BIRDS *of the* DARWIN REGION

Niven McCrie and Richard Noske

CSIRO PUBLISHING

National Library of Australia Cataloguing-in-Publication entry

McCrie, Niven, author.

Birds of the Darwin region / Niven McCrie and Richard Noske.

9781486300341 (paperback)
9781486300358 (epdf)
9781486300365 (epub)

Includes index.

Birds - Northern Territory - Darwin Region.
Bird watching - Northern Territory - Darwin Region.

Noske, Richard Alfred, author.

598.0994295

Published by

CSIRO Publishing
36 Gardiner Road, Clayton VIC 3168
Private Bag 10, Clayton South VIC 3169
Australia

Telephone: [+613] 9545 8555
Email: csiropublishing@csiro.au
Website: www.publishing.csiro.au

Front cover: Rainbow Pitta (photo: Niven McCrie)
Back cover: Rufous Owl (photo: Laurie Ross)

Edited by Joy Window
Cover design by James Kelly
Typeset by Thomson Digital
Printed by Ingram Lightning Source

Feb26_RP_ILS

Contents

Preface

Cities change more quickly in physical characteristics than natural environments. For this reason, in part, their wildlife is constantly changing. In the older, larger cities of Australia, where natural bushland tends to be confined to small remnants, the avifauna has been dramatically altered to one dominated by introduced species and a handful of adaptable natives.

Darwin is different. Firstly, there have been virtually no successful introductions or invasions of exotic birds. Secondly, despite its relatively small size by city standards, Darwin supports a diverse array of habitats, including a large corridor of woodland and riparian forest adjacent to Darwin International Airport and defence airbase, which virtually bisects the metropolitan area. Drainage has not been altered substantially, and the mangals (mangrove communities) that fringe the huge harbour and dissect the suburbs are still largely intact despite repeated attempts to reclaim them. This variety of habitats within Darwin and in its immediate surrounds explains, to a large extent, why the avifauna of the region is so rich.

Located in the north-west Top End, Darwin also differs profoundly from other Australian capital cities in having a monsoon-tropical climate in which the annual cycle of seasons is driven by rainfall, rather than temperature. The climatic extremes of the Wet and Dry seasons have been likened to those that characterise wet tropical rainforests and deserts, respectively, and the annual alternation of these seasons has a dramatic effect on the lives of the birds residing in the region, by altering the physical characteristics of their habitats and the food available to them. Additionally the warm conditions of the Dry season make a favourable environment for many birds that vacate their breeding grounds in southern parts of Australia for the winter. It is this combination of residents and visitors from myriad sources over the course of a year that makes the avifauna of the Top End so interesting, and worthy of study.

The beginnings of this book

In the early 1990s the authors instigated a three-year bird atlas project to map the distribution and determine the seasonality of birds of the Darwin region. Preliminary distribution maps were produced, but over the ensuing period both authors became preoccupied with other related projects, including the production of other books: one on the birds of Groote Eylandt[1] and the other a guide to bird watching in Darwin and the Top End.[2] Meanwhile, bird records of the Darwin region continued to accumulate, both from the authors and from numerous other sources, including data resources of the Northern Territory government, Eremaea and eBird. Two decades after the initial Darwin bird atlas project, almost 120 000 bird records had been amassed. These data form the basis for the maps, charts and text in this book.

But this book is more than a summary of bird distribution and seasonality. It is an account of what is known about birds of the region, from the authors' own field research, an extensive reading of the literature, and personal communications with numerous birdwatchers and ornithologists living in, and visiting, the region. We have deliberately minimised scientific jargon in the book, yet at the same time present an accurate portrayal of as many aspects of the ecology of the region's birds as currently available. Our aim was to provide information for birdwatchers at any level of experience, including students and other young people, while presenting material that will also be of value to ornithologists, ecologists, environmental and town planners, teachers and other professionals with an interest in our bird life.

Acknowledgements

This book could not have been written without an enormous amount of help from a great many people and we are most grateful to all who contributed to it, in whatever way. Our requests for assistance met with an exceedingly generous response, and for that we are indebted. If we have inadvertently omitted the names of any who provided assistance we tender our sincere apologies.

We have been fortunate in being allowed the use of several comprehensive datasets, the inclusion of which greatly enhanced the accuracy and value of the seasonality charts and distribution maps appearing in this book. Data from the NT Fauna Atlas was kindly supplied by Alaric Fisher and Ralf Koberstein from the Biodiversity Conservation Division and Natural Resources Division, respectively, of the NT Department of Natural Resources, Environment, the Arts and Sport, in Darwin. We are also very grateful to Richard and Margaret Alcorn for extracting an immense amount of data for us from Eremaea.

Hilary Thompson generously gave access to his large bank of personal records as well as providing a copy of his unpublished 'Annotated List of the Birds of the Top End', prepared jointly with Denise Goodfellow. For providing their personal survey data we are also grateful to Tony Hertog, Chris Healey, Ashley Carlson, Arthur and Sheryl Keates, Roger Jaensch, Mohd-Azlan J. Azad, and Heather Ryan. Gavin and Meg O'Brien kindly supplied shorebird count data and nesting records for the Darwin region.

The NT Birds newsgroup, courtesy of Yahoo, has proved to be an important source of information, going well beyond expectations when the first author set it up in 2002. The contributions of local birdwatchers in recording the details of sightings of particular species have been invaluable, particularly for the rarer species.

For their valuable comments on draft texts and additional data for specific groups of birds, we are especially grateful to Roger Jaensch (waterbirds), Arthur and Sheryl Keates (shorebirds), and Tony Hertog (raptors). Emails requesting information on one bird species or another were met with helpful responses from many people, including (in alphabetical order): Nigel Adlam, Gillian Ainsworth, Bryan Baker, Darryel Binns, Deb Bisa, Mike Carter, Rohan Clark, Stephen Debus, David Donato, Fiona Douglas, Lunar Eclipse, Johnny Estbergs, Keith and Lindsay Fisher, Sally Forsstrum, Dominic Funnell, Marc Gardner, Clive Garland, Stephen Garnett, Bastiaan Hensen, Mick Jerram, Ron Johnstone, Arthur and Sheryl Keates, Helen Larson, Tom McCrie, Clive Minton, Heather Moorcroft, Will Riddell, Danny Rogers, Andrew Spiers, Penny Steele, Colin Trainor, Fred van Gessel and David Webb. We are grateful to them all.

In this 'internet age', online data are becoming more readily available. We acknowledge access to the Australian Bureau of Meteorology website (www.bom.gov.au) from which we drew climate data, used both in writing the text and creating the climate chart used in the book's introduction. The maps used in this book include GIS data obtained from Open Street Map, Geoscience Australia, Australian Bureau of Statistics and the Collaborative Australian Protected Area Database (CAPAD).

Without photographs, this book would be less informative and much less attractive. Photographs are individually credited, but the authors expressly wish to thank the following for their generosity in contributing images: Anand Arya, Tony Ashton, Adrian Boyle, Ian and Jill Brown, Vince Bugeja, Nolan Caldwell, Ashley Carlson, Rohan Clarke, David Fisher, Marc Gardner, Mat and Cathy Gilfedder, Kirri Hardy, Bastiaan Hensen, Micha V. Jackson, Peter Kyne, Dr Chan Ah Lak, Judy Leitch, Tom McCrie, Scott McGregor, Ross Monks, Glenn Newton, Jon Norling, Tony Palliser, Belinda Rafton, Mike Reed, Will Riddell, Danny Rogers, Laurie Ross, Julie Sarna, Jim Sneddon, Brian Thistleton, Colin Trainor, Aiden Webb and David Webb.

We also express our gratitude to CSIRO Publishing. We are especially grateful to John Manger and Lauren Webb for accepting the immense amount of time the book took to complete and the frequently postponed deadlines, while continuing to provide encouragement and support throughout what has at times seemed an almost Herculean task. If, as is almost inevitable, errors have crept into the book, the responsibility for them is entirely ours. We are grateful to those who helped keep them to a minimum.

This book demanded a lot of hard work and cooperation from each author, but one or other of us did the lion's share of most of its components. Niven compiled and processed the database, constructed the maps and charts, collated the photos (and took many of them) and wrote the species texts for the non-passerines from Emus to Jaegers, as well as the vagrants and unconfirmed species. Richard wrote the Introduction, and the species accounts from the Pigeons and Doves onwards, including the passerines, except for the vagrants. The Preface and About this book sections were jointly written.

Niven would personally like to give his warmest thanks to his partner Judy Evans, for her unremitting support throughout the writing of the book. Her understanding and acceptance of the countless hours that he has been at the computer, neglecting chores and putting important things second, have been the single most crucial factor in completing much of his work.

Richard thanks his wonderful wife Rosalinda Isorena for her moral and gastronomic support during the many hours he was preoccupied with this book. He is also very grateful to Richard Luxton, owner of Coomalie Farm, for his tireless hospitality and interest in wildlife. Lastly, he pays tribute to the pioneers of Top End ornithology from the early explorers to the first chroniclers of Darwin's birds, especially David Crawford and Hilary Thompson.

Niven McCrie and Richard Noske

May 2015

Introduction

Text by Richard Noske

History of ornithology in the Darwin region

Under the watchful eye of the Larrakia indigenous people, Port Darwin was settled in 1869 by the British, 30 years after they first sighted the region. Originally christened Palmerston, the settlement was renamed 42 years later in honour of the most celebrated biologist of all time, although Sir Charles Darwin himself never came closer than Sydney. Although Northern Territory (NT) ornithology had begun three decades earlier with the extensive zoological exploration of the Port Essington district, it would be another century before the birds of Darwin received attention, with the month-long visit of Herbert Deignan in 1948, during the American-Australian Scientific Expedition to Arnhem Land.

The publication in 1967 of Storr's *List of Northern Territory Birds*[3] provided the first summary on the status and distribution of the birds of the NT, including the Top End, but it was not until 1972 that a detailed, annotated list of birds of the Darwin area was published.[4] This paper, covering the Humpty Doo and Marrakai regions as well as Darwin, provides a baseline for this book. Its author, David Crawford, was stationed at the Coastal Plains Research Station near Fogg Dam from April 1967 to January 1972. Crawford wrote nearly a dozen other papers on birds of the region, mostly concerning waterbirds, and including the first Australian records of Common Redshank and Asian Dowitcher.[5,6]

The early 1970s also saw the arrival of Hilary Thompson and John McKean, who were able to witness first-hand the devastating damage caused by tropical cyclone Tracy on Christmas Eve 1974, not only to the lives of Darwin's human inhabitants, but also to the vegetation, with its consequent effects on the birdlife.[7] Realising the significance of the newly built Sanderson (Leanyer) Sewage Ponds as a permanent, albeit artificial, wetland refuge for waterbirds and waders, McKean and Thompson embarked on monthly surveys of the ponds and their environs from 1974 to 1984. The data from these surveys later formed the basis for a series of important papers concerning the status and seasonality of gulls and terns,[8] cuckoos,[9] kingfishers,[10] cuckooshrikes,[11] ducks[12] and waders[13] in the Darwin region. Thompson's paper entitled 'Common birds of the Darwin suburbs'[7] was followed by *Common Birds of the Darwin Area*,[14] a guide to the identification and habits of 91 species, with brief descriptions of an additional 61 species.

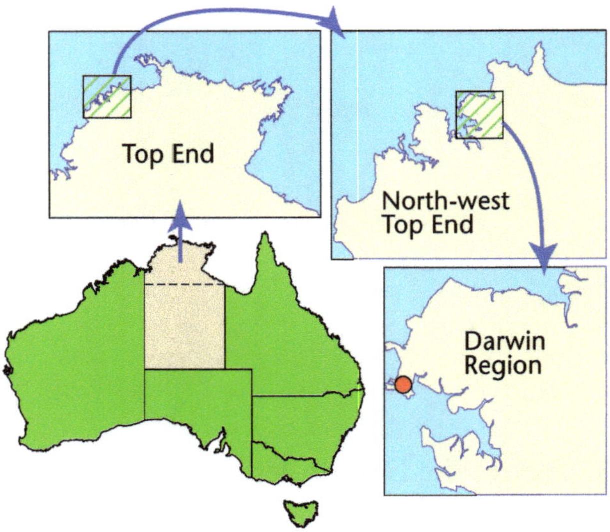

Fig. 1. Overview map showing the Northern Territory in Australia, with inset maps of the Top End, north-west Top End and Darwin region. Darwin city is indicated by a red dot.

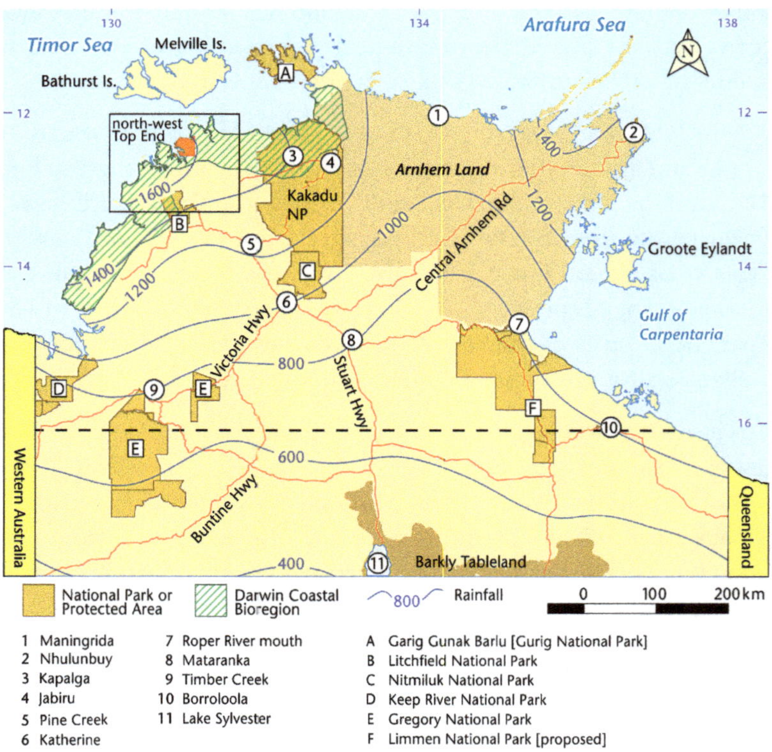

Fig. 2. Map of the Top End, as defined in this book as that area north of 16°07′E (shown as a dashed line). Keep River National Park and Borroloola lie approximately on its south-west and south-east boundaries, respectively. The map extends southward to show the north-western Barkly Tableland IBRA subregion, an important area for waterbirds.

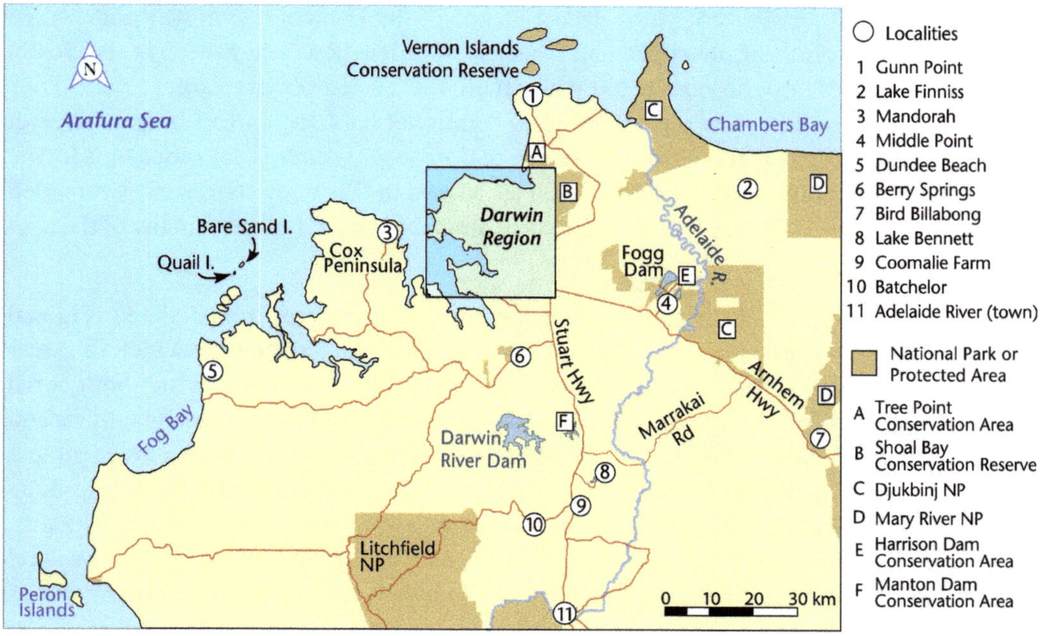

Fig. 3. Map of the north-west Top End.

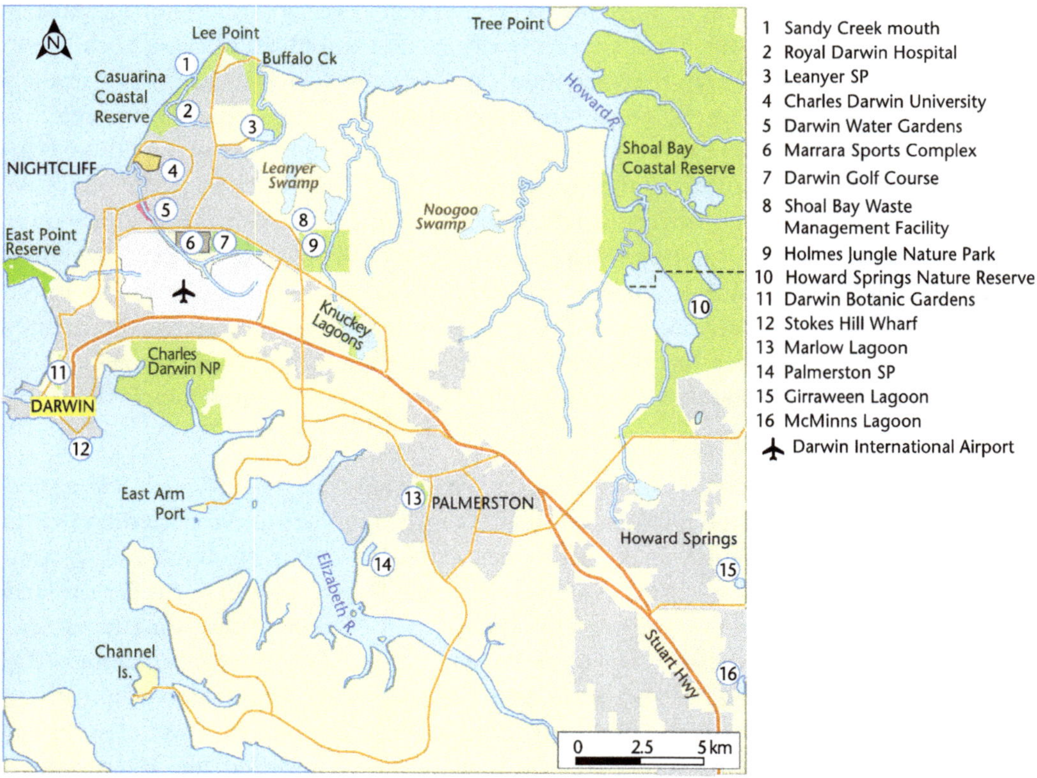

Fig. 4. Map of the Darwin region, showing localities mentioned in the text.

The first scientific survey of the terrestrial avifauna of the Darwin region was undertaken in 1986 by John Woinarski and colleagues from the Conservation Commission of the Northern Territory (CCNT), who conducted more than 1000 censuses over six weeks.[15] They repeatedly sampled 100 sites representing 20 terrestrial habitats, spread over an area of 20 × 20 km, centred on the Howard's Peninsula, most of which is encompassed in the north-eastern quadrant of the Darwin region as defined in this book. This survey provided valuable information on the relative abundance and habitat preferences of many of the species of the region.

Much of the literature concerning birds of the Top End emanates from Kakadu National Park (NP).[16,17,18,19,20] These studies provide vital comparative information by which to assess the status of Darwin's birds. In 1990, CCNT appointed Ray Chatto to conduct both aerial and ground surveys of waterbirds, seabirds and shorebirds, as well as marine mammals and reptiles, along the entire NT coastline, including the Darwin region. This ambitious project spanned 16 years in all, and led to the discovery of nationally and globally significant numbers of colonial nesting herons and terns,[21,22] and migratory waders,[23] on the floodplains, offshore islands and intertidal areas, respectively, of the Top End. Some of the breeding rookeries are surprisingly close to Darwin, and doubtless are the source of many waterbirds visiting our region.

Climate

Darwin lies in the monsoon or wet–dry tropics, which cover about 20% of the Australian continent, including the Top End, Kimberley region and much of Cape York. Here, in stark contrast to the temperate south of the continent, which holds most of Australia's human population, the seasons are driven by rainfall, not temperature. This region is uniformly warm to hot all year round. The difference between the mean maximum temperature (32°C) and mean minimum temperature (23°C) for Darwin is only 9°C. As expected, temperatures are lowest in June and July, coincident with the austral winter but, surprisingly, Darwin is hottest during the spring months of October and November, rather than the austral summer. Indeed, in terms of monthly maximum temperatures January and February are among the five coolest months, mainly due to the greater cloud cover of the Wet season. But while Darwin's temperature rarely exceeds 35°C, relative humidity at 9 o'clock in the morning averages 72–83% for six months of the year.[24]

With an average annual rainfall of 1700 mm, Darwin is easily the wettest Australian capital city.[24] More remarkable is the fact that about 90% of this rain falls in the period known as the Wet season, which comprises the six months from November through April, encompassing the Austral summer (see Fig. 5). Thus while the Wet Tropics in north Queensland typically receives some rain in every month, Darwin, like the rest of the Top End, experiences drought-like conditions for four to five months, from May through September. The transition period straddling the end of the Dry and start of the Wet season (September-November) is often known as the 'Build-up', during which the humidity and cloud cover steadily increases while temperatures remain high. Scattered localised, and often violent,

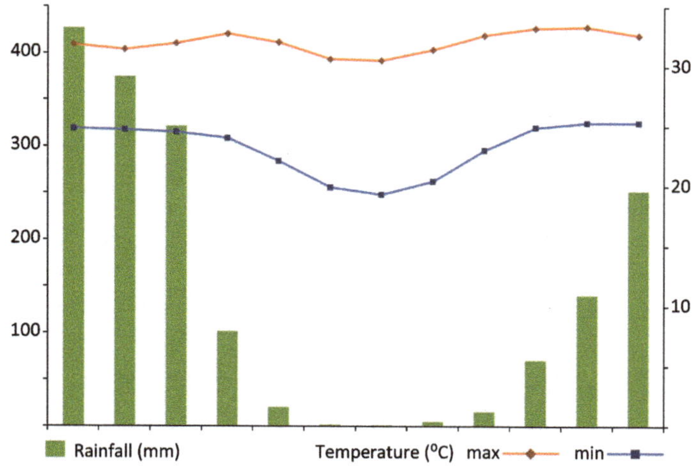

Fig. 5. Mean monthly rainfall and mean maximum and minimum temperatures, 1941–2012. Based on Darwin International Airport data from www.bom.gov.au.

convection thunderstorms, offer the only relief from this hot and humid weather, before the arrival of the Australian summer monsoon, which brings heavy rain across wide areas.[25] The timing of the onset of the Wet season, its duration and the total amount of rainfall, vary dramatically from year to year.[26,27] Over the 114 years from 1870 to 1993 Darwin's rainfall varied from 892 to 2644 mm in 1958 and 1974, respectively.[26]

Although the Wet season is sometimes considered to start in October, in this book we define it as the period from November to April, partly to enable easy comparisons of two halves of the year, and partly because the long-term average rainfall for October is less than half that of November.[26] Where relevant, we further divide the Wet and Dry seasons into three periods of two months: early, mid and late. While these divisions are somewhat arbitrary, and may in fact comprise different calendar periods in different years, they nevertheless provide the most convenient seasonal indicators for bird activity during the year.

Another important feature of the local climate is tropical cyclones. In the century between 1906 and 2006 there were 100 cyclones that passed within 400 km of Darwin, of which 10 passed within 50 km of the city.[24] Tropical cyclone Tracy, with its wind gusts of at least 235 km/h, is arguably the most significant tropical cyclone in Australia's history, accounting for 65 lives and the destruction of most of Darwin. While the most recent cyclone affecting Darwin at the time of writing was tropical cyclone Helen in January 2008, with maximum wind gusts of 102 km/h causing many trees to uproot or snap, and widespread disruption to power and communications, it was not nearly as severe as tropical cyclone Monica, which passed to the north of Darwin, but caused widespread damage to the Arnhem Land coast and hinterland in April 2006. While the effect of tropical cyclone Tracy on the vegetation in the region is evident in the even-aged stands of eucalypts in open forest around Darwin, two less intense cyclones that affected Darwin in 1980s appear to have had little effect on the vegetation.[28]

The avifauna and its responses to extreme seasonality

Darwin's avifauna in the Top End context

A grand total of 323 bird species has been confirmed for the Darwin region, including vagrants and introduced species. Another 11 species have been reported from the region, but remain unconfirmed. Of the confirmed species, 61 (19%) are here considered vagrants, and another four either escaped aviary stock or vagrants. That such a large proportion of the avifauna consists of very rare or unexpected species attests to both the proximity of Darwin to Indonesia and the alertness of the Darwin birdwatching community. Indeed, many species have been seen nowhere else in the NT (for example, Kentish Plover and Little Stint) or even Australia (Tricolored Grebe, Green Sandpiper and Christmas Frigatebird).

Of the 323 species in our region, 19 are largely endemic to the Australian monsoon tropics, including three species that are found only in the Top End and Kimberley regions (Rainbow Pitta, Silver-backed Butcherbird and Yellow-rumped Mannikin). While another 10 species are largely restricted to the monsoon tropics in Australia, they also occur in New Guinea and, in some cases, eastern Indonesia. An additional 33 species have a largely tropical distribution in Australia, but again also occur in New Guinea, Indonesia or even beyond. Although the NT has five endemic bird species, none has been recorded in the Darwin region, which is not surprising given that four of them (Black-banded Fruit Dove, Chestnut-quilled Rock Pigeon, White-throated Grasswren, White-lined Honeyeater) are confined to the sandstone escarpments and plateaux of western Arnhem Land and adjacent parts of Kakadu NP, over 200 km to our east. The fifth species, the Hooded Parrot, is largely restricted to woodland within 100 km of Katherine, but has occasionally been recorded as far north as Coomalie Farm, 45 km south of the Darwin region.

Waterbirds, shorebirds and seabirds

Almost a quarter of the birds that have been recorded in the Darwin region are associated with freshwater wetlands (Table 1). While the region has little in the way of freshwater wetlands itself, it is neatly wedged between two vast areas of floodplains, one starting around the Adelaide River just 30 km to the east, and the other, the Finniss River, about 60 km to the south-west. These floodplains, and those to their east and south, respectively, hold both nationally and globally significant numbers of colonial nesting waterbirds,[21] and are probably the source areas for many of the birds visiting Darwin. Indeed the largest of the 27 nationally significant waterbird breeding colonies in the Top End is located on the Adelaide River, only 20 km to the east of the Darwin region. This colony hosts between 20 000 and 30 000 birds belonging to 10 species, and represents the largest egret colony in Australia, with over 6000 Intermediate Egrets alone.[21] The floodplains and wetlands between here and the mouth of the East Alligator River support over 600 000 waterbirds of more than 50 species, including about 200 000 Magpie Geese.[29]

Table 1. Broad habitats used by birds recorded in Darwin.

Primary habitat	No. of bird species	%
Terrestrial	184	57.1
Freshwater wetlands	76	23.6
Shoreline	36	11.2
Marine	26	8.1
Total	322	

Similarly, Darwin is situated between the two richest areas for shorebirds in the NT, the most important of which comprises the mudflats, saltflats and wetlands around Fog Bay, only 60 km south-west of the Darwin region, where more than 38 000 shorebirds were counted during a single aerial survey in October 1995.[23] The second area is around the coast of Chambers Bay between the mouth of the Adelaide River and Point Stuart, where over 14 000 birds were counted from the air on one day in September 1993. Much of the coastline around the Darwin region, on the other hand, is lined with mangals rather than wetlands and consequently supports fewer shorebirds, though over 7000 birds were counted during a single survey between Lee Point and Tree Point, in the north-east corner of the region, in November 1995.[23] A total of 36 shorebird species has been reported in the Darwin region, representing 11% of the avifauna.

The Top End also boasts globally significant breeding colonies of terns, but they are concentrated in the north-east, mostly on islands off Arnhem Land and in the Gulf of Carpentaria.[22] The closest to Darwin is Seagull Island, just north of Melville Island and about 150 km north-west of Lee Point, which has the largest breeding colony of Greater Crested Terns in Australia, if not the world, while two sandy islands off Cobourg Peninsula, less than 200 km north-east of Lee Point, host colonies of three other tern species. These islands, or those further east, are probably the original source of Silver Gulls and the marine terns that visit, and possibly temporarily reside, in the Darwin region. These and other seabirds comprise about 8% of the confirmed species of the region (Table 1).

Landbirds

The Darwin region lies in the wettest part of the NT mainland, although the Tiwi Islands have the highest rainfall in the NT.[24] Progressing inland, mean annual rainfall (MAR) drops steeply from about 1700 mm in Darwin to 1300 mm at Mount Ringwood, about 90 km south-east of Darwin, and to 1100 mm at Pine Creek, a further 90 km south-east. The vegetation also changes along this north–south transect from eucalypt forest dominated by the evergreen Darwin Woollybutt and Darwin Stringybark to woodland dominated by partly or totally deciduous eucalypts like Northern Box, Salmon Gum and various bloodwoods.[30]

The avifauna also changes along this environmental gradient, but it is not until one reaches Pine Creek that the NT-endemic Hooded Parrot can be found year-round.

Fifty kilometres south-east of Pine Creek lie Yinberrie Hills, one of the most important known breeding areas of the Gouldian Finch.[31] A year-long study at this site revealed the strong seasonality of the avifauna at this latitude, where bird density increased 10-fold between June and December, and species composition changed demonstrably.[32] Thus 12 of the most abundant species in December were scarce or absent in June. Many of these species were nectar-feeders taking advantage of the blossoming eucalypts and other plants in the late Wet season. A further 40 km to the south is Katherine, 270 km inland from Darwin, where Yellow-tinted Honeyeaters and Apostlebirds are main-street birds, while the Australian Magpie reaches its northern limit just south of the town.

However, rainfall declines not only to the south, but also to the east of Darwin. For example, Point Stuart, 100 km east of Darwin, has a mean annual rainfall of 1555 mm, and Gunbalanya (Oenpelli), 140 km further east, 1400 mm.[24] On the Marrakai track, 40 km south-east of the Darwin region, four species that have never been reported from the Darwin region occur: the Chestnut-backed Buttonquail, Black-chinned Honeyeater, Jacky Winter, and occasionally Pictorella Mannikin.[2,4] Most of these species are more typical of drier regions, and are generally scarce north of Pine Creek. However, the Jacky Winter is resident on the paperbark-fringed floodplains at Coomalie Farm, 45 km south of the Darwin region, a habitat that also attracts large numbers of Diamond Doves each Dry season.

Movements

Excluding vagrants and introduced or escaped species, 258 bird species have been recorded in the Darwin region. Of this subtotal, 105 species (41%) appear to be predominantly resident, although only 88 of them have been confirmed as breeding in the Darwin region. The remaining 59% of the subtotal are predominantly mobile, being either long-distance migrants (25%) or 'visitors' that undertake small-scale, irregular or irruptive movements (34%; Table 2). That such a large proportion of the avifauna is mobile is testament to the extreme seasonality of the region. Indeed, while each species has been assigned to one category in Table 2, in reality many species in the region have both resident and non-resident populations.

Roughly two-thirds of the 64 migratory species that visit the Darwin region do so mostly during the Wet season or late Dry–early Wet transition period (Table 2). Most are shorebirds or waders that emanate from Siberia and northern China and apart from a handful of species, such as the Little Curlew and Oriental Plover, all can be found in coastal parts of southern Australia during the warm months of the year. However, a few East Asian breeding migrants are land birds (Oriental Cuckoo, Garganey and Eastern Yellow Wagtail) that rarely reach temperate latitudes in Australia. Although a large proportion of the populations of some shorebird species remain during the following Dry season, all of these Northern Hemisphere breeding species are here considered **Wet season migrants**. None of

Table 2. Status designations and seasonality of Darwin's birds according to reporting rates.

	Residents	Visitors	Migrants	Total
No obvious seasonality	77	22		99
Predominantly Wet	10	9	32	51
Predominantly Dry	18	52	21	91
Predominantly Build-up (Oct–Nov)		7	11	18
Total	105	90	64	259

these birds breeds in Australia. In contrast, three other Wet season migrants breed locally, then migrate in a north-easterly direction to New Guinea, where they spend at least part of the Dry season (Pacific Koel, Oriental Dollarbird and Torresian Imperial Pigeon). The increasing abundance of Torresian Imperial Pigeons during the Dry season in Darwin, however, suggests that the local population now consists partly of residents.

Approximately a third of the migratory species visiting the region are **Dry season migrants** (Table 2). Most are species that are known to migrate in a north-westerly direction from breeding grounds in south-east Australia,[33] which appears to be the source of some birds visiting Darwin, although there is no direct evidence for this supposition. Some populations of these species continue on to Indonesia, and their passage through Darwin is revealed by sudden increases in reporting rates during the Wet–Dry transition period. The Australian Pratincole, Sacred Kingfisher and Rainbow Bee-eater are examples of species in which some populations 'overwinter' in the Top End, while others continue further north. Several species that are predominantly Dry season migrants also have small breeding populations in the Darwin region (e.g. Rainbow Bee-eater, Black-faced Cuckooshrike, White-breasted Woodswallow), although it is not clear if these populations are resident. Thus the Bee-eaters of Darwin may comprise three separate populations: passage migrants to Indonesia, migrants that winter locally, and local, possibly sedentary breeders.

Seven additional species are classified as Dry season migrants because of their scarcity during the Wet season, and the lack of breeding records from the northern half of the NT. One exception is the Magpie-lark, which breeds widely in the NT, as well as in Darwin, yet the migratory population is so large and its arrival so predictable that the species warrants classification as both a migrant and resident. Significantly, the majority of Top End birds that feed on aerial insects are Dry season migrants, such as the Bee-eater, Tree Martin, White-breasted Woodswallow and Spotted Nightjar. The partial or complete evacuation of Darwin by these birds before the first rains suggests that aerial insects become scarce during the Wet season, or that they are more abundant elsewhere at this time. Although several aerial insectivores visit the region during the Wet season, they are either passage

Fig. 6. Knuckey Lagoons attracts large numbers of Magpie Geese during the Dry season even when the lagoons have all but dried out.

migrants that do not stay long in any one place (Pacific Swift and Oriental Pratincole), or occur in small numbers (Barn Swallow).

Of the 90 species classified as 'visitors', 58% visit mainly in the Dry season and 18% in the Wet or late Dry-early Wet transition period, leaving 24% that occur equally frequently in both seasons (Table 2). Of the 52 Dry season visitors, the largest group (44%) are the waterbirds. These include the common ducks and the iconic Magpie Goose, as well as herons, spoonbills and Brolga, which congregate, often in immense numbers, at permanent water-holes as the Dry season progresses (Fig. 6). Soon after the first rains break, these birds move to the vast floodplains to the east or south-west of the Darwin region to breed, taking advantage of the annual surfeit of aquatic plants and invertebrates.

The second largest group (29%) of Dry season visitors are birds of woodlands or grasslands, including the Red-tailed Black Cockatoo, large flocks of which are often attracted to recently burnt areas. Raptors such as the Spotted Harrier and Black-breasted Buzzard are also commoner at this time. Few Dry season visitors are forest-dwellers, but they include the Rufous Whistler and Olive-backed Oriole. While both of these species are migratory in south-east Australia, local birds are more likely visitors from more inland parts of the Top End, where they are known to breed.

Breeding seasons

While the mobility of many of Darwin's birds is an obvious response to the extreme climatic seasonality of the Top End, another much less appreciated response is found in their

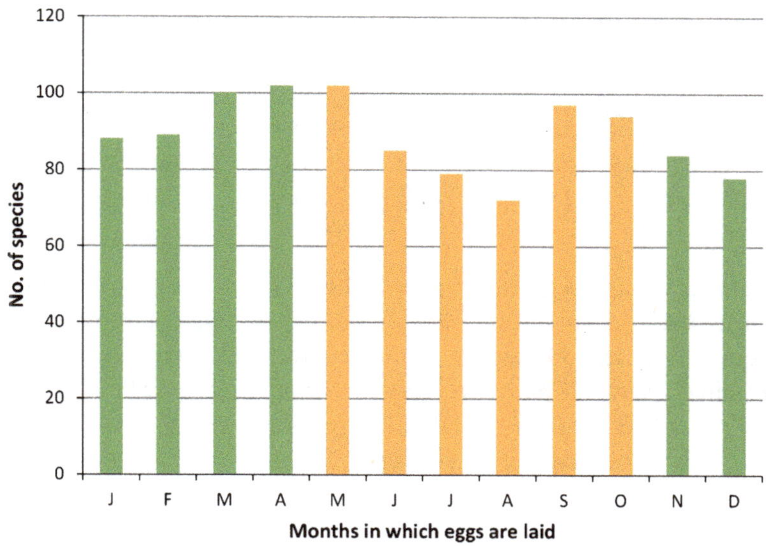

Fig. 7. The combined breeding seasons of 204 bird species of the Top End. Bars show the number of species breeding in each month of the year. Data are based on primary information sources.

breeding seasons. In southern Australia, and temperate climates worldwide, the great majority of birds breed in the warm spring and summer months, when most resources are at their peak. In stark contrast, the breeding seasons of birds in the monsoon tropics vary enormously, depending to a large extent on their feeding ecology.[34,35] The consequence of this diversity of breeding seasons is that birds of one sort or another can be found nesting in every month of the year (Fig. 7). In general, however, the numbers of nesting species are highest in the late Wet–early Dry transition period and late Dry season, and lowest in the mid Dry. Few species breed throughout the year, but a notable exception is the Rufous-banded Honeyeater, urban populations of which probably breed for longer than those in its natural habitats.[36]

Predictably, most waterbirds breed during the second half of the Wet season when water levels in wetlands are highest. For example, the iconic Magpie Goose, which builds its nests among grass and sedges on floodplains, usually lays its first clutches in February.[37] Colonial-breeding egrets and herons, which build their nests in coastal mangals and paperbark swamps, mainly breed from January to May, while cormorants and the Australian White Ibis breed somewhat later, finishing in July.[21] The breeding seasons of colonial-nesting terns are more variable than that of colonial waterbirds, occurring throughout much of the year, but peak in April–June and September–December.[22]

One of the few songbirds that breed almost entirely during the Wet season is the Rainbow Pitta, which is the sole species specialising on ground-dwelling invertebrates of monsoon rainforests in the Top End. This reproductive timing ensures that parents can feed their young on earthworms, which are driven to the soil's surface by rainwater.[38] The late Wet season is also the time when most grassland-dwelling birds, such as the cisticolas, Red-backed Fairywren, and Chestnut-breasted Mannikin breed, while grasses and sedges are tall and dense. Woodland-dwelling finches also breed during the latter half of the Wet season and first half of the Dry to coincide with the seed-set of perennial, then annual grass species.[39] Most Top End raptors breed mainly during the Dry season, presumably because their prey is easier to find and capture when there is least grass cover in wooded areas,[40] or in the case of species feeding on aquatic prey, when water levels are low.[41]

Unlike their temperate counterparts, many small woodland-dwelling songbirds of the Top End breed during the early to mid-Dry season, corresponding to autumn and winter.[35] For the honeyeaters, this clearly relates to the annual surfeit of nectar supplied by Dry season-flowering trees like Darwin Woollybutt, Darwin Stringybark, and Fern-leaved Grevillea.[42] However, most small leaf-gleaning insectivorous birds also breed at this time, possibly due to lower abundance or reduced activity of small arboreal insects during the Wet season, particularly during long periods of wet or overcast weather. Several mangal-dwelling insectivores have a biannual breeding season, the first season peaking in the Wet–Dry transition months of April and May, and the second in the late Dry, from September to November.[43,34] This unusual annual cycle appears to be a response to the biannual pattern of insect abundance in this habitat.[43,44]

Nevertheless, many medium-sized forest and woodland-dwelling songbirds breed in the late Dry season and early Wet, corresponding to the spring–summer breeding peak in southern Australia. With the exception of the Pacific Koel, fruit-eating birds also start breeding in the late Dry season, well before the main fruiting season of most monsoon rainforest plants.[45] While the very short, early breeding season of local Australasian Figbirds may be timed to take advantage of the emergence of cicadas, which provide protein for adults,[46] it is difficult to explain the lack of correspondence between the breeding seasons of the specialised fruit-eating pigeons and fruiting seasons. Clearly there is a need for more detailed studies of the nesting phenology, as well as moult cycles, of birds of Australia's monsoon tropics. Such information would help enormously in understanding how our birds have adapted to their highly seasonal environment.

Habitats of the Darwin region

Darwin's monsoon-tropical climate and coastal location have resulted in a mosaic of many distinct habitats which, while representative of the Top End as a whole, are atypically close to one another. During their rather brief stay in Darwin in 1968, collectors for the Harold Hall expedition remarked on this habitat diversity and its consequent avian diversity: 'four quite distinctive zones all held species that were rarely encountered in more than one of them, and yet this belt of coastal monsoon vegetation was barely two miles in depth'.[47] Indeed in a single 1-km transect near Buffalo Creek, for example, from the coast, one traverses tidal mudflats, beach, monsoon rainforest, mangal (mangrove communities), paperbark forest, eucalypt woodland, and depending on the time of year, grasslands or swamp. While each major habitat has its own suite of specialised bird species, their juxtaposition allows some mixing of bird species. In mangals, for instance, it is often possible to see species more typically associated with monsoon rainforest, because small patches of the latter habitat are embedded in the former, growing on cheniers above the high tide mark. The close proximity, and even intermingling, of habitats in and around Darwin may partly explain the adaptability of many birds to urbanisation, which is discussed later.

Since 1981, when Palmerston was being built, the combined population of this city and Darwin has almost doubled to over 103 000,[48] a quarter of which resides in Palmerston. Given this rapid population growth, it is hardly surprising that there has been some loss of bird habitats during the development of the metropolitan areas of Darwin and Palmerston, as well as in rural parts of the region. While collecting birds in Darwin during early 1948, Deignan witnessed the clearing of a 'fine stand of monsoonal forest … just behind the hostel at Nightcliff'.[49] Indeed, an analysis of aerial photographs taken between 1945 and 1990 revealed that 60% of the monsoon rainforest in Darwin was lost over those 45 years, of which 40% was attributable to clearing for urban development before 1974. The Nightcliff patch of rainforest was one such casualty, as all 51 ha were cleared, as was about 90% of the 110 ha of rainforest along Rapid Creek, 85% of the 73 ha in the Leanyer complex, and 77% of the 78 ha at East Point. Then in 1974 tropical cyclone Tracy caused massive damage to most of the remaining rainforest patches, but while some recovered through vegetative regeneration, others contracted or were lost completely due to intense fires fuelled by introduced Mission Grass which rapidly invaded patches opened up by the cyclone.[28]

Figure 8 shows the current distribution of urban, rural and natural areas in the region. The following is a brief description of the major habitats of the region, progressing inland from the coast, and the types of birds they support. Sites representative of each habitat are also mentioned.

Shoreline

Darwin's coastline is macro-tidal, with an enormous 6–7 m range twice daily during the spring tide period, which roughly coincides with the new and full moons each fortnight. During such tides, Darwin's coral reefs and vast intertidal mudflats (Fig. 9) are exposed for only a short time before being submerged again. The pace of water movement is much slower during days of neap tides, however, when the tidal range drops to 1–3 m. The mudflats, rocky reefs and beaches host numerous shorebirds, most of which are long-distance migrants that breed as far away as the Arctic Circle during the brief Northern Hemisphere summer, then undertake a round trip of 15 000–20 000 km or more each year to Australian shores during the Austral spring and summer. During low tides, these shorebirds disperse across the mudflats to forage for marine invertebrates, but as these become covered by incoming tides, they move to beaches or saltflats on higher ground, where they roost until the tides recede again. Lee Point and the mouth of Buffalo Creek are excellent examples of these habitats.

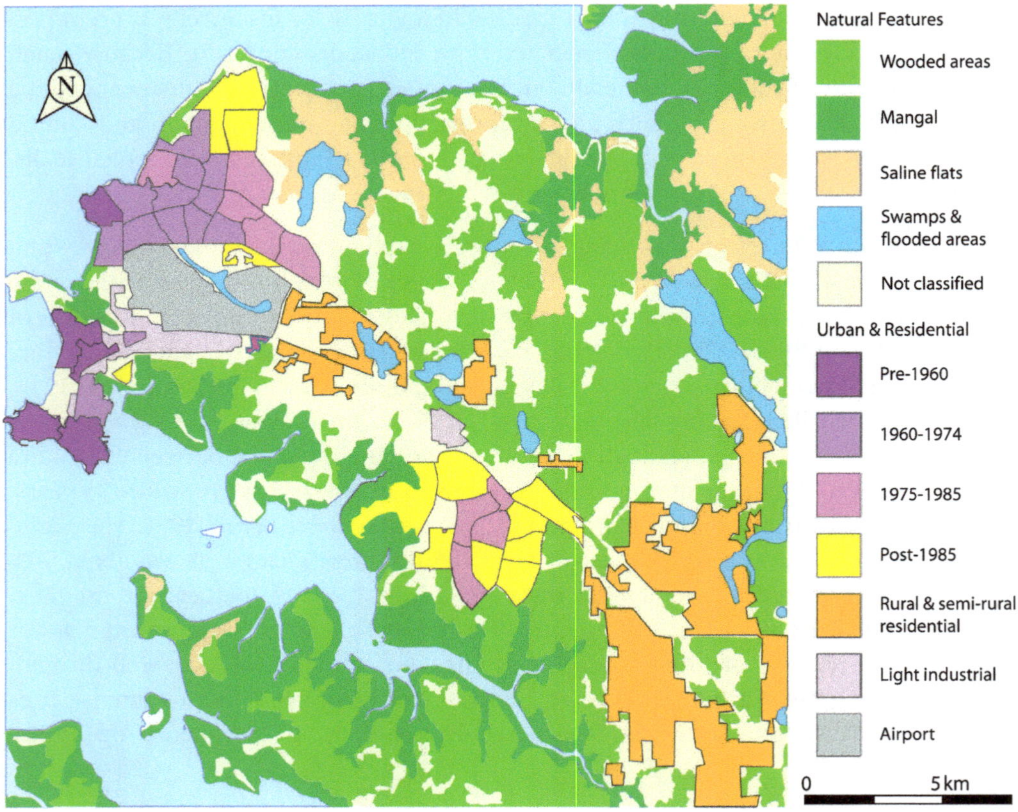

Fig. 8. Map of Darwin's physical environment, indicating broad vegetation types and built-up areas. The periods of suburban development have created differing ages of habitat, which affect the distribution of some bird species.

Mangals

Australia has more bird species that occur exclusively in mangals (mangrove communities) than anywhere else in the world. Paradoxically the Top End supports more of these birds than the Wet Tropics of north Queensland, despite having fewer mangrove plant species. Mangals cover over 4000 km of the NT coastline, constituting about one-third of the total mangal area of Australia, and Darwin Harbour supports the second most extensive mangal in the Top End.[50] With two large mangrove-lined creeks (Ludmilla Creek and Rapid Creek) dissecting the metropolitan area, and the vast mangals around the Harbour's East and Middle Arms, including a large proportion of the area of Charles Darwin NP, Darwin is undeniably the best city in Australia to see or study most of Australia's mangal-specialised birds. Accordingly a considerable amount of ornithological research has been conducted in these mangals, by both local and international scientists.[51,52,53,54] Like the mangroves that comprise mangals, mangal-specialised birds show zonation, some being largely restricted to the taller mangroves fringing tidal rivers and channels (Fig. 10), while others favour the stunted mangroves that surround barren saltflats (Fig. 11), which are a haven for nesting birds.[55] Several common mangal-dwellers also

Fig. 9. Tidal mudflats, like these south of East Point, are important feeding grounds for shorebirds.

Fig. 10. The seaward edge of mangals around the coast and along tidal rivers, such as Elizabeth River shown here, is usually dominated by dense Stilt-root Mangroves, often with a narrow fringe of widely spaced Star Mangroves.

Fig. 11. Barren saltflats, invariably surrounded by short Grey Mangroves, are scattered through the landward side of mangals.

inhabit monsoon rainforests,[19] which resemble mangals in having a closed canopy and a grass-free understorey.

Monsoon rainforests

Sometimes called monsoon vine forests, or simply monsoon forests, rainforests occupy only a tiny proportion of the land area of the Top End, occurring as small, scattered patches in a savanna-dominated landscape. Yet due to their diverse flora and complex structure, they have a rich avifauna, which includes several distinctive species.[15] A high proportion of the plant species of monsoon rainforests bear fleshy fruits that attract fruit-eating birds, which in turn act as dispersal agents of their seeds. Studies of the movements and patch size requirements of these birds took place in rainforests just east of the Darwin region.[56] Of the two broad types of monsoon rainforest, dry rainforests and vine thickets predominate in the Darwin region, occurring patchily around the coast on sand dunes and cheniers, and typically including many deciduous plant species.[57] In contrast, wet rainforests are associated with perennial streams and are dominated by evergreen plant species, typically including many palms (Fig. 12). Examples of the former can be found at East Point and in Casuarina Coastal Reserve (Fig. 13), while Holmes Jungle and Howard Springs

Fig. 12. Wet rainforest occurs mainly around perennial streams, and often includes dense stands of palms, such as these Carpentaria Palms at Holmes Jungle.

Fig. 13. Dry rainforest occurs patchily on dunes or cheniers around Darwin's coastline, often abutting mangals, and typically includes some deciduous trees. This photo shows the edge of the large dry rainforest patch in Casuarina Coastal Reserve.

Nature Reserve support examples of the latter. As mentioned above, monsoon rainforests are highly susceptible to invasion by weeds following fire, and much of the large patch at Kulaluk has been replaced with a monoculture of tall Coffee Bush.

Paperbark forest and riparian forest

Although a minor vegetation type in terms of total land area, paperbark (Melaleuca) forests and woodlands provide a vital source of food during the Wet season to a suite of locally mobile nectar-feeding birds,[16,58] as well as being the primary habitat of several other bird species (Fig. 14). Fringing the non-tidal, upper reaches of rivers, floodplains and other wetlands, paperbark forest is but one type of riparian forest. Other types of riparian forests are dominated by plant species typical of monsoon rainforest or eucalypt forest and, as they usually contain bird species characteristic of those habitats,[59] are treated separately in this book. Paperbark forests are usually inundated during the mid and late Wet season and, as few other plant species can tolerate annual flooding, they are typically composed solely of Broad-leaved Paperbarks or mixtures of paperbark species.[60] In Darwin this habitat is exemplified by Marrara Swamp, where large numbers of friarbirds, honeyeaters and lorikeets gather each year in January to feed on the copious nectar of the blossoming paperbarks.

Fig. 14. Paperbark forests, like this patch in Howard Springs, grow in areas that are flooded during the Wet season, and when in blossom attract many nectar-feeding birds by day, as well as flying foxes by night.

Eucalypt woodland and open forest

Often called savannas because of their understorey of tall grass, eucalypt woodlands and forests overwhelmingly dominate the vegetation of the Top End. The common tree species of this habitat in the Darwin region, and much of the Top End, are Darwin Stringybark and Darwin Woollybutt, which often occur together (Fig. 15). Both species flower in the mid Dry season, when they provide an important nectar source for honeyeaters and lorikeets.[42] Beneath the eucalypt canopy are smaller trees, often broad-leaved species such as Cocky Apple and the deciduous Kakadu Plum and Green Plum. On the seasonally inundated or poorly drained soils that occur around floodplains and creeks, eucalypts such as Ghost Gum and Long-fruited Bloodwood often dominate, with Pandanus and Fern-leaved Grevillea as typical mid-storey species. The ground is covered by grasses, both perennial and annual. Most conspicuous in the mid to late Wet season is Annual Speargrass, which reaches about 2.5 m in height before it is knocked down by the last storms of the Wet.

Seed-eaters, comprising cockatoos, parrots, doves and grassfinches, are the most distinctive component of the avifauna in the eucalypt woodlands. The monsoon-tropical woodlands are home to no fewer than seven species of grassfinches, three of which occur commonly in the Darwin region. This group, along with grass-dwelling insectivorous species such as cisticolas, are the birds most directly affected by the fires that burn across the Top End every year. Being the dominant habitat of the region, it was the eucalypt forests

Fig. 15. Eucalypt woodlands are the dominant habitat of the Top End, often called savannas because of their tall understorey of grass. In the foreground of this photograph can be seen the rough-barked Darwin Stringybark and, in the background, half white-barked Darwin Woollybutts and reddish-barked Smooth-stemmed Bloodwoods. The mid-storey typically comprises scattered broad-leaved trees and often Cycads (as shown here), while the ground is covered by dense Annual Speargrass. Note the charred tree trunks and, in the foreground, the lack of grass, indicating recent fire.

and woodlands that bore the brunt of tropical cyclone Tracy in December 1974. One legacy of this devastation is the even-aged stands of Stringybark and Woollybutt that abound around the outskirts of Darwin.

Floodplains and saline coastal swamps

Unlike the following wetlands, which are more or less permanent, these habitats are ephemeral, being inundated only during the Wet season. The Darwin region lacks the extensive floodplains of the Adelaide River and other large river systems further east, but small examples, such as Noogoo and Milners Swamps, exist around the upper reaches of the Howard River, Kings Creek and Mickett Creek. Largely covered by sedges and rushes during the Wet season, these swamps slowly dry out over the Dry season, when they are often burnt off, exposing the black cracking clay soils that form their substrate. Closer to Darwin, between Leanyer Sewage Ponds and Holmes Jungle, Leanyer Swamp consists of bare saltpans dissected by beds of sedges and salt-tolerant grasses which are home to many ground-dwelling birds (Fig. 16). Though now largely drained by a network of canals that were installed to reduce breeding habitat for mosquitoes, Leanyer Swamp used to be an important area for migratory waders because the shallow ponds formed after the first rains of the Wet season remained devoid of standing vegetation for longer than other wetlands, due to their high salinity levels.[61]

Lagoons and sewage ponds

Since they were constructed in 1971, Leanyer Sewage Ponds have attracted large numbers of ducks (Fig. 17), herons and terns, along with lesser numbers of shorebirds. Though uninviting in an olfactory sense, these artificial lagoons have proved a boon to birdwatchers because of their peculiar appeal to vagrants, especially from outside Australia. Leanyer SP has become known as a rarity hotspot, producing the first records of many species for the NT, along with some Australian firsts, and is probably the most reliable place in Australia to see Eastern Yellow Wagtails and Little Ringed Plovers. Smaller than Leanyer SP, Palmerston SP hosts fewer waterbirds, but the mangal and saltflats adjacent to the ponds provide extensive habitat for mangal-specialised birds and loafing waders, respectively. Set among rural blocks just beyond Darwin's easternmost suburbs, Knuckey Lagoons comprises four

Fig. 16. The seasonally inundated floodplains of the Top End are home to many breeding waterbirds from the mid-Wet season through to the early Dry, while the surrounding sedgelands and grasslands, such as these adjacent to Holmes Jungle, are habitat for grassland-dwelling birds, such as the two species of cisticolas, year-round.

Fig. 17. Leanyer Sewage Ponds attracts large numbers of ducks, mainly during the Dry season, as well as several rare Northern Hemisphere migrants.

roughly circular depressions that hold water for six to 10 months, depending on their depth (Figs 6, 18). They represent the most significant natural wetland, and along with Leanyer SP are the most important of all wetlands for waterbirds in the Darwin region.

Fig. 18. Fiddler's Lagoon, here shown in the early Dry season, is one of the four Knuckey Lagoons, located just outside the metropolitan area of Darwin.

References

1. Noske RA, Brennan GP (2002) *The Birds of Groote Eylandt.* Northern Territory University Press, Darwin.
2. McCrie N, Watson J (2003) *Finding Birds in Darwin, Kakadu and the Top End, Northern Territory, Australia*. Niven McCrie, Casuarina.
3. Storr GM (1967) *List of Northern Territory Birds.* Western Australian Museum, Perth.
4. Crawford DN (1972) Birds of the Darwin area, with some records from other parts of Northern Territory. *Emu* **72**, 131–148. doi:10.1071/MU972131
5. Crawford DN (1972) First record of Redshanks in Australia. *Emu* **72**, 112.
6. Crawford DN (1972) First Australian record of Asiatic Dowitcher. *Emu* **72**, 112–113.
7. Thompson HAF (1978) Common birds of the Darwin suburbs. *Northern Territory Naturalist* **1**, 7–12.
8. McKean JL (1981) The status of gulls and terns (Laridae) in the Darwin area 1974-80. *Australian Seabird Group Newsletter* **15**, 11–16.
9. Thompson HAF (1982) The status of cuckoos Cuculidae in the Darwin area Northern Territory, 1974–1980. *Northern Territory Naturalist* **5**, 13–19.
10. Thompson HAF (1984) The status of kingfishers and their allies (Coraciiformes) in the Darwin area, Northern Territory, 1974 to 1982. *Northern Territory Naturalist* **7**, 18–29.
11. McKean JL (1986) The status of cuckoo-shrikes (Campephagidae) in the Darwin area, Northern Territory, 1974–1984. *Northern Territory Naturalist* **9**, 8–14.
12. Shurcliff KS, McKean JL (1990) The status of ducks (Anatidae) in the Darwin area, Northern Territory, 1974–1986. *Australian Bird Watcher* **13**, 243–248.
13. Shurcliff KS, Thompson HAF (1993) Seasonal patterns of wader populations in Darwin, Northern Territory, 1974–1987. *Northern Territory Naturalist* **14**, 10–18.
14. Thompson H, Goodfellow D (1987) *Common Birds of the Darwin Area*. Sandpiper Productions, Winnellie.
15. Woinarski JCZ, Tidemann SC, Kerrin S (1988) Birds in a tropical mosaic: the distribution of bird species in relation to vegetation patterns. *Australian Wildlife Research* **15**, 171–196. doi:10.1071/WR9880171
16. Morton SR, Brennan KG (1991) Birds. In *Monsoonal Australia: Landscape, Ecology and Man in Northern Lowlands* (Haynes CD, Ridpath MG, Williams MAJ eds), pp. 133–149. A.A. Balkema, Rotterdam.
17. Morton SR, Brennan KG, Armstrong MD (1991) *Distribution and Abundance of Waterbirds in the Alligator Rivers Region, Northern Territory*. Supervising Scientist for the Alligator Rivers Region, Jabiru.
18. Brooker MG, Parker BS (1985) The birds of Kakadu National Park. In *The Kakadu Fauna Survey: An Ecological Survey of Kakadu National Park, Vol. 2* (Braithwaite RW ed.), pp. 327–388. Final Report to Australian National Parks and Wildlife Service. Division of Wildlife and Rangelands Research, CSIRO, Darwin.
19. Woinarski JCZ (1993) A cut-and-paste community: birds of monsoon rainforests in Kakadu National Park, Northern Territory. *Emu* **93**, 100–120. doi:10.1071/MU9930100
20. Woinarski JCZ, Press AJ, Russell-Smith J (1989) The bird community of a sandstone plateau monsoon forest at Kakadu National Park, Northern Territory *Emu* **89**, 223–231. doi:10.1071/MU9890223
21. Chatto R (2000) *Waterbird Breeding Colonies in the Top End of the Northern Territory.* Technical Report No 69. Parks and Wildlife Commission of the Northern Territory, Palmerston.

22. Chatto R (2001) *The Distribution and Status of Colonial Breeding Seabirds in the Northern Territory*. Technical Report No 70. Parks and Wildlife Commission of the Northern Territory, Palmerston.
23. Chatto R (2003) *The Distribution and Status of Shorebirds Around the Coast and Coastal Wetlands of the Northern Territory*. Technical Report No 73. Parks and Wildlife Commission of the Northern Territory, Palmerston.
24. Commonwealth of Australia Bureau of Meteorology (2014) Climate statistics for Australian locations http://www.bom.gov.au/climate/averages/tables/cw_014015.shtml.
25. McDonald NS, McAlpine J (1991) Floods and droughts: the northern climate. In *Monsoonal Australia: Landscape, Ecology and Man in Northern Lowlands* (Haynes CD, Ridpath MG, Williams MAJ eds), pp. 19–29. A.A. Balkema, Rotterdam.
26. Taylor JA, Tulloch D (1985) Rainfall in the wet-dry tropics: extreme events at Darwin and similarities between years during the period 1870–1983 inclusive. *Australian Journal of Ecology* **10**, 281–295.
27. Cook GD, Heerdegen RG (2001) Spatial variation in the duration of the rainy season in monsoonal Australia. *International Journal of Climatology* **21**, 1723–1732. doi:10.1002/joc.704
28. Panton WJ (1993) Changes in post World War II distribution and status of monsoon rainforests in the Darwin area. *The Australian Geographer* **24**, 50–59. doi:10.1080/00049189308703088
29. Chatto R (2006) *The Distribution and Status of Waterbirds around the Coast and Coastal Wetlands of the Northern Territory*. Technical Report No. 76. Parks and Wildlife Commission of the Northern Territory, Palmerston.
30. Wilson FA, Brocklehurst PS, Clark MJ, Dickinson KJM (1990) *Vegetation Survey of the Northern Territory, Australia*. Technical Report No. 49. Conservation Commission of the Northern Territory, Darwin.
31. Tidemann SC, Lawson C, Elvish R, Boyden J, Elvish J (1999) Breeding biology of the Gouldian Finch *Erythrura gouldiae*, an endangered finch of Northern Australia. *Emu* **99**, 191–199. doi:10.1071/MU99022
32. Woinarski JCZ, Tidemann SC (1991) The bird fauna of a deciduous woodland in the wet-dry tropics of northern Australia. *Wildlife Research* **18**, 479–500. doi:10.1071/WR9910479
33. Griffioen PA, Clarke MF (2002) Large-scale bird-movement patterns evident in eastern Australian atlas data. *Emu* **102**, 99–125. doi:10.1071/MU01024
34. Noske RA (2003) Monsoon madness: the upside-down breeding seasons of Top End birds. *Wingspan* **13**, 16–20.
35. Noske RA, Franklin D (1999) Breeding seasons of land birds in the Australian monsoon tropics: diverse responses to a highly seasonal environment. *Australian Biologist* **12**, 72–90.
36. Noske RA (1998) Breeding biology, demography and success of the rufous-banded honeyeater, *Conopophila albogularis*, in Darwin, a monsoonal tropical city. *Wildlife Research* **25**, 339–356. doi:10.1071/WR97070
37. Whitehead PJ, Saalfeld K (2000) Nesting phenology of magpie geese (*Anseranas semipalmata*) in monsoonal northern Australia: responses to antecedent rainfall. *Journal of Zoology* **251**, 495–508. doi:10.1111/j.1469-7998.2000.tb00805.x
38. Zimmermann U, Noske RA (2003) The breeding ecology of the monsoon rainforest-endemic Rainbow Pitta *Pitta iris*. *Emu* **103**, 245–254. doi:10.1071/MU02005
39. Tidemann SC, Woinarski JCZ (1994) Moult characteristics and breeding seasons of Gouldian *Erythrura gouldiae*, Masked *Poephila personata* and Long-tailed Finches *P. acuticauda* in savanna woodland in the Northern Territory. *Emu* **94**, 46–52. doi:10.1071/MU9940046
40. Frith HJ, Davies SJJF (1961) Breeding seasons of birds in subcoastal Northern Territory. *Emu* **61**, 97–111. doi:10.1071/MU961097

41. Corbett L, Hertog T (2011) Diet and breeding of White-bellied Sea-Eagles *Haliaeetus leucogaster* in subtropical river habitats in the Northern Territory, Australia. *Corella* **35**, 41–48.
42. Franklin DC, Noske RA (2000) Nectar sources used by birds in monsoonal north-western Australia: a regional survey. *Australian Journal of Botany* **48**, 461–474. doi:10.1071/BT98089
43. Noske RA (2001) The breeding biology of the Mangrove Gerygone *Gerygone laevigaster* in the Darwin region, with notes on brood parasitism by the Little Bronze-cuckoo *Chrysococcyx minutillis*. *Emu* **101**, 129–135. doi:10.1071/MU00013
44. Mohd-Azlan J, Noske RA, Lawes MJ (2014) Resource partitioning by mangrove bird communities in North Australia. *Biotropica* **46**, 331–340. doi:10.1111/btp.12108
45. Bach CS (2002) Phenological patterns in monsoon rainforests in the Northern Territory, Australia. *Austral Ecology* **27**, 477–489. doi:10.1046/j.1442-9993.2002.01209.x
46. Noske RA (1997) Short breeding season of Figbirds *Sphecotheres viridis* in Darwin, Northern Territory. *Corella* **21**, 44–47.
47. Hall BP (ed.) (1974) *Birds of the Harold Hall Australian Expeditions, 1962–1970: A Report on the Collections made for the British Museum of Natural History*. British Museum (Natural History), London.
48. Australian Bureau of Statistics (2014) Census QuickStats http://www.abs.gov.au/websitedbs/censushome.nsf/home/census?opendocument&navpos=10.
49. Deignan HG (1964) Birds of the Arnhem Land Expedition. In *Records of the American Australian Scientific Expedition to Arnhem Land*. (RL Specht ed.), pp. 345–425. Melbourne University Press, Melbourne.
50. Wightman G (2006) *Mangroves of the Northern Territory, Australia. Identification and Traditional Use*. NT Botanical Bulletin No. 31. Department of Natural Resources, Environment and the Arts, Palmerston.
51. Noske RA (1996) Abundance, zonation and feeding ecology of birds in mangroves of Darwin Harbour, Northern Territory. *Wildlife Research* **23**, 443–474. doi:10.1071/WR9960443
52. Sato NJ, Tokue K, Noske RA, Mikami OK, Ueda K (2010) Evicting cuckoo nestlings from the nest: a new anti-parasitism behaviour. *Biology Letters* **6**, 67–69. doi:10.1098/rsbl.2009.0540
53. Tokue K, Ueda K (2010) Mangrove Gerygones *Gerygone laevigaster* eject Little Bronze-cuckoo *Chalcites minutillus* hatchlings from parasitized nests. *The Ibis* **152**, 835–839. doi:10.1111/j.1474-919X.2010.01056.x
54. Mohd-Azlan J, Noske RA, Lawes MJ (2012) Avian species-assemblage structure and indicator bird species of mangroves in the Australian monsoon tropics. *Emu* **112**, 287–297. doi:10.1071/MU12018
55. Noske RA (2003) The role of birds in mangroves – pollination and insect predation. In *Proceedings: Darwin Harbour Region: Current Knowledge and Future Needs* (Working Group for the Darwin Harbour Advisory Committee ed.), pp. 74–86. Department of Infrastructure, Planning and Environment, Darwin.
56. Price OF (2006) Movements of frugivorous birds among fragmented rainforests in the Northern Territory, Australia. *Wildlife Research* **33**, 521–528. doi:10.1071/WR06029
57. Brock J (1988) *Top End Native Plants*. John Brock, Winnellie.
58. Woinarski JCZ, Connors G, Franklin DC (2000) Thinking honeyeater: nectar maps for the Northern Territory. *Pacific Conservation Biology* **6**, 61–80.
59. Woinarski JCZ, Brock C, Armstrong M, Hempel C, Cheal D, Brennan K (2000) Bird distribution in riparian vegetation in the extensive natural landscape of Australia's tropical savanna: a broad-scale survey and analysis of a distributional data base. *Journal of Biogeography* **27**, 843–868. doi:10.1046/j.1365-2699.2000.00439.x

60. Franklin DC, Brocklehurst PS, Lynch D, Bowman DMJS (2007) Niche differentiation and regeneration in the seasonally flooded *Melaleuca* forests of northern Australia. *Journal of Tropical Ecology* **23**, 457–467. doi:10.1017/S0266467407004130
61. Crawford DN (1980) Saline coastal swamp in Northern Territory as a habitat for waterbirds. *Emu* **80**(1), 36–38.

About this book

The Darwin region

In this book the Darwin region is defined as the area bounded by 12°19′S in the north, 12°35′S in the south, 130°49′E in the west and 131°05′E in the east. This square of 16′ × 16′ is roughly 29 × 27 km, covering an area of about 800 km^2. The nearest well known localities to these boundaries are Lee Point in the north, Howard Springs NR and McMinns Lagoon in the east, Channel Island road in the south, and East Point in the west.

The data

Almost 120 000 records comprise the dataset from which the distribution maps and seasonality charts were created. The relative contributions of the major sources of data are shown in Fig. 19. Those records for which locality information was imprecise, or considered to be inaccurate, were eliminated from the dataset used for the maps. Records of national rarities were retained if there was sufficient supportive evidence, even if they had not been submitted for appraisal by the Birds Australia Rarities Committee (BARC). Examples of these are relatively few, and the majority of vetting concerned rarities at a regional level. Many of these were likely the result of observers making assumptions about bird distribution based on experience elsewhere, so species documented as, for example, Grey Fantail or Little Eagle may have been Northern Fantail and immature Brahminy Kite. Some misidentifications may have remained in the dataset, but it is unlikely that their number is consequential. Further notes on how the data were used in preparation of the maps and charts are provided below.

The average annual number of records per decade in the dataset show a steady increase since 1960, while the average number of species rose abruptly during the late 1960s and 1970s, then approached a plateau during the 1980s (Fig. 20).

Species accounts

The species accounts following this introductory section form the core of this book. In addition to the text, most accounts comprise a brief profile of the species concerned, a map of its distribution in the Darwin region, and a chart showing its seasonal occurrence. However, for species that have been recorded fewer than six times or in fewer than three grid

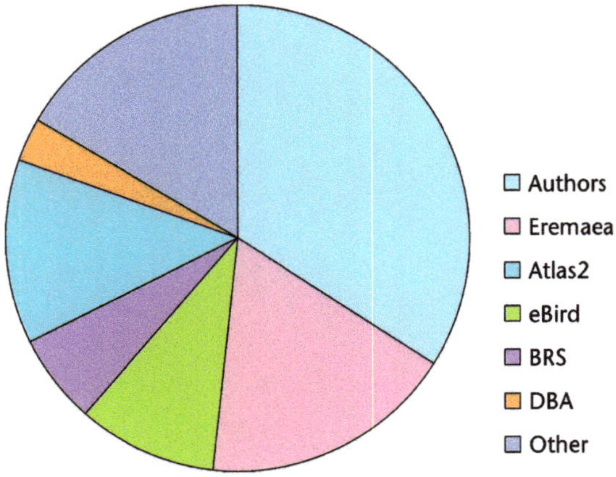

Fig. 19. Pie chart showing the main sources of data. Eremaea data supplied by R & M Alcorn; Atlas 2 and BRS data supplied by NT Department of Natural Resources, Environment, the Arts and Sport; eBird Basic Dataset, Version: EBD_relAug-2013, Cornell Laboratory of Ornithology, Ithaca, New York, August 2013; DBA data from original Darwin Bird Atlas project. 'Other' includes data obtained from various individuals.

cells, the locations and months of those records are given in the text, in lieu of a map and chart, respectively.

For English and scientific names, and the sequence of species, we have followed the International Ornithological Committee (IOC) World Bird List, version 5.1,[1] in recognition of its being an up-to-date authority on the taxonomy of all bird species. It is no longer tenable, for example, to refer to Australia's cicadabird as simply 'Cicadabird', when there are 13 species of cicadabirds distributed from the Philippines to the Solomon Islands.

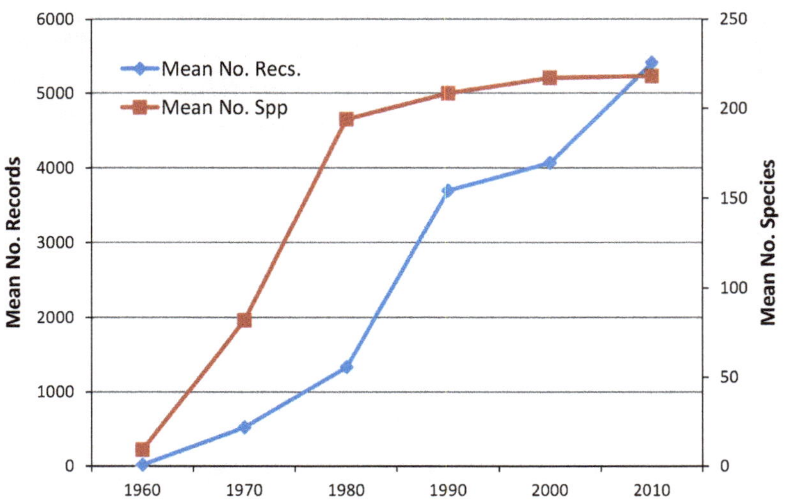

Fig. 20. Average annual numbers of records and species in the Darwin region dataset per decade since 1960. Data for final decade incomplete.

Species profiles

Heading each species account is a concise summary of the species' global geographical range, and its habitats, status and breeding season within the region.

Range broadly describes the species' global distribution, highlighting those that are endemic to the Australian monsoon tropics. The purpose of this information is not to provide details of everywhere it may be found, but rather to place its local distribution into a wider context, including breeding and non-breeding range where appropriate. Because many species recorded in the Top End occur in Indonesia, and distributional details of birds in Indonesia are not readily available to most people, we have sometimes specified particular islands occupied by birds. We hope that this will stimulate interest in the birds of eastern Indonesia, which is one of the most poorly known regions in the world. Additionally some Australian birds migrate to Indonesian islands, and surveys are needed to establish the extent of this migration.

Habitats list the major Top End habitats in which the species occurs using the categories identified in the Introduction section of this book. As habitats are usually described in more detail in the main text for each species, the habitat types provided here are very broad.

Status provides a summary of the relative abundance and seasonality of the species in the Darwin region. Traditional terms such as 'common' and 'rare' are used to imply both abundance and spatial distribution. *Residents* are here defined as those species or populations in which the majority of adults are thought to be sedentary, spending their whole lives in a relatively small area. Thus residents are expected to breed in the region.

In contrast, *migrants* are species that are known to undertake long-distance movements from their breeding grounds to wintering grounds every year. Most such birds do not breed in Australia, but there are some exceptions, as discussed in the next section. In Australia many species appear to have both resident and migratory populations, often in the same area. The remaining species are classified as *visitors* because they do not breed in the region and appear to undertake smaller scale seasonal movements within the Top End.

Breeding refers to the species' breeding season, which is traditionally defined as the months in which eggs are laid. As the dates of egg-laying are rarely known, they are usually extrapolated from records of the dates on which nestlings or fledglings were observed. Much of the breeding season information published in field guides and non-scientific bird books is imprecise and, in the case of the Top End, often erroneous due to unquestioning repetition of information from the early literature, which suffers from inaccuracies of provenance and limited sampling. In this book, therefore, we have used original unpublished data of our own or provided by other observers, as well as published data from primary sources, and tried to limit the use of secondary information sources. Where breeding data are not available for the Darwin region (DR), we have indicated whether they refer to the Top End (TE) or Northern Territory (NT). Additionally, 'not in NT' indicates Australian breeding birds that are not known to breed in the NT, while 'not in Australia' is used for non-breeding migrants to Australia.

Distribution maps

For each species a map is provided to show both the extent of its distribution and its relative reporting rate within the region. Unlike traditional bird atlas records, the data used for the species maps in this book were not submitted as individual surveys, but derive from a range of sources, as previously described. In order to calculate reporting rates for map grid cells, all records from a particular cell on the same date were considered to form a single survey. The following simple formula was then applied, using surveys containing three or more records:

$$\text{cell reporting rate} = \frac{\text{sum of records of species for cell}}{\text{sum of surveys for cell}} \times 100$$

The reporting rates for each cell were then compared with the maximum reporting rate of the species (in cells with 20 or more surveys) to provide a 'relative reporting rate', which was used to determine the size for each dot on the map. Again, a simple formula was used:

$$\text{relative reporting rate} = \frac{\text{cell reporting rate}}{\text{maximum reporting rate}} \times 100$$

As might be expected, the number of surveys in each grid cell varied (Fig. 21). In several parts of the region, particularly in the remote north-east, access is limited both by physical barriers such as a lack of roads, particularly all-weather roads, and by entry restrictions, if the land is private property or belongs to the Australian Government Department of Defence. In such cells all species recorded will have a high reporting rate, simply because of the low number of surveys. In order to reflect this bias in reporting rates, the dots on each map have been assigned two colours. Dots in those cells for which fewer than 20 surveys were available are coloured grey, while those for which there are 20 or more surveys are black. Figure 22 illustrates how the dot sizes and colours are represented on the maps.

Seasonality charts

Charts are used to show monthly fluctuations in the relative frequency of records of each species. The number of bird surveys and records shows a gradual rise over the Dry season, particularly from June onwards, followed by a dramatic drop at the start of the Wet (Fig. 23). The former probably partly reflects an increase in the number of observers over the Dry season, as this is the most popular time for interstate birdwatchers, as well as other tourists, to visit. However, it also probably reflects a rise in survey effort by local birdwatchers, as the late Dry heralds the arrival of migratory shorebirds on the coast and of large aggregations of waterbirds on permanent water bodies. The lower survey effort over the Wet season probably relates to an increase in inclement weather, as well as reduced access to some areas in the second half of the season due to flooding. The latter also introduces a spatial bias to sampling. While it is impracticable to filter the data to eliminate

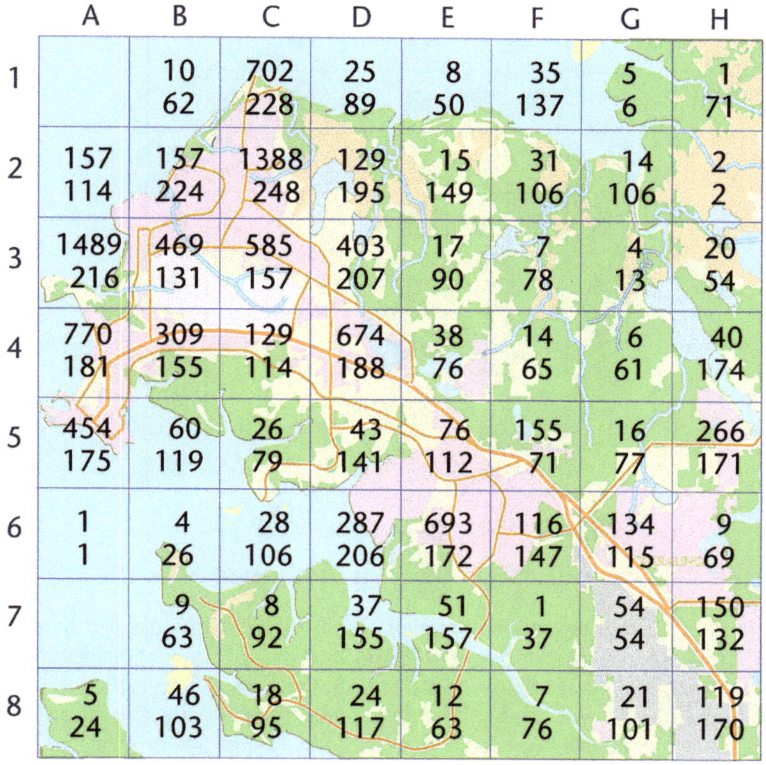

Fig. 21. Map of Darwin region grid cells showing number of surveys (upper value) and number of species (lower value) for each cell. The physical features in the background (based on Fig. 7) provide context for the uneven spatial distribution of surveys. Note the lack of surveys in the north-east corner (due to difficulties of access).

these sources of bias, we have endeavoured to compensate for the seasonal bias in sampling by calculating a reporting rate for each species, based on the following simple formula:

$$\text{monthly reporting rate for species A} = \frac{\text{sum of records of species A in month}}{\text{sum of surveys for month}} \times 100$$

To further highlight seasonality, different coloured bars are used to represent Wet season months and Dry season months, although it should be noted that these seasons grade into each other during the Build-up and Wet–Dry transition periods (see Introduction: Climate). It should also be remembered that the charts are based on the frequency of records rather than counts, so they do not necessarily reflect the numerical abundance of the species over the year. Figure 23 shows a chart for a typical Dry season migrant.

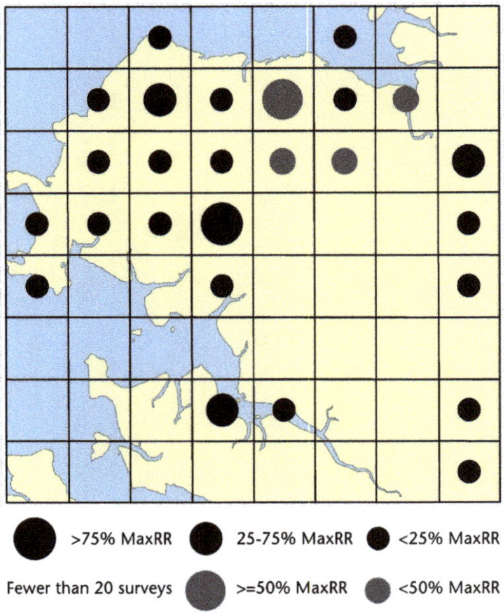

Fig. 22. Sample distribution map, with an explanation for the dot colours and sizes. With a MaxRR value of 10% for example, a large black dot indicates that the species was reported in that cell from 7.5 to 10% of surveys, while a medium-sized dot indicates a reporting rate of 2.5 to 7.5%, and a small dot, less than 2.5% reporting rate. Grey dots represent cells with fewer than 20 surveys, with a large grey dot in this example representing a reporting rate of 5% or more, and a smaller grey dot less than 5%.

In addition all 64 grid cells were assigned to one of two broad geographical categories (coastal or subcoastal), and to one of three landscape categories (urban, rural or natural). Chi-squared tests were performed on these data for many species for the purpose of

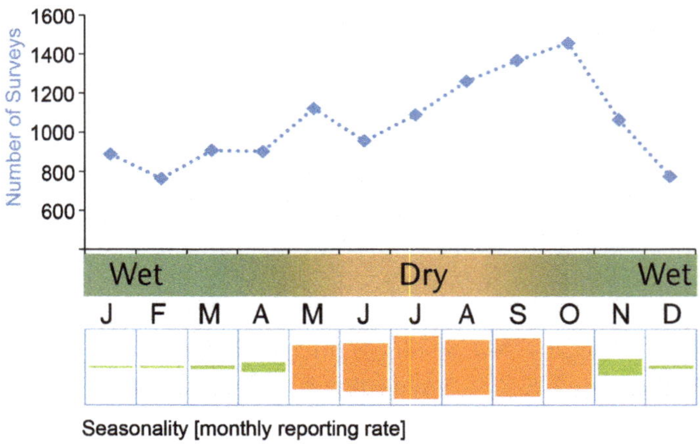

Fig. 23. Seasonality chart (below) of Australian Pratincole, a typical Dry season migrant, showing the coloured bars used to indicate Wet season and Dry season reporting rates. The line graph (above) shows the number of surveys for each month. The graduated band, not present on charts in the species accounts, gives a more realistic representation of the seasons.

objectively describing their pattern of distribution. These tests compared the frequency of records in each cell for the species concerned with their expected frequency according to the cell coverage of each category. Yate's correction factor was applied in the case of the geographical comparison, where there are only two categories.

Reference

1. Gill F, Donsker D (eds) (2015) IOC World Bird List (v 5.1). doi:10.14344/IOC.ML.5.1

Abbreviations

CR: Coastal Reserve

DR: Darwin region

NG: New Guinea

NR: Nature Reserve

NP: National Park

NSW: New South Wales

NT: Northern Territory

Qld: Queensland

SA: South Australia

SP: Sewage Ponds

Tas: Tasmania

TE: Top End

Vic: Victoria

WA: Western Australia

Species accounts

Emu
Dromaius novaehollandiae

RANGE: mainland Australia. HABITATS: mainly woodland and grassland. STATUS: rare visitor. BREEDING: May–July (NT).

Judy Leitch

Though there are scattered records of Emus throughout the Top End, the species is rarely seen near Darwin. The first report was in open forest near Howard Springs, in December 1967,[1] and a 'half-grown' bird was also observed near there in August 1973.[2] More recent observations include single birds seen in the McMinns area in 1978, at Karama in December 1995, and in the Howard Springs area in December 1998 and June 2010.[3]

There has been a single published report of breeding in the Darwin region,[4] though there are no details provided. The sighting of two chicks along the Wangi Road in Litchfield NP, 80 km south of Darwin, in July 1973,[2] and an adult with one chick in that area in December 2008 provide the nearest evidence of breeding to Darwin. Further out, single males with up to five chicks were also sighted at Kapalga, Kakadu NP, in June, July and December in three years between 1978 and 1993.

There was an apparent drop of up to 80% in the reporting rate for Emus in the NT during the 20 years 1981 to 2001.[5] While this may overstate the actual decline, the species is listed as Vulnerable in the NT, indicating a reduction of over 30% in the last 10 years. The major threat is considered to be fire,[6] which contributes to a reduction in the amount and type of food available, and may reduce breeding success through the destruction of eggs.

Magpie Goose
Anseranas semipalmata

RANGE: largely tropical Australia and NG. HABITATS: wetlands, floodplains, cultivated areas. STATUS: very common visitor (mainly Dry season), but resident in TE. BREEDING: Feb–May (TE).

The Magpie Goose is an archaic species that possibly evolved around 66 million years ago, at the time of dinosaur extinction, and is more closely related to the

Niven McCrie

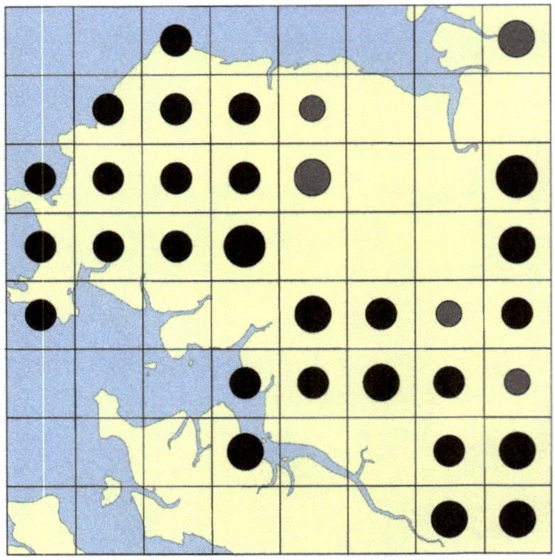

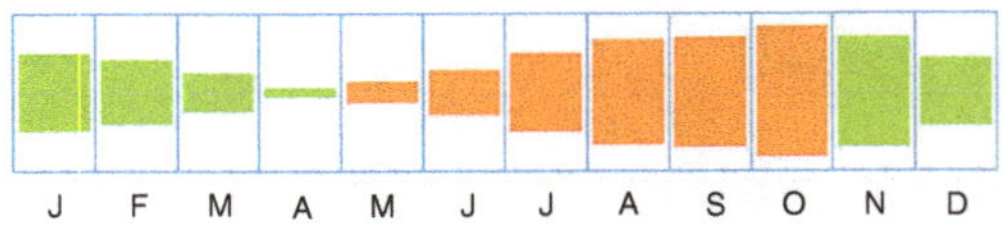

South American screamers than to any other birds. The sole member of the family Anseranatidae, it is unique in many respects, and holds the status as Australia's most studied waterfowl. Unlike other ducks and geese, Magpie Geese do not moult their flight feathers simultaneously after breeding, and so are able to fly year-round.[7]

Magpie Geese occur in suitable habitat throughout the Top End in vast numbers. In the mid-1980s[8] the population of Geese was estimated to be 2.3 million birds for the coastal plains of the western Top End, but more recent estimates suggest the population is much lower, at about one million.[9] From 1958 to 1971,[10] Goose numbers in the Top End declined by as much as 80%. Initially the Water Buffalo was blamed,[1] deep channels caused by their tracks thought to cause premature drainage of nesting habitat. By 1980 numbers had recovered to levels comparable to those before the decline,[10] and subsequent research, between 1980 and 1988 at Kapalga, Kakadu NP, showed that buffalo trampling and grazing influenced the location of nests, but not their numbers. It was found that the decline was more likely due to a combination of fire damage to floodplains and poor rains. Indeed, nesting studies on the Mary River during 1988–93[11] revealed that the density of nests fluctuates markedly from year to year, almost failing completely in El Niño years, and depends on the time of the onset of the Wet season. Perhaps ironically, breeding can also fail if late floods occur and drown the nests.[11]

Typically, Magpie Geese in the Top End move seasonally between wetlands with wild rice beds on floodplains when breeding during the late Wet, to wetlands with spike-rush during the Dry season, where they use their hooked bills to dig up bulbs. Although Magpie Geese may move considerable distances under adverse conditions, satellite telemetry tracking of 10 Geese in Kakadu NP[12] indicated that when adequate resources are available, they do not move far, with the maximum linear distance any of the Geese travelled from the site of release being 114 km. Indeed, a Goose banded near Humpty Doo in 1953 had its band recovered after the bird was shot some 23 years later, only 7 km from where it was banded.

In the Darwin region, both reporting rates and counts are highest in October, with high counts generally continuing into the first week of November. Numbers then drop markedly as birds move out to feeding areas opened up by early Wet season rains, to prepare for the breeding season, and between March and June, Magpie Geese are almost completely absent from the region. Numbers are greatest at Knuckey Lagoons, typically peaking in October, and between 1995 and 2010 the mean October count at the site was over 1500 birds. The highest count, of 6000 birds, occurred there in October 1990 and there were 4000 present in October–November 2002 and October 2003. There have also been several counts in excess of 1000 birds at Holmes Jungle swamp, but otherwise much smaller numbers, generally not more than the low hundreds, occur around other small wetlands, in orchards, or on open grassed areas such as golf courses and parklands.

Counts over the years 1974–86,[13] covering six sites in the Darwin region, showed an average annual total of 5527 birds, but there was a significant difference between 1979, when the maximum total was 12 853, and 1986, when only 1423 were counted. This was considered a cyclical phenomenon, rather than a long-term population decline. The timing of Goose movements also varies between years, depending on floodplain water levels. In 2001–02, for example, there was an earlier than usual influx into the Darwin region, with high numbers in early August. The highest December count was also in that year, suggesting that wetlands were still dry, possibly leading to a delay in breeding.

Magpie Geese are the most sought after waterfowl by hunters in the Top End, and comprise some 80% of total waterfowl taken. About 90% of recreational hunting takes place at Howard Springs Hunting Reserve and Shoal Bay Coastal Reserve within the Darwin region, and at Lambell's Lagoon Conservation Reserve and Harrison Dam Conservation Reserve not far outside of the region. In 1984 some 58 000 geese were shot in these areas, representing about 10% of the regional population, but in 1985, this was reduced to 17 000.[14] In 2001[15] an estimated 33 000 birds were taken with 985 permits, and in 2008, there were about 1600 permits issued, suggesting a possible total harvest in excess of 50 000 birds that year. Estimates of numbers harvested by Indigenous people in the NT vary markedly, with one estimate of up to 288 000[16] per year and other estimates from less than 60 000[17] to about 150 000. It is also estimated that at least an additional 20 000 birds die as a result of wounding by hunters, or are illegally shot by crop growers. Geese are known to ingest the toxic lead shotgun pellets that accumulate in wetlands, using them for grinding up food in the gizzard, and it has been shown that the ingestion of just one or two pellets is sufficient to cause the death of an adult Goose. The use of toxic shotgun pellets has been banned throughout Australia, but in the NT the ban applies only to hunting under permit, therefore excluding Indigenous shooters. While the ban will significantly reduce the levels of toxic shot in wetlands, at least those controlled by the NT Parks and Wildlife Commission, the effects of lead shot already deposited may continue to be felt by birds for some time.

The late Dry season influx of Magpie Geese into Darwin, to alternative feeding habitats, is driven by the drying up of subcoastal wetlands, as ground becomes too dry for birds to dig up wetland plants. It also appears to be partly driven by hunting pressure, as birds flee wetlands during September when the hunting season gets underway. Mango growers in areas near hunting reserves also blame the hunting on the influx of birds to their orchards, where they can cause considerable damage to fruit, with losses of up to one-third of crops.[18] However, it may be that the birds have become reliant on the fruit for food, and the timing of their arrival in orchards may be more about the stage of fruit development than anything else. Birds also move into mango orchards adjacent to lagoons, where they stand under trees seeking shade, not only to eat fruit. Historically, the Magpie Goose has been known for its putative role in the failure of rice farming in the Top End, at Humpty Doo, but this has been shown to be greatly exaggerated.[19]

Although they attain sexual maturity by 24–36 months of age, male Magpie Geese typically do not breed until their fifth year. Polygyny is the norm in Magpie Geese, with the typical breeding unit comprising one male and two females. These trios are considered stable, though there is no certainty that they are lifelong. All three birds participate in building the nest – a large mound of floating vegetation – and in incubating the eggs. Both females lay in the same nest, which contains up to 16 eggs, although it is more usually about 9.[20] After 24 or 25 days of incubation, the young hatch and within a day are able to swim off with their parents, feeding mainly on highly nutritious Wild Rice.

The most important breeding areas in the NT are on the floodplains to the south-west and east of Darwin, the nearest being around the Finniss River and Adelaide River, respectively.[8] Consistent with the lack of major floodplains within the Darwin region, no nesting has been reported closer to Darwin than Fogg Dam, where they breed irregularly in very small numbers.[21]

Spotted Whistling Duck
Dendrocygna guttata

RANGE: Philippines, e. Indonesia, NG. HABITATS: wetlands. STATUS: vagrant. BREEDING: not in NT.

The Spotted Whistling Duck was first recorded in Australia in 1995[22] at Weipa, north Queensland. The number of reports, and birds, have been increasing, and there have been several breeding records at Weipa and elsewhere on Cape York since 2001. It appears that the species is extending its range in Australia, gradually moving further south in Queensland, and it is not unlikely that its movements have also been westward, into the NT. The first report of the species in the Darwin region, and the NT, was of a single bird seen at Leanyer SP from late December 2011 until mid-February 2012.[23] A second record, also of a single bird at Leanyer SP, occurred in December 2012. A subsequent sighting of a Spotted Whistling Duck at Fogg Dam in March 2013[24] may have been the same bird.

The presence of at least 20 birds on Melville Island in July 2014[25] suggests the possibility that the species may also colonise the Top End.

Plumed Whistling Duck

Dendrocygna eytoni

RANGE: n. and e. Australia. HABITATS: wetlands, grassland. STATUS: uncommon visitor, mostly in the Dry. BREEDING: Feb–Apr (NT).

Ashley Carlson

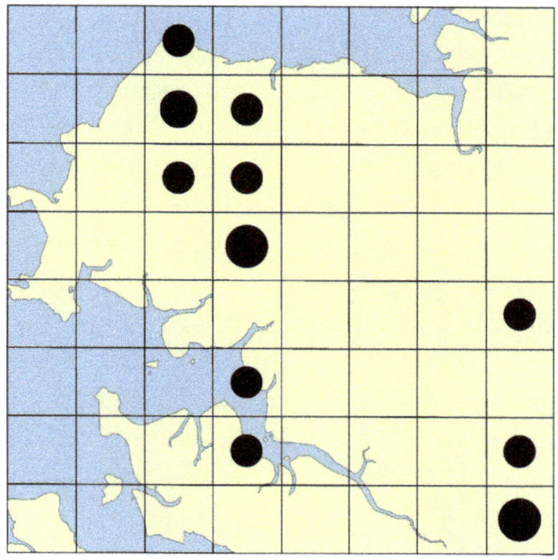

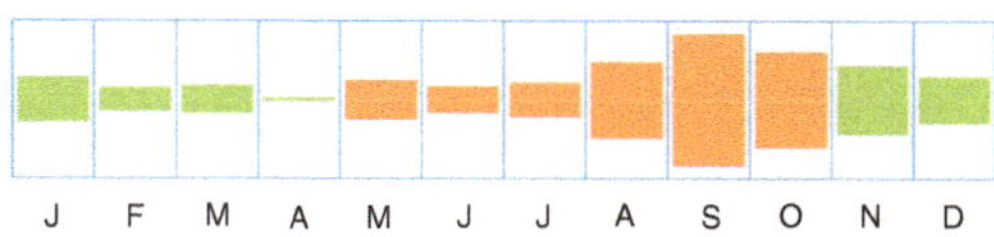

Having more in common with swans and geese than the 'true' ducks, the Whistling Ducks have only a single moult each year. The male and female, which apparently mate for life, show little or no sexual dimorphism and share the nest building, incubation of the eggs and care of the young. While there are clear similarities between Plumed and Wandering Whistling Ducks, one of the more significant differences is in feeding; while Wandering feed almost entirely in water, and commonly during the day, Plumed feed mostly on land, and at night.

Far more widely distributed within Australia than the Wandering Whistling Duck, the Plumed is much less numerous in the Top End than Wandering. It is most typical of savanna wetlands such as in the Barkly Tableland where roosts of more than 10 000 regularly occur, whereas Wandering is relatively uncommon and often absent in the savanna.[26] There are few confirmed breeding records in the Top End, and although some 50 000 Plumed Whistling Ducks, generally in large groups including one recorded flock of up to 15 000 birds,[27] may be present in Kakadu NP at the end of the Dry season, they are thought to originate from subtropical breeding grounds.[28]

The species occurs only rarely in saline wetlands on and near the coast,[27] and is a somewhat irregular visitor to the Darwin region, possibly due to the limited amount of their preferred open grassland habitat. In the years from 1980 to 2013, there were seven in which the species was not recorded in the region, and around Humpty Doo it was not recorded in two out of five years from 1967 to 1971, although there were over 1500 at Fogg Dam in October 1970.[1]

While Plumed Whistling Ducks have been recorded in all months except May, the species is most frequently reported from September through November. Across the Top End as a whole, the highest monthly frequency of records and numbers counted were from March to October.[27] It is interesting that October, when numbers peak in the Darwin region, is at the end of the peak months for the Top End as a whole. It may be that the Darwin region is something of a last resort, perhaps as suitable habit contracts, and birds move out of the Top End, with some staging for a short time in the Darwin region. Alternatively, it may be that hunting pressure is a significant factor, as the season begins during September, and this species, like the Wandering Whistling Duck, is regularly hunted. Counts during the 1990s and 2000s showed the highest mean number was in October, but there was considerable variation in numbers between years. From 1974 to 1986, the maximum number counted in the Darwin region was 170 birds, in November 1978,[13] but in October 2002 1250 birds were recorded at Leanyer SP, and 1000 were present at McMinns Lagoon.

Plumed Whistling Duck breeding in the NT normally takes place inland, south of the Top End, from February to May, but a pair with eight young were seen on Batchelor SP, 80 km south of Darwin, in April 2005,[27] and during the 1950s clutches were found in the months between February and May on wetlands between Darwin and the western edge of Arnhem Land.[29]

Wandering Whistling Duck
Dendrocygna arcuata

RANGE: largely tropical Australia, NG, Philippines and Indonesia. HABITATS: wetlands. STATUS: common resident. BREEDING: Jan–Apr (TE).

Bastiaan Hensen

Of the two species of Whistling Duck regularly occurring in the Darwin region, the Wandering is by far the more abundant, and counts in the low thousands are a regular occurrence at Leanyer SP during the late Dry season. However, these numbers pale into insignificance against the peak total of 400 000 estimated for the Kakadu region during surveys from 1981 to 1984.[28] Wandering Whistling Ducks prefer wetlands with deep water, where they feed by dabbling or diving, reportedly up to 3 m, to strip the flowers or seeds from aquatic plants.[30] Deep water with nearby banks that provide loafing and roosting areas are particularly favoured,[28] and this

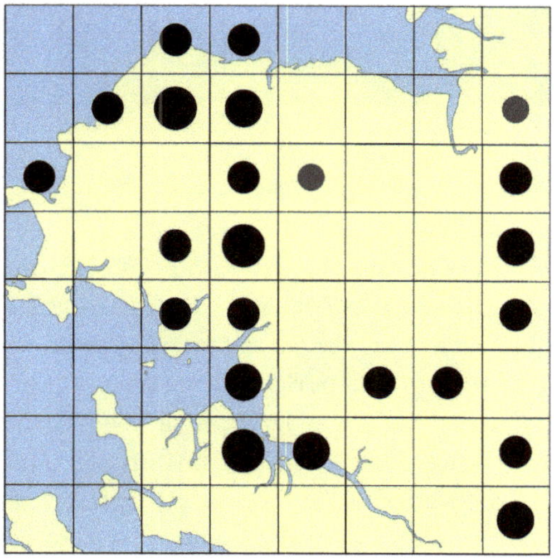

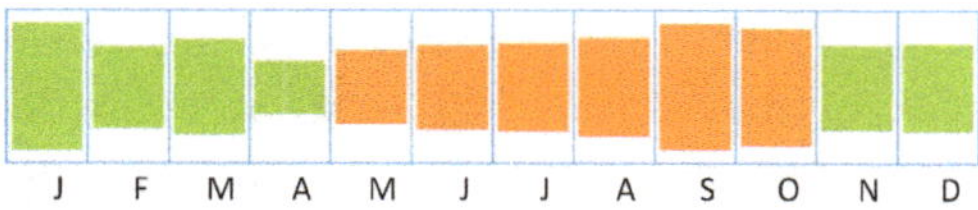

may explain why the abundance of this species in the Darwin region is highest at Leanyer SP.

Both the numbers of birds and the number of records in the Darwin region increase over the Dry season to a peak in September–October, with numbers falling significantly in November or December. A second, sharp peak in January perhaps indicates that birds from outside of the region pass through on their way to breeding areas. The maximum number reported for the region is 6500 birds, at Leanyer SP in November 1974.[13] However, that number appears atypical and, from counts at the site during the period 1995 to 2008, the maximum was 4500 birds in early December 2002, and the next highest was 3500 in early November 1999. These counts produced yearly maximum numbers three times in September and November and twice in December and January. The December high counts were in years of drier than normal conditions, suggesting that birds were perhaps delaying breeding, and using the Darwin region as a refuge. However, the region is also part of the birds' regular annual movements, and the presence at Leanyer SP during the Dry seasons from 1981 until 1987 of a partially albinistic bird, likely the same individual, provides some indication of site fidelity in the Dry season. By contrast, at Knuckey Lagoons, counts were on average only 20% of those at Leanyer SP, with the maximum number counted being 750 in August 1999. Outside Leanyer SP, the largest number recorded in the Darwin region was 2000 at Holmes Jungle Swamp, in January 1987.[31]

Whistling Ducks of both species comprise about 10% of waterfowl taken by hunters in the NT. No recent data are available, but in 1984 this amounted to about 6000 birds of which about 90% were taken within 100 km of Darwin.[14]

The most important breeding area is on floodplains of South Alligator River, where an estimate of up to 30 000 birds have been reported nesting.[32] This is clearly not a regular occurrence, however, and other surveys have put breeding numbers at colonies in the Top End in the hundreds rather than thousands, and it is considered that birds move into the Top End from breeding grounds elsewhere, although it is not known where. Most birds using the Darwin region probably move to the Adelaide River or Daly River floodplains to breed,[27] but some breed locally. The 6–8 eggs, laid on the ground, generally in dense vegetation and usually some distance from water, are incubated by both parents for 28–30 days. After the last duckling has hatched,[27] the brood moves to the water. Newly hatched broods of seven or eight ducklings have been recorded at Leanyer SP from March to May[26] and at Knuckey Lagoons in April.[33]

Black Swan
Cygnus atratus

RANGE: largely e., se. and sw. Australia. HABITATS: wetlands. STATUS: vagrant. BREEDING: Mar–Aug (NT).

Kirri Hardy

Although among the smallest of the world's swans, the Black Swan's neck is proportionately longer than in any other, giving it a uniquely elegant silhouette. While the majority of swan species moult at a regular time each year, the Black Swan does not, as it ties its moult period to weather conditions, making it able to breed at any time of year when conditions are favourable. The Black Swan was first reported in the Top End in October 1972, when two were found near Marrakai Station on the Mary River.[2] Over the following 15 years, however, nine additional sightings were made in the Top End.[13] The sole record from the Darwin region is of an adult bird at Leanyer SP in December 1979,[2] perhaps more than coincidentally following a record from Fogg Dam in July of that year. There were also three at Beatrice Lagoon in July 1986, and three, probably the same birds, at Fogg Dam, the following November.[34] The most recent report for the Top End was a single bird at Leaning Tree Lagoon in January/February 2011.[35] The only breeding record for the Top End concerns a pair of birds that nested at Oenpelli, western Arnhem Land, in March 1980.[4]

Freckled Duck
Stictonetta naevosa

RANGE: largely inland s. and e. Australia. HABITATS: wetlands. STATUS: rare visitor. BREEDING: Mar–May (NT).

Mainly occupying wetlands in central eastern and southern Australia, the Freckled Duck occasionally occurs in more northerly regions; this includes occurrence in shrubby savanna wetlands, where breeding may occur, and in open lakes (artificial or natural) where birds may be seeking drought refuge. At Alice Springs SP, it has

Bastiaan Hensen

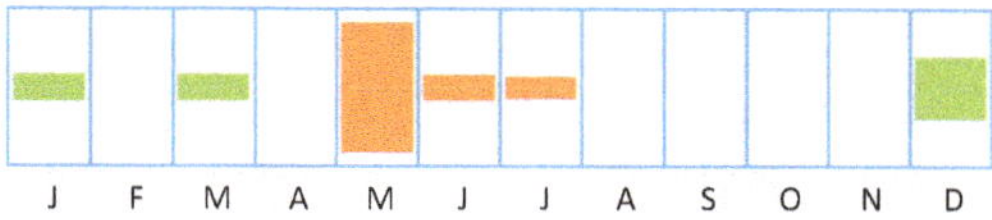

been recorded in all but two years between 1985 and 2013, but it is only infrequently found north of there. During surveys conducted in five years between 1993 and 2002, however, the species was recorded throughout the Barkly Tableland wetlands, mostly in Wet seasons, and breeding was also confirmed.[36] It was first reported for the Top End in the Alligator Rivers region in 1902,[20] but there appear to be no reports from the Darwin region until 1994, when one was found at Leanyer SP. Subsequent records, also from Leanyer SP, were in January 2003,[21] and December 2012.[37] This bird was present for only a few days, then from late March until early July 2013 one was also seen, but it could not be ascertained if these sightings involved the same individual. The latter record was in a period when irruptions of the species were noted in other coastal areas of Australia, with birds reported in unprecedented numbers in south-east and north-east Queensland, and one even reported from Sydney Harbour. Irruptions tend to occur when inland drought follows a series of exceptionally wet years when the naturally fluctuating population has been markedly augmented by inland breeding.[26]

Raja Shelduck
Tadorna radjah

RANGE: tropical Australia and NG. HABITATS: terrestrial and estuarine wetlands. STATUS: moderately common resident. BREEDING: Jan–Jun (TE).

Surely the most confiding of local waterfowl, the Raja Shelduck is a common sight on Darwin's ovals, parks and golf courses, and even on suburban street verges, as well as on wetlands, estuaries and sewage ponds. Although its range in Australia has contracted considerably since the early twentieth century, Raja Shelduck is still abundant in the Top End, and its national conservation status is considered Least Concern. It is also one of the most vocal of native ducks, with the rattling call of the females and hoarse whistle of males commonly heard around Darwin wetlands. It exploits a broader range of habitats than most other duck species, inhabiting mangals and tidal mudflats, as well as freshwater wetlands and rivers. Of the waterfowl breeding in the Top End, Raja Shelduck and Green Pygmy Goose are the only species that nest in tree hollows, and they are often to be seen perching in trees or branches overhanging water. Though not as abundant as other species of ducks, the highly conspicuous Raja Shelduck was the most frequently recorded duck species across the Top End in both ground and aerial surveys during the 1990s.[38] Estimates from

Niven McCrie

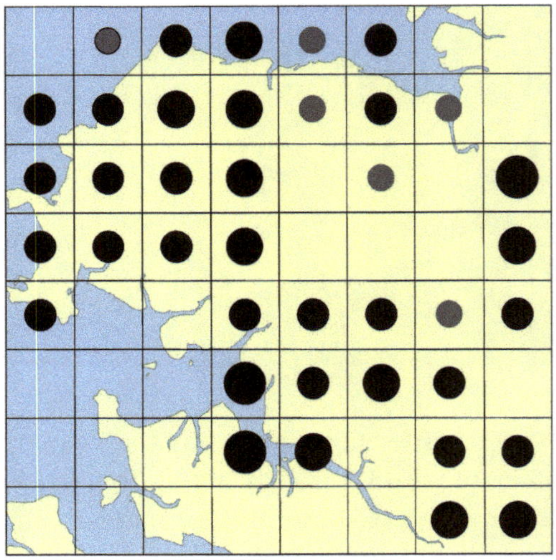

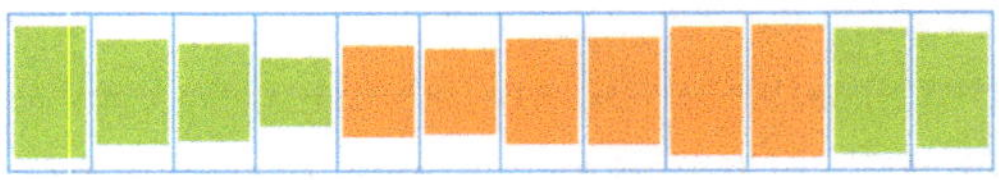

those surveys put the Top End population at about 7000, while surveys between 1981 and 1984 estimated a peak population of about 20 000 for the Alligator Rivers region.[28] Similarly, in the Darwin region it was recorded more often and from more cells than other waterfowl species.

The frequency of records in the Darwin region increases gradually after May until a peak in October, then decreases through the Wet season, apart from a curious spike in January, possibly representing birds on passage to breeding sites.

From 1974 to 1986 Raja Shelduck was the fourth most abundant species of waterfowl on and around Leanyer SP, with an average annual total of 3930 birds. Annual totals in that period fluctuated markedly, with a ninefold fall from 7718 in 1982 to 903 in 1986. Monthly counts during this period showed highest numbers occurring in November, and lowest numbers in May, with the maximum count (3500 birds) in November 1982. In contrast, counts at Leanyer SP over the 14 years from 1995 to 2008 show that monthly averages remained high from December through March. In the period 1995 to 2004, counts in excess of 100 birds were made there in 38 of 83 months, and included all months except June, for which the maximum was 78 birds. Three counts of 600 birds, the largest concentrations for the region, and all at Leanyer SP, occurred in February, September and December.

Around Darwin Golf Course in Marrara, weekly counts from July 2006 to June 2007 rose gradually from August to a peak of 102 birds in December. One to several pairs, or families, regularly visited a Nightcliff lawn beside the mangrove-fringed Kulaluk Bay from 2002 to 2008; visits occurred throughout the year, with a maximum of 70 birds in February 2007. Movements are likely to be regional, as within the NT the species is not known to occur in significant numbers south of the Top End.

Australian Shelduck
Tadorna tadornoides

RANGE: mainly se. and sw. Australia. HABITATS: terrestrial and coastal wetlands, grassland. STATUS: vagrant. BREEDING: not in NT.

Kirri Hardy

Largely restricted to the south-west and south-east of the continent, and rarely straying to the tropics, the Australian Shelduck has only once been recorded in the Top End. An immature male bird was present at Leanyer SP from 16 March 2012 until at least early July 2012.[39,40]

Pink-eared Duck
Malacorhynchus membranaceus

RANGE: whole of Australia. HABITATS: chiefly shallow wetlands. STATUS: irregular Dry and early Wet season visitor. BREEDING: not in TE.

Laurie Ross

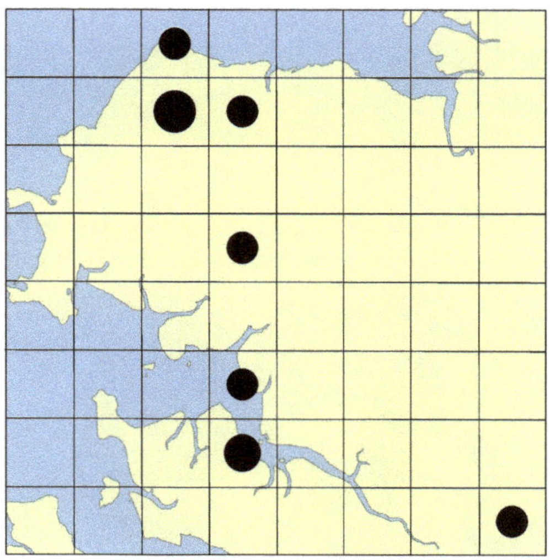

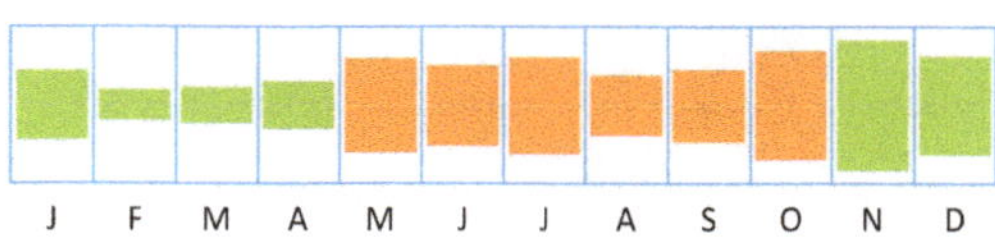

The Pink-eared Duck is considered moderately common in southern NT,[41] yet there appear to be no references to its occurrence in the Darwin region before 1978, although it was subsequently recorded in 26 of the 35 years from then until 2013. Given the scarcity of records in the region before 1978 and their relative regularity since then, and even considering that some birds may have been overlooked, it appears that this species has increased its non-breeding range in recent decades. Sewage ponds near Melbourne and Adelaide attract large aggregations of Pink-eared Ducks, and the creation in 1971 of the Leanyer SP has provided this type of habitat that was not previously available. Indeed, the greatest densities, and the most frequent reporting of the species, are from that site. Pink-eared Ducks may begin arriving early in the Dry season, from April, and by late May or June there are usually birds present if there are going to be any for that year. Surveys across the Top End during the 1990s[27] also yielded few or no records from December to March but, oddly, the highest frequency and abundance occurred in April.

Averaged monthly counts at Leanyer SP[13] during 1974–86 showed a peak in June, with the highest count of 350 birds occurring in June 1985. Over this 10-year period, annual totals of monthly counts varied from six birds in 1983 to 968 in 1980, with an average of 208 birds. In contrast, from 1996 to 2007, monthly average counts rose markedly in June, but did not peak until October, which had an average count of 224 birds, with the highest count of 500 in September and October 1998. During 2013, there was an unusually large influx of waterfowl to coastal areas, apparently Australia wide. At Leanyer SP, numbers of Pink-eared Duck built up from 17 in late March[42] to at least 450 by June.[37]

Major breeding strongholds of this species are the Channel Country, Murray–Darling Basin and inland areas south-west of Australia's east coast. Nearer Darwin, breeding has been recorded in the upper reaches of Sturt Creek, south-west of the Top End, and in the Barkly Tablelands.[26] As wetland habitat is not widely available every year in these parts of the NT, it appears unlikely that breeding occurs annually in these areas.

Maned Duck
Chenonetta jubata

RANGE: throughout Australia but largely excluding tropical n. and extreme arid zones. HABITATS: lightly timbered areas near water. STATUS: vagrant. BREEDING: not in TE.

Niven McCrie

Three birds at Knuckey Lagoons from September to November 1994, and two birds at Leanyer SP in September 1997, appear to be the only records of Maned Duck from the Darwin region. The greatest number recorded for the Top End was 10 birds near the South Alligator River in October 2001. Other Top End sightings, all of individuals, are from Moon Billabong on the Wildman River in July 1979;[4,43] Lake Finniss, about 70 km east of Darwin, in October 1994; Adelaide River sewage ponds in July 2013; and Bird Billabong[44] in September 2013. Given the overall relatively high abundance of the species and its expansion in range, including into cities, coincident with agricultural development in Australia, it is perhaps surprising that few have been reported from the Darwin region.[26]

Green Pygmy Goose
Nettapus pulchellus

RANGE: tropical Australia, NG and rarely e. Indonesia. HABITATS: wetlands. STATUS: moderately common resident. BREEDING: Jan–Mar (TE).

Niven McCrie

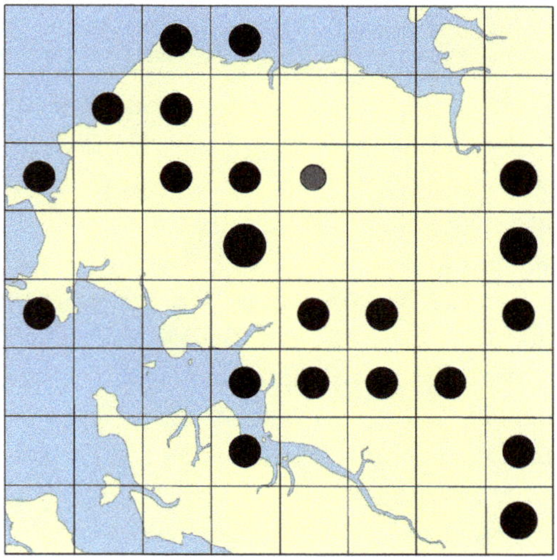

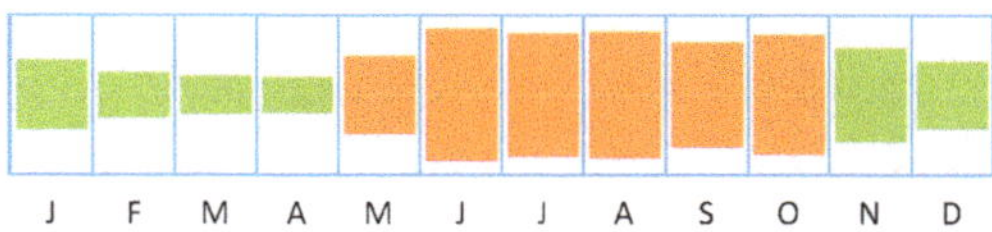

Pygmy Geese, the smallest of the world's waterfowl, are, despite their name, considered to be perching ducks, a classification they share in Australia only with Maned Duck. The largely tropical-restricted Green Pygmy Goose shows a clear association with deep lagoons, lakes and floodplains covered with floating aquatic vegetation. On these water bodies it feeds mainly on the flowers and seeds of waterlilies and aquatic plants on or up to about 30 cm below the water's surface. The highest reporting rates of the species are from McMinns and Knuckey Lagoons. It is only an infrequent visitor to Leanyer SP, probably because of limited aquatic vegetation there.

Unlike most of the waterfowl in the region, the Green Pygmy Goose shows a relatively stable spread of records throughout the Dry season from May to November. More in line with other local waterfowl, however, is the noticeable drop at the beginning of the Wet season, most likely due to the birds' dispersal to small inland wetlands to breed.[27] The small increase in January may represent birds on passage to those areas. This pattern was also evident during surveys of the Top End during the 1990s,[27] with the number of records and abundance of birds being higher between May and October than between December and April.

At Knuckey Lagoons, average monthly counts were also highest from June through October, with a maximum count of 320 in mid-October 1996, but the second highest number, around 200 birds, was in June and July 1997. Large numbers in the early Dry season, such as an estimated 800 at Fogg Dam in July 1970,[1] suggest post-breeding recruitment of young birds. In combination, these observations indicate that the Green Pygmy Goose is sedentary for much of the year, and dispersing or becoming inconspicuous during the Wet season when they breed.

In 1970 Green Pygmy Goose populations in the Fogg Dam area reportedly declined by as much as 90%[45] compared with the previous two years and, although not verified, this phenomenon may have been the case also in the Darwin region. The three previous Wet seasons, particularly 1969, had below average rainfall, and water levels had fallen to the extent that Fogg Dam was unable to support the birds. It was considered that this resulted in mass mortality, but no dead birds were found, and many may have found refuge at Manton Dam, a large permanent body of water about 40 km south-west of Fogg Dam. Additionally, the above average rainfall in the Katherine area in the 1969 Wet season may have opened up alternative, temporary wetlands to which birds may have fled. Within a year of the decline at Fogg Dam, numbers there were back to more than 70% of those pre-1970, further suggesting that birds had moved to alternative refuges, as breeding is unlikely to fully account for such a rapid increase.

At Knuckey Lagoons greater numbers of Green Pygmy Geese are to be found in the deeper, more densely vegetated central area than the shallow water near the edges of the swamp. On the edges birds tend to feed in pairs rather than groups, and regularly chase away other Green Pygmy Geese that infringe on their feeding area. Usually the intruder leaves after the aggressor swims swiftly towards it, but occasionally there are more aggressive confrontations with birds jumping up at each other. These confrontations, however, are not of the intensity reported for birds approaching the breeding period, when males may engage in sustained aggression, sometimes springing out of the water, trilling loudly and buffeting competitors.[20] In contrast, they often feed very close to other species, such as Pacific Black Duck and Hardhead, without incident.

A Brown Goshawk was seen diving on and striking a low-flying Green Pygmy Goose about 1 m above the water at Manton Dam, in July 2002.[46] Both birds dropped to the water, but the half-submerged goshawk soon flew off without the Green Pygmy Goose, which resurfaced briefly 15 m away, then dived again.

Nests are built in tree hollows, although there are unsubstantiated reports of nesting on swamp vegetation. Clutch size is not known with certainty; however, broods in Darwin ranged from four to nine, the latter perhaps unusually large, although no reliable data on typical clutch sizes for the species[30] are available. There are no reports of nests in the Darwin region, but breeding evidenced by the presence of young has been recorded on several occasions. Broods at Knuckey Lagoons in late April 1993[26] and 1995, March 1997[33] and a brood at Marrara Swamp in March 1998,[47] suggest the laying season to be largely in January and February, with dependent ducklings present during March and April.

Pacific Black Duck
Anas superciliosa

RANGE: whole of Australia, NG and NZ. HABITATS: wetlands. STATUS: common, largely Dry season visitor. BREEDING: Dec–May (TE).

Niven McCrie

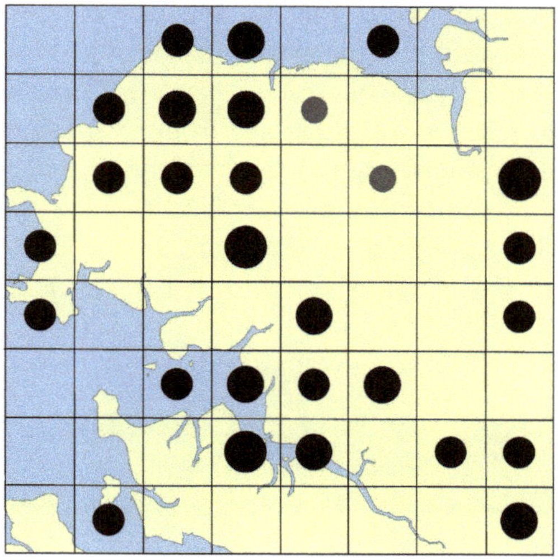

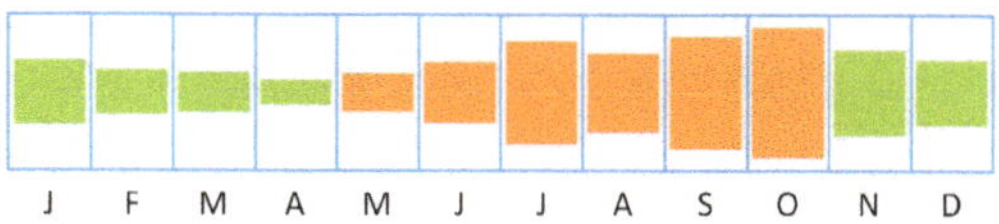

Though recorded in all months of the year in the Darwin region, the Pacific Black Duck is scarce during the major part of the Wet season and early Dry season. The species is known to be dispersive, and birds occurring in the Top End perhaps originate from inland eastern or southern Australia.

From ground surveys in the early 1990s, the Top End population of Pacific Black Duck was estimated at 17 000.[38] Between 1981 and 1984, however, the peak population of the Alligator Rivers region was estimated to be around 50 000, but this estimate was derived from both ground and aerial surveys.

From 1974 to 1986[13] it was the third most abundant species of waterfowl on and around Leanyer SP, with an average annual total of 4267 birds. Annual totals fluctuated markedly from 162 in 1986 to 7547 in 1979. The 1979 total was largely boosted by the unusually early, and unusually large, influx of some 2000 birds in July, many which may have been recounted in following months. Monthly counts during this period showed a similar pattern to the Wandering Whistling Duck, except that numbers rose more quickly during the early Dry season, peaking in October, and were already low by December. In the period 1995-2008 monthly totals were found to be highest in September-October at both Knuckey Lagoons and Leanyer SP. There is a sharp decline in November or December, shortly after numbers peak. In 1996, for example, numbers dropped from 1300 birds on 13 October to 200 by 4 November. The increase during the Dry season is, however, not as marked. Counts in excess of 100 birds have been made from late May, although July and August are typically when the larger influxes occur. Maximum counts at Knuckey Lagoons in this period were 2000 in October 1996 and 1300 in October 2000; at Leanyer SP, the maximum count was 650 in September 1998.

Breeding records in the Top End are few and scattered, but clearly not through lack of nesting habitat. Pacific Black Ducks nest in a variety of situations, from scrapes on the ground to tree hollows, often some distance from the water, and there is no shortage of these situations. In the early 1970s, broods were recorded in the East Alligator River area from January to May, and in May 1987 at Kapalga, near the South Alligator River, an adult with 10 young ducklings was disturbed on a vehicular track. Within the Darwin region, the only known breeding record is of a brood of newly hatched ducklings at Leanyer SP in March 1979.[13]

Pacific Black Duck comprise about 5% of waterfowl taken by hunters in the NT - in 1984 this amounted to about 4000 birds, of which probably 90% were from within 100 km of Darwin.[14]

Grey Teal
Anas gracilis

RANGE: whole of Australia, NG, e. Indonesia. HABITATS: wetlands, occasionally estuarine or marine waters. STATUS: moderately common, largely Dry season visitor. BREEDING: Apr–May (one record).

David Webb

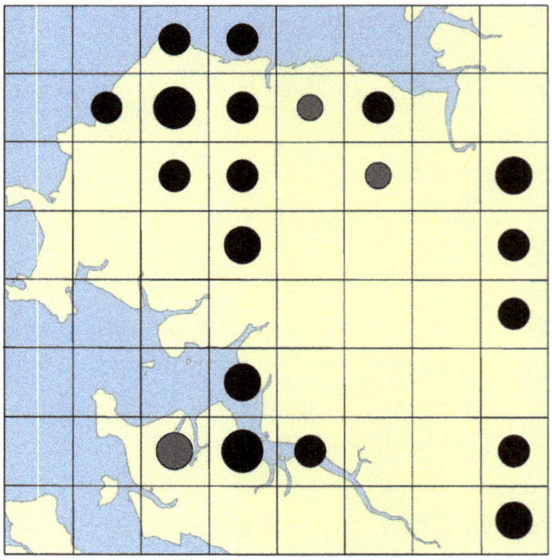

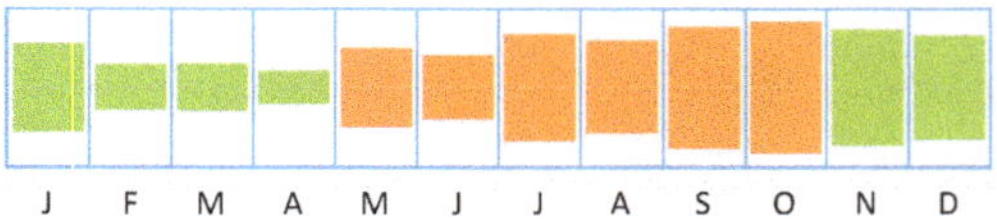

Archetypal nomads, with populations and individuals showing little regularity in the direction of their movements, most of the Grey Teal that visit the Top End probably originate from areas at least several hundred kilometres to the south,[28] perhaps the Barkly Tablelands, where the species is the second most frequently observed breeding waterbird.[26] Although it has been postulated that Grey Teal are able to locate newly created wetland habitat from some distance away using olfactory cues,[48] they also have strong spatial memory, and given that wetland habitat is available in the Darwin region throughout the year, and that the majority of birds occurring in the region are to be found at Leanyer SP, it is not unlikely that it is the latter that guides them there. The size of the influx varies dramatically from year to year, and is almost certainly related to environmental conditions elsewhere in Australia. Of the 84 000 birds counted in surveys across the Top End during the 1990s,[27] over 62% (52 000) were recorded in 1992, a year of intense drought in southern Australia, and 17% in 1999.

Seasonality in the Darwin region shows a similar pattern to that of the Pacific Black Duck, with the number of records peaking in October, and declining over the Wet season until February–March, when there are few records. It has been recorded at fewer sites than Pacific Black Duck, and, also contrasting with that species, the reporting rate at Leanyer SP is higher than at Knuckey Lagoons.

Grey Teals may arrive suddenly in the region in large numbers. Similarly, their departure tends to be quite sudden, and almost complete, with birds typically leaving the region during November and only small numbers remaining between December and March. Totals of monthly counts around Leanyer SP from 1974 to 1986 revealed that numbers of Grey Teal fluctuated more dramatically than those of other common duck species, from 49 in 1974 to 5186 in 1980. The average annual total was 2214 birds. The seasonal pattern of abundance also differed somewhat from the other common ducks in this area, the highest and lowest numbers occurring up to three months earlier than in the latter, in September and February, respectively. A decade later, from 1995 to 2008, mean monthly counts at Leanyer SP were highest in June in at least three years, declining sharply in July. Maximum counts were 1470 in June 1991 and 1450 in June 1999. While some birds are usually present at Leanyer SP in all months, with averages exceeding 100 birds from April to November, the species is typically absent from or in small numbers at Knuckey Lagoons until July, with the maximum numbers occurring in October, such as 500 in 1996 and 200 in 1999.

The only breeding record for the Darwin region, and seemingly for the Top End, is of a brood of nine recently fledged ducklings seen on McMinns Lagoon in mid-May 2004.[27]

Chestnut Teal
Anas castanea

RANGE: subtropical and temperate Australia. HABITATS: wetlands. STATUS: vagrant. BREEDING: not in NT.

Bastiaan Hensen

Normally found well south of the tropics, the Chestnut Teal has been recorded only four times in the Top End. Two specimens were collected from near Darwin in 1958,[20] although no precise locality is available. The other records, all from Leanyer SP, occurred in June 1990,[21] October 2003[49] and July 2013.[21,50]

Northern Pintail
Anas acuta

RANGE: breeds ne. of Europe, Asia and America, winters in Africa, s. Asia and C. America; occasionally se. Asia and NG. HABITATS: wetlands. STATUS: vagrant. BREEDING: not in Australia.

Anand Arya

Prior to the discovery of an adult male Northern Pintail at Knuckey Lagoons[51] in October 2003, the only record for Australia was a bird seen near Perth in July 1985.[30]

The day after first being seen at Knuckey Lagoons, the Northern Pintail moved a short distance to another lagoon, only about 700 m distant, and two days after that appeared at Leanyer SP, where it stayed until the end of December. During its two-month stay at Leanyer SP it moulted into winter plumage.

Garganey
Anas querquedula

RANGE: breeds temperate Eurasia; winters Africa, s. and se. Asia. HABITATS: wetlands. STATUS: rare Wet season migrant. BREEDING: not in Australia.

Dr Chan Ah Lak

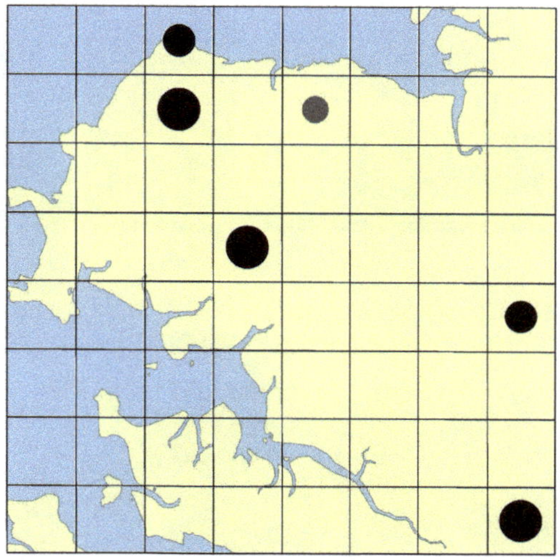

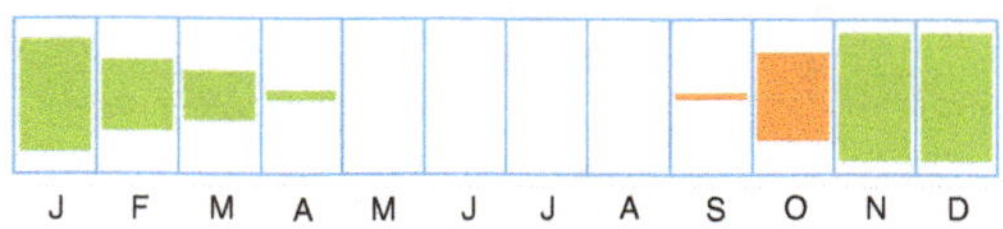

First recorded for Australia in 1896,[20] when two birds were shot at a lake in Victoria, there were only two further Australian records of Garganey, also from Victoria, until the species was found at Fogg Dam, in October 1967,[1] and observed there and at nearby Harrison Dam until January 1968. In Darwin, the species was first noted at Leanyer Swamp in late December 1969,[1] with subsequent records from Leanyer SP, where a male in eclipse plumage was present in March 1975[2] and a male in breeding plumage in April 1978.[52] Following those records, Garganeys were recorded in Darwin in 20 of the 26 years from 1981 until 2006, but with the exception of three birds seen near Katherine in October 2008[53] there were no other NT records between 2006 and 2014. The earliest arrival date is 13 October, and the latest date 10 April.[52] The maximum count for the region (125 birds including many males in breeding plumage) occurred at Leanyer SP in March and April 1991.[21] Numbers gradually built up from 12 in mid-January[54] to 80 by mid-February and peaked at about 125 from mid-March until early April. The majority of records, however, involve only one to three individuals, but other high counts include 14 at Knuckey Lagoons in November 1988, nine at McMinns Lagoon in November 1993[26] and 16 at Leanyer SP in January and February 1995.[26]

Garganey are reported most frequently from Leanyer SP and Knuckey Lagoons, but have also been seen at Palmerston SP, and in January 1996 two Garganey were seen at a swamp near Micket Creek, during a Wet season when none were recorded at the more regularly visited localities.

Feeding on aquatic plants and animals, Garganey are often near or with large groups of Grey Teal and Pacific Black Duck, when they can be difficult to detect. However, Garganey are readily identified in flight, and it is unlikely that many birds have been overlooked in recent years in the Darwin region. However, during aerial and ground surveys in the 1990s, at least 66 Garganey, including a maximum single count of 20, were sighted at nine locations, as far east as the mouth of the East Alligator River, Kakadu NP, and as far west as the Daly River mouth. A Garganey was found at Lake Woods, in the Barkly Tableland, in May 1993,[55] and the species appears to be a regular visitor to the Kimberley region, Western Australia, where groups of up to 20 have been recorded.[56]

Hardhead
Aythya australis

RANGE: whole of Australia, NG, rarely e. Indonesia. HABITATS: wetlands. STATUS: uncommon, largely Dry season visitor. BREEDING: not in TE.

Niven McCrie

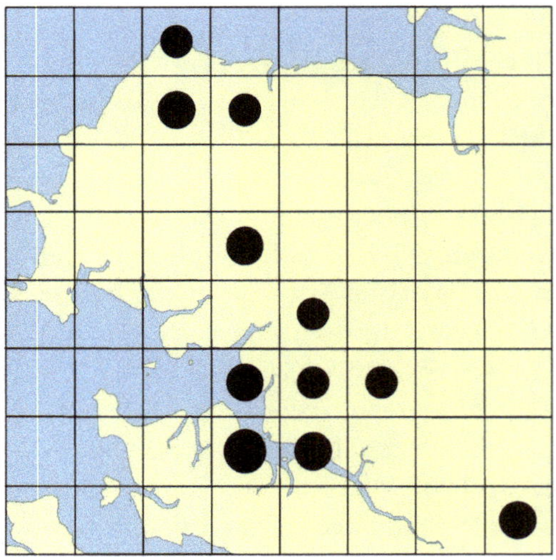

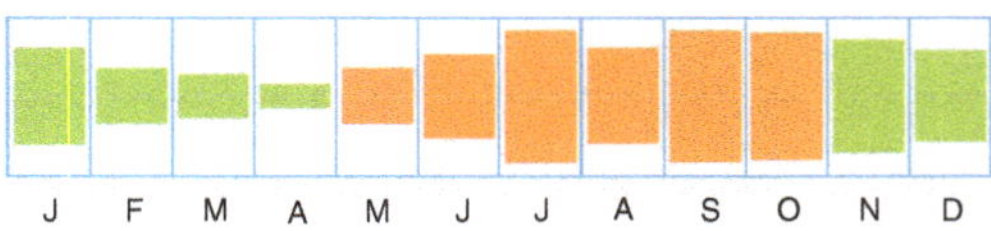

Apparently named by taxidermists because of difficulties in processing its head, rather than because of any physiological cranial feature,[57] the Hardhead is Australia's only member of the pochards, a diving duck group.

Puzzlingly, there appear to be no records of Hardhead in the Darwin region before the end of 1976,[13] though it was reported from Fogg Dam at least as early as 1969, with over 60 there in August 1970,[1] and its distribution had been described in 1966 as 'throughout the Territory'.[58] The completion of the sewage ponds at Leanyer in 1971 provided new habitat, yet the absence of early records from Knuckey or McMinns Lagoons is curious, and it may indeed be that the occurrence of the species in the region is a relatively new phenomenon. It has been recorded in the Darwin region in most but not all years since 1976, generally in small numbers in relatively deep water on lagoons and swamps, though with sometimes large numbers at Leanyer SP, where 600 were recorded in late July 2003. There are records for the Darwin region from all months, but it is scarce during the mid to late Wet season. Surveys from 1974 to 1986 showed average monthly counts peaking in November, after a smaller spike in July, with the total of monthly counts during this period fluctuating markedly from nine birds in 1974 to 1312 in 1980. The average annual total was 432 birds, but about 500 were counted in December 1979[34] alone. In contrast to the period 1974–86, between 1990 and 2010 mean monthly counts peaked in August, at more than three times that of the November mean, even though monthly reporting rates in November were significantly higher than in August. The difference in monthly abundance is perhaps not surprising however, given that the first 10 years of counts immediately followed the species' first appearance in the region, and a pattern had not become established.

Hardhead are not known to breed in the Top End, but they occur in substantial numbers, up to thousands, in the shrubby blacksoil wetlands of internal draining catchments in the Barkly Tableland and upper reaches of Sturt Creek, and breed there in at least small numbers.[26] In the Channel Country, in the south-west NT, Hardheads are quick to exploit floods, and arrive in large numbers, with some confirmed breeding. These occurrences external to the Top End, in temporary wetlands, may help explain the origins of birds that reach the Darwin region, especially after clusters of wet years in the inland such as during the mid-1970s and the years 2010 and 2011. Establishment of permanent reservoirs in tropical Australia, notably Lake Argyle, on which tens of thousands of Hardhead occur, has also possibly contributed to the recent increase in the Top End.[26]

Blue-billed Duck
Oxyura australis

RANGE: se. and sw. Australia. HABITATS: wetlands. STATUS: vagrant. BREEDING: not in TE.

Julie Sarna

The Blue-billed Duck typically makes only short distance movements, between breeding and non-breeding wetlands but, like other duck species, will undertake longer distance movements during periods of adverse conditions or to exploit temporarily available habitat.[30] The species' occurrence in the Darwin region, however, surely represents one of its longest movements, with the nearest regular breeding areas some 2500 km south, and the nearest regular non-breeding area about 1500 km south.[30] The single Darwin record was an eclipse-plumaged male at Leanyer SP, in late March 1985.[59] Interestingly, this was also the first record for the NT, yet the species was subsequently recorded in Alice Springs and surrounding areas in 14 of the 24 years from 1989 to 2012.[24]

Orange-footed Scrubfowl
Megapodius reinwardt

RANGE: e. Indonesia, NG and tropical (sub-)coastal Australia. HABITATS: monsoon rainforest, mangrove edges, parks, gardens. STATUS: common resident. BREEDING: poorly known; includes late Dry and early Wet seasons.

Niven McCrie

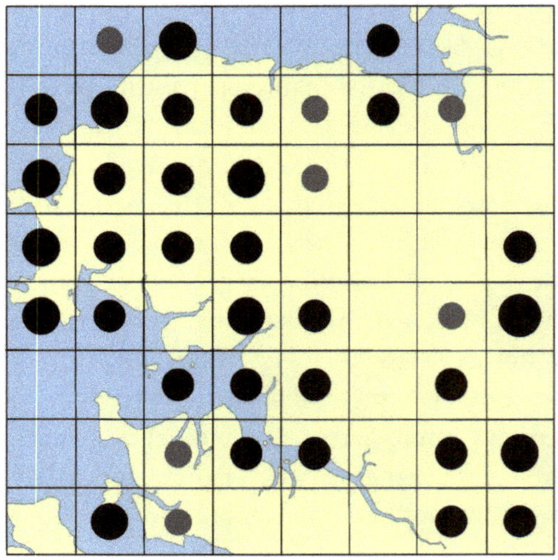

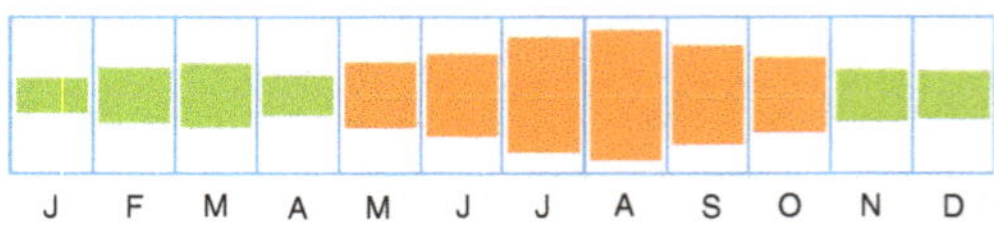

Orange-footed Scrubfowl is the most widespread of the world's megapodes, or mound-builders, a family of birds that bury their eggs in mounds of decaying vegetation or sand, or lay them in burrows, to be heated by microbial decomposition or the sun, rather than incubate them with body heat. Megapode chicks are able, indeed are compelled, to care for themselves from the moment they hatch. For Orange-footed Scrubfowl chicks, their first task, which can take two to three days, is to dig their way through some 2 m of soil and decomposed plant matter before they reach the outside world. They are able to fly from the time of hatching, and there are local observations of chicks in Darwin suburbs, in areas far from any nesting mounds, that are assumed to have flown there.

Prior to the mid-1980s, Orange-footed Scrubfowl were reported to be scarce in the Darwin region, and restricted to just a few local patches of monsoon forest and mangrove fringes.[60] Since then, however, they have greatly expanded their range and numbers, becoming regular sights on Darwin's parklands, roadsides and nature-strips, and, to the annoyance of some Darwin residents, in household gardens, where the birds scratch up mulch as they forage, or use it to construct mounds.[61] In suburban Coconut Grove, for two or three weeks during June 2006, up to 10 birds roosted together in a garden tree,[62] though they are more typically seen singly or in pairs. There are no measurements of densities; however, 10 birds were seen foraging alongside 1.5 km of road during the early morning, at Casuarina CR.[21]

Scrubfowl do much of their foraging during the early morning or late afternoon, and at those times are most likely to be seen well away from cover. As the day heats up, Scrubfowl tend to move into cooler, more densely vegetated areas, where they may rest in trees, or continue feeding, their noisy scratching on the forest floor clearly audible as they forage for snails, beetles, earthworms or other invertebrates.[63] When disturbed, Scrubfowl generally run into denser cover, but they will also often take flight, landing high in a tree not far from the disturbance. In some instances however, Scrubfowl seem particularly tame and will venture onto house verandas or into picnic areas, even when people are numerous, and there are records of birds entering buildings.[64]

Nesting mounds of the Scrubfowl vary in size, with large ones spanning around 20 m and reaching 3 m in height.[65] Size is dependent on location and on age, as previously used mounds are added to, often over many years. At East Point[65] mounds in coastal vine thickets with sandy soils averaged 0.8 m in height and 3 m in diameter, while mounds in monsoon forest, composed of coarse gravel and leaf litter, averaged 2.5 m in height and 10 m in diameter. Decomposition of leaf litter is the major source of heat for incubation in monsoon forest, whereas in the more open vine thickets, solar heat contributes.[65] Mounds have been found in eight of Darwin's suburbs,[21,64,66] and although there was an apparent reduction in the number of mounds between surveys in 1998[64] and 2002,[66] it appears that numbers have increased again. The highest urban density of mounds is in Ludmilla, where four nest mounds, of varying size and ages, including two in gardens, were built within a radius of about 500 m.[67]

Surveys conducted on Darwin radio suggest that the majority of property owners are comfortable with Scrubfowl mounds in their gardens,[66] but there is anecdotal evidence of householders destroying mounds on their property. Other than the hospitality of the property owner, the abundance and success of nest mounds in suburban gardens depend on the availability of leaf litter[65] and the absence of cats and dogs,[66] which may harass the adult birds or kill the young.

Despite the abundance of Scrubfowl and their mounds in the region, little is known of the timing of egg-laying, or the hatching of chicks, but the majority of hatchlings and young birds have been reported during the period from the beginning of September until the end of April.[34,43,44,65,68,69]

Brown Quail
Coturnix ypsilophora

RANGE: e. Indonesia, NG; whole of Australia except arid region. HABITATS: grassland, sedgeland, woodland, cultivated areas. STATUS: moderately common resident. BREEDING: Feb–May (TE).

The common quail of the Top End, Brown Quails are less secretive than other quail or button-quail in the area and, indeed, they can be quite tame, with birds coming to forage among people.[70] Reporting rates in the Darwin region are highest at Casuarina CR,

David Webb

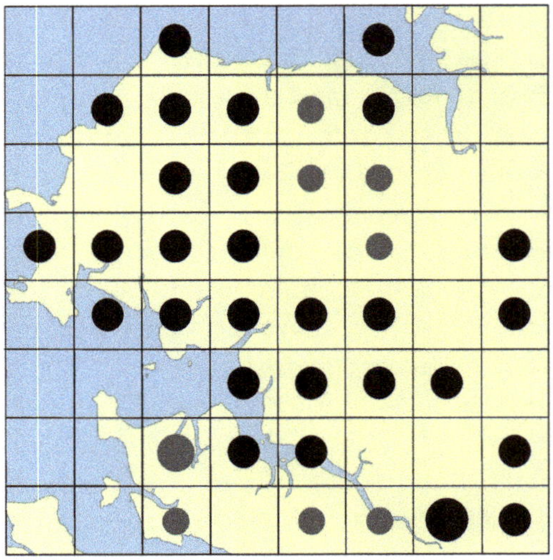

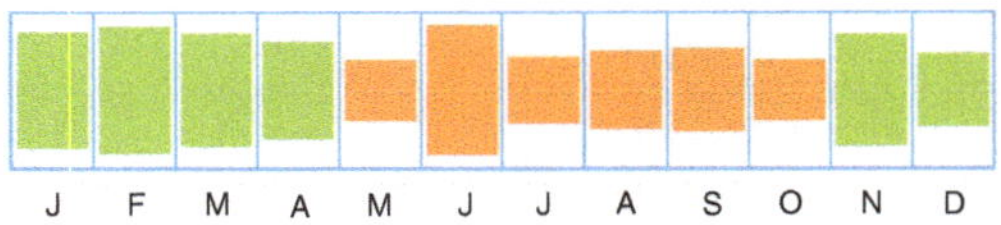

Charles Darwin NP and Howard Springs, but the species is widespread, occurring in most areas where there is open woodland with grass cover, and patchily around swamps and the grassy verges of saltflats bordering mangroves. Most observations are of birds along roadsides, particularly during the early morning or late afternoon as small groups venture into the open to forage. Although generally seen in small coveys of up to 10 birds, during September 1984, 60–70 birds were seen at Holmes Jungle, and 40 birds were reported near Palmerston SP.[34]

Brown Quails in the Darwin region appear to be largely resident, though in the Humpty Doo area,[1] during the late 1960s and early 1970s, most records were from July to November, with few sightings from December to April. These observations may have been made on floodplains, which would be inundated during the Wet season, but provide suitable habitat during the Dry.

Perhaps surprisingly, no nests have been reported from the Darwin region, but sightings of young birds suggest that breeding is widespread. Pairs with chicks were observed near East Arm in June 1968, near Berrimah in March 1970,[1] at Robertson Barracks, near Palmerston, in May 1997[33] and at Darwin International Airport in April 2004.[71]

At Kapalga, Kakadu NP, flying young were present at one site in May 1990, and at two sites in May 1991. A nest containing eight eggs found near Katherine during September 2004 is the only report of breeding in the Top End for the latter part of the year.[72]

King Quail
Excalfactoria chinensis

RANGE: s. and se. Asia, NG, tropical and e. Australia. HABITATS: swampy grassland, sedgeland. STATUS: rare visitor. BREEDING: Jan–May (TE).

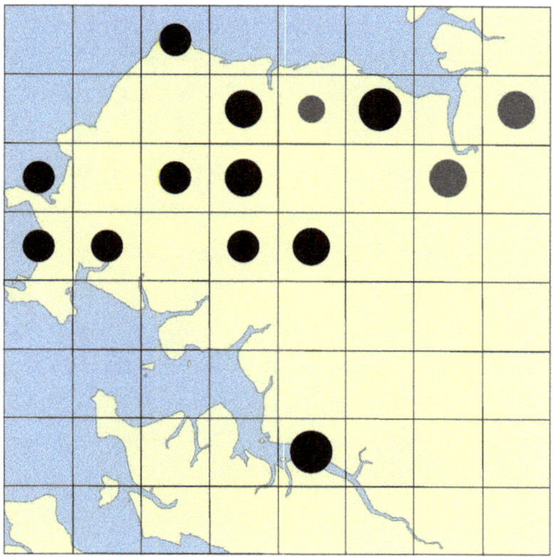

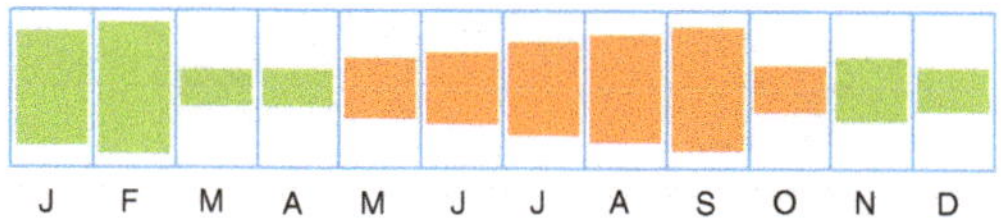

The smallest of the world's quails, and the most widely distributed, King Quail occurs only patchily in the Top End, where it is largely restricted to the wetter north-west. Although described as 'moderately common' on dry sedge-plains and rice stubble in the Humpty Doo-Fogg Dam region from June to December in 1967,[1] the following two years produced few records. In the Darwin region also, the species is sporadic. It was reported in only 21 of the 35 years from 1977 to 2012, and with the exception of 1998, when there were three reports, all in September, there have been no more than two records in any year. However, it is likely to be more common than records suggest as, even when present, the birds are difficult to find, preferring to squat or run through dense cover rather than fly when disturbed. Although there are too few records to show a valid pattern of seasonality, it may be that the apparent drop in the number of records in the Darwin region during the Wet season is a result of birds moving away as wetlands, their favoured habitat, become flooded.

The majority of records come from low-lying wetlands, such as Leanyer Swamp, Holmes Jungle and Shoal Bay, but birds have also been reported from grassy mangrove edges in Charles Darwin NP, and in the Ludmilla Creek area.

There has been no confirmed breeding in the Darwin region, and indeed the only evidence of breeding anywhere in the Top End was the discovery in May, of an unknown year, of a clutch at Oenpelli,[29] in west Arnhem Land.

Streaked Shearwater
Calonectris leucomelas

RANGE: breeds on the East Asian mainland, from Japan to Russia, migrating to seas off NG, Arafura Sea and South China Sea. HABITATS: pelagic. STATUS: vagrant. BREEDING: Mar–Sep (TE).

Laurie Ross

The first record of Streaked Shearwater in Australian seas was the sighting of large numbers in seas from Cape Wessel to about 400 km east of that point in November 1971,[73] but it has since been shown to occur in significant numbers in the Arafura Sea and northern Gulf of Carpentaria.[74] In summer 1983, it was found to be 'numerous' between Darwin and Nhulunbuy,[75] and in February 1984 a raft of about 150 were also seen on waters about 300 km north-east of Darwin.[76] A live bird captured at Jabiru in December 1989,[34] likely to have been blown in by tropical cyclone Felicity, was the first inshore record for the NT. Given that the species is present in waters off the Top End during the cyclone season, it is somewhat surprising that there are not more records of birds along NT coasts, as was the case following tropical cyclone Gertie in Broome, in December 1995,[77] when large numbers of Streaked Shearwaters were forced inshore. The only inshore reports of the species from the Darwin region are of a bird seen sitting on the water at Lee Point in early January 1986,[78] and another at Buffalo Creek in August 2000,[23] also 'loafing' on the water. A specimen that was donated anonymously to the NT Museum has no details of locality or date, but may have been collected locally.[76]

Wedge-tailed Shearwater
Puffinus pacificus

RANGE: tropical and subtropical Pacific and Indian Oceans. HABITATS: pelagic. STATUS: vagrant. BREEDING: not in NT.

www.wildlifeimages.com.au/Rohan Clarke

Breeding in dense colonies on islands off the coast of north-west and eastern Australia, Wedge-tailed Shearwaters have been observed in the Darwin region only during the Wet season, particularly during stormy weather when birds are more likely to be forced inshore. An exhausted bird was found at Rapid Creek in early February 1973,[79] following a week of gale force winds from the north-west. This bird died and was prepared as a specimen for inclusion in the NT Museum. A beach-washed bird was collected in February 1977, from 'a Darwin beach', but no further details of the locality were provided.[34] Single birds were recorded at East Point in January of 1974, 1975 and 1986.[34]

The nearest known breeding is in a small colony on West Island, Ashmore Reef,[80] about 845 km west of Darwin. Birds there breed during the Wet season, leaving the area from early April and returning from August. As all Darwin region records have coincided with the breeding season they may represent non-breeding birds or breeding birds caught in cyclonic winds when out foraging as the species is known to forage up to 500 km from their breeding colonies.

Short-tailed Shearwater
Puffinus tenuirostris

RANGE: breeds s. Australia; winters mainly in n. Pacific Ocean. HABITATS: pelagic. STATUS: vagrant. BREEDING: not in NT.

www.wildlifeimages.com.au/Rohan Clarke

Australia's most numerous seabird, the Short-tailed Shearwater normally leaves its breeding colonies on islands off Tasmania and the south-east mainland in April–May, and arrives a month later in Japan. Their well-documented trans-Pacific migration route is well east of the NT coastline and, although the species has been recorded near Cairns,[30] typically it does not occur further north in Australia than south-east Queensland. The only record of this species for the Darwin region involved a single bird seen and photographed at Stokes Hill Wharf in Darwin in January 2011.[81] Another Short-tailed Shearwater was seen about 10 km off the north coast of Melville Island in May 2014,[37] perhaps suggesting that further surveys may reveal that the species is more frequent in the Timor and Arafura Seas than has been thought. Prior to these two observations, the only record for the NT pertains to two birds found on Groote Eylandt in 1999: one dead, the other rehabilitated and released.[82]

The migration of Short-tailed Shearwater is almost entirely around eastern Australia, so its occurrence in and near the Darwin region is remarkable. It has recently been found, however, that small numbers occur from April to June in the northern Straits of Malacca, Malaysia,[83] and in May 2011 up to eight birds were observed moving east on the Singapore Strait during squally weather.[83] This is during the period of the species' northward migration, and shows that at least some birds must move west, though it cannot be certain at which point they do so.

Wilson's Storm Petrel
Oceanites oceanicus

RANGE: breeds in Antarctic region; migrates n. to Indian and Atlantic Oceans. HABITATS: pelagic. STATUS: moderately common off TE coast, mainly in Dry season. BREEDING: not in NT.

www.wildlifeimages.com.au/Rohan Clarke

Although there are no inshore records of Wilson's Storm Petrel from the region, birds are encountered quite regularly within about 30 km from Darwin's shores. In March 2013[50] the species was reported to be numerous around Fenton Patches, an area about 24 km north-west of Darwin, where artificial reefs have been constructed for recreational fishing. A local fishing boat skipper, on having the birds pointed out by an accompanying birdwatcher, reported seeing 'petrels', undoubtedly this species, fairly regularly about an hour out from shore.[37]

Tricolored Grebe
Tachybaptus tricolor

RANGE: e. Indonesia and NG. HABITATS: wetlands. STATUS: vagrant. BREEDING: not in Australia.

Niven McCrie

Formerly considered to be a race of Little Grebe, the Tricolored Grebe may instead be more closely related to Australasian Grebe.[84] It was first recorded at Leanyer SP from 22 September to 30 October 1999, then nine years passed before it was recorded again, at the same site, with an individual being sighted regularly from September to December 2008. The following year, one was present at Leanyer SP from October to December. Given its similarity to the Australasian Grebe this species could easily have been overlooked in the past, and may be a more frequent visitor than these few records suggest. The dark – rather than yellowish – eyes, and the lack of white in the primaries appear to be the best field marks for separating it from Australasian Grebe.

In the Darwin region, and tropical Australia generally,[30] the Australasian Grebe breeds from December to April. While birds of this species are beginning to moult into breeding plumage in September, the majority are still in non-breeding plumage by the end of that month, and even as late as mid-November.[21] All observations of Tricolored Grebe in Darwin, however, have been of birds in breeding plumage, and it appears that this species, unlike Australasian Grebe, may retain this plumage throughout adult life. Young have been reported on Timor-Leste in April, suggesting egg-laying in March and April, but breeding apparently occurs at any time during the year,[85] which may account for the absence of a 'non-breeding' plumage.

In the Wallacean region of Indonesia, Tricolored Grebe is a locally numerous breeding resident of lakes, ponds, swamps and rivers[86,87] and, indeed, in surveys conducted in Timor-Leste between July 2002 and December 2004[87] it was the second most frequently recorded waterbird. It is therefore not an unlikely visitor to northern Australia.

Australasian Grebe

Tachybaptus novaehollandiae

RANGE: Australia, NG, e. Indonesia. HABITATS: mainly wetlands. STATUS: common, mostly Dry season visitor, with some breeding residents. BREEDING: Dec–Apr (TE).

Laurie Ross

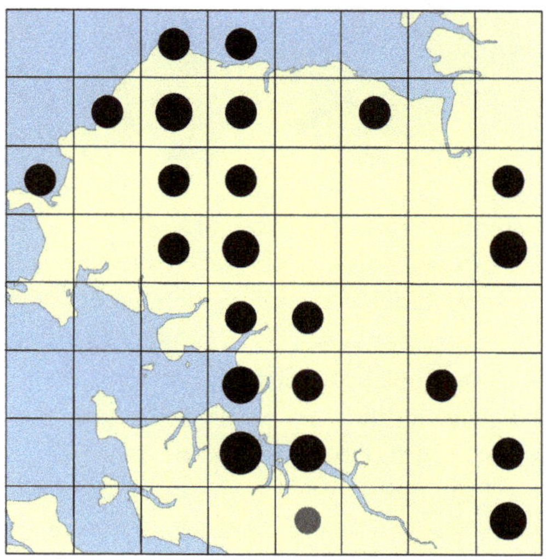

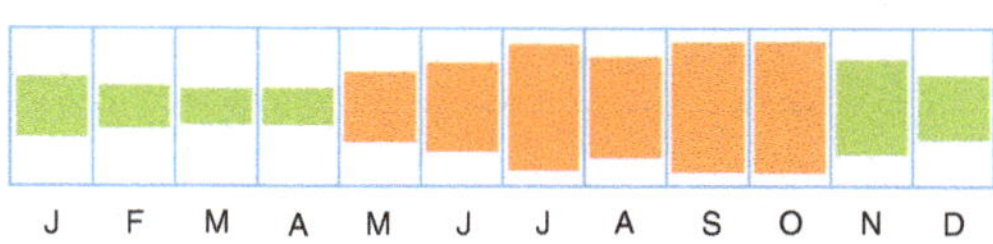

While the Australasian Grebe is a common resident in the Top End, it is clearly seasonal in the Darwin region, with the majority of birds dispersing to nest on large water bodies as they fill with rainwater at the start of the Wet season. Birds start returning as early as February or March, when many adults are still in breeding plumage, and juveniles are in large numbers, suggesting that breeding is widespread across the Top End. Maximum numbers normally occur between September and December before birds move out of the Darwin region again. The greatest number recorded in the region was 650 birds at Leanyer SP in July 1979.[31] Following this unusually early peak, numbers dropped off quite rapidly, with only 200 still present in September, but built up again, reaching 500 birds in November.

Although breeding normally takes place in the Wet season, from December to April, there is one record of a nest with eggs in August 1984 at Palmerston SP. Nests have also been found at Knuckey Lagoons and McMinns Lagoon. On one lagoon at Kapalga, Kakadu NP, 10 nests were found at various distances from the shore, all probably containing eggs, in March 1986.[43]

Hoary-headed Grebe

Poliocephalus poliocephalus

RANGE: whole of Australia and NZ. HABITATS: freshwater and estuarine wetlands. STATUS: vagrant. BREEDING: not in TE.

Bastiaan Hensen

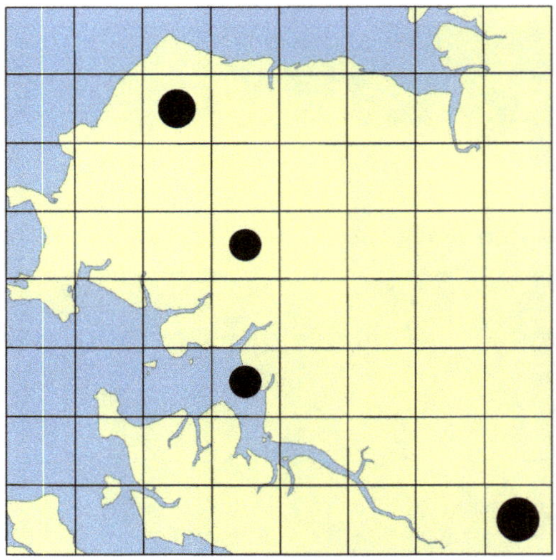

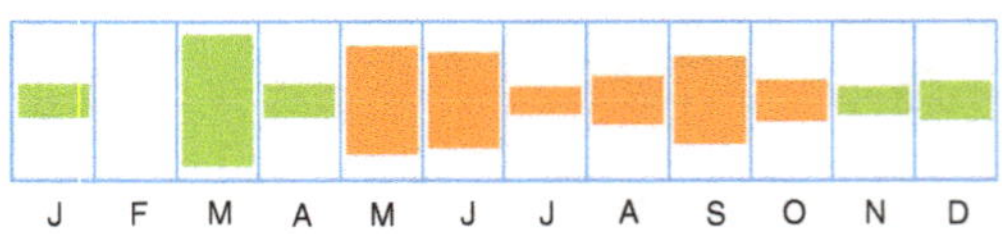

Although regularly seen in southern NT, the Hoary-headed Grebe rarely reaches the Top End, and has only been reported in the Darwin region since June 1974[34] when one was found at Leanyer SP. Seventeen birds, the maximum number recorded in the region, were reported there in June 1976,[34] with the only other reports of multiple birds being seven at McMinns Lagoon in August and September 1990, and two at Knuckey Lagoons in October 2011.[88]

Further out of Darwin there are sporadic records, usually of only one or two birds, including at Fogg Dam,[1] Adelaide River SP, Kakadu NP[89] and Bird Billabong.

Great Crested Grebe
Podiceps cristatus

RANGE: Worldwide except Americas. HABITATS: wetlands, coast. STATUS: vagrant. BREEDING: not in TE.

Adrian Boyle

The cosmopolitan Great Crested Grebe was not reported in the NT until November 1941, when one was seen at Newcastle Waters, about 600 km south of Darwin.[41] The first report from the Darwin region was in January 1979, when a single bird was seen and photographed at Leanyer SP.[34] This coincided with an influx of the species into the Top End between June and December 1978, perhaps partly due to movement of birds out of western New South Wales.[4] Five adult birds were recorded also at Leanyer SP in November 1995,[90] and although there have been no records in the Darwin region since then, Great Crested Grebe has been frequently recorded on Darwin River Dam, one of the very few large bodies of fresh water within 50 km of Darwin. The species may be a regular Dry season visitor in small numbers to other large water bodies in the Top End, as there are reports from a lake on Delamere Station,[21] south-west of Katherine, and lakes in the Barkly region,[26] as well as from northern Western Australia.[91]

Black-necked Stork
Ephippiorhynchus asiaticus

RANGE: Pakistan to tropical and subtropical Australia. HABITATS: wetlands, occasionally tidal mudflats and beaches. STATUS: moderately common resident. BREEDING: Mar–Jun (TE).

Niven McCrie

In contrast to its declining status in the southern parts of its range, the Black-necked Stork is still moderately common and widespread in wetlands in the Darwin region, though not in the large numbers that occur just outside of Darwin in areas such as Fogg Dam.

The frequency of records drops in the late Wet season, then increases again as the Dry season progresses, partly due to the species' tendency to congregate around drying or receding wetlands and billabongs where fish are trapped.

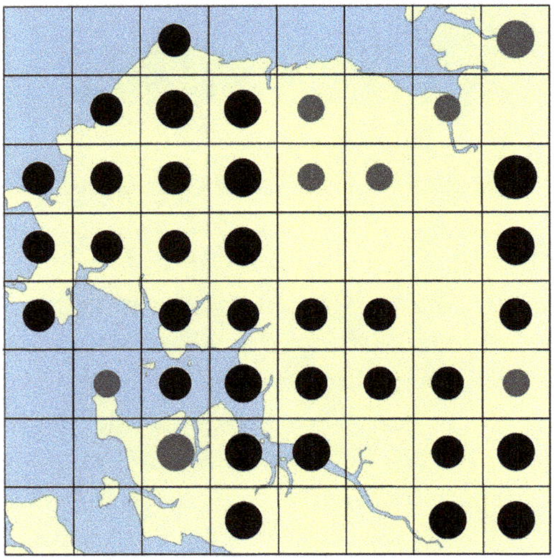

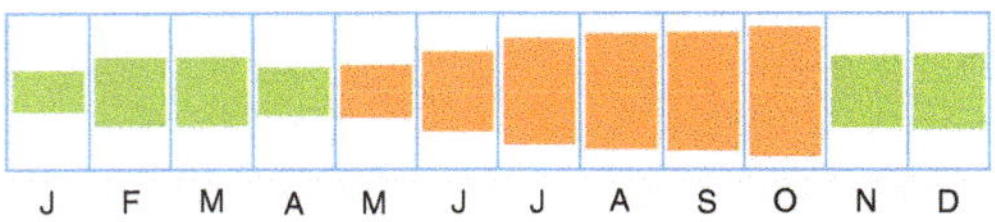

Normally found singly or in small groups of two to four birds, comprising adults and juveniles, larger groups are occasionally seen. While the maximum count in the Darwin region appears to be only 12 birds at Holmes Jungle in mid-October 2011, there were 257 present at Mamukala, in Kakadu NP.[92]

Aerial counts of the Top End during the 1990s totalled 1800 birds, though the population in the Alligator Rivers region alone was estimated in the early 1980s to be up to 1500 birds.[28]

Fish, especially eels, tadpoles and insects, are the major food items of Black-necked Stork, but they will prey on turtle hatchlings as they emerge from the sand, and one bird seen foraging in open forest in the Darwin region was thought to be hunting rats.[1]

While dark-plumaged juveniles have been recorded in the Darwin region in all months, there appears to be no evidence of the species breeding here, and a nest containing two chicks at Beatrice Hill,[93] just south of Humpty Doo, in March 1979 is the nearest confirmed nesting. Most nests in the Top End have been found to have eggs or small young from March to May; however, a few late breeding or re-nesting pairs lay clutches as late as July, at the time most other young have fledged.[38]

Australian White Ibis
Threskiornis moluccus

RANGE: Australia, e. Indonesia, NG, NZ. HABITATS: wetlands, estuaries, mangal, urban parks. STATUS: moderately common resident and migrant. BREEDING: Mostly Mar–Jul (NT).

Viewed as an urban pest in some Australian cities, mainly due to its habit of scavenging food in public places and waste disposal sites, numbers of Australian White Ibis in the Darwin region do not appear to have increased to the same extent as they have elsewhere. Nevertheless, up to 400 birds have been counted at the Shoal Bay Waste Management Facility.

While the Australian White Ibis occurs in the Darwin region year-round, the frequency of records increases steadily over the mid-Dry season, peaking in the late Dry before falling sharply. The species is widespread in the region but the highest reporting rates are from Howard Springs NR, where birds scavenge from picnickers and rubbish bins. Reporting rates are

Niven McCrie

also high at Knuckey Lagoons, where it forages near the margin of shallows. On the coast, these birds forage both on tidal mudflats and inside mangrove forests at sites like Buffalo Creek and Nightcliff. At Leanyer SP, where the largest number recorded was 80 in June 1979,[93] White Ibises often stand along the concrete walls of the intake channels, perhaps feeding on cockroaches as do Pied Herons.

An estimated 34 300 Australian White Ibises breed among 39 colonies around the Top End coast and adjacent floodplains, the largest colonies being to the south-west of Darwin, including one of 4000 birds near the mouth of the Daly River.[38] Colonies

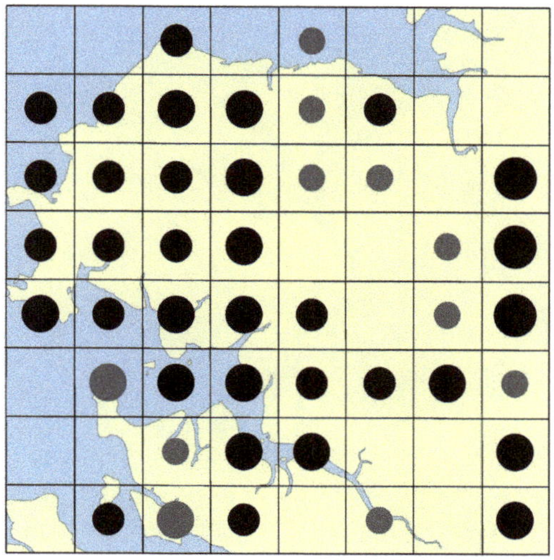

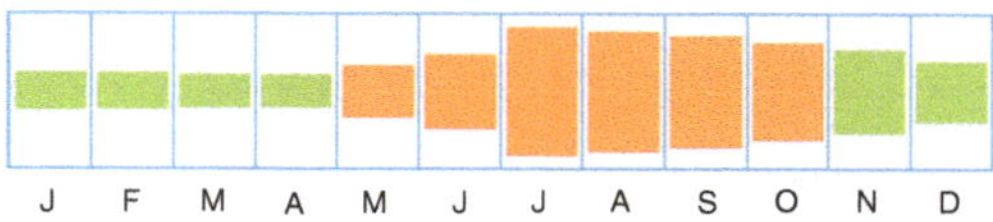

were equally distributed between coastal mangroves and paperbarks on floodplains slightly inland, in both cases mixed with other breeding waterbirds. However, when nesting on their own they often nested in Common Reed. Breeding takes place from January to at least August, but is mostly concentrated between March and July.[38] Breeding has not yet been confirmed in the Darwin region, but in October 1983 several birds with partly completed nests were seen at Howard Springs.[34]

Straw-necked Ibis
Threskiornis spinicollis

RANGE: whole of Australia. HABITATS: floodplains, grassland, open woodland, cultivated areas and parklands. STATUS: common Dry season migrant. BREEDING: mainly Feb–May (NT).

A long-distance migrant from southern Australia, the Straw-necked Ibis is a characteristic sight on parklands, ovals and wetlands in the Darwin region throughout the Dry season. Young birds banded near Kerang, Victoria, were recovered near the Roper River,[41] south-east of the Top End, a movement of some 2500 km. While this suggests that Straw-necked Ibises in the Top End originate from south-east

Niven McCrie

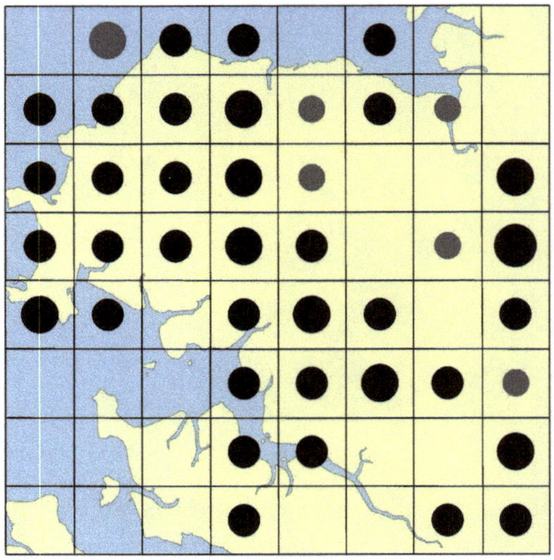

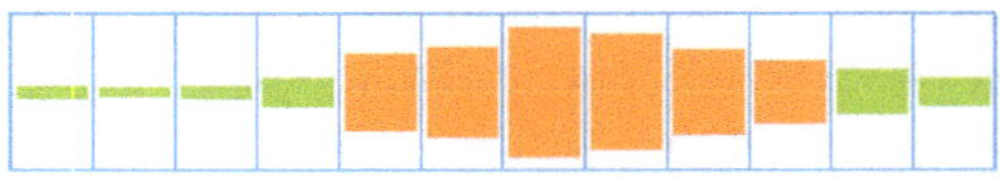

Australia, it is also possible that many derive from breeding colonies on lakes in the Barkly Tableland. Particularly following major flood events, large numbers, including a maximum of 5500 pairs, have been found nesting there.[94] The lakes have, however, not been surveyed annually, and it is not known how regularly the species breeds there in significant numbers.

In the Darwin region the first Straw-necked Ibises, often including many juveniles, begin arriving in April or May. After reaching a peak in August, the frequency of records starts to decline, presumably as the birds begin the long haul back south, and few birds are seen in the region between December and April.

Occasionally flocks can be seen in woodland or even open forest, foraging on burnt ground immediately after a fire, and for up to four weeks afterwards, moving out as the grasses sprout and grow.[95] Although widespread in the region, the highest reporting rates and greatest numbers are from Knuckey Lagoons, where up to 300 were counted in July 1967.[96] Other large flocks include a May record of 160 on a sports oval in Fannie Bay, and 120 at Shoal Bay in August 1973.[34] In the Alligator Rivers region, the Dry season population during the early 1980s was estimated to about 80 000 birds, making it at least temporarily one of the most numerous waterbirds in that region.[28]

Glossy Ibis
Plegadis falcinellus

RANGE: cosmopolitan, including whole of Australia. HABITATS: wetlands, floodplains. STATUS: uncommon Dry season migrant. BREEDING: Jan–May (TE).

Much less frequently recorded than the other two ibis species in the Darwin region, the Glossy Ibis is the latest to arrive in any numbers. Although not consistent between years, there is usually an influx from August through October, followed by an exodus at the start of the Wet season. In the late 1960s and early 1970s also, the species was

Nolan Caldwell

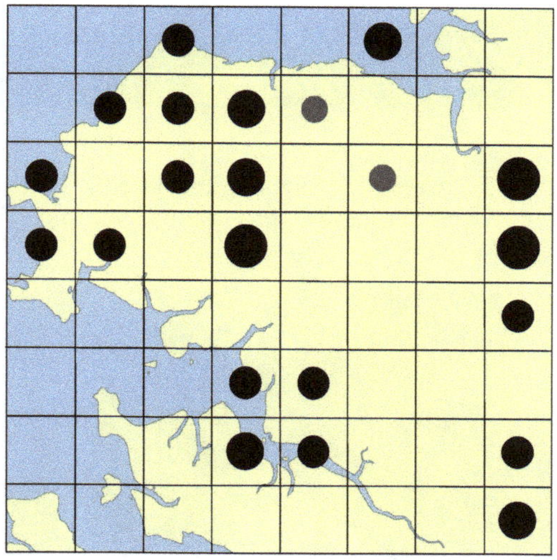

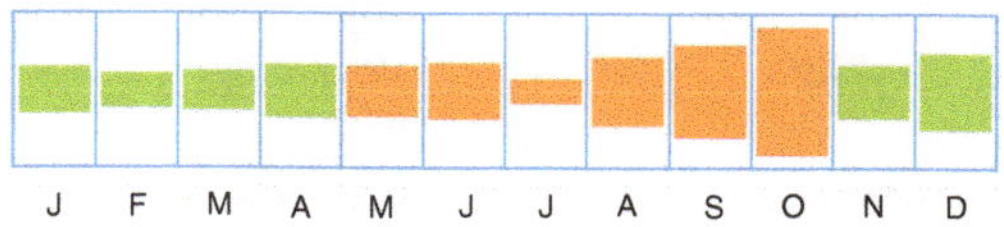

found to be scarce in the Darwin–Humpty Doo region from November to May.[1] In the Alligator Rivers region during the early 1980s, the population peaked at an estimated 60 000 in September–October, but numbers dropped off markedly during the Wet.[28] Ground surveys across the Top End in the 1990s, which recorded an estimated 35 500 birds, revealed a similar pattern, with numbers dropping off in November, as birds migrated to breeding areas, and returning in March.

The highest reporting rates are from wetlands around Howard Springs, Holmes Jungle, Knuckey Lagoons and McMinns Lagoon. There, as at other wetlands, birds forage for invertebrates in shallow water or soft mud. Some Glossy Ibises have also been seen feeding in parks with Straw-necked and Australian White Ibises.

Generally seen in quite large groups throughout its range, the species may be found in flocks of up to 5000 birds in the Top End, but in the Darwin region, it is often seen singly, or in small groups up to about 10 birds. The largest counts have been 600 birds at Holmes Jungle Swamp in September 1977,[31] and 200 birds at Knuckey Lagoons in October 2003.[21]

The only known breeding colony of this species in the Top End is in the mangroves on the eastern side of the Keep River estuary, near the Western Australian border, about 350 km south-west of Darwin. This large, mixed species colony contained about 5000 Glossy Ibis in 1999.[38] Eggs were present in mid-March but by the end of June all breeding had ceased.[38] Breeding has also been recorded in the Barkly Tableland, where over 3700 nests were found in March 2006,[26] and from Kununurra,[26] in north-west Western Australia.

Royal Spoonbill
Platalea regia

RANGE: mainland Australia, but largely excluding arid areas of SA and WA; NG, Indonesia, NZ. HABITATS: wetlands, floodplains, mangal. STATUS: moderately common visitor. BREEDING: Apr–Jul (TE).

Mat and Cathy Gilfedder

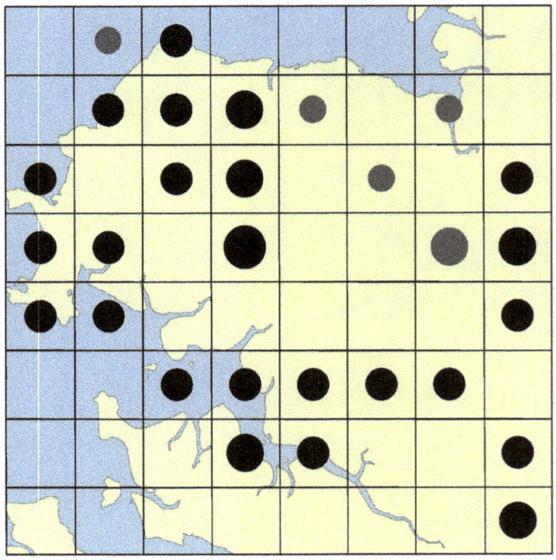

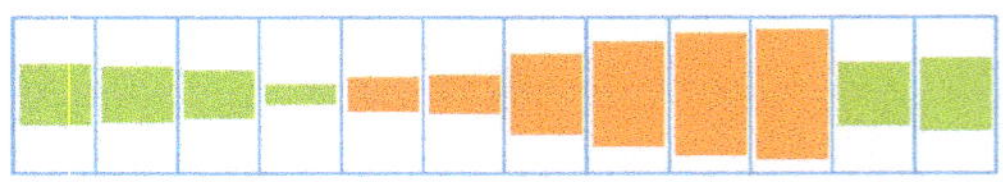

Not observed in the Darwin region as often as the two larger ibis species, the Royal Spoonbill is nevertheless present throughout much of the year in local wetlands, though scarce in the late Wet and early Dry seasons. As with many other waterbirds, such as the Intermediate Egret and Black-necked Stork, records of this species increase steadily during the latter half of the Dry season as birds move from temporary swamps to areas of permanent water. After October, the number of records drops sharply, indicating that these birds move out of the region, probably in response to food or habitat shortages as wetlands dry up, rather than to breed, since nesting does not begin for another four months. The relatively few sightings from March to June, coinciding with the breeding season, probably pertain to non-breeders.

Birds are generally present on the fringes of swamps, and while flocks of up to 400 have been recorded in the Top End,[27] they are found only in small numbers, typically fewer than 10, in the Darwin area.

Although a few Top End breeding colonies of the Royal Spoonbill were known in the late 1970s, intensive surveys around the coast and adjacent floodplains during the 1990s revealed 25 colonies containing about 5500 birds. The majority of these colonies are to the south-west of Darwin, the largest, containing 700 birds, on the Finniss River floodplain. Larger colonies are situated on floodplains with nests in paperbarks, while colonies in mangroves along the banks of tidal rivers[38] tend to be small. Royal Spoonbills start breeding later than other waterbirds in the region, the first eggs being laid in March and the last in July. In May 1977[34] four nests, though with only one bird sitting, were discovered at Fogg Dam, but nesting has not been reported there, nor closer to the Darwin region, since then. The number of birds occurring in the Alligator Rivers region during the early 1980s was estimated to be 35 000.[28] Given that this number is almost seven times the estimated Top End breeding population, it is possible that the latter is augmented by visitors from southerly breeding populations during the Dry season.

Yellow-billed Spoonbill
Platalea flavipes

RANGE: whole of Australia. HABITATS: wetlands, floodplains. STATUS: rare Dry season visitor. BREEDING: Jul (NT).

Laurie Ross

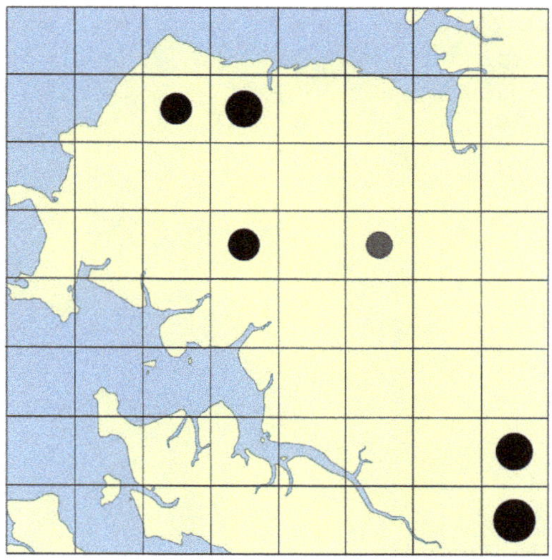

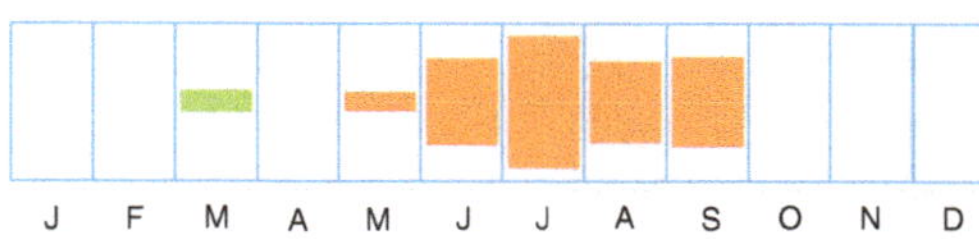

Unlike the Royal Spoonbill, the Yellow-billed Spoonbill is a rare non-breeding visitor to the Top End. It was recorded nine times in the Darwin–Humpty Doo region between 1969 and 1977, all but one record being of single birds; however, 30 birds, the greatest number reported in the Top End, were present at Fogg Dam in April 1989.[97] Apart from a few records in the late Wet season, this species visits almost exclusively during the Dry season. In the Alligator Rivers region[28] too, it is present only from May to November. The majority of Darwin region records since 1978 are from Knuckey Lagoons, but two birds were also seen regularly at Leanyer SP during the Dry season of 1996, the last recorded for our region.

In the Top End, breeding has only been recorded at Victoria River Downs, over

300 km south of Darwin, in July 1976.[98] Further south, a total of nine pairs were found breeding in various lakes in the Barkly wetlands in March 1993.[26] There are breeding colonies in north-east Western Australia, but whether birds that venture into the Top End derive from there, or from the large colonies in south-west Queensland, New South Wales and Victoria, is not known.

Black-backed Bittern
Ixobrychus dubius

RANGE: Australia, but sporadically distributed in coastal and near coastal regions; s. NG. HABITATS: wetlands with dense cover. STATUS: vagrant. BREEDING: not in NT.

Ian and Jill Brown

Secretive habits and small size combine to make the Black-backed Bittern the least known of Australia's herons. The first record in the NT was of an adult male bird collected at Point Charles Lighthouse, about 20 km west of Darwin, after it struck a window of the lighthouse during a storm.[41] No date is available for the collection, but the collectors were present in the area during 1902.

The only confirmed record in the Darwin region was in January 2001, when two birds, an adult male and a female or juvenile, were seen at Holmes Jungle.[21] Reports from Fogg Dam, however, suggest that the species is at least an irregular visitor, rather than a vagrant, to the Top End. Confirmed recent records include a male bird found in December 2012[99] and two birds, a male and a female, in February 2013.[100,101]

Egg-laying in southern Australia has been recorded from October to December, and following breeding, as wetlands dry up, birds are thought to be migratory, with some moving to New Guinea. The Darwin and Fogg Dam records of pairs coincide with the reported breeding period in southern Australia, and are perhaps suggestive of breeding, but there has been no evidence to confirm this.

Black Bittern
Dupetor flavicollis

RANGE: tropical and e. Australia, s. and se. Asia, NG. HABITATS: wetlands, estuaries. STATUS: scarce breeding resident and non-breeding visitor. BREEDING: Sep–Feb (NT).

Aiden Webb

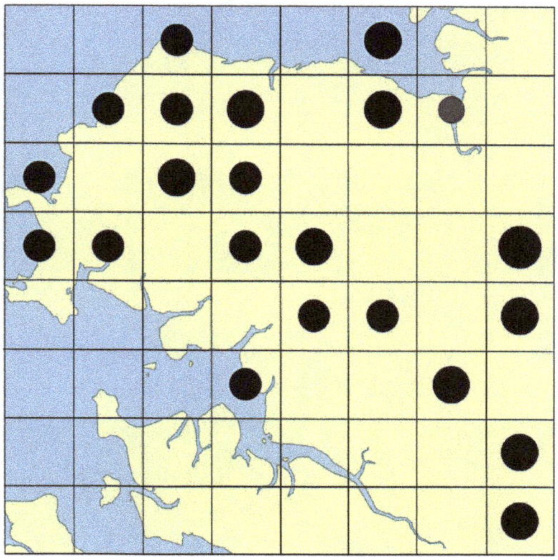

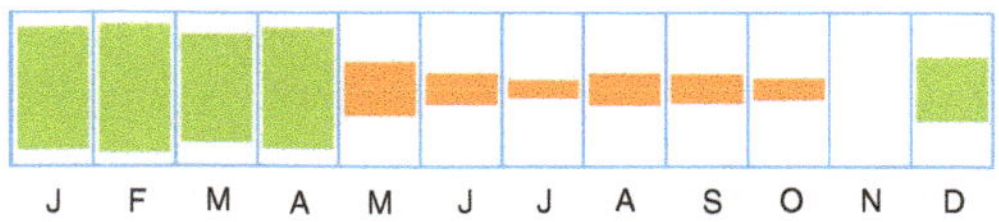

Despite its wide range, the Black Bittern remains a poorly known species in Australia, doubtless due to its secretive and generally solitary nature. In the Darwin region, fewer than 60 records were collected over the years 1985 to 2013, and although they were reported from several areas, only four sites - Leanyer SP, Leanyer Swamp, Knuckey Lagoons and Howard Springs - had more than five records.

In marked contrast to other herons and egrets, records of Black Bittern were much more frequent in the Wet season than in the Dry, with about two-thirds of records occurring from November to April. However, during Top End surveys in the 1990s, although the species was recorded on only 13 occasions, all were between February and October, mostly from near Darwin or in the southern Gulf of Carpentaria. The major local habitats of the species are mangroves and paperbark swamps, as well as sedgeland and reeds at the edges of swamps or rivers fringed by rainforest. As wetlands become drier and the reeds at their edges die, birds appear to move to riverine habitats,[21] including small streams in monsoon forest. The use of mangroves may thus be limited to the Wet season, when they breed. Although most records are of single birds, over 30 birds were counted in a few hours along the treed banks of the lower Finniss River,[102] about 50 km south-east of our region.

Reports of breeding are few. In mid-January 1916,[70] three nests were discovered in mangroves on the King River, Arnhem Land, with two containing four eggs each, and the other still under construction. In Kakadu NP, one nest with three chicks was found in monsoon rainforest on 20

February 1982.[103] Within the Darwin region, a bird was flushed from a nest containing two eggs in mangroves near Palmerston on 11 February 1987. It may be that egg-laying was incomplete, as the typical clutch is reported to be three to five eggs. An adult and fledged young were recorded in small mangroves at Doctor's Gully in November 1995,[104] and three unfeathered chicks were found near Darwin[27] in October 2001, though a precise location was not provided. These reports collectively suggest a breeding season extending from September to February or later.

Nankeen Night Heron
Nycticorax caledonicus

RANGE: whole of Australia, Indonesia, NG. HABITATS: wetlands, estuaries. STATUS: uncommon resident. BREEDING: Mar–Jun (TE).

Ashley Carlson

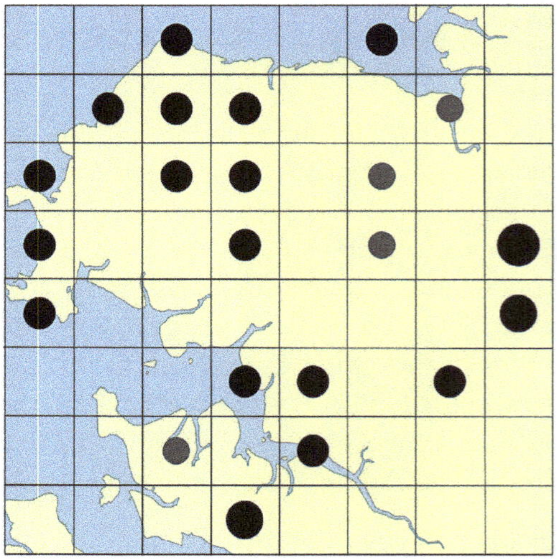

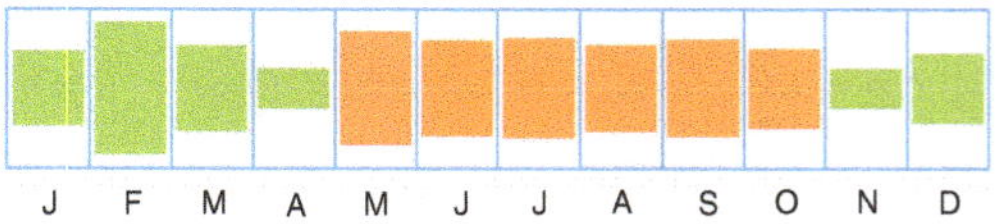

Roosting during the day in loose colonies under the thick canopy of mangroves, paperbarks or rainforest trees, the nocturnal Nankeen Night Heron emerges on dusk to hunt small fish, frogs, crustaceans and insects along rivers or around floodplains and billabongs.

Nankeen Night Herons were reported as 'numerous' at Holmes Jungle in July–August 1964,[105] and up to 100 birds were recorded in mangroves near Buffalo Creek in November 1976 with a similar number there in August 1977.[31] Most records in the Darwin region, however, are of one or two birds, and only occasionally up to five. The species is common at Fogg Dam, where birds can be seen foraging along the causeway at night.

A Nankeen Night Heron breeding colony discovered near the mouth of the South Alligator River in 1981 was, for many years, thought to be the only one in the NT. That colony held up to 2000 nests, with most containing half-grown young in mid-March.[106] During the 1990s[38] up to 19 000 breeding birds were recorded, spread over 26 colonies in the Top End, suggesting that the Top End breeding population is significant in a national context. The largest Top End colony held about 4000 birds,[38] but those nearest Darwin, on the Adelaide River, contained fewer than 100 birds. The majority of colonies are located in mangroves. Breeding takes place from January to August, but most is from March, the peak egg-laying time, to June.[38]

Striated Heron
Butorides striata

RANGE: pantropical; tropical and subtropical Australia. HABITATS: coast, estuaries. STATUS: uncommon resident. BREEDING: Sep–Oct (DR).

Niven McCrie

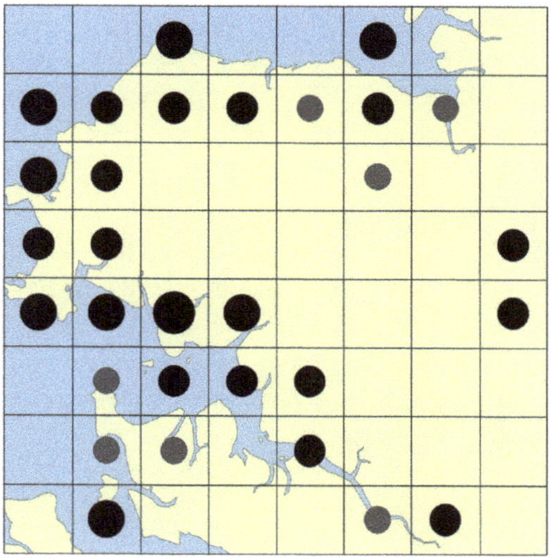

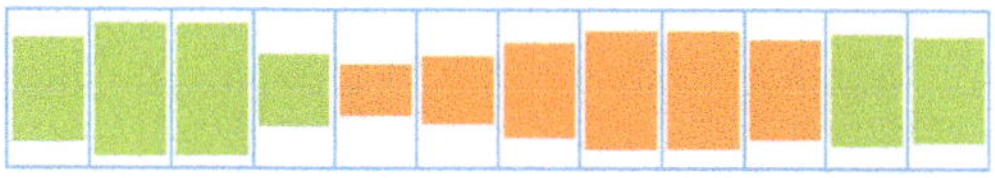

Unlike elsewhere in its worldwide range, where it may be found over 4000 m above sea level, the Striated Heron in Australia is largely associated with intertidal areas, hence the old name 'Mangrove Heron'. With some 23 to 30 subspecies recognised, depending on whose taxonomy is followed, the Striated Heron has greater geographic plumage variation than any other heron species.[107] Top End Striated Herons belong to the race *stagnatilis*, which was first collected at Port Essington and which has two colour forms, the commoner grey form being paler than the race found along the eastern seaboard. Rarely seen around Darwin, the other form is rufous on the underparts.[30]

The species occurs on almost any reefs and mudflats around Darwin's coastline, although it is rare to see more than two or three birds together. Reporting rates are highest on the reefs at Nightcliff and East Point, though they are also frequently seen some way from the coast, such as along tidal creeks and rivers and even on golf courses. The chart shows some level of seasonality; however, this may be due to survey bias, with the species being recorded more frequently during the period when shorebirds are more regularly surveyed at shared habitats.

Although the species normally hunts by standing and waiting, or stalking and diving, Striated Herons are known to use bait, such as picnic scraps, to lure fish close enough for them to catch. A possible instance of such behaviour was observed at Yellow Water, Kakadu NP, where one bird was watched apparently attempting to bait fish by placing a dragonfly on the water surface,[108] though it did not result in any successful catches during the period of observation. The species has also been seen hunting from boat mooring ropes in Darwin Harbour.

Not dependent on the flooding of seasonal wetlands, Striated Herons in the Darwin region commence nesting up to five months earlier than other herons breeding in the Top End. This species normally breeds in single pairs;[38] however, loose colonies of up to five nests have been reported. In the mangroves at Nightcliff, the closest distance between two active nests was 19 m, but 33 m was more typical.[47]

Of 31 active nests, 81% were built in Stilt-root Mangroves, and the average height was 4.27 m above the ground.[47] Two was the normal clutch of Darwin birds, though elsewhere in Australia three or four eggs are laid, and outside Australia up to eight. The young spend about four weeks in the nest or nest tree after hatching, but after three weeks young birds will move off the nest, and adopt a near vertical bittern-like stance if the nest is approached, or sometimes drop into the water or onto the mud, and swim or run away. In September 2006 a possible pre-copulatory display was witnessed, in which an adult bird gave at least six low booming growl-like calls ('hwow') at 2–3 minute intervals. Moreover, its facial skin and legs were strongly suffused with red, appearing a deep salmon-pink.[47]

Javan Pond Heron
Ardeola speciosa

RANGE: se. Asia, Indonesia. HABITATS: wetlands. STATUS: vagrant. BREEDING: not in Australia.

Niven McCrie

A common small heron over much of Indonesia, as far east as Sumba and Flores, the Javan Pond Heron was unknown in Australia until one was found in Darwin in early March 2007.[109] The bird was first discovered at a rainwater puddle on the edge of a large open field in Casuarina CR near the mouth of Rapid Creek. After a few days the bird had moved to a small creek at the edge of the reserve, where it was present until last seen in mid-March.

This species is considered to be entirely sedentary in its normal range,[107] so its occurrence in Darwin is unexpected.

Eastern Cattle Egret
Bubulcus coromandus

RANGE: all continents except Antarctica. HABITATS: wetlands, grassland. STATUS: common non-breeding resident. BREEDING: Nov–May (TE).

Niven McCrie

Eastern Cattle Egrets were introduced to Derby, Western Australia, in 1933, but the discovery in 1948 of a large number near Oenpelli, Arnhem Land,[110] represents their first documented natural occurrence in Australia. Despite its late arrival into the NT, the Eastern Cattle Egret now appears to be the most numerous breeding egret species in the Top End, after the Intermediate Egret,[38] with a breeding population estimated at over 30 000 birds in 1994.[38] It occurs in suitable habitat throughout the Darwin region, and in all months, though frequency of records gradually increases from February, through the late Wet season

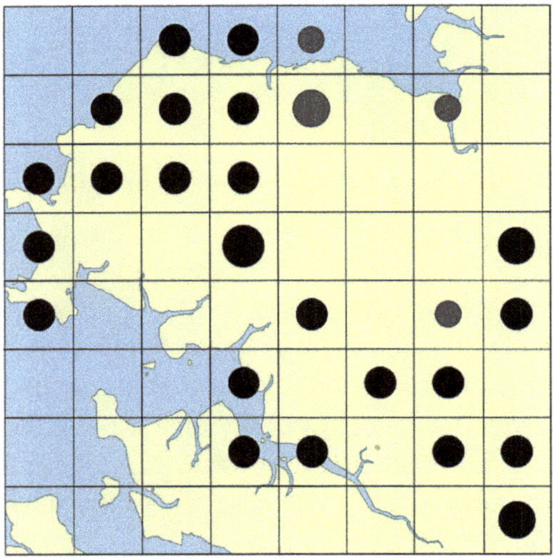

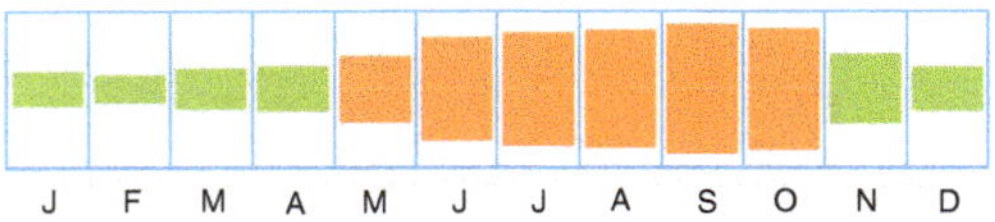

and Dry season, peaking in September–October, before dropping at the start of the Wet. Numbers in the Darwin region also increase during the Dry season and decline at the beginning of the Wet season.

Typically Eastern Cattle Egrets are found in groups of up to about 20 birds feeding in dry grassland, and although up to 250 were reported near Leanyer SP in January 1979,[31] groups of this size in the region are very uncommon. Birds are most commonly seen in paddocks and wetland edges, often alongside horses or cattle and occasionally with goats, pigs, Magpie Geese, or, in some areas outside of Darwin, buffalo, but they have also been seen in woodland and tropical forests.[28] The species occurs widely in the region, with the highest reporting rate from Knuckey Lagoons, which also provides the highest mean counts of the species.

Eastern Cattle Egrets begin breeding earlier than other colonial nesting heron or egret species[29] in the Top End, mostly in November or December, a timing that may have evolved to take advantage of the maximum development of the grassland. It appears, however, that many birds in the Darwin region delay breeding to coincide with the breeding of other heron species,[38] as birds with bright orange breeding plumage are seen around Darwin mainly from December to February, although there is an anomalous record of a bird seen in advanced breeding plumage in July.[21] Many birds move out to breed during December, some earlier, while those in breeding plumage that remain later are likely to be late breeders, who time their breeding to coincide with that of the other heron species.

Breeding colonies are mostly south-west of Darwin, in the area between the Finniss and Moyle Rivers, and east of Darwin, between the Adelaide River and Murgenella Creek. The majority of nests are built in mangrove trees, in colonies with access to both large floodplains and large numbers of cattle or buffalo.[38] At least four Eastern Cattle Egret colonies hold in excess of 3000 birds, with the largest known colony, on the lower Adelaide River, comprising more than 10 000 birds.[29]

While some migration has been shown for birds in eastern Australia,[111,112] there is no evidence of this in the Top End, other than movement to and from heronries.

White-necked Heron
Ardea pacifica

RANGE: Australia, except very arid regions; NG. HABITATS: wetlands, floodplains. STATUS: uncommon visitor, mainly Dry season. BREEDING: Jan (NT).

Mat and Cathy Gilfedder

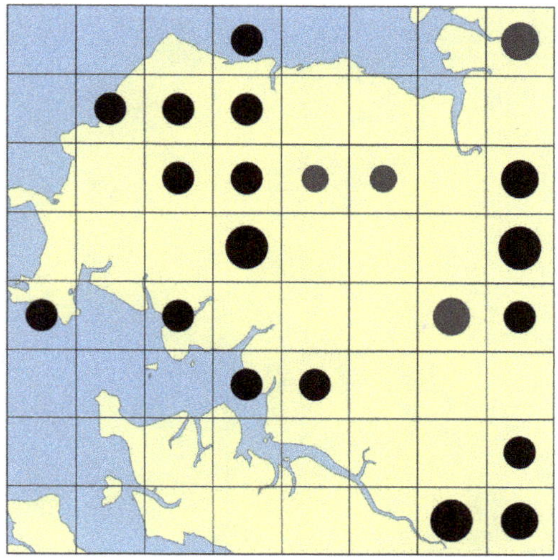

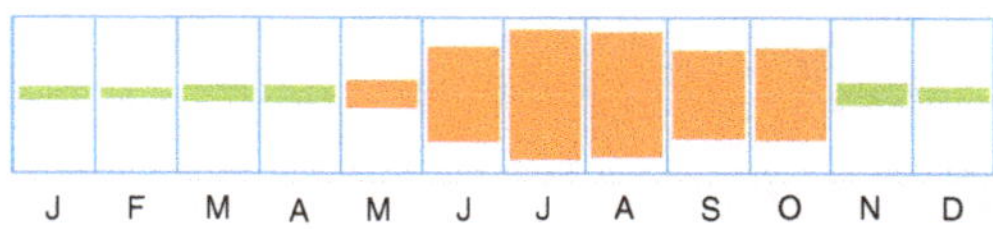

Although frequently present on floodplains and the larger bodies of water in the Humpty Doo region, such as at Fogg Dam and Beatrice Lagoons, the White-necked Heron is a somewhat uncommon visitor to the Darwin region. The species has been recorded throughout the year, but records are much more numerous during the Dry season than the Wet.

Typically only one or two birds are recorded at any time, although five birds were seen on a drained pond at Leanyer SP in June 1991,[31] and up to 12 were present at Knuckey Lagoons in July and August 1999.[21]

The only report of breeding in the Top End is from the Adelaide River floodplain in January.[29] Yet in the Alligator Rivers region[28] alone, the population may be as high as 3000 birds, clearly suggesting a large movement of birds into the Top End from inland south-eastern Australia,[4,28] or perhaps from the Barkly Tableland, where small numbers breed.[113]

The favoured habitat of White-necked Heron is shallow water with a depth of less than 7 cm, usually in marsh or wet grass where there is a clear view of its surroundings.[114] Food is typically small aquatic prey,[115] and primarily includes amphibians, lizards, mussels, shrimp, crayfish, insects, small birds and mammals, and occasionally carrion.[114]

Great-billed Heron
Ardea sumatrana

RANGE: tropical Australia, se. Asia, NG. HABITATS: estuaries, rivers, vegetated wetlands. STATUS: uncommon resident. BREEDING: Feb–May (TE).

Mat and Cathy Gilfedder

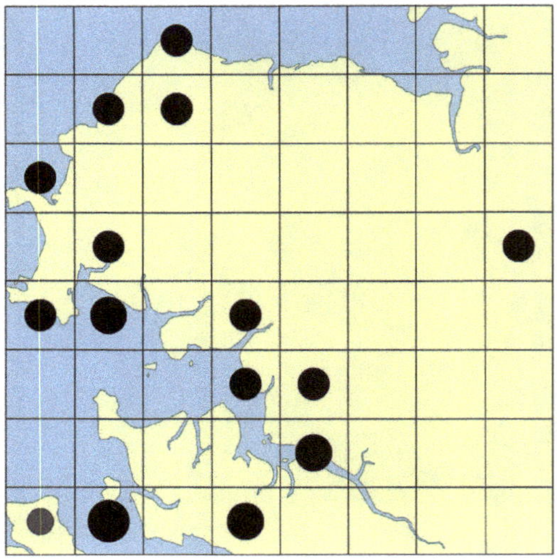

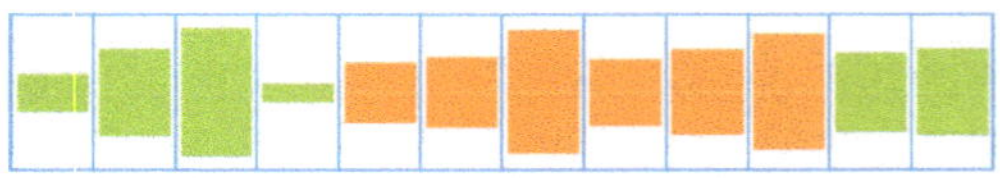

Among the largest of the world's herons, the widespread tropical Great-billed Heron is resident in the Top End. Densities of three birds per kilometre have been reported near the mouth of the Wildman River, about 120 km east of Darwin,[116] and although there are no measures of abundance for the Darwin region, it is likely that densities are lower than this, as the species appears to be generally uncommon. It has, however, been observed in many of the accessible mangrove areas in the region, even near public boat ramps, and it can be assumed to occur in the large areas of suitable habitat that have not been surveyed because of difficulties in access. Moreover, its secretive nature is likely to cause under-reporting. Typically seen on the shores of Darwin Harbour, and along estuaries, often among Stilt Mangroves, Great-billed Herons also occur considerable distances upstream along larger rivers such as the Roper, Katherine and Victoria Rivers, and on subcoastal wetlands, particularly where there are extensive areas of paperbarks and Freshwater Mangrove.[30]

During high tide Great-billed Herons move to higher ground away from the water, or perch in larger mangroves, and birds have also been seen taking larger prey to a perch. When flushed, they will often fly only a short distance, land on mud and walk into the mangroves.[107]

Variations in the number of monthly records during the year are probably due to the birds' increased visibility at particular times of year, rather than to movements in and out of the region.

Unlike many herons, the Great-billed Heron breeds solitarily, in simple pairs, rather than in colonies. Their massive stick nests are generally constructed on a horizontal limb of a mangrove, up to 10 m or more above ground, or sometimes water, and may be used for many years.[117] Few nests of this bird have been found, however, doubtless due to the species' solitary nature and the difficulty of accessing areas it inhabits. The first nest to be described was found 'near Darwin' in February 1843, but it has since been suggested that this was more likely the nest of Black-necked Stork.[30] A nest with one chick was found overhanging a tidal creek 'near Darwin' in March 1982. The young bird adopted a bittern posture when approached.[107] Juvenile birds are occasionally seen around Darwin Harbour,[21,118] and may be the result of local breeding, although juveniles are known to disperse.

Elsewhere in the Top End, a bird was found sitting on a nest in a Darwin Woollybutt at the junction of the McKinley and Mary Rivers in early May 1981[43] and two fledglings were seen begging food from an adult on the edge of the river at Douglas Hot Springs in mid-August 1987.[43] As chicks are thought to take at least two months to become fully feathered, these records indicate that egg-laying occurs at least from February to May.

Great Egret
Ardea alba

RANGE: Australia, s. and e. Asia, NG and NZ. HABITATS: tidal and freshwater rivers, wetlands, foreshores, estuaries. STATUS: moderately common resident. BREEDING: Dec–Apr (TE).

Niven McCrie

The Great Egret utilises a broader range of habitats than other egret species and, reflecting this, it has been reported more widely in the Darwin region than any of the other egrets. High reporting rates from Knuckey and McMinns Lagoons suggest that these areas provide the preferred habitat, although birds also can be found on foreshores, along estuaries and at sewage ponds.

Maximum counts of Great Egret have been at the Holmes Jungle wetland, with about 80 in June 1974,[31] and 45 in October 1996. It is generally the least numerous of the white egrets in the Top End, and in the Darwin region[21] most sightings involve

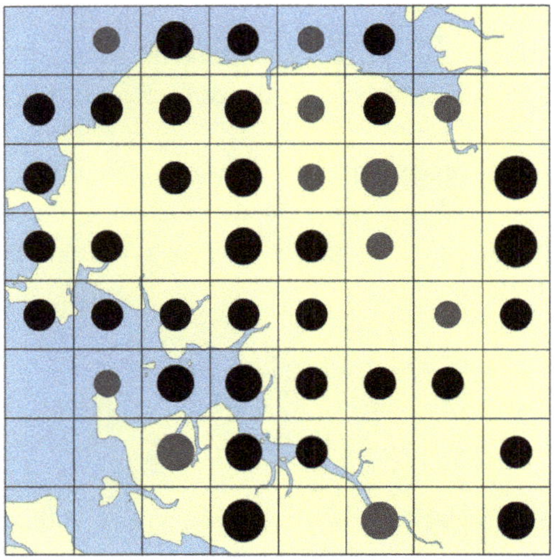

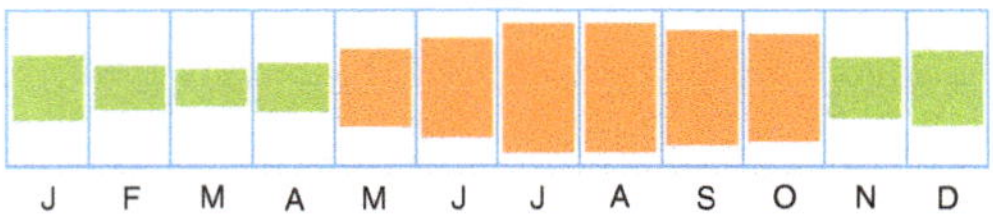

fewer than about 10 birds. As with other waterbird species in the region, numbers are generally higher during the Dry season; however, breeding takes place over a shorter period than in most other Top End waterbirds, so the drop in recording frequency is somewhat short lived.

Numbers of Great Egrets breeding in the Top End are reported to be higher than in colonies elsewhere in Australia, with as many as 31 000 birds nesting across 29 heronries.[38] Birds breed in mixed egret colonies around the coast of the Top End, largely in mangroves at the mouths of rivers and creeks, or along banks of rivers near to the coast.[38] Peak egg-laying occurs from late January to early February, while young leave the nest from mid-April to late June.

Intermediate Egret
Egretta intermedia

RANGE: whole of Australia, Asia and Old World. HABITATS: wetlands. STATUS: moderately common resident. BREEDING: Jan–May (TE).

Niven McCrie

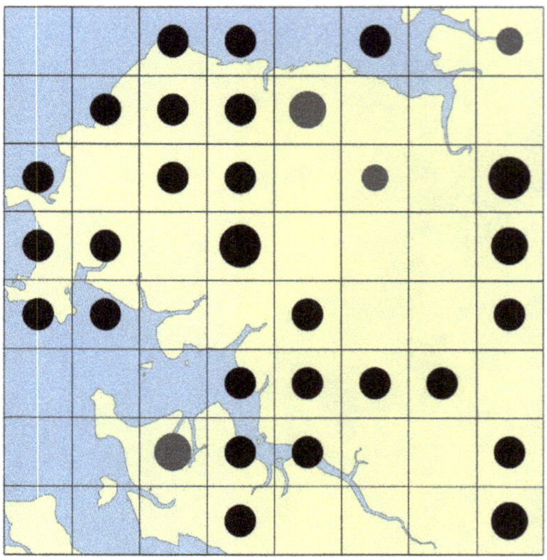

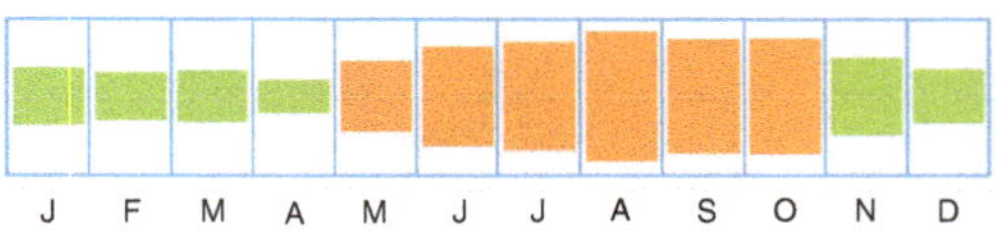

As with the Great and Little Egrets, the Top End seems to be the most important region in Australia for breeding colonies of the

Intermediate Egret. As many as 93 000 birds nested annually during the 1990s at 26 Top End heronries, with up to 6000 at a single site.[38] This species was the most frequently recorded and most abundant of all egret species recorded during the 1990s surveys, although Eastern Cattle Egrets, which were under-counted, may have been in numbers as great or even greater. Surveys on the Magela Floodplain in Kakadu between 1981 and 1984 found it to be the third most abundant waterbird, behind Magpie Goose and Wandering Whistling Duck, and about 10 times as numerous as either Little Egret or Great Egret.[28] In the Darwin region, however, it was recorded less frequently than the Little Egret, and from fewer sites than either that species or Great Egret. The highest reporting rates were from Knuckey and McMinns Lagoons. Although the species occurs in large numbers on sewage farms elsewhere,[30] there are relatively few records of Intermediate Egret from Leanyer SP, as there is very little suitable habitat there.

While the species may occur in thousands at Fogg Dam and on the Adelaide River floodplain, counts of 100 at Leanyer Swamp in December 1977,[34] 50 at Holmes Jungle Swamp[34] and 120 at Knuckey Lagoon in September 2000[21] are typical maximum numbers for the Darwin region. Birds are present throughout the year but numbers drop off from as early as November, as birds move out to breeding colonies. From May, following breeding, numbers increase again, with a peak in abundance usually during August and September.

Intermediate Egrets are much less likely to feed on the coast or in saline wetlands than either Great or Little Egrets,[38] and the scarcity of records from Lee Point, for example, where both Great and Little Egrets are regularly seen, reflects this. Instead, Intermediate Egrets typically forage in the shallow water at the edges of billabongs and sedgelands, though they will also walk on matted vegetation or lily pads in deeper water, exploiting areas not used by other egret species,[28] and occasionally feed well away from the water in dry grass.[21] When hunting, birds often stand motionless, and then dart forward to take prey, usually from the water surface or from vegetation, but also at times submerging their head. Intermediate Egrets are often seen individually or in small groups feeding or resting, but roost communally at night.[116]

The majority of Intermediate Egrets in the Top End are very likely the result of local breeding, with movement only between breeding colonies and Dry season refuges.[28] As with the other egret species, Intermediate Egrets mostly nest in mangroves along the coast and downstream sections of tidal rivers. The nearest colony to the Darwin region, which is also the largest in the Top End and contained an estimated 6000 birds in 1994, is in a tributary off the Adelaide River.[38] The peak egg-laying period for the Intermediate Egret is from late January to early February, but some occurs in late March or even early April. Fledglings leave the nest from mid-April to late June.

Pied Heron
Egretta picata

RANGE: Tropical Australia, NG and e. Indonesia. HABITATS: wetlands, floodplains, sewage ponds, rubbish dumps. STATUS: very common Dry season visitor. BREEDING: Dec–May (TE).

Niven McCrie

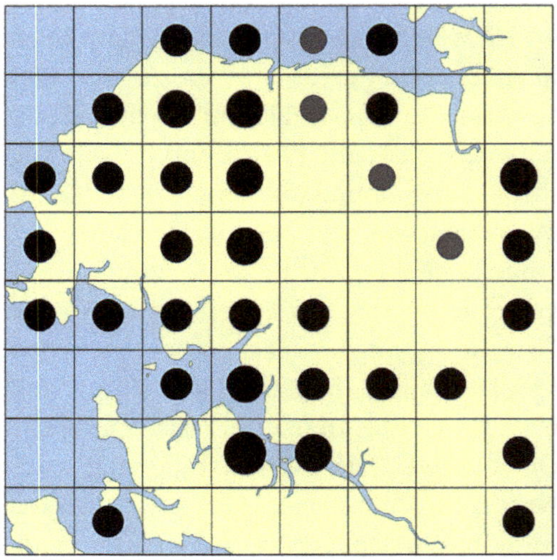

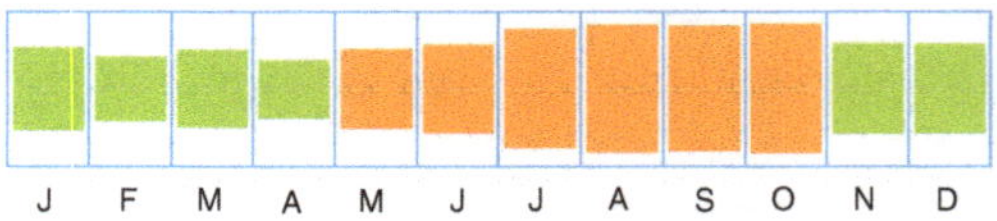

Like the Great-billed Heron, the Pied Heron is restricted to the tropics but, unlike the former, the diminutive Pied Heron occurs widely, and is the most abundant of the herons and egrets in the Darwin region. Counts conducted at Leanyer SP from 1985 to 2010[21] showed numbers of birds to be highest from May to November, followed by a sharp decline. While numbers are low during the period December to February, there are still a few birds present, and it may be that first year, non-breeding birds are joined by other birds using the Darwin region as a staging post en route between Dry season refuges and breeding colonies. The proportion of first-year birds appears to be high during late March and April, and it may be that these non-breeding birds move in while adults are at their breeding colonies.

Large concentrations - up to 500 birds - occur at Leanyer SP, but the Shoal Bay Waste Management Facility appears to hold the highest numbers, with in excess of 1000 birds counted there in October 1999 and September and November 2010. At Holmes Jungle Swamp, adjacent to the rubbish tip, over 500 birds have been counted and it appears that birds move regularly between this site and the rubbish tip. Up to 500 birds have also been reported at Knuckey Lagoons, although large numbers there are uncommon, and it is more usual to see only small numbers around the water's edge. The Top End breeding population was estimated during the 1990s at 22 600 birds, yet in the Alligator Rivers region alone, an estimated 50 000 birds are present during the Dry season,[28] suggesting that the breeding population may be significantly augmented by birds emigrating from other areas.

Plumages vary markedly between adult and young, with the latter having an entirely white head, and from a distance these can be mistaken for White-necked Herons. However, there have also been a small number of sightings of red morph birds, which have much of the white replaced with colour, varying from buff to rich rufous.

Aquatic insects form the major part of the Pied Heron's diet, but small fish and frogs are also taken.[107] Pied Herons usually feed in shallow water, gleaning insects from vegetation, though they may also hunt in deeper water. When standing and feeding in deeper water, birds are rarely seen to submerge their heads, preferring to glean from the surface.[21] They may also feed in water too deep to stand in, and in such cases they rely on perches from which to hunt, and vegetation, fences, posts or even the backs of Pelicans[21] may be utilised. At Leanyer SP birds can be seen lined up along the concrete banks of the sewage inlet channels, watching for, then lunging to capture, cockroaches[21,107] as they emerge from the closed channel into the open. Pied Herons appear to display more intraspecific aggressive behaviour than other herons in the region. Particularly as feeding resources contract, and birds forage closely together, aggressive behaviour is often seen. Birds face each other, and if neither bird submits and moves off, they may jump into the air to a metre or more, flapping their wings and stabbing at each other.

In the Darwin region, as dusk approaches, large flocks of Pied Herons move out from Shoal Bay Waste Management Facility, Leanyer SP, Holmes Jungle, Knuckey Lagoons and other daytime feeding areas, to fly to their roosting sites. They move in a south to south-easterly direction, small groups coalescing into larger flocks as they approach their roosting site in mangroves, mainly on Catalina Island,[119] in Middle Arm. Monthly counts[119] of birds going to roost, from November 2012 to October 2013, ranged from 1700 birds in February, to an amazing 5895 birds in November, the latter being significantly higher than previously estimated for the region. Very few birds remain near feeding areas; however, some have been observed arriving at Knuckey Lagoons just on dusk and may overnight there in dense vegetation. At daybreak the birds return from their roost, flying in large loose flocks - some over 1500 birds.[21] At Knuckey Lagoons up to 500 have been seen to arrive from the direction of the roosting sites around dawn, forage for a short time, then leave again within about an hour after sunrise.

Pied Herons breed extensively in the Top End. During surveys in the 1990s, 23 breeding colonies, containing up to 3000 Pied Herons, were found in the Top End. These colonies invariably contained other waterbird species, and nearly all colonies[38] were found in mangroves along the coast or short distances up tidal rivers. Extreme dates of egg-laying are 20 January and 13 April, with peak laying from late January to early February. Young birds started leaving their nests by mid-April, and the last broods were gone by late June.[38]

White-faced Heron
Egretta novaehollandiae

RANGE: whole of Australia; NG, Indonesia, NZ. HABITATS: wetlands, occasionally estuaries. STATUS: uncommon visitor, largely Dry season. BREEDING: Dec–May (NT).

Niven McCrie

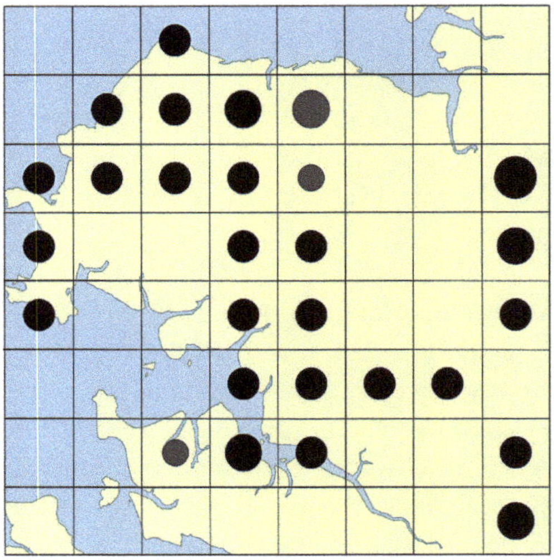

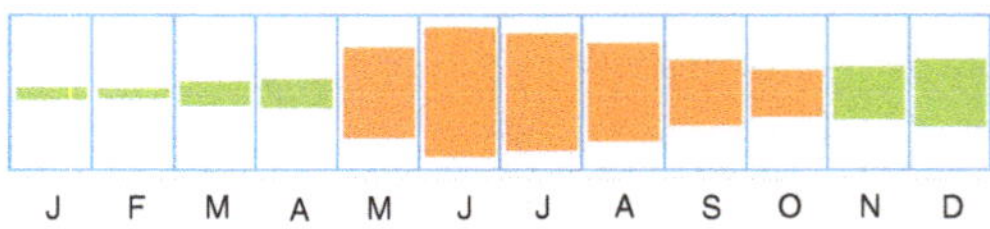

Like the White-necked Heron, the White-faced Heron is predominantly a Dry season visitor to the Top End, though, unlike the former, it does not seem to disappear from the Darwin region until well after the start of the Wet season. An increase in records in May indicates a quite sudden influx, presumably deriving from south-east Australian breeding populations, which are known to migrate to the tropics during autumn, followed by a gradual decline through the mid and late Dry seasons. The December spike may be partially a result of late southward movement through the Darwin region from Indonesia. The same sharp rise in both frequency and abundance of these birds during May was found during surveys which covered the entire NT coast over the years 1990–2006.[27] In these surveys, 80% of records were of birds in groups of five or fewer, but in the south-east of the Top End there were observations of groups comprising 100 to 350 birds.

The White-faced Heron has been reported more widely than the White-necked Heron, and is perhaps able to make use of smaller areas of habitat, occurring around the shallow fringes of swamps and lagoons, occasionally in sewage works or in temporarily flooded pools on roadsides. While it is almost always seen singly, 17 birds were recorded at Leanyer SP in July 1976.[31]

The White-faced Heron will sometimes nest in colonies; however, it is more often a solitary nester, and usually builds its nest high in a tree, and sometimes on artificial structures. In April 2003, an adult and three chicks were observed in a nest atop an electricity pylon adjacent to the Adelaide River floodplain, about 5 km south of the town.[21] The still downy chicks suggest egg-laying to have been in early March. There appear to be no breeding records closer to Darwin, although there are reports from further north on the Adelaide River floodplain,[29] and it occurs in some other parts of the Top End.[28] An estimated population of about 2000 birds for the Alligator Rivers region[28] was derived from surveys between 1981 and 1984, so it is clear then that many of the birds appearing in the Top End move in from elsewhere, probably from the species' major breeding areas in south-eastern Australia.

Little Egret
Egretta garzetta

RANGE: whole of Australia, Asia and Old World. HABITATS: wetlands, estuaries, shores, sewage ponds. STATUS: moderately common resident. BREEDING: Jan–May (TE).

Niven McCrie

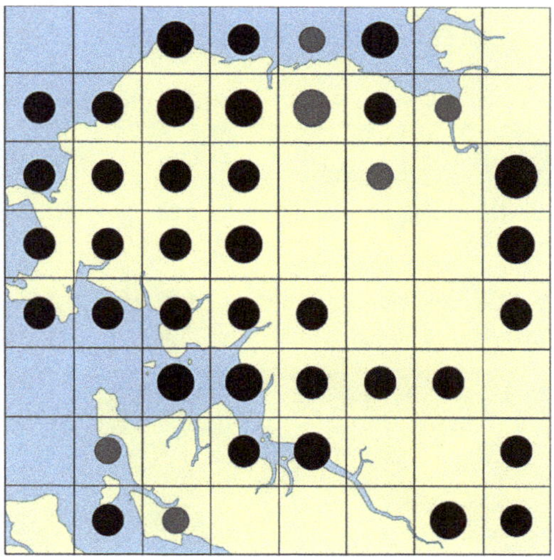

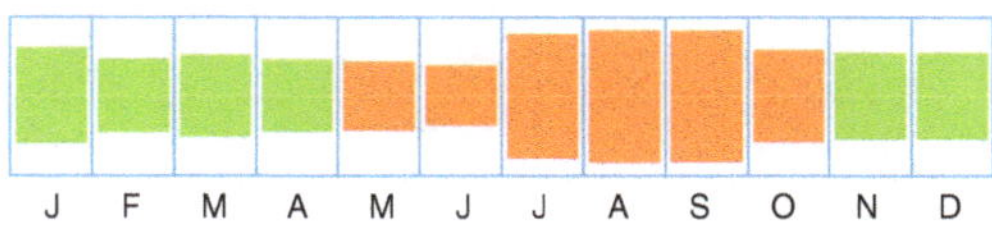

With an estimated breeding population of around 18 000, the Little Egret is the least abundant egret species in the Top End. Yet these numbers indicate that the Top End is the most significant Little Egret breeding region in Australia. In the Darwin region, it was the most frequently recorded species among the egrets. This reflects the range of habitats it utilises, including tidal mudflats on shorelines and along mangrove creeks, floodplains, sewage ponds and temporary waterholes. Little Egret concentrates around channels and open water more than the other egret species,[27] and birds are sometimes seen among Eastern Cattle Egrets, foraging in flooded paddocks.[21]

Though some birds are present throughout the year, reporting rates indicate that many move out of the region from late January, and return during the early part of the early Dry season. This absence coincides with their known breeding season, so it appears that birds present in Darwin during that time are non-breeding birds.

Maximum numbers occur from October to January, and lowest numbers from February to May. Like Great Egrets, Little Egrets are usually found in groups of less than 10,[27] but up to 100 were recorded around the Holmes Jungle Swamp in January 1975.[31] It may be that birds from outside of Darwin, at least in some years, use the Darwin region as a staging area before moving to breeding colonies in late January and February.

Some evidence suggests that the Top End hosts birds from south-east Asia, as well as local breeding birds. Since 2005 there have been numerous sightings and photographs taken of one to three individuals of the nominate subspecies *garzetta*, which occurs in Africa, Europe and Asia.[120] Most reports have come from Leanyer SP, but these birds have been seen also at Buffalo Creek, Knuckey Lagoons and Fogg Dam.[21]

As with the other egrets, most breeding colonies are in mangroves along the coast or tidal rivers. In 1993, the largest single colony, containing around 4500 birds,[38] was located in a tributary off the Adelaide River, less than 40 km from Knuckey Lagoons. Peak laying for Little Egrets is from late January to early February, and young are present through March and April, with the earliest fledglings leaving the nest by about mid-April, and the last by late June.[38]

Pacific Reef Heron
Egretta sacra

RANGE: Australia, e. Asia, NG, NZ. HABITATS: tidal mudflats, reefs, estuaries. STATUS: moderately common resident. BREEDING: Aug–Dec (NT).

Niven McCrie

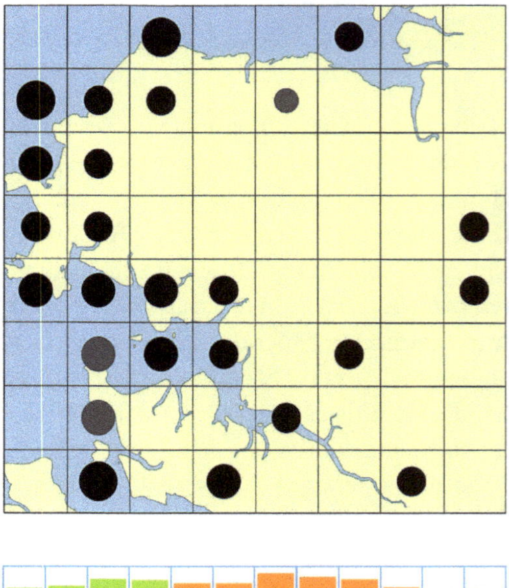

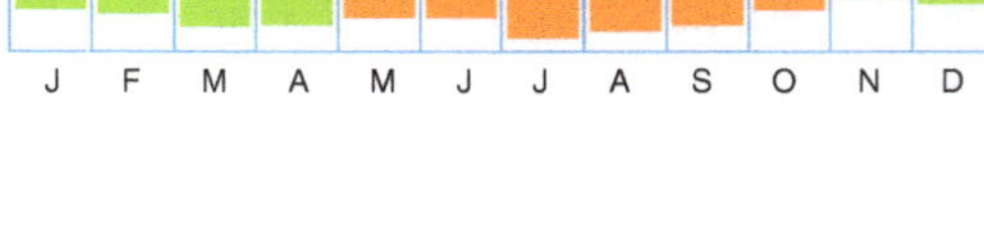

There are two colour morphs of Pacific Reef Heron, with some interbreeding between the two. While birds of southern Australian are invariably of the dark grey morph, the white morph becomes increasingly common towards the tropics, at least in the east. In apparent contradiction to that trend, however, dark birds outnumber white ones in the Darwin region by about four to one, or possibly more. Both colour morphs are frequently seen together, and there seems to be no specific habitat preferred by either morph, though in Darwin Harbour white birds apparently feed predominantly on fish, while dark birds mostly feed on crustaceans.[107]

Although restricted to coastlines and islands, the Pacific Reef Heron is moderately common, and can be seen on reefs, mudflats and sandy beaches, or roosting in mangroves, and even on mooring ropes and boats in Darwin Harbour. The highest reporting rates are from Casuarina CR, Lee Point and along the Elizabeth River, but there are records from most coastal localities, and from Leanyer SP. Seasonality data indicate a possible movement of birds to breeding areas out of the region during the Wet season, although it is also possible that birds move to breeding sites within the region. In either case, the rise in records during the Dry season would result from birds returning to feeding areas, augmented to some extent by young.

A total of over 4000 birds was found in the Top End during coastal surveys in the 1990s.[38] The species normally occurs singly or in small groups, but there were nine records during those surveys of 100 or more, including an estimated 500 on a small island 'near Darwin'.[27] All of the large groups were recorded in the period late August to early October, in Flatback Turtle nesting areas, and it may be that the birds gather there to feed on hatchling turtles.[38]

In contrast to other egrets, peak breeding activity of this species is mainly on islands and takes place from August to December.[38] In February 1998 a bird carrying a stick in its bill was seen flying from the shore of Darwin Harbour apparently landing under Stokes Hill Wharf,[21] but confirmation of nesting was not obtained. A report of the species 'nesting colonially under a pier' in Darwin Harbour[107] apparently also refers to this wharf, but there is insufficient detail to substantiate the record.

In October 2003 a white and a grey morph bird were photographed together at a nest at East Point.[121] Although the nest is empty, it is undoubtedly that of Pacific Reef Heron,[102] and provides the most substantial evidence available of the species nesting in the Darwin region.

Australian Pelican
Pelecanus conspicillatus

RANGE: whole of Australia and NG, vagrant to e. Indonesia. HABITATS: terrestrial and coastal wetlands. STATUS: moderately common. BREEDING: Mar–Sep (TE).

There are few breeding colonies of the Australian Pelican anywhere in northern Australia,[4] but a colony of up to 6000 birds, located on North Peron Island, 120 km south-west of Darwin[38] – the only permanent colony in the NT – is one of the largest permanent colonies of this species in Australia. However, the colony is active

Niven McCrie

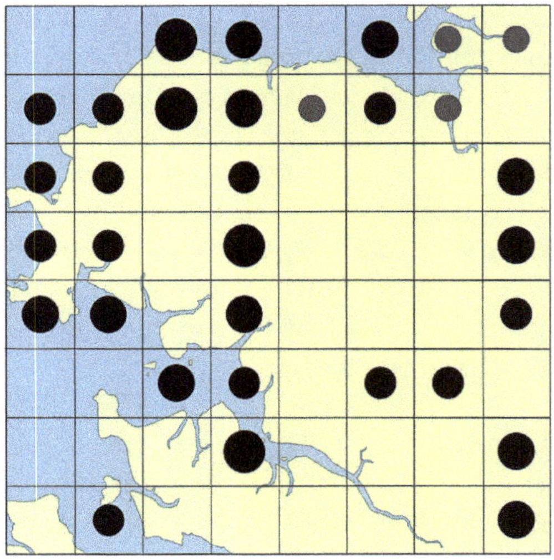

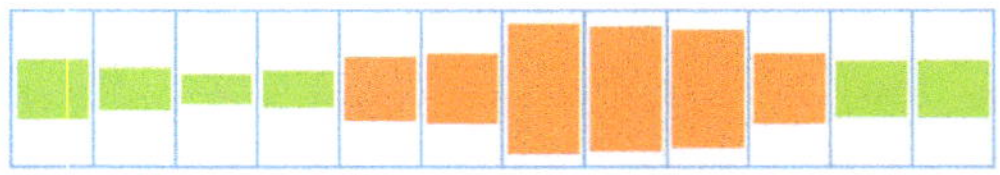

during the Dry season, from March to September, when numbers are at their peak in the Darwin region, so it is likely that these birds derive from elsewhere. Indeed, up to 50 000 individuals are thought to inhabit the Alligator Rivers region[28] of Kakadu NP alone, and clearly cannot originate from that colony.

An estimated 700 pelicans at Harrison Dam in late July 2013,[122] and over 400 at the nearby Beatrice Lagoons in August 2007,[123] surpass any counts in the Darwin region, where peak numbers rarely exceed 100 birds, and probably reflect the extent of suitable habitat available. Numbers normally build up during the Dry season, particularly in the period June to September, as birds move from drying wetlands inland towards the coast, and it is in this period that peak numbers are normally found. The considerable annual variation in abundance is probably due to conditions inland, with higher numbers following good breeding seasons.

Lesser Frigatebird
Fregata ariel

RANGE: mainly tropical Indian and w. and central Pacific Oceans. HABITATS: pelagic, coast. STATUS: uncommon Wet season (usually) visitor. BREEDING: Mar–Sep (Ashmore Reef).

Superbly adapted for life on the wing, with the lowest wing-loading of any birds, many frigatebirds take advantage of the continual thermals over our warm tropical seas and may spend the entire non-breeding season in the air. Most frigatebirds encountered in the Darwin region are Lesser Frigatebirds, though two other frigatebird species have

www.wildlifeimages.com.au/Rohan Clarke

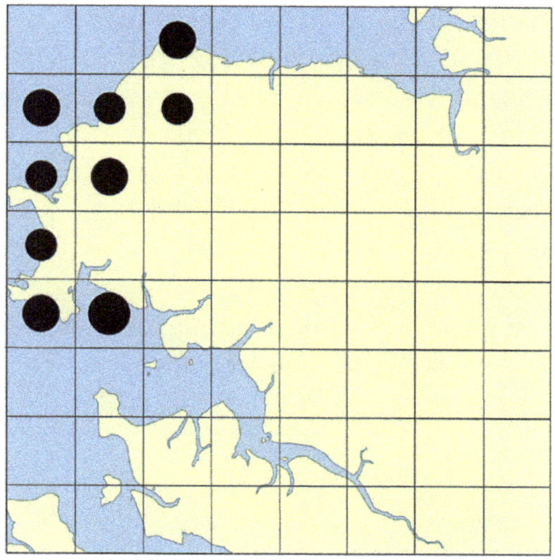

also been recorded. Lesser Frigatebird was reported in the region in 21 of the years between 1986 and 2014, invariably during Wet seasons, and usually but not always associated with strong windy conditions. Although generally seen singly or in groups of up to about 10 birds, 80 were seen at Lee Point in 1986[43] and groups of over 50 were seen over the suburbs in January 2010 and 2011.[21]

Lesser Frigatebirds are conspicuous in flight from a considerable distance, so they are unlikely to be overlooked, and indeed it may be assumed that most occurrences are reported. It is not surprising then that there are reports from throughout the suburbs including Casuarina and Palmerston shopping centres, Parap, Nightcliff and Fannie Bay, but in January–February 1973 birds were sighted up to 16 km inland from Darwin harbour.[79] In 2011, groups of 50 or more were reported from 10 to 12 January, but by 13 January few were reported. A tropical low formed in the eastern Indian Ocean on 10 January, and intensified into tropical cyclone Vince two days later.[124] The sudden appearance of the Frigatebirds at this time, followed by the rapid drop in numbers, demonstrates clearly the association between weather and the species' occurrence in the Darwin region. It also suggests that the birds originated from around Ashmore Reef, either blown in by strong winds or, aware of the low pressure system, moving ahead of it to seek shelter. The timing makes the latter explanation more likely. Moreover, having very little webbing on toes and non-waterproof feathers which if waterlogged can make it impossible to take flight, frigatebirds are ill-adapted to being on the water, and would benefit from avoiding the prospect of being forced onto the water by stormy conditions.

At Buffalo Creek, Lesser Frigatebirds have been observed harassing Silver Gulls and waders, and one was seen to take food from a Crested Tern.[40] The famed frigatebird kleptoparisitism – or piracy – only accounts for a small percentage of Frigatebird food intake, with most of their diet comprising squid, snatched from the surface of the sea, or flying fish, taken just above the surface of the water.[125] Numerous birds also associate with trawlers to feed on cast-off catch during the prawn fishing season.[118]

During Dry seasons in at least three years flocks of between 500 and 2000 Lesser Frigatebirds have been seen soaring over islands in Fog Bay.[126] The birds were flying from the south and west, at an altitude of over 100 m, in the late afternoon. Boat operators have also reported a roost of the species at the mouth of Bynoe Harbour. The nearest known breeding colonies are on the Wellesley Islands, just over the border in Queensland, about 1000 km from Darwin,[127] and Ashmore Reef,[30] about 800 km from Darwin, where over 4000 birds were present in April 2010.[128] The birds breed on Ashmore from March to September[91] but each pair breeds only every second year. Thus the large numbers in Bynoe Harbour during the Dry season are possibly non-breeders from Ashmore.

Christmas Frigatebird
Fregata andrewsi

RANGE: breeds Christmas Island; non-breeding birds wander largely se. Asia. HABITATS: pelagic. STATUS: vagrant. BREEDING: not in NT.

www.wildlifeimages.com.au/Rohan Clarke

Breeding only on Christmas Island, the Christmas Frigatebird's movements outside the breeding season are not well known, though it appears to be largely restricted to Christmas Island and south-east Asia during that period. There are records in two years for the Darwin region, and these appear to represent the only confirmed records for mainland Australia. In January 1974, up to two birds were seen and photographed on different days and by different observers[129] during a period of north-westerly gales. A Christmas Frigatebird in late juvenile or early second cycle plumage was seen and photographed at Nightcliff, in January 2014,[130] also during severe weather conditions.

Great Frigatebird
Fregata minor

RANGE: Mainly Indo-Pacific. HABITATS: pelagic. STATUS: rare Wet season visitor. BREEDING: Not in NT.

www.wildlifeimages.com.au/Rohan Clarke

Great Frigatebirds are rarely seen inshore except during periods of cyclonic weather, when they may use mangroves for shelter, and may even be blown short distances inland.[30] The first acceptable record for the Darwin area was of five birds over Lee Point in January 1968, during the rough weather of cyclone Bertha.[1] In January 1974, there was a storm-driven influx of frigatebirds, including Great, Lesser and Christmas Frigatebirds, into the Darwin region, and Great Frigatebirds were reported from Nakara, Casuarina, Buffalo Creek and Lee Point.[129] Other records include sightings from Lee Point in January 1983, East Point in March 1985 and, following storms produced by cyclone Hector in January 1986, Great Frigatebirds were reported at Coconut Grove, Lee Point and Myilly Point.[34,43] An adult female seen at Nightcliff in March 2001[21] appears to be the most recent sighting.

The largest number reported in the Top End was a flock of 24 birds that were observed at sea between the Peron Islands and Darwin in May 1982.[34] The nearest breeding colony is on Ashmore Reef, where 40 active nests were found in 2010, and there are indications that the numbers of breeding birds there are increasing.[128]

Brown Booby
Sula leucogaster

RANGE: pantropical. HABITATS: pelagic. STATUS: uncommon visitor, mainly late Dry and Wet seasons. BREEDING: not in TE.

The only booby to be recorded in the Top End with any regularity, the population of Brown Boobies was estimated during the 1990s to be about 3000 birds. However, most Brown Boobies in the Top End occur in northern areas of the Gulf of Carpentaria,[127] and they are largely absent, or in very small numbers, in the south-west Top End, including the Darwin region, probably because the waters are shallower and more turbid.[127] Records are most frequent during the late Dry and early Wet seasons and, although the species is not restricted to a set breeding season, this suggests a

www.wildlifeimages.com.au/Rohan Clarke

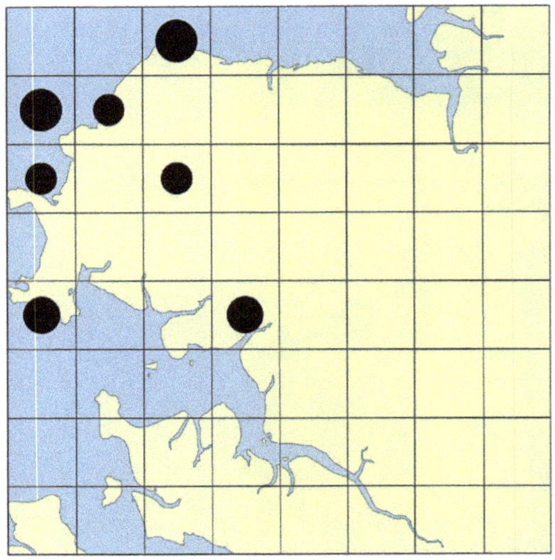

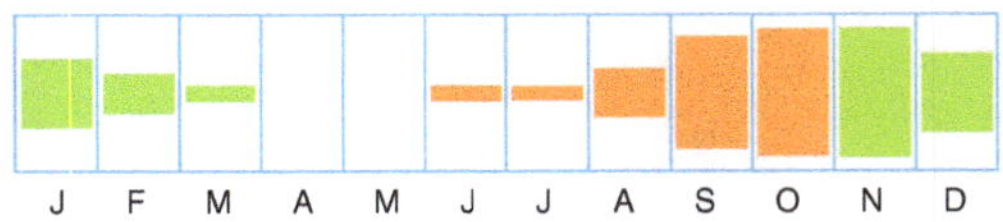

dispersal of juveniles and/or an exodus of adult breeding birds at that time of year. Sightings mostly refer to solitary birds, although occasionally groups of birds are recorded, particularly during stormy periods, when they may be forced inshore, and are occasionally observed sitting on rocks, trees or exposed structures on or near the shoreline. At East Point during July 1964 there were sightings of up to 30 or more birds perching on a cable over the water, remnants of an anti-submarine installation from World War II.[96] This observation is significant not only in being the greatest number recorded together for the region, but for the fact it was at a time when the species is rarely recorded. Most records come from the Nightcliff and Buffalo Creek-Lee Point areas. In north-west Western Australia the Brown Booby is most abundant 18–36 km offshore[30] and observations at sea suggest that the situation is similar off the NT coast.

There are no confirmed breeding sites on the NT coast,[127] with the closest known breeding colonies being on Booby Island off the Kimberley coast, Western Australia, about 750 km west of Darwin. Immature or juvenile birds are most often reported in the Darwin region in January, and although this fits with the reported main season at both colonies mentioned, such plumage cannot be taken to indicate first-year birds, as it is not until their third year, when the birds typically return to colonies to breed for the first time, that they attain their adult plumage.

Little Pied Cormorant
Microcarbo melanoleucos

RANGE: whole of Australia, NG, Indonesia and SW Pacific STATUS. HABITATS: wetlands, coastal waters. STATUS: moderately common. BREEDING: Feb–June (NT).

Laurie Ross

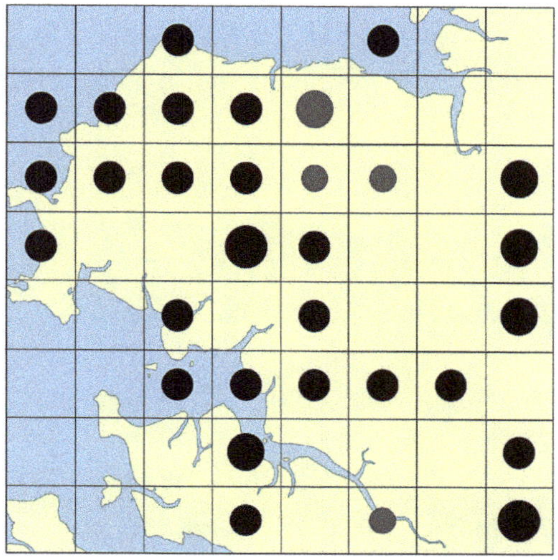

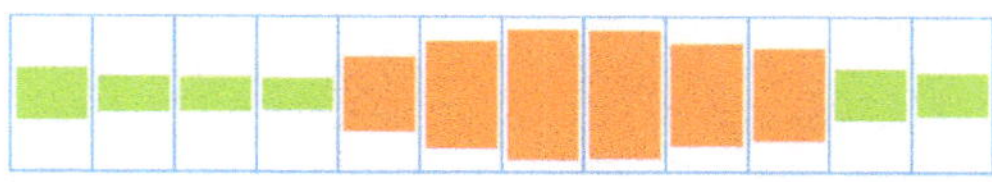

Although less numerous in the Top End than Little Black Cormorants, Little Pied Cormorants are more widespread. From a small sample of gut contents in Kakadu NP,[131] the diet of Little Pied Cormorant was found to comprise about 54% fish and 34% invertebrate prey, while the diet of Little Black was found to be 99% fish. Not being entirely dependent on fish for their diet, Little Pied Cormorants are able to take advantage of smaller pools of water,[28] and spread across a broader area. Their tendency to hunt alone rather than in flocks like the Little Black[131] also means their habitat requirements are less restrictive, thus making more areas available to them.

In the Darwin region, Little Pied Cormorant is most numerous, and most frequently reported, on freshwater wetlands, particularly at Knuckey and McMinns Lagoons, where the reporting rates are more than double those from elsewhere. Small numbers can be seen fishing at Buffalo Creek and in other estuaries, or along the coast, and individual birds can often be found perched in trees along small creeks in the region. Although the maximum number reported was 70 birds at Leanyer SP in May 1976,[30] the majority of counts in excess of 50 birds have been in the period August to October, with numbers dropping during the Wet season as birds move to breeding areas. Much larger numbers occur in the Fogg Dam and Adelaide River regions, particularly late in the Dry season as wetlands are drying out, and hundreds of Little Pied Cormorants, along with large numbers of egrets and other waterbirds, are concentrated at the remaining pools.

Although no breeding has been reported in the Darwin region, the Little Pied

Cormorant is the most numerous breeding species of the cormorant and darter group in the Top End, with at least 18 500 estimated to nest around the coast and adjacent floodplains.[38] The majority of the 36 breeding colonies so far discovered are in the north-western Top End, with the closest to Darwin, in mangroves near the mouth of the Adelaide River, holding over 600 birds when surveyed in March 1993. In large mixed species colonies such as this one, which include many egrets, egg-laying occurs mostly in February and March, while most young have left their nests by late May. In colonies dominated by cormorants and darters, on the other hand, Little Pied Cormorants tend to breed later, and are often still on eggs in early May.

Little Black Cormorant

Phalacrocorax sulcirostris

RANGE: Australia, NG, Indonesia, NZ. HABITATS: wetlands, coastal waters. STATUS: common, mostly Dry and early Wet season visitor. BREEDING: Mar–Jul (TE).

Laurie Ross

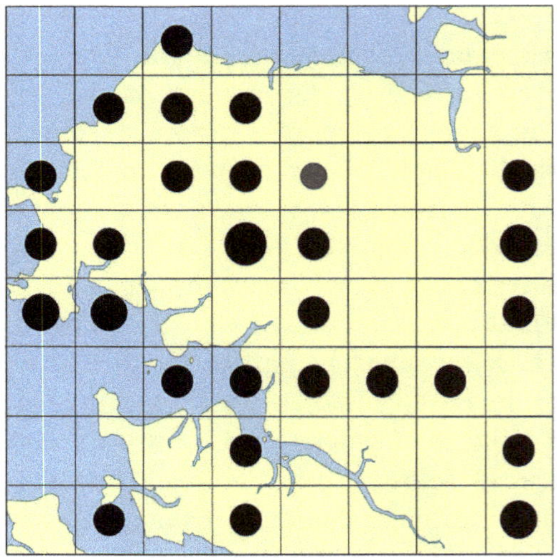

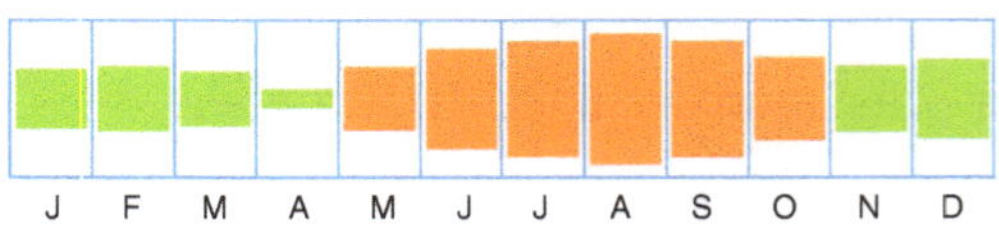

More dispersive than the Little Pied Cormorant, the Little Black Cormorant is also more seasonal, with the number of records increasing steadily over the Dry season, and declining during the mid-Wet. Numbers also rise during the Dry season, and typically peak in August or September, although 130 birds, the maximum single count for our region, were seen in mangroves near Leanyer SP in February 1978.[34] Reporting rates are highest at Knuckey and McMinns Lagoons, but are somewhat lower than for Little Pied Cormorant. There are relatively few records in estuaries or along exposed coast line, but small numbers - up to about 10 birds - can be seen in Frances Bay, the wharf areas, and in some other parts of Darwin Harbour.

Little Black Cormorants hunt socially for small fish, which comprise up to 99% of their diet, and form larger social groups than do Little Pied Cormorant, often numbering several hundred. Their apparent dependence on fish for food means Little Black Cormorants are more likely to need deeper, more open water than Little Pied.[131] At McMinns Lagoon, 12 birds were apparently following a flock of Australian Pelicans as they fished.[31]

No breeding has been reported in the Darwin region, but a minimum of 8500 Little Black Cormorants are estimated to breed in the Top End. Most breeding colonies are located in the north-west of the Top End, particularly in the area between the Moyle and Finniss Rivers.[38] Colonies tend to be on floodplains or along subcoastal rivers, and may be small, comprising only cormorants, or large mixed colonies with egrets, spoonbills and other waterbirds. Birds are reported to be on eggs from late February to early May, with young seen from mid-April to late June. As with Little Pied Cormorant, those in colonies dominated by egrets nested earlier, and those in colonies with the later nesting species, such as Royal Spoonbill, also tended to breed later.

Australian Pied Cormorant
Phalacrocorax varius

RANGE: Whole of Australia and NZ only. HABITATS: mainly marine, but also wetlands. STATUS: uncommon, non-breeding. BREEDING: Mar–Jul (TE).

Nolan Caldwell

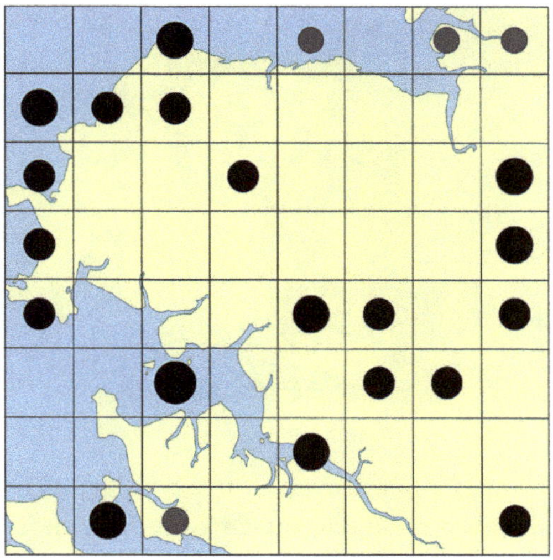

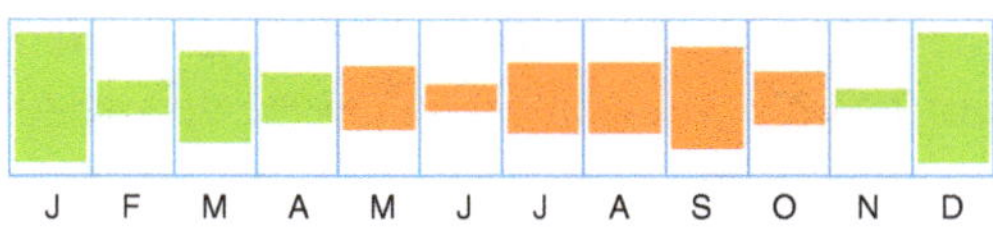

In the Darwin region, Australian Pied Cormorants are generally seen singly, and small numbers have been recorded in all months. Exceptionally, there were 150 at Leanyer SP in December 1978[34] and 20 on dead mangroves near Ludmilla Creek in February 1983.[31] In recent years there have been frequent records of up to 15 or more roosting on the shore or feeding offshore from Nightcliff and Buffalo Creek, and the records appear to be increasing. Records are widespread, but the highest reporting rates from the more frequently visited areas are from Buffalo Creek–Lee Point and Casuarina CR.

The only reported breeding within the Darwin region is an undated record of 12 pairs breeding on Haycock Reach, in Middle Arm, about 8 km south of Palmerston;[30] however, it is not known whether the breeding was successful.

An estimated 4000 Australian Pied Cormorants breed in colonies around the Top End. The majority of colonies, and the largest number of birds, occur in the south-west corner of the Gulf of Carpentaria,[38] with only two small colonies in the western Top End. Most colonies are in mangroves on, or close to the coast, and about half of the reported colonies have comprised only Pied Cormorants. The breeding season extends from about late March to July, with birds laying from late March, and some still with eggs in late May.[38]

Great Cormorant

Phalacrocorax carbo

RANGE: cosmopolitan, including whole of Australia and NG. HABITATS: wetlands, coastal waters. STATUS: vagrant. BREEDING: May (NT).

Kirri Hardy

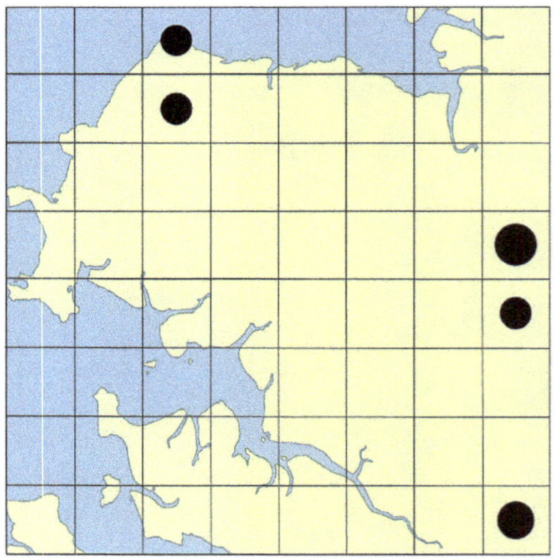

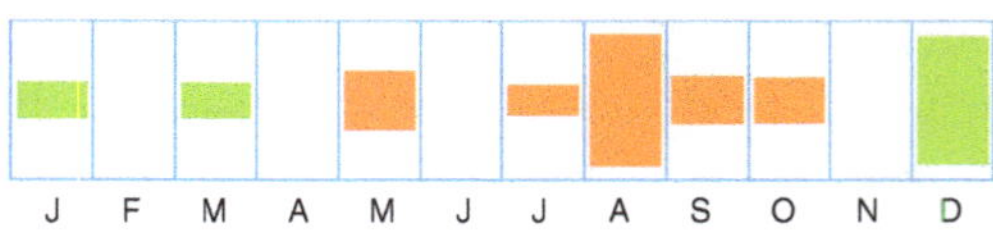

Though a familiar sight in the wetlands of southern Australia, the Great Cormorant is rare in the Top End. Indeed, the species' presence in the NT was not confirmed until 1972, when two birds were recorded about 20 km south-west of Berry Springs. There have since been numerous observations of the species at nearby Darwin River Dam, and although there has been no evidence of breeding, it appears at least to be a regular visitor there.

Reports from the Darwin region have been only sporadic, and even then it is possible that some reports of the species have involved misidentified Little Black Cormorants. At least five birds were seen at Dinah Beach in December 1983,[31] but otherwise all reports have been of single birds. Of the 13 records up until 1992, almost half were from Knuckey Lagoons. Away from there, records include a bird observed flying offshore at Lee Point in October 1978,[31] one along a creek behind the beach at Ludmilla[132] and a bird at Leanyer in November 1976.[2]

Elsewhere in the Top End, there are several records from the Katherine[2] and Victoria River[133] regions, but only one record from the Alligator Rivers region.[28] Five birds were also reported at Kapalga and another on Seven-mile Lagoon, Wildman River in August and September 1977.[2]

The main breeding distribution of Great Cormorant is south-east Australia, with the only confirmed breeding in the NT at Lake Woods, in the Barkly Tableland, where 25 pairs were found nesting in May 1993.[26]

Australasian Darter
Anhinga novaehollandiae

RANGE: Whole of Australia, NG and e. Indonesia. HABITATS: estuaries, wetlands, rivers. STATUS: uncommon. BREEDING: Mar–Jul (TE).

Niven McCrie

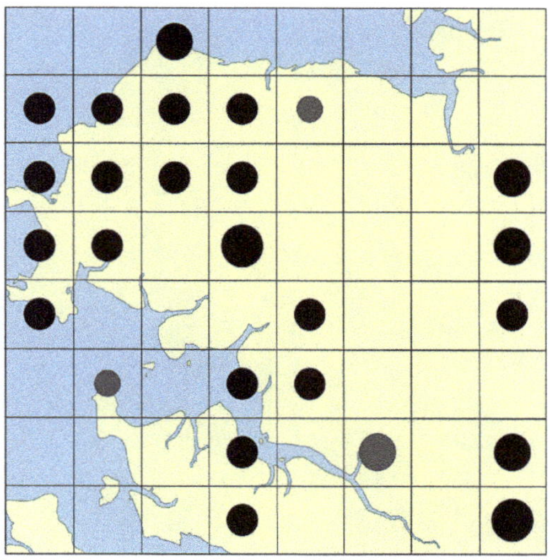

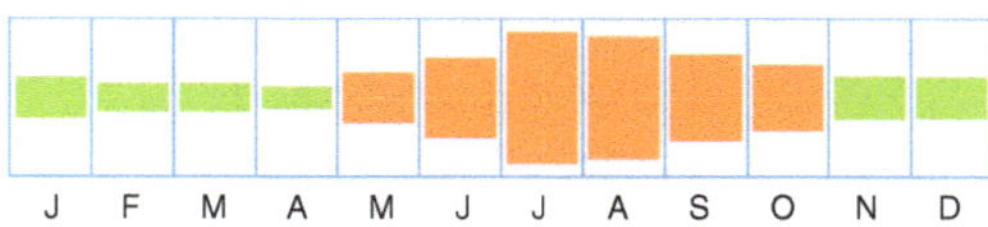

While the Australasian Darter occurs throughout the year, records are most frequent during the mid-Dry, at the end of the species' breeding season. The highest reporting rates in the Darwin region are from McMinns and Knuckey Lagoons. Both lagoons meet the birds' requirements of smooth open water for feeding, and while McMinns has fringing trees for perching and drying wings, in the case of Knuckey Lagoons wooden posts provide a suitable alternative.

In the Darwin region, there are rarely more than 10 birds present at a site at any one time. However, at Beatrice Lagoons, near Fogg Dam, over 100 were recorded in June 1997.[21] The birds were spread out on small 'islands' among groups of Little Black

Cormorants, or feeding in the water, and there were continually birds in flight, circling the lagoon complex or flying across the Arnhem Highway to deeper pools on the other side of the road. In the Alligator Rivers region, where there is significant seasonality with most individuals present in the Dry season, the population has been estimated at about 2000 birds.[28]

An estimated 4700 Darters breed in 27–30 colonies in the Top End, most in the north-west, and chiefly within about 40 km of the coast.[38] Colonies are located inland from the coast, in paperbarks around floodplains and billabongs, with smaller groups of birds often nesting along mangrove-lined coastal creeks. Although eggs were recorded at colonies from February to July, with some colonies still active into August, most breeding is concentrated between March and July.

Eastern Osprey
Pandion cristatus

RANGE: Australia, Indonesia, NG. HABITATS: coast, larger rivers, occasionally wetlands. STATUS: uncommon resident. BREEDING: Mar–Aug (DR).

Colin Trainor

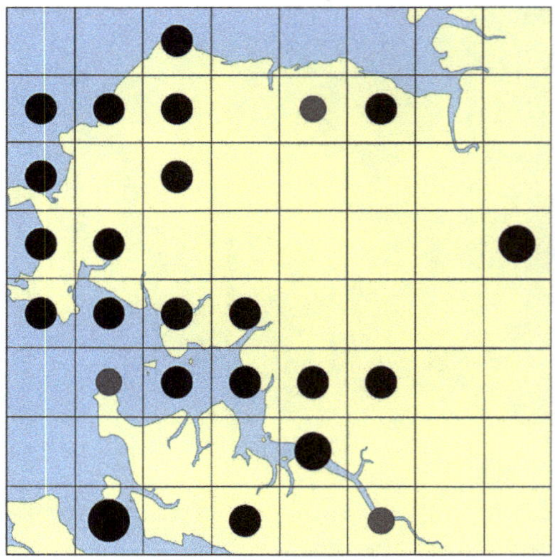

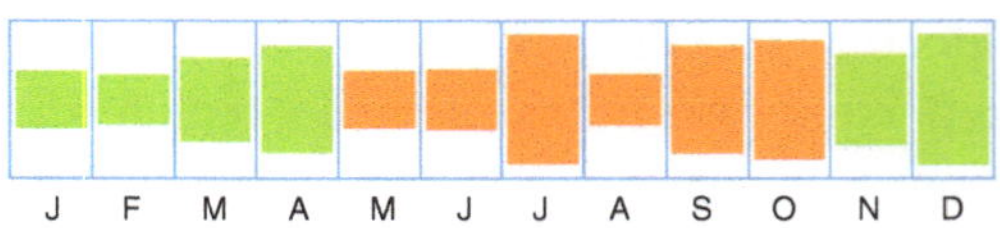

Following a taxonomic revision in 2008,[134] the Osprey went from being a near-globally distributed species to two species, Western and Eastern Ospreys, with the latter being the one that occurs in Australia.

The Eastern Osprey is a sedentary, breeding resident in the Darwin region, and variations in monthly reporting rates probably reflect the conspicuousness of their behaviour over the year. The small spike around April, for example, may be associated with increased activity of breeding birds and, given their local breeding season of May to

August, the increase in records from September to December may be due to recruitment of young birds.

During aerial surveys of the coastal Top End in the 1990s, the Eastern Osprey was the fifth most frequently recorded species, though they were less common in the western half of the Top End than in the east, where, in many areas on the western side of the Gulf of Carpentaria, active nests were less than a kilometre apart.

Ospreys have been seen hunting in several areas around Darwin's coast, but most feeding observations refer to birds carrying food, perhaps to nests. They have also been seen consuming prey at sites near the point of capture, such as on reefs or even on street lights along coastal roads, but there are no reports that identify food items to species level. Ospreys seen hunting at Vestey's Beach were presumed to be the owners of a nest at Stuart Park, just over 1 km away, but those seen hunting at East Point would be about 5 km from the nearest known nest.

Although there have been no systematic searches for nests, at least 10 nest sites have been found in the Darwin region. Ospreys are well known for building nests on artificial structures, and this appears to be the main practice in the region. Nests have been constructed on communication masts at Shoal Bay and along Lee Point Road, on electricity pylons at Dinah Beach and along the Elizabeth River, on a high antenna at the Casuarina Campus of Charles Darwin University, and in the city, on a television transmission tower. Natural sites are also used, with nests in trees at Casuarina Beach, and near the mouth of Ludmilla Creek.[3]

The television tower nest was used over several years, with observations of adults feeding chicks in mid-July of 2010[3] and 2011.[3] The nest at Charles Darwin University was constructed during March 2005, but there was no evidence of it being used until May 2010, when a bird was seen sitting. Eggs have been observed in the Top End from April to August, with most clutches in May,[27] while young nestlings have been observed in April, and large young in May, suggesting that eggs may be laid as early as late February. As eggs take about 35 days[135] to hatch, and the young up to 60 days to fledge, the eggs must have been laid in the television tower nest between late April and mid-May.

Black-shouldered Kite
Elanus axillaris

RANGE: whole of Australia. HABITATS: grassland, woodland, floodplains. STATUS: uncommon resident. BREEDING: Feb–Oct (TE).

The range and numbers of Black-shouldered Kites have increased since European settlement,[135] probably as a result of an increase in suitable habitat created through the clearing of forest and woodland for agriculture. The species is nevertheless uncommon in the Top End, including the Darwin region, where it is largely restricted to the grassy flats around Leanyer Swamp, Holmes Jungle swamp, Knuckey Lagoons

Nolan Caldwell

and Shoal Bay. Birds have also been seen at Palmerston and Lee Point, and as near to the city as Fannie Bay, Bayview Haven and Dinah Beach,[1,136] but such reports are erratic.

There are records of Black-shouldered Kites in the region covering all months, but reporting rates fall to their lowest level at the onset of the Wet season, before increasing again during the mid to late Wet season. While lower reporting rates may be expected during the Wet season, due to increased difficulty in capturing ground-dwelling prey, the February–March rise is puzzling, but is perhaps related to detectability. There may be some local movement from inundated flats to areas where they are more readily detected and, with frequent heavy rain at that time, birds may

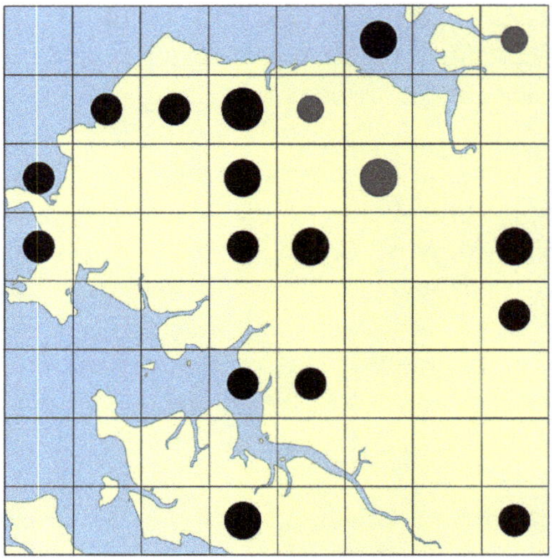

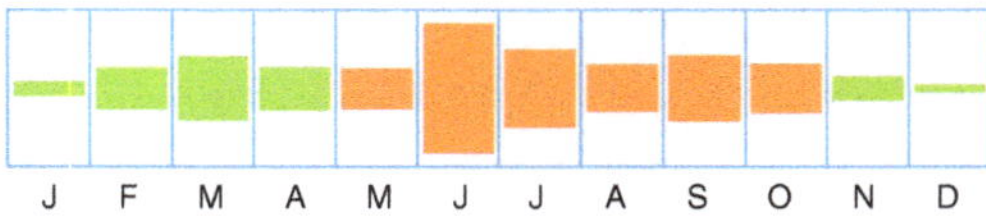

need to hunt more during the middle parts of the day, and as a result become more frequently seen.

Although juvenile birds are occasionally seen in the region, there are few breeding records. In March 2001[21] a pair was seen building a nest by the Berrimah power substation, close to a tree in which a Black Kite was nesting, but it is not known if successful breeding occurred. Breeding has occurred at Middle Point, near Fogg Dam, and it may be that it occurs there regularly. Further out of Darwin, at Coomalie Farm, and at Kapalga, Kakadu NP active nests were found in May and March, respectively, and two pairs were seen with fledglings at the latter site in early November.[43]

Letter-winged Kite
Elanus scriptus

RANGE: inland Australia. HABITATS: dry, timbered watercourses and sparsely vegetated grassland. STATUS: vagrant. BREEDING: Feb (TE).

Judy Leitch

Letter-winged Kites are well known for their irruptions into areas from which they have been absent for extensive periods, or into new areas. Such events occur following years with higher than average inland rains, which lead to a marked increase in the abundance of prey, particularly Long-haired Rats. After breeding continuously while food is abundant, large numbers of Letter-winged Kites are forced to disperse when numbers of these rats decline again.

Major irruptions of Letter-winged Kites occurred during the years 1951–53, 1966–70 and 1976–77.[135] During the first of these, there were records from north-west Western Australia, but whether any ventured into the Top End is not known, as there is an almost complete lack of data from that period. None were recorded during extensive bird surveys of the north-west Top End from 1967 to 1971,[1] clearly indicating that the irruption of the late 1960s did not extend that far. During the Dry season of 1977, however, the irruption extended well into the Top End, precipitating the first records of the species for the Darwin region, with birds reported from Fannie Bay, Lee Point, Leanyer Swamp and Leanyer SP.[2] In the years following that irruption, breeding occurred at Kapalga, Kakadu NP, in 1979,[43] at Jabiru in 1979 and 1980[135] and at Middle Point, near Fogg Dam, during February 1979.[43] Another irruption occurred at this locality during the late Dry season of 1994, when 60 or more birds, and several nests, were present. No feeding or hunting observations were reported but Letter-winged Kites were observed on several occasions around dusk, moving towards Fogg Dam. It is likely the birds preyed on Dusky Rats, which are normally abundant on floodplains of that area,[137] although numbers vary annually, apparently related to the amount of rainfall in April, which has a major effect on breeding conditions.[138] A sharp fall in numbers of these rats from 1995 to 1998[138] may have caused the birds to disperse, and perhaps led to some deaths through starvation. Nevertheless, up to six Letter-winged Kites were still seen in that area until late November 1996, and sporadic reports of one or two birds until 1999 may also relate to remnants of that colony.

Square-tailed Kite
Lophoictinia isura

RANGE: mainland Australia excluding very arid regions. HABITATS: open forest and woodland. STATUS: rare, mainly Dry season migrant. BREEDING: not in NT.

Mat and Cathy Gilfedder

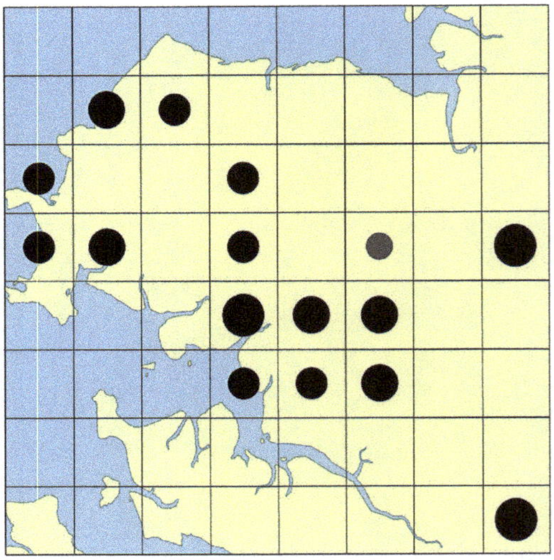

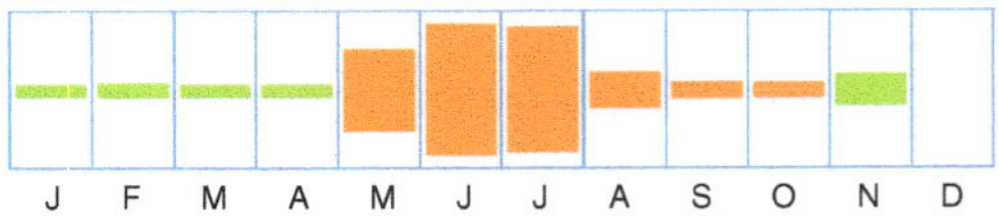

Typically seen gliding gracefully just above the tree canopy in search of small birds and foliage insects in the trees, Square-tailed Kites, apparently the majority of which are first-year birds, migrate during the austral winter from southern Australia to the Top End, where they spread themselves sparsely across woodlands and open forests, often close to the margins of dry floodplains, rivers or mangroves. There are no reports of breeding in the NT and, indeed, this is the only diurnal raptor for which this is the case.

The Square-tailed Kite was first reported from the Darwin region in January 1979,[135] when a bird was seen at Marlow Lagoon, Palmerston, and it was recorded in 19 years between then and 2013. Reports tend to occur in clusters, suggesting that they are re-sightings of a single bird, so records are more likely to indicate the length of time a Square-tailed Kite stays in the region, and the area it covers, rather than the species' abundance. Records are widespread in the region, but there appears to be a rather disproportionate number of records from urban areas. This, however, can be readily explained by the birds' visibility as a relatively low flying raptor, and because of its rarity, incidental observations are almost always reported. Reporting rates indicate that Square-tailed Kite is largely a Dry season visitor, but in addition to the few Wet season records in the Darwin region, the species has also been seen in November–December in Kakadu NP in three separate years.[43]

There have been no reported feeding observations in the Darwin region, but at Coomalie Farm, 45 km south of the Darwin region, a Square-tailed Kite was seen diving on a Weebill nest, and flying off with one or two chicks, the nest left in tatters, and at Kapalga, in Kakadu NP, one was seen feeding on a Spotted Tree Monitor.[43]

Black-breasted Buzzard
Hamirostra melanosternon

RANGE: largely arid and semi-arid Australia. HABITATS: dry floodplains, grassland and woodland. STATUS: uncommon Dry season visitor. BREEDING: Jun–Oct (TE).

Laurie Ross

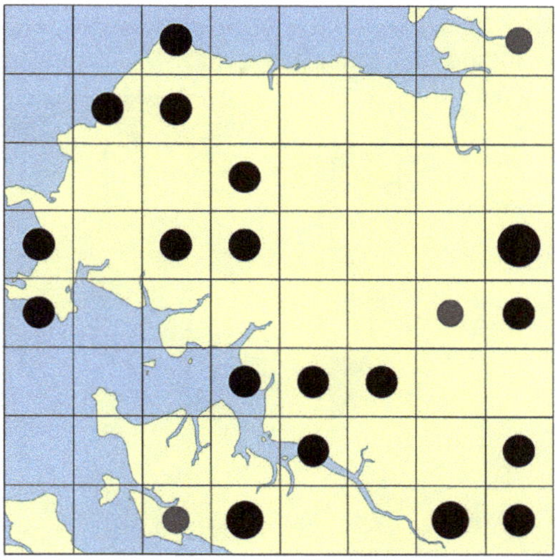

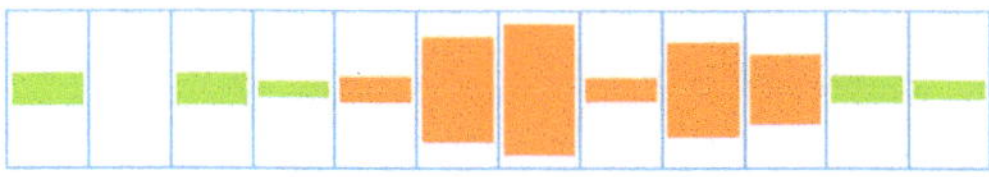

As with other uncommon or rare raptors, most observations of Black-breasted Buzzard in the region are reported, and these show the species to be almost exclusively a mid-Dry season visitor to the region, with a sharp reduction in reporting rates in August, probably the result of dispersal to breeding areas. Sightings are widespread, as would be expected for a soaring, easily recognised raptor, though reporting rates are highest in eastern parts of the region, around Palmerston, Howard Springs and Humpty Doo, where the woodland habitat is not dissimilar to that in which it is recorded throughout much of its range.

Black-breasted Buzzards are famed for their technique of dropping rocks on Emu eggs in order to crack them to gain access to the yolk, but their diet consists mainly of mammals, small to medium-sized birds, nestlings, reptiles and carrion.[135] Observations of feeding in the Darwin region have, however, only involved reptiles and carrion, with identified items of the latter including Water Buffalo, Tawny Frogmouth[2] and Agile Wallaby.

Breeding occurs sporadically in the Top End, but the nearest report to Darwin is a nest constructed in a Darwin Woollybutt, about 4 m above the ground, found adjacent to the Arnhem Highway, about 90 km to the south-east. At Kapalga, in Kakadu NP, nests were found in three years,[43] and all were built 15–18 m above the ground in trees close to a running creek. Observations of chicks or eggs at these nests indicate that the eggs were laid in July or August, though a second nest in one year had a bird sitting on eggs or small chicks in early November. Disgorged pellets under one nest contained the remains of a Red-collared Lorikeet.

Pacific Baza
Aviceda subcristata

RANGE: tropical and subtropical Australia, NG and e. Indonesia. HABITATS: forest and woodland including mangal, urban areas. STATUS: uncommon resident. BREEDING: Sep–Feb (DR).

Niven McCrie

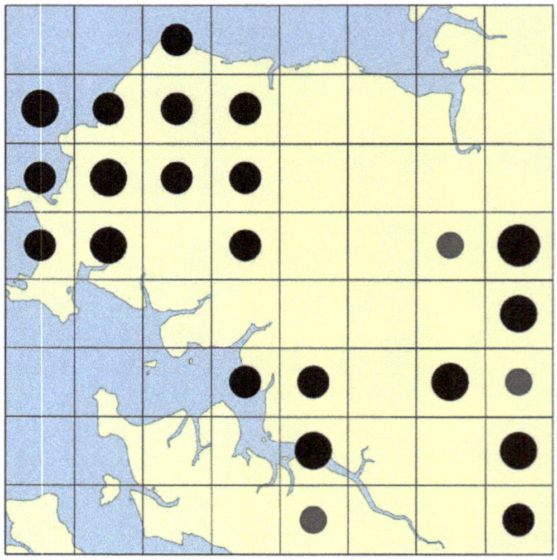

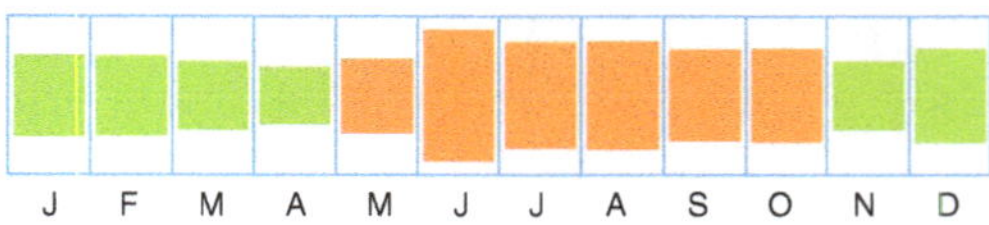

Pacific Bazas in Australia's north-west differ somewhat from those in eastern Australia in being slightly smaller and having darker underpart barring. At times those features have been used as a basis for assigning it subspecific status; however, the differences have been shown to be clinal and, in the case of measurement, apparently insignificant.[135]

In addition to forested areas such as Casuarina CR, Holmes Jungle and Howard Springs NR, Pacific Bazas have been reported throughout Darwin's suburbs, mostly in flight, and rather more frequently in the older suburbs of Parap, Ludmilla and Nightcliff. The largest groups seen in the region, comprising up to six birds, have also been seen in flight over the suburbs.[21,139,140] There is little variation in monthly reporting rates throughout the year, as the species is a breeding resident, but an apparent increase at the beginning of the Dry season may be due to the addition of juveniles to the population, as more sightings of groups occur in the period April to August also.

Foraging observations show that local Bazas, like those elsewhere, feed on stick insects and preying mantises. At Buffalo Creek, a sub-adult bird was observed to tumble through the foliage of a Black Wattle at about 10 m height, then perch with a large green praying mantis clenched in its right foot, resting against a branch. It pulled the head off, dragged out the gut from the exposed end of the thorax, then pulled off each of the forelegs in turn, dragging out tissue from the exposed base, subsequently discarding the remains of these. Most foraging is done in trees and shrubs but one Baza was seen to take a stick insect from the ground, then fly to a perch in a tree to consume it.[141]

The first reported evidence of breeding in the region appears to be of an adult with two fledglings at Sadgroves Creek, in January 1990,[34] but the first nest was not found until February 1999. This nest, a shallow bowl of sticks lined with green leaves, containing nearly fledged young, was located in the Darwin General Cemetery in Jingili. Since then, however, nesting has been observed across a broad area, including in the suburban areas of Nightcliff,[142] Casuarina and Tiwi,[93] as well as in mangroves at Kulaluk and Sadgroves Creek, and the forested areas of Holmes Jungle, Howard Springs NR, Black Jungle, and Shoal Bay. At Ludmilla Creek, a pair was observed building a nest in mangroves in early October 2007.[47] The female was seen sitting on the nest during November, and by 30 November, young were present.[47] A nest found at Manton Dam, about 50 km from Darwin, held eggs in mid-November 1986, and another nest, found there in mid-January 2003, contained two near-fledged young.[3] Aerial display flights, where the pair of birds repeatedly dive and then climb again, have been observed in September, October and April. These are generally considered to be pre-nesting displays, although no breeding has been reported in the region around April.

Little Eagle
Hieraaetus morphnoides

RANGE: whole of Australia. HABITATS: woodland, grassland, open country. STATUS: rare, mainly Dry season visitor. BREEDING: reputedly May–Jun (TE).

During the 1960s, the Little Eagle was described as 'common and widespread in every part' of Cobourg Peninsula,[143] but survey data both before[110] and since then[43,144] suggest that while there may be small irruptions in some years, the species is at best an uncommon Dry season migrant into any part of the Top End. Measurements from specimens[143] collected on Cobourg Peninsula align those birds with populations in eastern Australia, indicating that Top End migrants derive from that part of the country.

In the Darwin region, the Little Eagle is scarce and irregular, with reports from only 20 of the 36 years from 1977 to 2012. Other than in 1990, when records covered

Scott McGregor

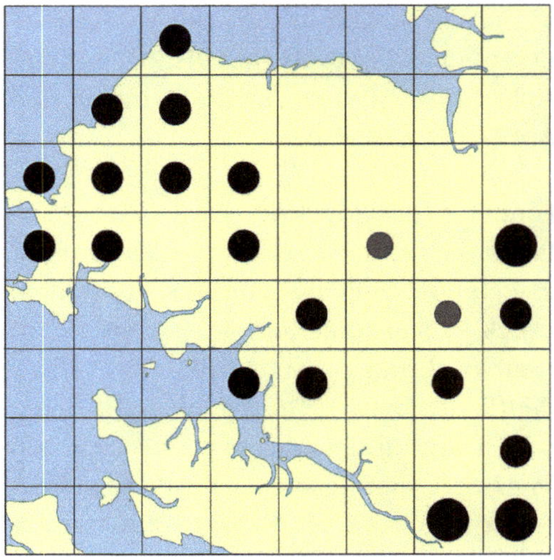

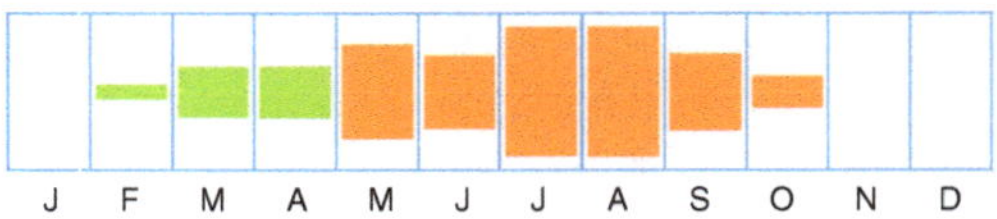

all months from March through September, the species was reported in no more than four months of any year.

Sightings of Little Eagle show a more disjunctive distribution than any other raptor in the region. It is most frequently seen in rural areas, where suitable habitat is extensive, but otherwise sightings have been largely confined to the urban area and its immediate surroundings. While the presence of resident observers in the urban area increases the likelihood of a soaring bird such as this being seen, it is probable that some reports of Little Eagles refer instead to immature Brahminy Kites, which appear quite similar when viewed in flight. The true status of the species, at least in the urban area, is therefore somewhat uncertain.

Little Eagles prey mostly on vertebrates, but occasionally insects are eaten. Near Lee Point, a Little Eagle was seen plunging into Casuarina trees, in the manner of a Pacific Baza, to pluck prey, probably a stick insect, from the outer foliage.[145] Although there have been no other feeding observations within the Darwin region, one seen perched at Kapalga was clutching a snake, possibly a Common Tree Snake, about 1 m in length.[43]

A Little Eagle was seen to fly from its nest, in a Darwin Woollybutt near Humpty Doo, in June 1991,[93] although the contents of the nest were not seen. A subsequent sighting of a Little Eagle sitting on a nest in a River Redgum near Katherine, during May 2002,[146] appears to be the only other breeding record from the western Top End.

Wedge-tailed Eagle
Aquila audax

RANGE: whole of Australia, NG. HABITATS: open country. STATUS: scarce visitor. BREEDING: Jun–July (TE).

Will Riddell

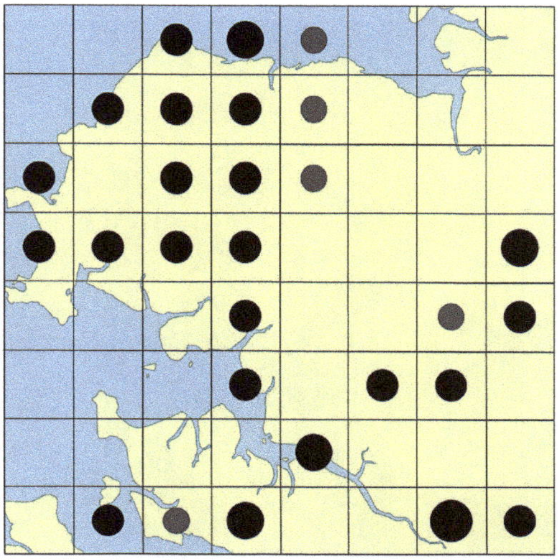

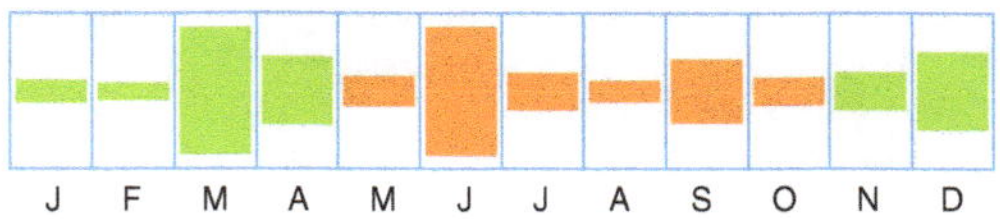

Sightings of Wedge-tailed Eagles in the urban area, all involving single birds in flight, have come from Marrara, Winnellie, Berrimah, Bayview Haven and Leanyer. The species is seen more frequently in rural areas, both as a consequence of the availability of suitable habitat for hunting, and because of the greater abundance of road-killed animals due to higher vehicle speeds. Identified road-killed prey includes Agile Wallaby, Gould's Goanna,[34] and a dog.[21] Two birds were seen chasing an Agile Wallaby on the Channel Island road, making several stoops, but were unsuccessful, due to thick undergrowth.[43]

There are breeding records from two widely spaced locations in the Darwin region. A nest in a Ghost Gum at Noogoo Swamp, near the Howard River in 1977, contained a large chick, which was banded. Another pair is reported to have nested along the Channel Island road in 1990. A pair was seen performing tumbling displays over Tiger Brennan Drive in October 2003,[147] but this is not necessarily an indication of courtship.

Red Goshawk
Erythrotriorchis radiatus

RANGE: tropical and subtropical Australia. HABITATS: woodland. STATUS: very rare visitor. BREEDING: Jul–Sep (TE).

Niven McCrie

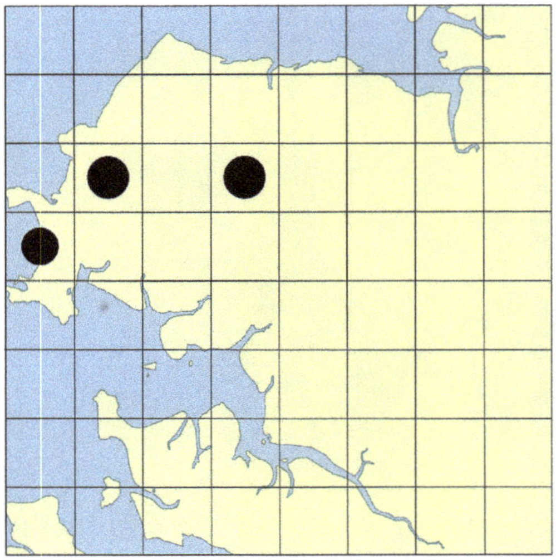

The Top End population of Red Goshawk was estimated from surveys during the late 1980s to be about 200 pairs.[148] While this number is higher than elsewhere, and shows the Top End to be the species' stronghold, Red Goshawk is infrequently encountered anywhere in its range and, for many birdwatchers, their only sighting is at known nest sites.

In the Darwin region, single birds were reported from Lee Point during May 1982,[31] Parap during September 1986 and May 1990,[21] and Berrimah in July 1990. The nearest confirmed nesting to Darwin is from Gunn Point, which falls just outside of our region. The Goshawks breeding there, and their offspring, are likely the source of Darwin records, as the nest was known to be active during the 1980s, and its abandonment in the early 1990s coincided with the last reports of the species in the Darwin region.

Eggs collected in late May 1902, and labelled 'Port Darwin',[149] may have come from elsewhere in the Top End, as that name is known to have been used by early collectors for locations at least as far away as Pine Creek, about 220 km south of Darwin.

Grey Goshawk
Accipiter novaehollandiae

RANGE: tropical and e. Australia, NG. HABITATS: rainforest, mangal, riverine forest, urban areas. STATUS: uncommon resident. BREEDING: Egg-laying Mar–Sep (DR).

Will Riddell

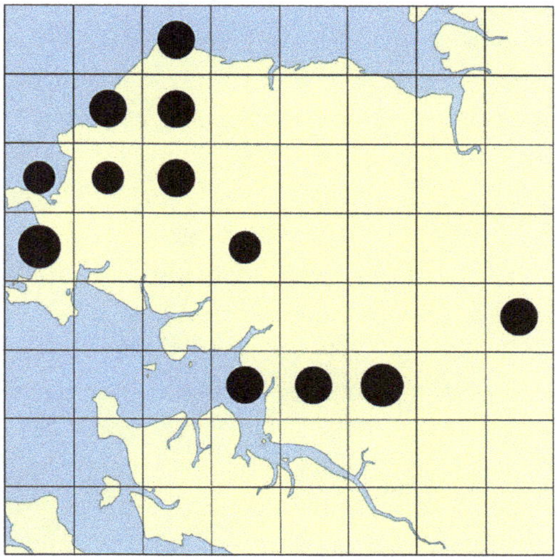

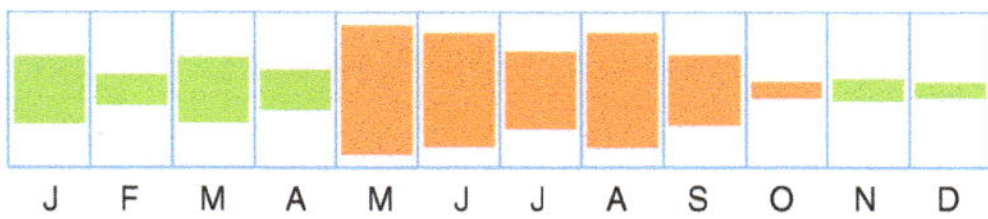

This largely coastal species occurs in two colour morphs, grey and white. In the Top End, the white morph appears to predominate in riverine areas, being regularly seen along the Adelaide, Mary, South Alligator and Daly Rivers, as well as their adjacent billabongs, such as Yellow Water.[150] In the Darwin region, however, although white morph birds have been seen, most are of the grey form. It has recently been shown that Grey Goshawks in the Top End and Kimberley differ in juvenile plumage from those on the east coast, most noticeably by the bolder streaking, or speckling, on the breast.[21,91,151]

While the majority of records are from areas near monsoon forest or mangroves, which are used for roosting and hunting, there are many reports of birds flying over the suburbs, and they are occasionally seen perched in garden trees or on street lighting or power lines.

Monthly reporting rates fluctuate markedly during the year, perhaps unusually so for a breeding resident. However, rather than reflecting presence or absence from the region, variations may come from changes in conspicuousness. The higher reporting rates occur during the species' local breeding season, and may be due to the extra hunting activity required to raise young, as well as the addition of juveniles to the population.

Throughout its range, the diet of Grey Goshawk is reported to be mostly mammals, but also birds, including nestlings, reptiles and occasionally carrion.[135] In the Darwin region, the only evidence of Grey Goshawk interest in mammals has been their presence near Flying Fox colonies,[146] but no predation has been confirmed, and it appears that birds and reptiles may form the greatest part of their diet. Grey Goshawks have been seen pursuing Red-collared Lorikeet and Bar-shouldered Dove, and feeding on Common Tree-snake, White-bellied Mangrove Snake, Northern Water Dragon and Gilbert's Dragon.[43,152,153] The single observation of scavenging involved a Goshawk picking up carrion from the road in front of an oncoming car.[154] Along the Daly River, south of the region, an adult Grey Goshawk was seen to capture a large live Cane Toad, but as the goshawk then dragged the toad into deep reeds, it is not known whether it was consumed.[155]

There are breeding records of Grey Goshawks from the late Wet season through to the late Dry season, with chicks reported in April, July, August and September.[40,156] Nests have been found in mangroves adjacent to the urban area, and it is likely that breeding occurs also in the extensive mangal elsewhere in the region. At least two nests were built in garden trees in Tiwi and another in Leanyer, and it may be the proximity of those suburbs to mangrove areas that has influenced their location. One of the Tiwi nest trees was removed for safety reasons however, and it may be that the pair from that nest was then responsible for the nest at Leanyer, as the two were not seen to be active concurrently. A pair of Grey Goshawks that successfully raised young at a nest in mangroves at Ludmilla in April 2012 raised a second brood in November of that year, the first confirmed record of double-brooding by the species.[153]

Brown Goshawk
Accipiter fasciatus

RANGE: Australia, e. Indonesia and NG. HABITATS: woodland, floodplains. STATUS: moderately common resident; common Dry season migrant. BREEDING: Jul–Dec (TE).

Two subspecies of the Brown Goshawk occur in the Darwin region. The resident race *didimus* is restricted to the tropics, while the race *fasciatus*, which occurs Australia wide, is a Dry season migrant to the Top End, probably from south-eastern Australia.[157] The former differs from the latter race in being about 10% smaller and somewhat paler, though the plumages of both adult and immature birds of each race differ only subtly.[135]

The fall in monthly reporting rates after a sharp peak in June may indicate passage through Darwin to islands to the north, or dispersal of birds to other areas of the Top End, but there is no corresponding peak during the late Dry season, suggesting that southward migration occurs over a broader front.

Although widespread across the region, including urban gardens and streets, the

Laurie Ross

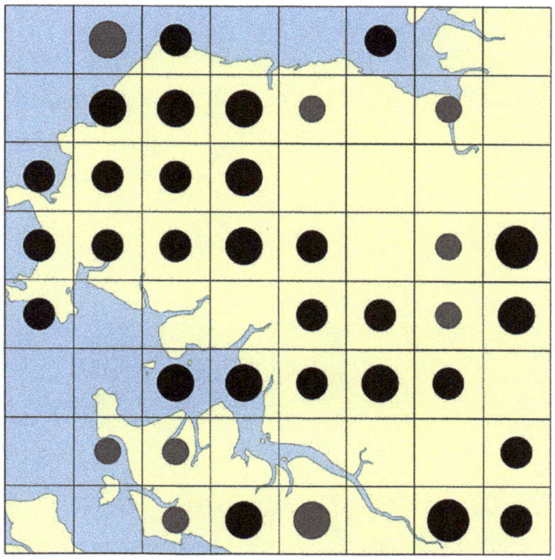

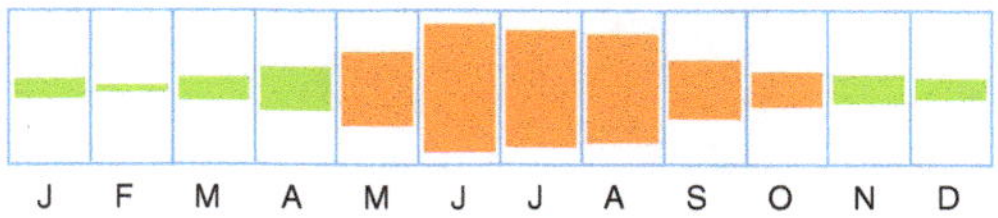

highest reporting rates coming from Knuckey Lagoons and Leanyer, Holmes Jungle and Howard Springs swamps, all areas that feature wetlands and open grass.

The Brown Goshawk's diet consists mainly of birds, small mammals and reptiles, insects and occasionally carrion.[135] The only mammalian prey reported in the region is Dusky Rat,[31] but avian prey has been observed frequently. In addition to the apparently regular taking of caged birds, these include the Magpie-lark, Red-backed Buttonquail, Spotted Nightjar, Black-faced Cuckooshrike nestlings and the eggs of a Bar-shouldered Dove.[158,159,160] The nightjar was captured on a road at night, and Brown Goshawk has also been seen hunting insects by the light of car headlamps at night. The species occasionally tries for larger prey, and one was seen to attack and kill an Eastern Cattle Egret,[161] and others attempt to capture Raja Shelduck[158] and Orange-footed Scrubfowl.[67] A juvenile Brown Goshawk at Bees Creek was seen tapping its foot in leaf litter, perhaps trying to flush invertebrates or lizards.[162] Although not seen to do so in the Darwin region, a Brown Goshawk, together with Brown Falcons, was observed hunting prey fleeing from fire near Katherine.[3]

In 2009 a pair of Brown Goshawks nesting in suburban Millner raised two broods of two fledglings, the first confirmation of the species raising two broods in a single breeding season.[158] The same pair nested successfully the following year, and attempted a second brood, but on that occasion were unsuccessful, although they were again successful raising two broods in 2013.

The few historical breeding records in the Darwin region include a nest containing three large chicks near Royal Darwin Hospital in November 1987,[43] and an observation of a bird sitting on a nest in woodland near Tracy Village in early August 1999. However, it is likely that breeding has been much more extensive, as searches in mid-September 2014 revealed at least nine nests containing brooding birds.[163] Breeding records from Kapalga[43] in Kakadu NP indicate that egg-laying occurs there in September and October.

Collared Sparrowhawk
Accipiter cirrocephalus

RANGE: Australia, NG. HABITATS: woodland. STATUS: scarce resident. BREEDING: Sep–Oct (TE).

Nolan Caldwell

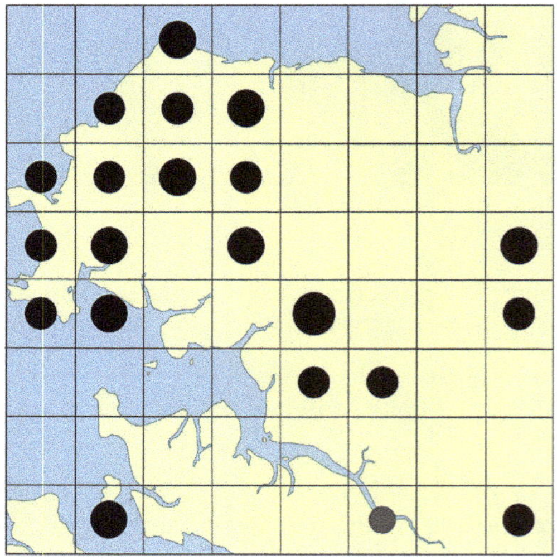

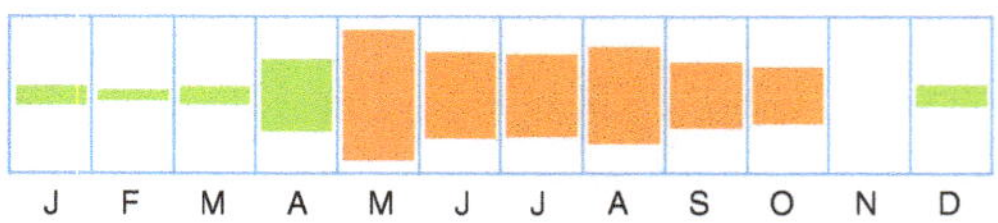

Although apparently more common in forests south of Darwin,[164] the Collared Sparrowhawk is uncommon in the Darwin region. Owing to possible confusion between this species and the smaller form of the Brown Goshawk the true status of the Sparrowhawk remains somewhat obscure.

Collared Sparrowhawks prey chiefly on small birds, mammals, reptiles and insects.[135] One was seen feeding on a wader, probably a Sharp-tailed Sandpiper,[21] and there are local observations of Sparrowhawks pursuing a Peaceful Dove, and various honeyeaters and woodswallows. Like the Brown Goshawk, this species is known to prey on captive birds. One was observed in a hen-house in Tiwi, apparently trying to capture Bar-shouldered Doves that were attracted to the seeds on the ground. At Charles Darwin University's Casuarina campus, Sparrowhawks have been observed hunting Northern Bent-winged Bats emerging from a stormwater drain in which the bats roosted.[165,166]

The only confirmed Top End breeding record is from McArthur River, where a nest with eggs was found in late September 1913. Active nests were reported from Elsey Station in September 1944[167] and Kakadu NP in October 1981,[43] but no details were given about the stage of breeding. A pair of Collared Sparrowhawks was seen mating in October 2011[168] in urban Katherine, but no further evidence of breeding was seen. In July 2011 a Collared Sparrowhawk was observed at a nest at the edge of suburban Leanyer,[40] but by August the nest, which had not yet been confirmed as belonging to the species, appeared to have been abandoned.[40]

Swamp Harrier
Circus approximans

RANGE: Australia, NG, s. Pacific Is. HABITATS: wetlands, floodplains, grassland, cultivated areas. STATUS: scarce Dry season migrant. BREEDING: not in NT.

Scott McGregor

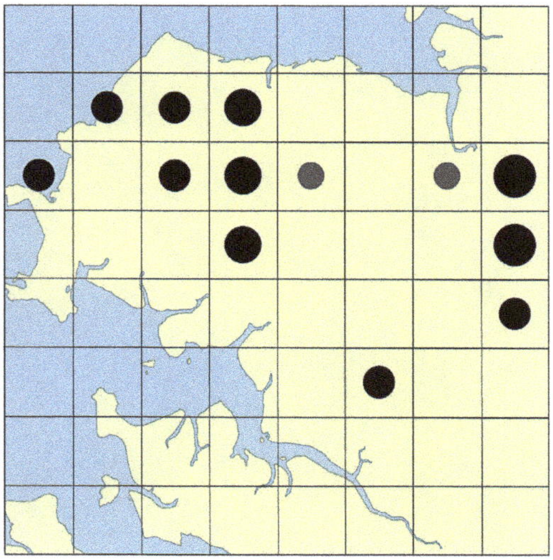

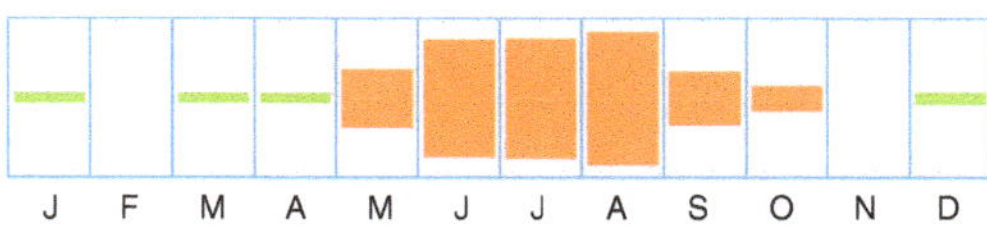

The Australian breeding range of Swamp Harrier is restricted to the south-east and south-west mainland, and Tasmania. The south-west population is not considered to be migratory, but the south-east, and particularly Tasmanian, populations are known to be strongly so.[135]

There are occasional records of Swamp Harrier in the Darwin region during the Wet season, but it is clear that May, the recognised beginning of the Dry season, sees the arrival of migrants into the region, and October, the recognised end of the Dry season, sees their departure. Reflecting the species' affinity for terrestrial wetlands, the highest reporting rates are from Holmes

Jungle and Leanyer Swamps, Knuckey Lagoons and the Shoal Bay wetlands. There is a slight increase in reporting rates as the Dry season progresses, and as with waterbirds, it may be that some Swamp Harriers move towards the coast as wetlands further inland begin to dry up. Numbers in the region are always small, and reports rather erratic, and it may be that some observations represent birds that have wandered into the region temporarily from the more extensive wetlands outside of the region, such as Fogg Dam.

Spotted Harrier
Circus assimilis

RANGE: Australia, e. Indonesia. HABITATS: open country, grassland, cultivated areas. STATUS: rare, mainly Dry season visitor. BREEDING: very rarely; Jun (NT).

Vince Bugeja

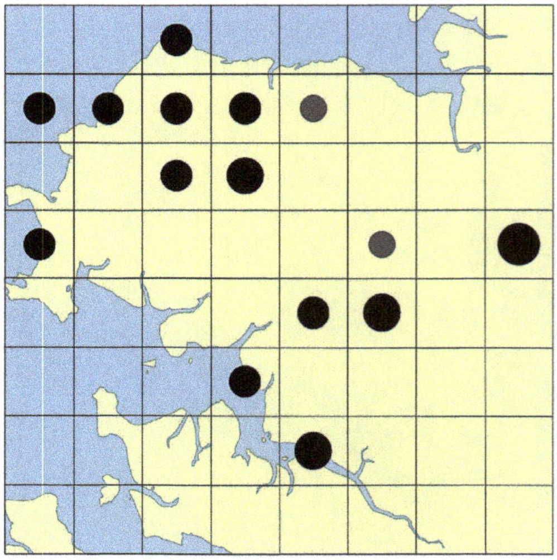

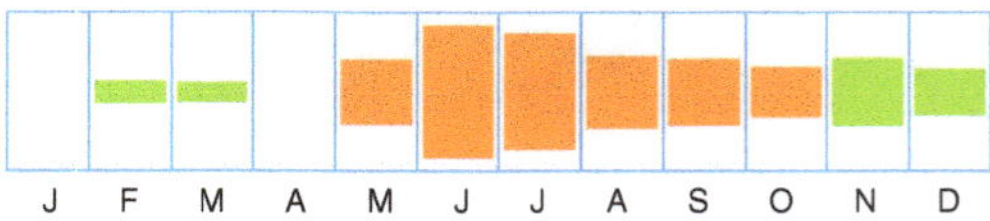

The only harrier species that nests in trees, rather than on the ground, Spotted Harriers breed mostly in subtropical and temperate Australia, and while there is strong evidence that eastern populations migrate northward through the inland to spend their winters in the tropics, the origin of Top End visitors is uncertain.

The June peak in records suggests a possible passage movement through Darwin, and while the species occurs in eastern Indonesia, populations there are mainly resident, with no clear evidence of migration from Australia.

Within our region, Spotted Harriers were observed in only 10 of the 21 years from

1979 to 1999, but then in 10 of the next 15 years from 2000 to 2014. The species is more regularly sighted in the Humpty Doo area, and around the floodplains of the Adelaide River,[1,21] though even there they are still uncommon.

At Victoria River Downs, in the south-west Top End, Spotted Harriers are reported to be present year-round,[98] but no breeding has been recorded. Indeed, there is little evidence of Spotted Harriers breeding anywhere in the Top End, and only one instance has been confirmed. An active nest, situated in a Ghost Gum on the edge of a floodplain at Kapalga, Kakadu NP, was seen during July 1979. Two adults were near the nest, which was thought to contain small nestlings.[43] In late August 2009, a food-begging juvenile bird and an adult were seen along a road near Maningrida, in western Arnhem Land, suggesting local breeding.

Black Kite
Milvus migrans

RANGE: Africa and Eurasia to se. Asia, NG, whole of Australia except sw. HABITATS: woodland, coast, wetlands, sewage ponds, refuse dumps, urban areas. STATUS: very common Dry season migrant, and scarce resident. BREEDING: Feb–Oct (TE).

Niven McCrie

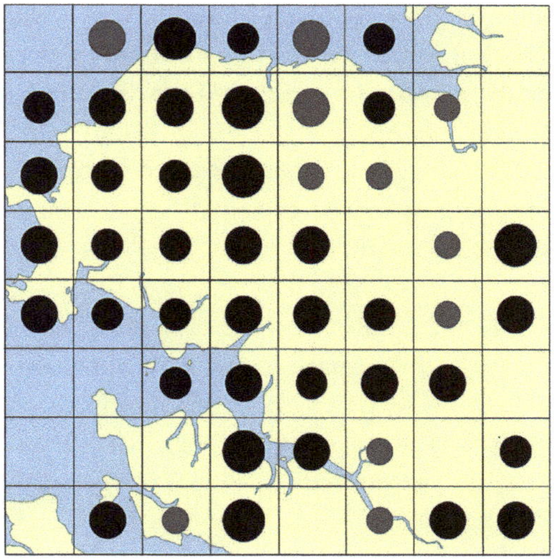

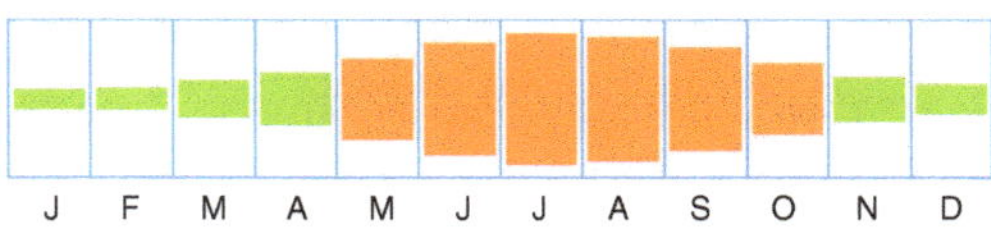

If any bird typifies the mid to late Dry season in the Darwin region, it is the Black Kite. The third most frequently recorded bird in the Darwin region, and probably the world's most numerous raptor, it is not unusual to see hundreds of Black Kites soaring above Darwin's outer suburban and rural areas at that time of year.

Black Kite occurs throughout the region, and in most habitats, but the greatest densities are generally found at Shoal Bay Waste Management Facility and the adjacent Holmes Jungle floodplain, where at least 1700 birds were seen in August 1978,[34] with similar numbers in August 1996 and July 1997.[21] Just before dusk most of the birds from there, and probably nearby areas, move east to woodland at Holmes Jungle, where night-time roosts may consist of hundreds of birds, often with 40 or more to a tree.[21]

From a peak over the mid-Dry months from June to August, numbers gradually fall as the Wet season approaches, and from December, the species all but disappears from the Darwin region. In years where the Wet season arrives later, however, the birds stay longer, or move only a short distance inland, but in other years few may be seen even as far south of Katherine, some 300 km south of Darwin. The kites normally start returning to Darwin during March and April, although in some years, particularly when there has been late Wet season monsoonal activity,[43] they remain scarce until well into June.

While the heat of the day sees Black Kites soaring on thermals, in the early morning, before these thermals form, hundreds of birds can sometimes be seen loafing on beaches or mudflats at Buffalo Creek and Kulaluk Bay.

The Black Kite is probably most noted for its predatory attraction to fires, and is sometimes called Fire Hawk because of this. Tens or hundreds of kites, coming from up to 20 km away, gather at fires to pounce on fleeing insects, reptiles, small mammals and birds, the kites sometimes approaching flames so closely as to singe their feathers.[95] At the fires, animals are concentrated in a relatively small area, physically exposed as they flee the flames and, in their panic to escape, are less likely to be aware of predators. Moreover, it may be that some become narcotised by the carbon monoxide or dioxide produced during the fire, making them even easier to capture.[95] While there have been observations suggesting that Black Kites carry burning sticks from one location to another in order to create new fires,[169] it seems more likely that this is an accidental outcome of an attempt to capture prey, with the kite grabbing, then dropping, the burning stick immediately it becomes aware of what it is. Soon after the fire has died down, the kites leave the site and may soar high on thermals in search of new fires.[95] Black Kites are not restricted to fires in hunting live prey. They have been seen collecting insects in flight with their open bills, or with their talons, then eating them as they fly, and there are observations of them feeding on Soldier Crabs at Lee Point beach,[71] and hunting Dusky Rats on the South Alligator River floodplains.[43]

As can be expected of an accomplished scavenger, Black Kites feed on a wide range of food items, which they secure in a variety of ways. They will take scraps from around public areas, and at schools they will dive to collect scraps, often despite harassment by children.[21] They have been seen attempting to steal food from a Brown Goshawk,[162] as well as from other Black Kites. Kites frequently feed on carrion along roads, and many are killed by vehicles as a result.

Although most Black Kites leave the coastal Top End to breed further inland, small numbers nest in the Darwin region. Nests were reported from Leanyer in October

1981, April 1986 and October 1986, and at Lee Point in October 1989.[31] A nest with young was found on a cement hopper near Fort Hill Wharf in September 1984, and a nest with eggs was at the Berrimah Power Substation in June 1999.[21]

Whistling Kite
Haliastur sphenurus

RANGE: whole of Australia; NG. HABITATS: all except rainforest. STATUS: very common resident. BREEDING: Mar–Oct (TE).

Nolan Caldwell

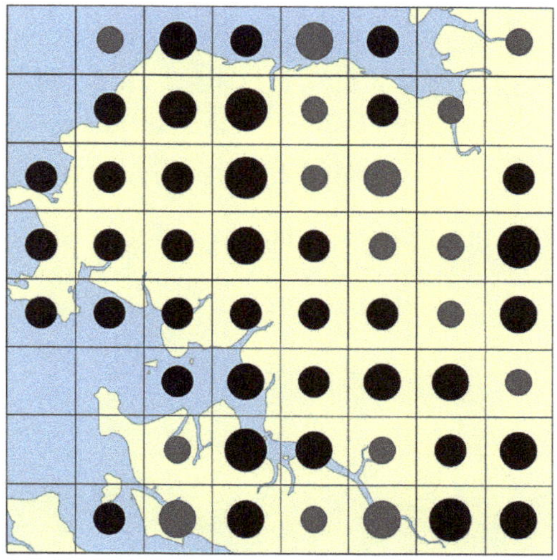

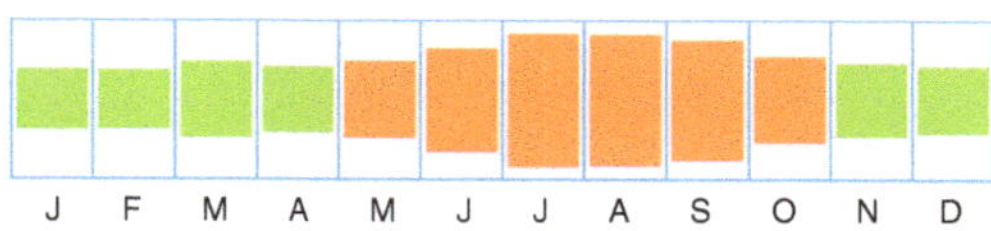

Like the Black Kite, the slightly larger Whistling Kite is a scavenger, but unlike the former species, which largely disappears during the Wet season, the latter is resident in the Darwin region. Although there is a small increase in records during the Dry season, there appears to be little evidence of large-scale movement, although an adult banded at Humpty Doo in July 1979 was found less than 10 months later some 2400 km away near Bourke, New South Wales.[135]

Although probably more numerous during the Dry season, Whistling Kite appears more noticeable during the Wet season when, with the near complete absence of Black Kites, it is by far the most common raptor seen in skies over the Darwin region. Of an estimated 1000 kites at the Shoal Bay Waste Management Facility in late September 2010, over 50% were Whistling Kites, and the remainder Black Kites. By late November counts yielded less than 200 birds, of which 90% were Whistling Kites.

Whistling Kites feed more on dead birds or fish than do Black Kites, and are often seen picking at the remains of road-killed animals or at scraps at fishing spots and picnic areas.[60] However, they also hunt grasshoppers and small rodents, and have been observed to run along a beach to catch Soldier Crabs.[170] They have been seen to steal, or attempt to steal, prey from egrets and Silver Gulls,[28] as well as from other Whistling Kites.[21] Despite their larger size, Whistling Kites are subordinate to Black Kites when the two species feed together, and allow the latter to feed first.[170] Although not as numerous at fires as the Black Kite, they do hunt at them, and stay up to three weeks after the fire has passed.[95]

Breeding in the Darwin area is likely to be more regular than records suggest. A nest was found near Leanyer SP in August 1974[31] and another with two chicks on the edge of saltmarsh at Shoal Bay in early November 1996.[96] Nesting activity was also observed at Humpty Doo in February and May although the years were not specified.[34]

In June 2003 a bird was found sitting on a nest near McMinns Lagoon while in June 2008, a pair was building or renovating a nest, possibly that of an Osprey, on a pylon in the water by the Channel Island bridge. At Marrara Swamp, an adult was seen perched beside a nest on 21 March 1998, but the contents of the nest were unknown. These breeding records, and others from elsewhere in the Top End, suggest laying occurs from at least March to October.

Brahminy Kite
Haliastur indus

RANGE: s. and se. Asia, NG, tropical and subtropical coasts of Australia. HABITATS: coast, estuaries, wetlands, urban areas. STATUS: common resident. BREEDING: May–Oct (DR, TE).

While beaches, reefs and tidal rivers are its natural haunts, this species is frequently encountered in Darwin's suburbs, as well as in the older suburbs of Palmerston. The spread of records shows that the species is resident, and, as some nest sites are used repeatedly, it is likely that these birds are largely sedentary. The small spike in records during January may result from the addition of juveniles into the local population. Though normally seen soaring or resting singly or in loose groups of two or three, at Leanyer SP, particularly in January and February, birds may be seen in large groups, perched on fences and other structures. Of 12 birds at the sewage ponds in early January 1996, six were adults and three were in typical brown juvenile plumage, with streaking on the breast and spotting on the wings, while the remaining three showed traces of immature plumage, with some brown over the white on the head and breast. By May the number of birds there dropped to six, comprising two adults, one sub-adult with some brown on the crown, and three very brown juveniles with no signs of paler underparts or underwings. Similarly, in 1997, 19 birds were

Niven McCrie

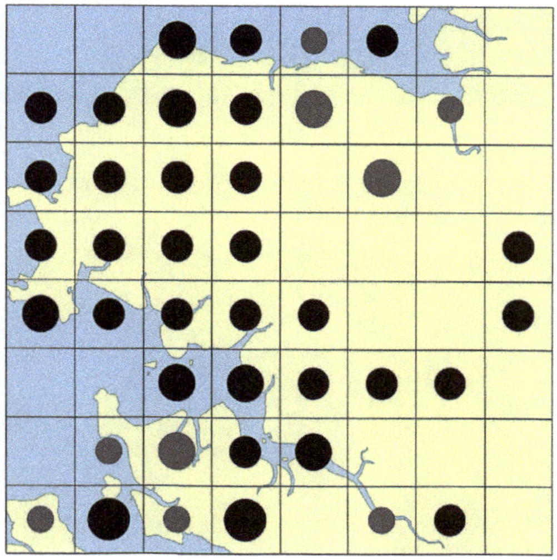

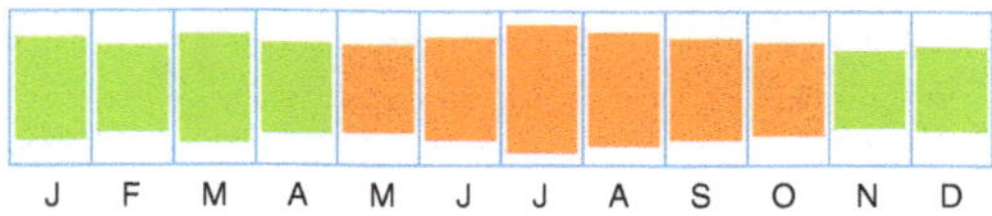

observed in late January, but by June only six remained. In March 2001, 15 birds included 10 juveniles.

Brahminy Kite is primarily a scavenger, but it will also take live fish, crabs and other marine life along the shoreline and in shallows, and reptiles and insects from the land. One was seen to plunge into a Carpentaria palm and was subsequently observed dismembering a small, honeyeater-sized bird,[171] and another attempted, unsuccessfully, to snatch Brown Goshawk chicks from their nest. This indicates that at least occasionally Brahminy Kites, while not in the same league as falcons and goshawks, will opportunistically prey on live birds.[172]

Nesting is apparently widespread in the Darwin region, though few data are available. Birds have been observed carrying nesting material in their bills in Malak and Tiwi in early March and early April, respectively. One pair of birds was known to nest in Frangipanni Street, Nightcliff,[173] over several years. A Kite was brooding on a nest in a paperbark tree in a caravan park near the mouth of the Howard River in early August 2007, while in October 2008 another was sitting on a nest built on an electricity pylon close to the entrance to Charles Darwin NP. A nest behind the jail in Fannie Bay had a near fully grown juvenile in September 2011,[174] while on Dum In Mirrie Island, Bynoe Harbour, a nest in mangroves had two large chicks as early as July 1977.[43]

Bird surveys of the Top End coastline during the 1990s showed that the highest density of Brahminy Kite nests was along the south coast of the Van Diemen Gulf to the east of Darwin. In one aerial survey in early September at least 10 pairs were found nesting in a 125 km section of coastline, though nests were up to 18 km inland.

White-bellied Sea Eagle
Haliaeetus leucogaster

RANGE: coastline around whole of Australia, s. and se. Asia, NG. HABITATS: coast, estuaries, large rivers, wetlands. STATUS: moderately common resident. BREEDING: May–Jun (TE).

Nolan Caldwell

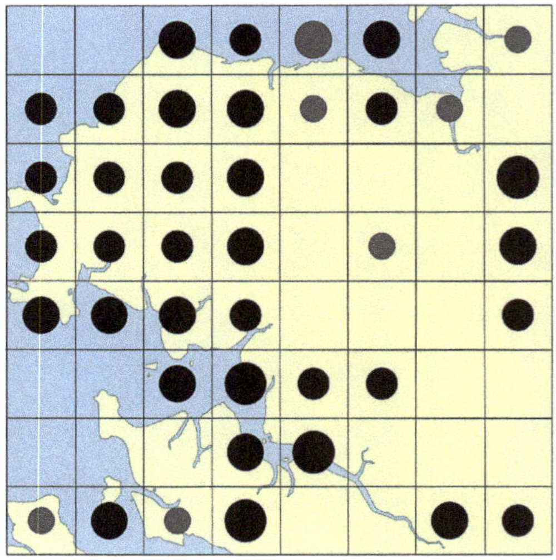

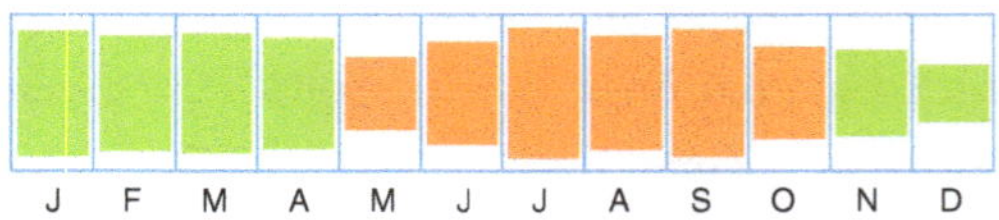

White-bellied Sea Eagles are present in the Darwin region throughout the year, though there is a small increase in reporting rates during the Wet season. Colour-marking of birds at Kapalga, in Kakadu NP, suggests that territorial adults may move long distances,[135] and this could account for the apparent influx.

Sea Eagles are opportunistic feeders and, in addition to hunting a variety of vertebrates and invertebrates, will also eat carrion, and in the Darwin region have been seen feeding on road-killed animals or birds, even on busy roads.[175] They have been observed stealing fishing bait,[136] and stealing the carcass of a Wandering Whistling Duck that Whistling Kites were actively feeding on.[21] At Leanyer SP and, to a lesser extent, at Palmerston SP, Sea Eagles prey on Wandering Whistling Ducks, perhaps taking advantage of injured or sick birds. Sea Eagles will also prey on flying foxes, particularly during the Dry season, and up to five birds have been seen at one time, hunting at a large flying fox colony at Kapalga.[43]

White-bellied Sea Eagle nests have been found throughout much of the Top End coastal and subcoastal areas, with nests constructed in trees, on rocks and on the ground. The highest nest densities were found along the Mary River, east of Darwin, where 21 active nests were found,[27] and on Cobourg Peninsula, where there were reported to be 19 regularly used nest sites.

The Top End breeding season is surprisingly short. During surveys of the Top End coastline in the 1990s, when some 40 nests were found,[27] eggs were recorded between early May and mid-June, downy young in July, and well-feathered young from mid-July to mid-September. On Buffalo Creek[43] eggs were present in a nest during July 1992.

Breeding success in the Darwin region has not been studied, but was quite low at Kapalga, where only two of 13 nesting pairs in 1981 reared young. Of the nine pairs that produced eggs, six failed during incubation or shortly after, and one chick died from pox virus.[176]

Australian Bustard
Ardeotis australis

RANGE: mainly inland and tropical Australia. HABITATS: grassy plains, open woodland. STATUS: rare Dry season visitor. BREEDING: Sep–Mar (TE).

Kirri Hardy

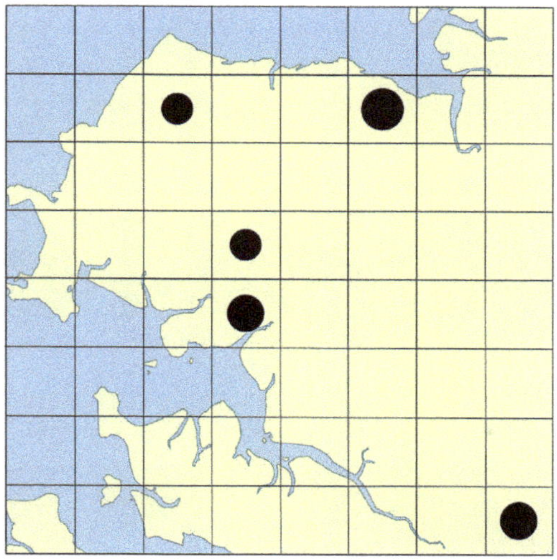

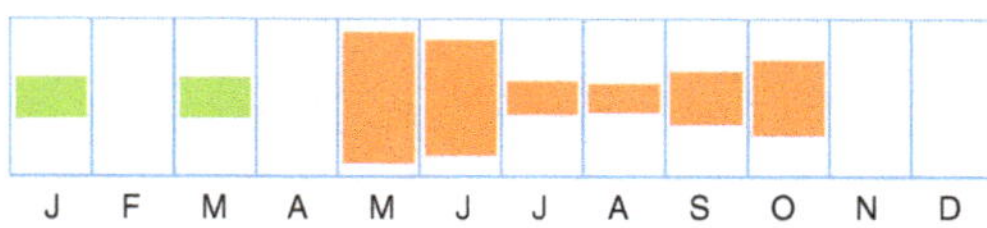

During 1942–43 Australian Bustard was reported as 'fairly common' in suitable habitat around Adelaide River, and 'many thousands' were shot for food by troops stationed in the area.[177] The species declined markedly during the 1980s and 1990s, and though this was more dramatically evident in the southern NT, numbers in the Top End also fell.[178] Hunting is still considered to be a threat but woody weed intrusion, which reduces the amount of suitable habitat, also appears to have had an impact on the species' numbers.[178] Other causes for the reduction of numbers in the Top End are Cane Toads, which compete with the Bustard for food, and fires, which also reduce habitat and food resources and, if late in the season, destroy eggs. However, the Bustard has to some extent taken advantage of these threats, and will feed on Cane Toads, apparently immune to the toads' toxins,[178] and during the late Dry season Bustards often move into recently burnt areas,[178] scavenging insects and small vertebrates killed by fire,[178] and at times to attend fires, to prey on insects and small vertebrates fleeing the flames.

Unlike those in areas of high rainfall variability, Bustards in the monsoonal Top End are considered to be largely sedentary,[178] and it is generally the younger males, without established display territories, that disperse further afield. The Douglas-Daly region, about 160 km south of Darwin, has high densities of Bustards relative to regional and national standards, and some of the birds seen in the Darwin region may be the result of post-breeding dispersal from there. There are 16 records for the Darwin region, all since 1978, and of the 14 that are dated, eight have been in the period May-July, which fits with post-breeding dispersal, and three have been during September and October, suggesting that these birds may have been on passage to breeding areas. Thirteen of the 16 sightings occurred between 1978 and 1998, and three during the period 2010-13. The majority of records are from open grassland, with sightings near Lee Point, Berrimah and Humpty Doo, but there is one sighting from Leanyer SP, in June 1987, and one was seen to land on the bank of a billabong at Bees Creek, presumably to drink, in January 2013.

Buff-banded Rail
Gallirallus philippensis

RANGE: s. and se. Asia, Indonesia, NG, Australia, NZ, sw. Pacific islands. HABITATS: wetlands, coast. STATUS: moderately common visitor or mobile resident. BREEDING: not confirmed in NT.

David Webb

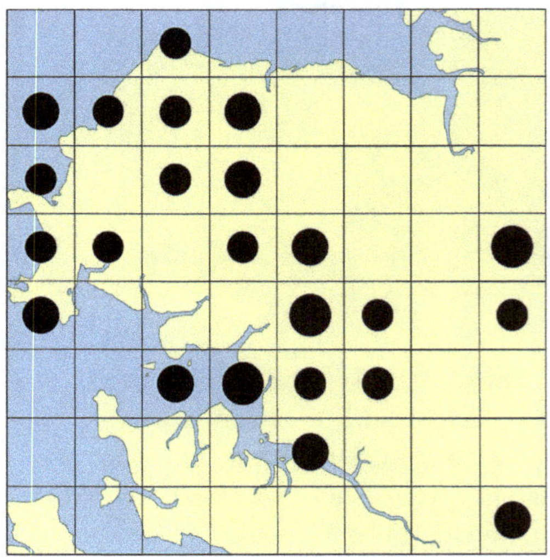

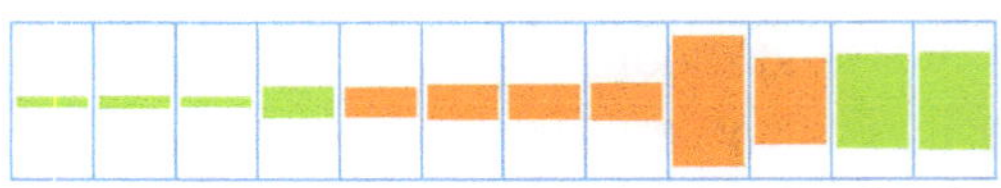

Buff-banded Rails are typically seen around wetlands, along shorelines including reefs, beaches and mangrove fringes, and have also been seen perched atop isolated, near fully submerged mangroves during high tide. They are becoming more frequently seen in urban parks and gardens, and can be quite confiding, appearing in small courtyards of housing units or office buildings, and occasionally in city streets.[3] Generally seen individually, records of larger groups, including up to 40 at Holmes Jungle in October 2002,[21] are all from the late Dry season, and it may be that birds move towards the coast as suitable habitat further inland contracts at this time. However, there are few records during the Wet season, and while this is likely in part to result from increased difficulty of detection,[26] the late Dry season concentrations may represent birds gathering before dispersal from the Top End, to breeding areas.

Buff-banded Rail was reported in the Darwin region every year from 1977 through 2013, with the exception of five of the years from 1988 until 1998. Indeed, during 1988 and 1989, there were no reports of any crake species. It is possible that the draining of Leanyer Swamp, and the consequent loss of habitat, during the later 1980s was at least part of the cause for this.

A young bird found in a Wulagi garden during September 2011 is the only suggestion of breeding for the region, although no information on its age is available. However, a dead adult bird was also found in the garden, suggesting that the young may have still been dependent, thus strongly supporting the possibility of local breeding.[179]

Lewin's Rail
Lewinia pectoralis

RANGE: Lesser Sundas, NG, e. and se. Australia. HABITATS: wetlands, often under tall cover. STATUS: vagrant. BREEDING: not in NT.

Ross Monks

Lewin's Rail was first recorded in the NT in June 1983, when a juvenile male was found in riverine forest at Casuarina CR.[180] The bird had been injured, apparently by a cat, and was released into a large planted aviary at the former Yarrawonga Zoo, near Palmerston. It died the following day, and was prepared as a museum specimen. Traces of the sheaths of flight feathers suggest that while the bird may have been capable of flight, it is unlikely to have flown any great distance, implying that it was the result of local breeding.[180] During September the following year, and in April 1986, Lewin's Rails were again reported, on the basis of calls, from near Holmes Jungle.[34] There have been no records since then, but this species can be extremely cryptic, inhabiting 100% closed cover, and is easily overlooked.[26] Lewin's Rail occurs in New Guinea, and has been recorded from Flores in south-east Indonesia, so it is not surprising that it would also occur in the Top End.

Pale-vented Bush-hen
Amaurornis moluccana

RANGE: Philippines, e. Indonesia, NG and tropical–subtropical Australia. HABITATS: wetlands. STATUS: uncertain – uncommon visitor, possibly small numbers resident. BREEDING: not in NT.

Vince Bugeja

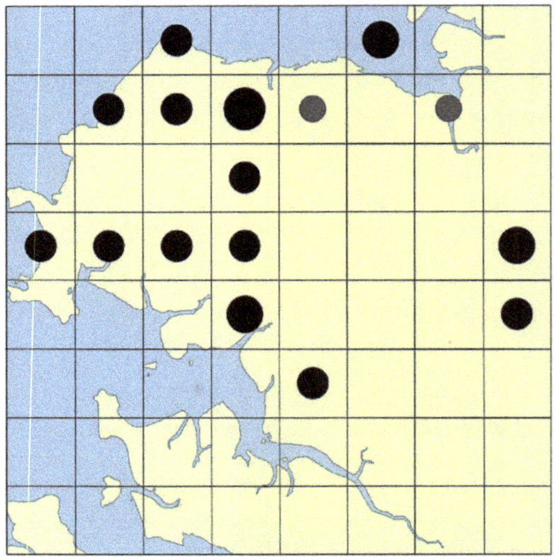

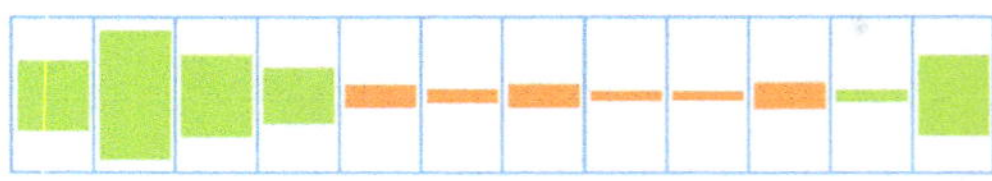

It is fortunate that the Bush-hen's calls, which consist of shrieks, piping and clicks, are so distinctive, as otherwise this secretive bird, which typically confines itself to dense vegetation, would go virtually undetected. Gaining views of the birds is particularly difficult, and an observation of a pair emerging from woodland adjacent to Leanyer Swamp, to bathe and drink at a rainwater puddle, is one of few reports of birds remaining in the open for extended periods.

Although there are records of Bush-hen from only five of the years from 1973 until 1999, it was reported in nine of the 12 years from 2000 until 2011, but whether the species' occurrence in the Darwin region is increasing, or, as is more likely the case, birds went undetected in previous years, is uncertain. Other than in 1986, when there were records covering seven months, no year had more than four months in which the species was recorded, and in six years there was only one month represented. There is evidence that some east coast birds migrate across Torres Strait, but otherwise nothing is known of regular movements. The scarcity of Top End records during the Dry season suggests movement out of the region during that time. However, it may also reflect lower detection rates as, outside of the breeding season, Bush-hens are reportedly more active and call more in overcast conditions, which occur only rarely during the Dry season.

The majority of observations have come from Leanyer Swamp and the wetlands at Holmes Jungle, which provide expanses of suitable habitat, but which also receive much attention from birdwatchers. Birds have also been reported from Charles Darwin NP, Buffalo Creek, Fannie Bay, Winnellie, Marlow Lagoon, Palmerston SP, Howard Springs, Shoal Bay, and Middle Arm.[34]

The breeding season of Bush-hen in eastern Australia is from October to March or April, the period when reporting rates in the Darwin region are highest, yet the only suggestion of local breeding was an immature bird, caught in a mammal trap on the edge of forest at Holmes Jungle, in mid-May 1986.[47] Indeed, the only confirmed breeding reported in the Top End is from near Pine Creek, in 1979, where a nest containing eggs was found in mid-February,[181] and another, containing three eggs and two recently hatched chicks, was found in early March.

Baillon's Crake
Porzana pusilla

RANGE: Europe and n. Africa, Asia; Australia, mainly e., se. and sw. mainland. HABITATS: wetlands with dense aquatic vegetation. STATUS: rare Dry season visitor. BREEDING: none in NT.

David Webb

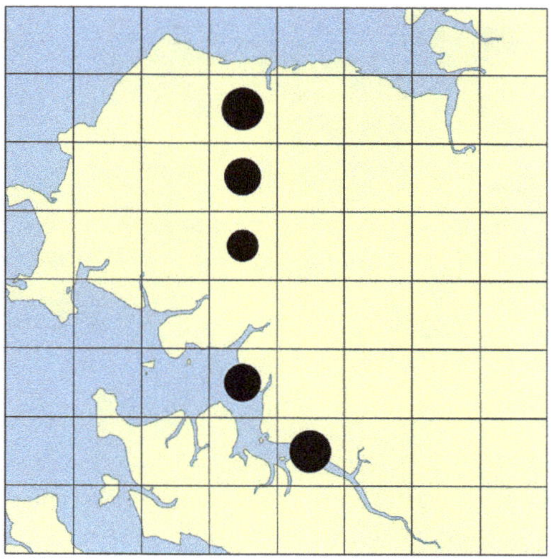

Globally the most widely distributed species of all crakes and small rails found in Australia, occurring as far afield as Europe, the Baillon's Crake is nevertheless the least reported of the three crakes occurring in the Darwin region. The species was first recorded in the NT during July 1974, when there was an influx into the Top End. Up to 20 birds were seen at Kapalga, near the South Alligator River,[135,182] and three birds, subsequently noted to have stored body fat consistent with preparation for migration,[182] were collected as specimens. The first records for the Darwin region followed in September of that year, with the observation of at least five birds, and the collection of a specimen, at Holmes Jungle swamp[182] and additional sightings at Leanyer Swamp and Knuckey Lagoons, during September and November. There were subsequent records from Holmes Jungle in September 1984, and Palmerston SP in 1985, but otherwise, despite occasional sightings at Fogg Dam, there have been no further observations in the Darwin region.

During surveys on the Magela floodplain in Kakadu NP from 1981 to 1984,[28] densities of up to 11 birds per square kilometre were found in dense aquatic vegetation, composed of Hymenachne, Spike-rushes and wild rice, over relatively deep water.[28] Records were largely confined to the months April through June, though a sharp fall in the latter month is suggestive that birds disperse, or perhaps continue further north, as numbers in southern Australia do not typically increase until September, when the local breeding cycle begins.

Spotless Crake
Porzana tabuensis

RANGE: Australia, NZ, NG, Philippines, e. to the Marquesas and Pitcairn Is. HABITATS: wetlands with emergent vegetation. STATUS: rare visitor. BREEDING: Sep–Jan (Aus).

Jon Norling

Inhabiting dense cover, and easily overlooked, the furtive Spotless Crake was recorded only once during Top End surveys in the early 1990s, and it was not found during surveys of the Alligator Rivers region between 1981 and 1984. The species tends to forage in dense cover,[3,26] venturing into the open less frequently than the other small crakes, making it even more difficult to detect. There have been more records in the Fogg Dam region, probably as a result of the number of observers visiting the area, and because suitable habitat is easily accessed.

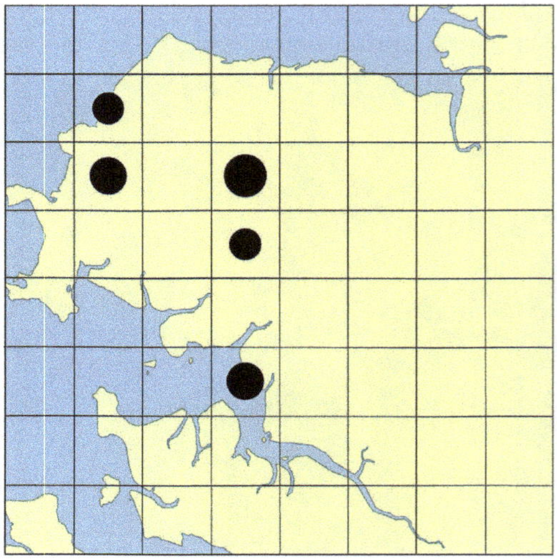

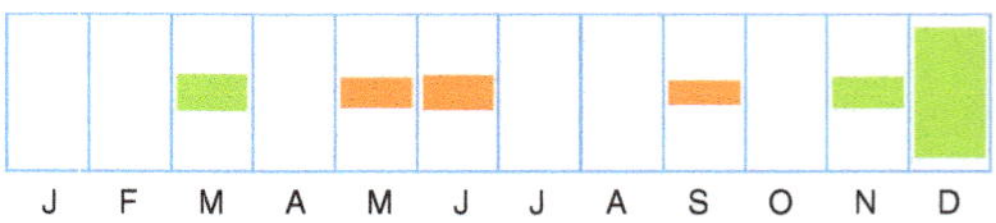

The Spotless Crake was first reported in the Darwin region in September 1982,[34] when a specimen was collected from near Casuarina. There were several sightings at Holmes Jungle swamp during September 1984.[34] During December 1993,[21] three birds were observed on densely matted vegetation at Palmerston SP. At least 10 birds, but possibly up to 20, were present in tall dense *Typha* at Holmes Jungle swamp during June 1994,[24] and while this may have been due to an influx at that time, it is probably more likely due to the survey method, and may indicate that numbers are higher than other records suggest. Surveys based on calls, targeting permanent habitat such as reed swamps, would give the most useful assessment of status. All sightings since then, at Holmes Jungle swamp,[26] Knuckey Lagoons[183] and McMinns Lagoon,[3] have been of single birds.

No breeding has been reported in the Top End; however, the peak egg-laying period of the species in eastern Australia occurs during the months September to December,[135] a period during which birds are present in the Top End.

White-browed Crake
Porzana cinerea

RANGE: se. Asia, NG, tropical Australia. HABITATS: vegetated swamps and other wetlands. STATUS: uncommon visitor late Dry and Wet seasons; some resident. BREEDING: Dec–Feb (DR).

Kirri Hardy

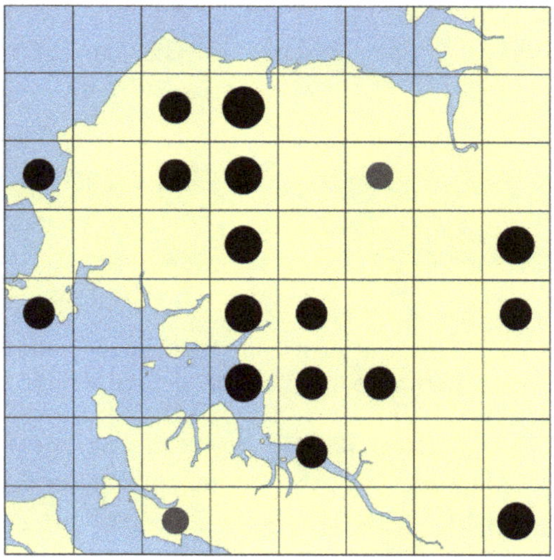

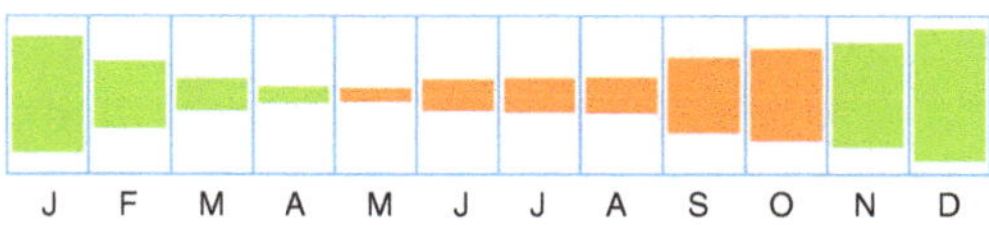

Despite its being relatively common in parts of its range, and less secretive than some other crake species, less is known about White-browed Crake than other crakes found in Australia. It has been suggested that the species arrives during the late Dry season, breeds, then moves northwards,[1,4] and while variations in reporting frequency in the Darwin region support this to some extent, at least small numbers remain throughout the year.

Though still more often heard than seen, White-browed Crakes are not normally as shy as the other crakes that occur in the region. Typically observed walking across water lilies or feeding along muddy swamp edges,[21,64] sometimes several metres from cover, White-browed Crakes will also emerge above the reeds, either having climbed the stems or flying short distances between patches of reeds. They have also been reported to perch in vegetation, or to run along branches of fallen trees,[135] but at Noogoo Swamp an adult was seen perched 3 m above ground in an isolated mangrove, and near Palmerston SP[21,47] one was seen flying to a perch some 5 m up, also in a mangrove.[21] Most often found in ones or twos, at least 12 birds were seen on floating matted vegetation at Palmerston SP during October 2005. While it is not unusual to encounter parties of up to six birds, basing counts on calls can be unreliable, as when two birds are calling it often sounds like a greater number are present.[26]

The frequency of records of White-browed Crakes rises markedly in the late Dry season, a period coinciding with the onset of the local breeding cycle. However, few records of breeding have been reported in the Darwin region, and it may be that while nests or chicks go unseen, breeding related activity, including increased frequency of calling, makes the adults more easily detected. Those few confirmed breeding records comprise nests holding five eggs at Knuckey Lagoons, in February of 1982 and 1986, and a recently hatched chick at Palmerston SP in January 1985.[34] These records suggest egg-laying in November or December, however, two half grown young seen at Fogg Dam in August 1991,[127] would indicate egg-laying in June or July. On the Thai–Malay Peninsula,[184] egg-laying is reported from January, February, July and August, and it may be that a bi-modal breeding season is typical of the species.

Chestnut Rail
Eulabeornis castaneoventris

RANGE: tropical Australia (Top End and Kimberley); Aru Islands. HABITATS: mangal. STATUS: uncommon resident. BREEDING: Sep–Feb (DR).

The Chestnut Rail is probably the most sought-after of the mangrove-endemic birds by birdwatchers, but despite being common in the region - at least nine Chestnut Rails were seen along less than 2 km of Buffalo Creek, during August 2008,[40] for example - it is also one of the more difficult to see. Its loud pig-like grunting is a characteristic sound of the mangal at almost any time of the day, but birds venture out of cover for short periods only, to forage on the muddy banks or bathe in pools created by the receding tide. When they do emerge from mangroves, the rails may forage along the banks of rivers and creeks in full sun, apparently undisturbed by passing boats and their human passengers, but the situation inside the mangal forest is very different, and the species appears shy and wary when approached by land.

Extensive degradation of mangroves for development in areas close to the city threatens the species' existence there, although birds may still be seen at Stuart Park and Bayview Haven, and downy chicks were seen in the latter area as recently as February 2009. A dead rail found on a suburban roadside in November 2011[21] had undoubtedly been transported

Laurie Ross

on a vehicle bumper for some distance, and although there is no certainty of its origin, the mangal fringing the busy road in Winnellie and Stuart Park is one of the few places a rail is likely to encounter traffic at speed. Indeed, a Chestnut Rail was seen dashing across a road, neck outstretched, through heavy traffic there.[185]

Nests[21,47,186] have been most commonly reported in mangals dominated by the Spurred (or Yellow) Mangrove, with some built in pure stands of Stilt-rooted Mangrove, and in mixed species stands. The Chestnut Rail typically builds a platform nest of bulky sticks, with a poorly defined bowl, below or rarely in the lower part of the canopy of mangroves, generally 1-3 m from the ground. Nests are frequently supported by collapsed leaning saplings, which additionally offer a walkway, eliminating the necessity for flight in order to access the nest. Chestnut Rails rarely if ever fly, and little more than wing flapping as a bird makes a hasty retreat from the water's edge to the cover of mangroves has been observed. Indeed, in the late Wet, after breeding, it

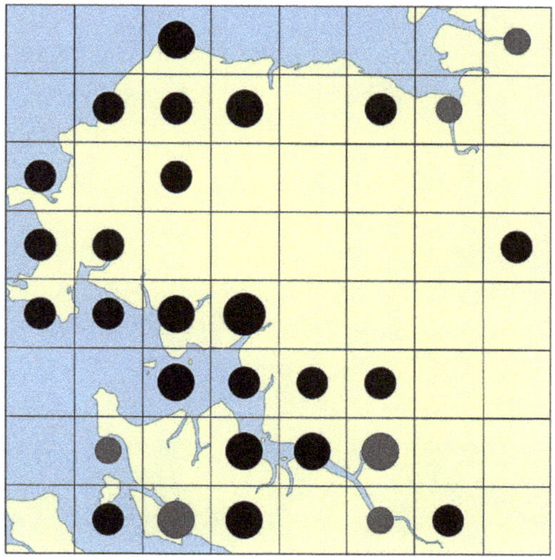

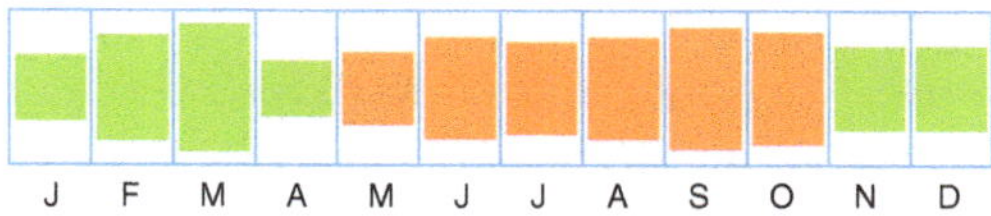

undergoes a complete primary moult and is incapable of flight due to loss of all flight feathers.[187]

Along the Blackmore River, 14 nests were reported in an 18 ha mangal block, suggesting a density of one breeding pair to 0.4 km of river bank on one side. Five of these nests were active in January 1995, with three containing two eggs, and the others, three and four eggs.[186] Active nests have also been found in October,[47] and chicks or dependent young have been reported in October,[188] December,[67] January[189] and February.[37]

Chestnut Rails may favour mangals where the banks of tidal rivulets are steep,[190] although it is not clear why. Perhaps it relates to the density of crabs, its main food. Rails feed, sometimes at night,[187] by probing crab burrows or chasing exposed crabs along the ground. The crab, once captured, is usually tapped against a prop root before being swallowed.

Calls are often given antiphonally, where one individuals' calls follow those of another, in a series lasting up to 15 seconds. Contact notes include a loud *chek chek*[187] and occasional Emu-like grunts.

Few observations of aggressive behaviour have been observed, though two adults were seen squabbling over a large crab, with one subsequently snatching it from in front of the other, then running with it into the mangroves,[21] and on another occasion a Rail foraging at the landward edge of mangroves was seen to chase away an intruding Swinhoe's Snipe.[21] At Middle Arm, a Chestnut Rail, with the feathers on its back erect, was seen 'hassling' a Saltwater Crocodile.[136]

Purple Swamphen
Porphyrio porphyrio

RANGE: Africa and Europe through Asia to NG; whole of Australia. HABITATS: wetlands. STATUS: rare late Dry season visitor. BREEDING: Feb–Apr, Jun–Jul (TE).

Laurie Ross

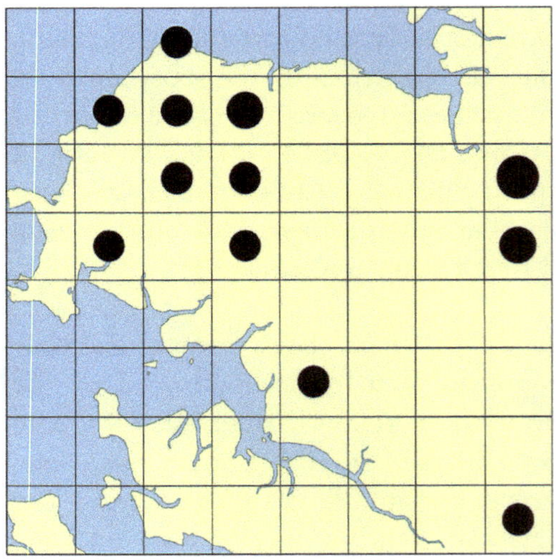

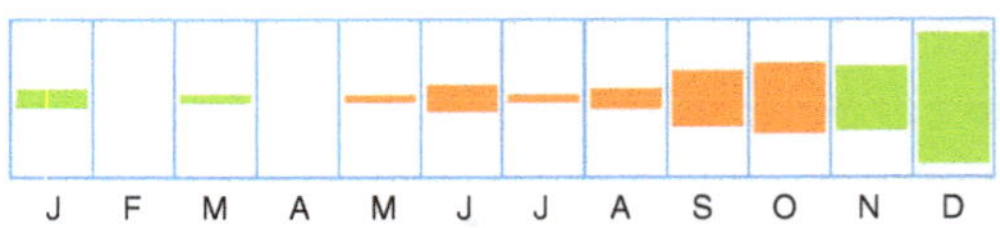

The Purple Swamphen is rarely recorded in large numbers in the Darwin region, where a maximum of 40 were present at Holmes Jungle Swamp in December 1984,[34] although 220 were seen in mango orchards near Fogg Dam,[21] in November 2002. Reporting rates are highest at Leanyer Swamp and Shoal Bay wetlands. Over 1000 birds were seen in a tight group at Kapalga in May 1982,[43] and there were 2000 birds seen in the south-west Top End during June 1999. Surveys of the Alligator Rivers region from 1981 to 1984 found Purple Swamphens restricted largely to Boggy Plain, a backswamp of the South Alligator River, where mean monthly counts peaked at almost 3000 in June, and at 1830 in September. On the Magela floodplain, the only other wetland where the species was recorded during those surveys, numbers peaked in September, which had a mean count of 344. In contrast, surveys during the 1990s failed to find any birds during September, and the number of observations during April, for which there are no records in the Darwin region, outnumbered those in any other month by a factor of almost three.

Although no nesting has been observed in the Darwin region, it occurs at least irregularly in some areas of the Top End. The nearest is a report of young at Opium Creek, about 100 km east of Darwin, in March 1988,[191] and they were found nesting between February and July 1957, in the floodplains between Darwin and western Arnhem Land.[27] Further east, about 500 birds, and an unknown number of nests, were seen among spike-rushes on the Murgenella Creek floodplain during aerial surveys in April 1994,[38] and nesting has also been reported from the Cobourg Peninsula.[27] Local breeding records cannot account for Swamphen numbers in the Top End, indicating that birds originate from elsewhere, probably from the species' stronghold in eastern Australia. Moreover, the great fluctuation in numbers, along with years during which the species appears to be absent, suggests that the Swamphen is dispersive, perhaps irruptive, rather than a regular migrant.

Eurasian Coot
Fulica atra

RANGE: from Africa and Europe through Asia to NG; whole of Australia. HABITATS: wetlands. STATUS: uncommon Dry or early Wet season visitor. BREEDING: Nov–Jan (TE).

Niven McCrie

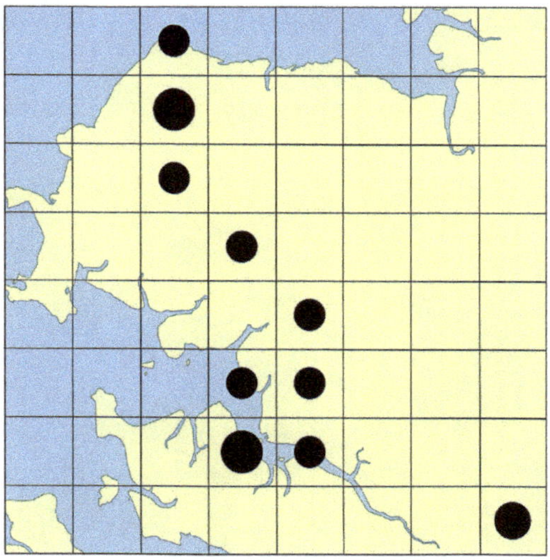

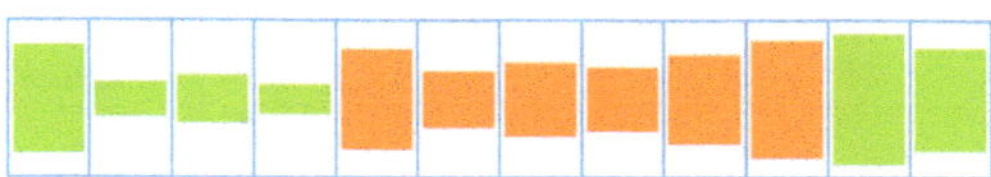

The Eurasian Coot was not recorded during aerial surveys of the Alligator Rivers region from 1981 to 1984, but during ground surveys of the Magela floodplain, small numbers were counted in three months between May and November, with an average of about 350 birds in September. During surveys across the Top End in the 1990s,[27] the greatest number of birds were in the south-west, where one group of 5000 birds was found in June 1999. Otherwise, apart from a count of 500 birds near

Murgenella Creek in Arnhem Land, Coot were only seen in small groups, with the majority of records from in or just east of Darwin. The surveys recorded birds in all months but April, May and November, with multiple records only in July and August.

In the Darwin region, the presence of Eurasian Coot appears to be a recent phenomenon, quite likely related to the construction of the Leanyer SP during the early 1970s, which created suitable habitat that was not previously available. The highest reporting rates, and numbers of Coot, are at those ponds, where maximum counts of 200 and 142, in December 2002 and January 1979, respectively, compare with the highest count outside of the ponds, of 50 birds, recorded in October 2002 at McMinns Lagoon. Analysis of records in the Darwin region for the period 1978 until 2013 reveal the Coot's nomadic nature. During those 35 years, there were two years with records in eight months, eight years with no records, eight years with only one month represented and four years in which there were records in two months.

No breeding has been recorded in the Darwin region. There is a report of breeding during the early Wet season at two sites on the Cobourg Peninsula,[27] but no further details are available. Otherwise, the observation of 59 pairs nesting in wetlands on the Barkly Tablelands in 1993,[192] is the only confirmed breeding in the NT.

Sarus Crane
Grus antigone

RANGE: s. Asia, n. Australia. HABITATS: grassland, wetlands. STATUS: vagrant. BREEDING: not in NT.

Adrian Boyle

The only record of Sarus Crane for the Darwin region was in December 1978, when two birds were seen near Holmes Jungle Swamp.[34] Other than undated records of three birds at Fogg Dam and five birds at Berry Springs,[135] the presence of two birds at Yellow Water, in Kakadu NP, during October and November 2013, is the only additional report of Sarus Cranes in the northern Top End.

The small number of sightings in north Western Australia,[56] together with those records known from the Top End, indicate some westward dispersal, albeit irregular, from the species' Queensland stronghold, and it may be that some Sarus Cranes, possibly in mixed flocks with Brolgas, are overlooked.

Brolga
Grus rubicunda

RANGE: n. and e. Australia, s. NG. HABITATS: wetlands, floodplains, grassland. STATUS: uncommon Dry season visitor. BREEDING: Sep–Mar (TE).

Nolan Caldwell

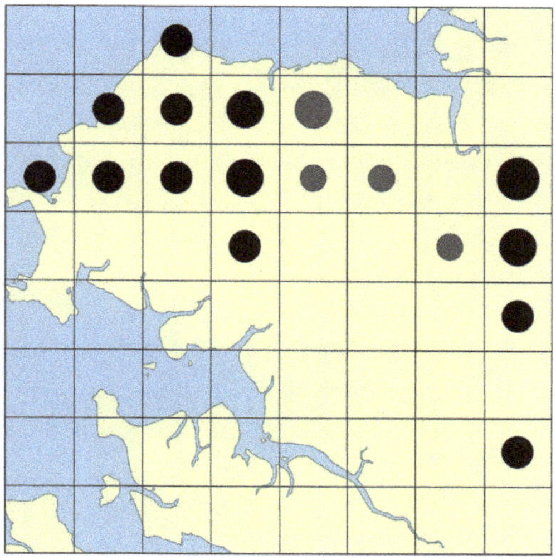

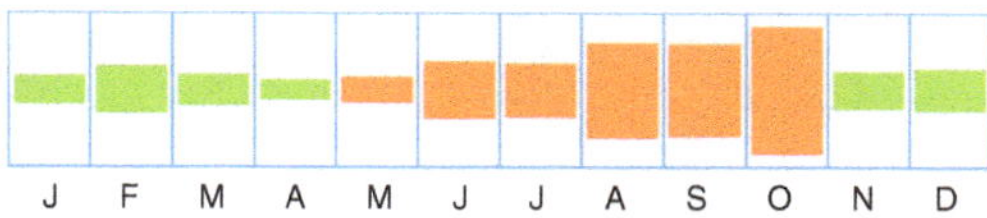

Estimates made from aerial surveys during 1981–84 suggest that up to 24 000 Brolgas may inhabit the Alligator Rivers region during the Dry season,[28] and although generally seen in small numbers - about half of all sightings during surveys in the 1990s were of one or two pairs - flocks of up to 5000 have been seen, largely in west Arnhem Land.[27] During the 1990s surveys, the greatest abundances of Brolgas were found from July to September. In the Darwin region, however, numbers consistently peak in October, before a dramatic fall in November, suggesting that birds are gathering in preparation for their departure. Many of these birds will make their way to breeding grounds, but as male birds are reported to breed only from their fourth or fifth year, and females from their third, it appears likely that at least some birds may simply be dispersing to habitat created by early Wet season rains. While numbers in the Darwin region fall well short of the large flocks seen outside of the region, groups in excess of 100 birds are not unusual, and there are several reports of 300 to 400 birds. With the exception of one of these reports, which was from the Howard Springs Hunting Reserve, all are from the Holmes Jungle wetland, and all, with the exception of one September record, are from October.

Of 14 Top End locations where Brolgas were found nesting during surveys in the 1990s, nine were in the area from the Daly

River to Murgenella, about 100 km south-west and 200 km east north-east of Darwin, respectively.[27] While breeding has been reported from other areas of the Top End, including a nest containing two eggs found on swampy land near Lake Bennett, about 60 km south-east of Darwin, during January 2008, the total number of breeding birds is small, and most Brolgas found in the Top End clearly originate from elsewhere. The peak of egg-laying for Top End breeders occurs in March, but has been reported in all months except for the periods July–August and November–December.[29,30]

The only breeding records for the Darwin region are of a nest containing two eggs, found at Holmes Jungle Swamp in January 1982,[31] and an observation of two adults with one young at Leanyer Swamp in July 1974. The presence of a few pairs during the Wet season in some years suggests that more breeding may occur locally, but the areas where this is likely to occur are largely inaccessible to observers at that time of year.

Red-backed Buttonquail
Turnix maculosus

RANGE: e. Indonesia, NG and mainly tropical Australia. HABITATS: grassland, woodland. STATUS: uncommon, probably resident. BREEDING: mid-Wet season (TE, historical).

Brian Thistleton

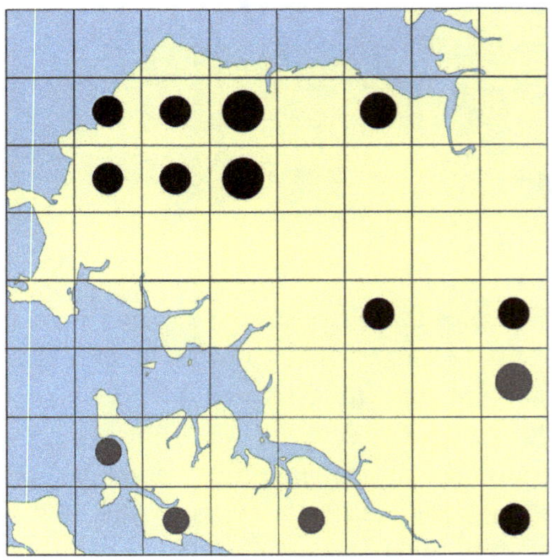

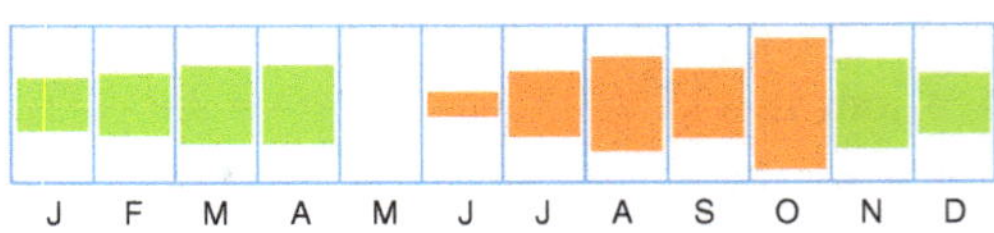

In contrast to the true quails, the female buttonquails are larger and more brightly coloured than the males. Female buttonquails also differ from males in having deep booming calls, which they use to initiate courtship, and it is often this feature that alerts observers to their presence. As is typical anywhere, many of the buttonquails flushed from grassland or woodland in the Darwin region provide only brief views as they fly a short distance and drop back into the grass, and so remain unidentified. The majority of those where identification has been considered certain, however, are

Red-backed Buttonquail. While these birds are most often seen close to water, in the grassy edges of streams, lagoons or floodplains, they also occur in open woodland. Up to five have been seen together,[193] but most reports are of one or two birds seen as they flush. Typically shy, birds are occasionally seen in the open, and one rather bold individual was seen in a garden in suburban Tiwi, foraging on seed spilled from a bird feeder.[194]

The majority of records come from grassy wetlands at Leanyer Swamp, Knuckey Lagoons, Shoal Bay and Holmes Jungle. Reporting rates are highest at Holmes Jungle; however, the grassland there is often burnt out during the Dry season, and buttonquails may be forced into the less accessible grassland beyond, although some may return after the fire has passed to forage on the burnt ground. As well, feral dogs have frequently been seen at the site, and it is likely that these disturb, and perhaps kill, some buttonquails.

Red-backed Buttonquails show no obvious seasonal pattern, other than a sharp increase in observations during October, in the late Dry season. In Kakadu NP, Red-backed Buttonquails are said to be more numerous in the Wet season,[103] apparently due to the movement of birds to wetter areas as suitable habitat further inland contracts.[30] Although there is no evidence to suggest that numbers are higher in the Darwin region during the Wet season, the October peak may indicate an influx, either from further inland or from surrounding savanna woodland, with birds subsequently dispersing over local floodplains or moving further afield. Although no nesting has been recorded in the Darwin region, its occurrence is not unlikely, and the October peak may be a result of birds becoming more easily detected as they increase activity, and call more frequently, before the breeding season.

Notably, with the exception of two records in 1993, there appear to be no reports during the 10-year period 1988–97. While this indicates a possible reduction in numbers, it may be that birds were seen, but not reported, due to uncertainty of identification.

Red-chested Buttonquail
Turnix pyrrhothorax

RANGE: n., e. and se. Australia. HABITATS: woodland, grassland, floodplains. STATUS: rare Dry season visitor. BREEDING: Jan–Feb (NT).

The Red-chested Buttonquail is widespread across the Top End to western Arnhem Land and south to Katherine, though its occurrence is patchy and records are sporadic. In the Darwin region, the species appears to be a rare visitor, with observations only from the Dry season, perhaps partly a result of post-breeding dispersal, or of birds moving to wetter areas to escape harsh conditions inland.

The first report of the species in the Darwin region appears to have been in July 1974, when six were observed at Leanyer Swamp. One was also seen there during September of that year, but there were no

Adrian Boyle

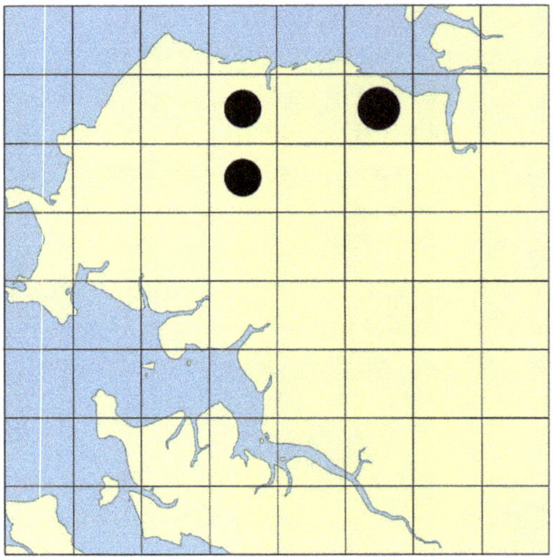

further reports until one was seen, also at Leanyer Swamp, in July 1986.[31] Although reported to be numerous in grassland at Holmes Jungle[146] during the late Dry Season of 1995, records are sporadic, with more years of absence than those with records.

There has been no recorded breeding in the Top End, though it has been recorded in Borroloola, south-east of the Top End, and in the Barkly Tableland.

Bush Stone-curlew
Burhinus grallarius

RANGE: Australia, s. NG. HABITATS: woodland. STATUS: common resident. BREEDING: May–Nov (DR).

A comment that 'about 50 pairs could be located during the course of a day's wandering',[177] relating to the Adelaide River area during 1942–43, may no longer be quite accurate, but unlike in southern Australia, Bush Stone-curlew is still common throughout the Top End. Its distinctive calls are a feature of Darwin nights and, on dark overcast days or in rainy weather, calls may be heard throughout the day.

The Bush Stone-curlew is resident throughout Darwin's suburbs, roosting during the day in parks and reserves, and coming at night to ovals, grassland and other open spaces including car parks and roadsides to feed. During the Wet season birds often gather around streetlights, where they hunt insects. Birds are also frequently present on saltpans, such as those near East Point and occasionally in mangroves on the suburban coastline. While birds typically roost in wooded areas or well-vegetated parks, individuals or small groups may roost under isolated trees in otherwise bare paddocks.

Bush Stone-curlews are generally seen as pairs or small groups, perhaps comprising families; however, densities can be high in favoured roosting areas. Up to 70 birds have been seen roosting in the front yard of

Niven McCrie

a semi-rural house block near Knuckey Lagoons, and more than 20 birds were at a regular roost beside Rocklands Drive, near Darwin Hospital. At this site birds tended to roost during the day on bare ground, on the shady side of trees. They appeared to be undisturbed by the regular, often heavy vehicular traffic and the frequent passing of walkers. When aware of an observer most birds simply stood still or lay down, though occasionally they moved slowly to more sheltered positions. Towards evening, as they became active, they also became

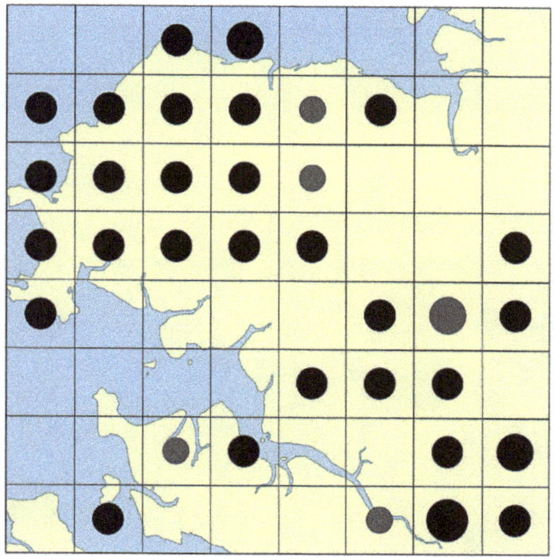

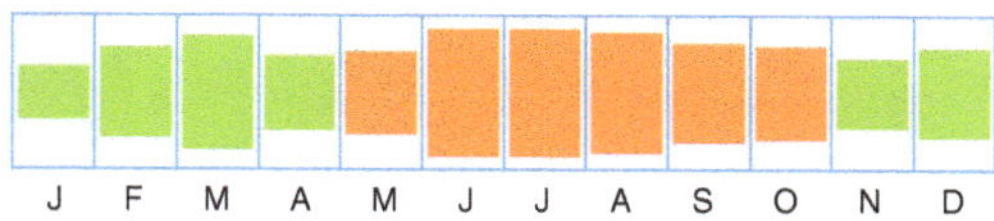

warier, generally running and sometimes flying into deeper cover when disturbed. The site has since been cleared and replaced by a housing development. A few birds are occasionally still seen on the nature strips and small parkland areas there, but the majority of birds have clearly been displaced.

Breeding is widespread throughout the region, with nests or young reported from Nakara, Nightcliff, East Point, Tiwi, Moil, Berrimah and Knuckey Lagoons, and is likely also to occur in other suburban areas. There are reports of young from early June and July, though breeding occurs mainly in the latter part of the Dry season, with eggs from August, and young from September to November.

Beach Stone-curlew
Esacus magnirostris

RANGE: from the Andamans, through se. Asia to Indonesia, NG, sw. Pacific islands, Australia. HABITATS: coast. STATUS: uncommon resident. BREEDING: Mar–Dec (TE).

Niven McCrie

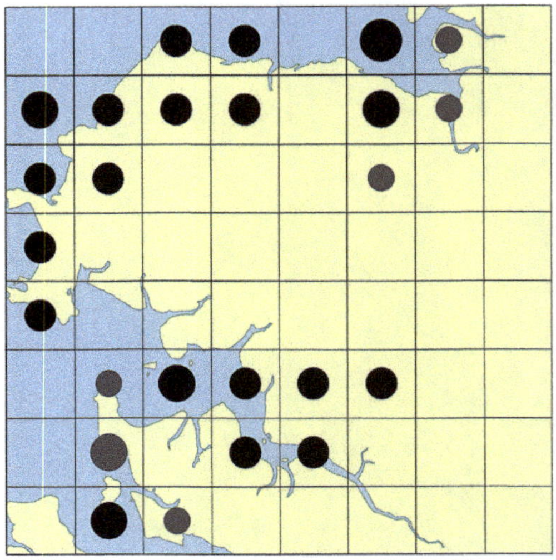

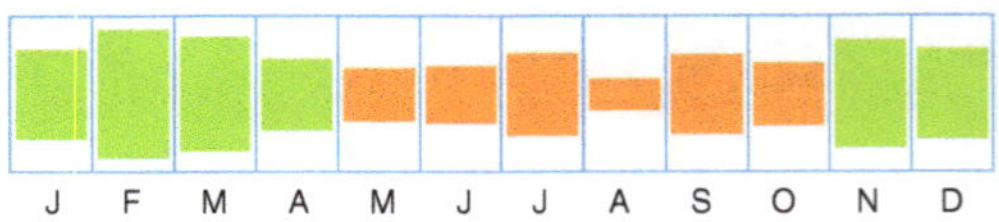

One of the largest shorebirds in the world, with a massive bill, which it uses for feeding on crabs and other marine invertebrates, and a unique, distinctive black and white facial pattern, the Beach Stone-curlew is unmistakeable. Nevertheless, it can be easily overlooked, even when foraging in relatively open habitats, as its largely grey-brown plumage blends in with dark mudflats, and particularly reefs. The species occurs around the entire Top End coast and most offshore islands, and a minimum of 700 birds has been estimated for the Top End coast.[195] An aerial survey of Darwin and Bynoe harbours in November 1994 located 43 individual birds,[195] and the population on islands in Fog Bay has been estimated at 25 individuals.[126]

The Beach Stone-curlew is most commonly seen singly or in pairs, and occasionally in groups of three, perhaps comprising an adult pair with an immature. During the day, or when the tide is high, birds are typically seen roosting in the shelter of mangroves, and sometimes standing around on open mudflats, such as near East Point, and adjacent to Palmerston SP. While tide levels significantly influence when birds feed, Beach Stone-curlews are most frequently seen foraging during the late afternoon, and will often be out on reefs until at least sunset. The extent to which they feed nocturnally is not known, but birds will congregate on beaches during the Flatback Turtle hatching season to prey on the hatchlings as they emerge from the nest burrows at night.[195]

The Beach Stone-curlew is rather sparsely distributed in the Darwin region, and possibly becoming less common as coastal development takes place. Birds are occasionally seen near the city area, at sites such as Cullen Bay, Doctor's Gully and Stokes Hill Wharf, but this is becoming less frequent, and the coastal areas of Fannie Bay, East Point, Nightcliff and near Lee Point appear to be the species' stronghold in the urban area.

Nesting takes place from the late Wet season to the late Dry season, with eggs seen - or inferred by the presence of chicks - from March to December. Confirmed nesting within the Darwin region is scant, though reporting rates indicate that the species is resident in the region. It may be that nests are overlooked, or that some birds, at least those from areas heavily frequented by people, including birdwatchers, move to more isolated sites to breed. The only records of nesting on the urban coastal area come from East Point, where a nest containing one egg was found in October 2006, and a few days later the pair of birds was seen with a small chick.[196] An adult was also observed apparently sitting on a nest there during late May and early June 2011, and in early July two adults were seen with a near fully grown juvenile in early July,[3] suggesting that breeding had taken place there. A nest containing a single well-camouflaged egg, found at Gunn Point[197] in December 2006, comprised a collection of a few dried roots and sticks, placed above the high-tide mark, but nests may also be little more than shallow depressions in the sand. A nest containing one egg was reported in early October 1998 near the Elizabeth River,[47] beyond Palmerston, and fledged young have also been recorded at Palmerston.[21]

On Quail Island, just outside of the region in Bynoe Harbour, four nests were found in June and July,[47] and in other parts of the Top End, recently fledged young were recorded in September and December, a small chick in July and single eggs in March and September.[195]

Pied Oystercatcher
Haematopus longirostris

RANGE: almost entire coastline of Australia. HABITATS: coast. STATUS: uncommon resident. BREEDING: May–Oct (TE).

Laurie Ross

Data from aerial and ground surveys during the 1990s suggested a minimum population of 1700 Pied Oystercatchers in the Top End. Most were recorded on the coast, but some were seen roosting on wetlands just in behind the coast, apparently due to bad weather or very high tides. Birds were generally spread around the coast and islands in pairs, but in some places they formed quite large roosting flocks, with around 30 records of flocks in excess of 50 birds. Only one of these large groups was in the western Top End, in Fog Bay.[195]

Pied Oystercatchers are frequently observed around coastal areas of Darwin,

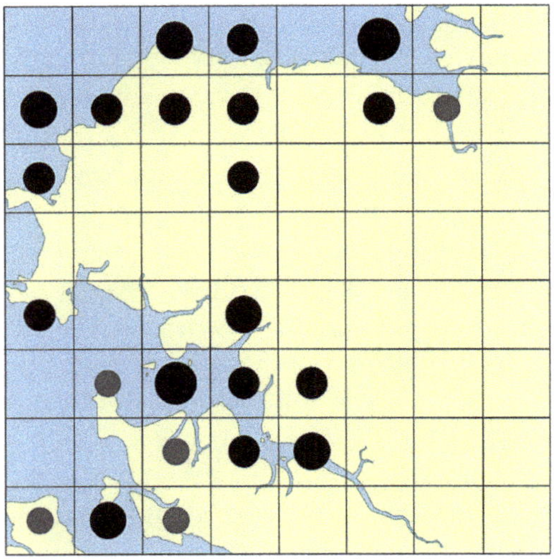

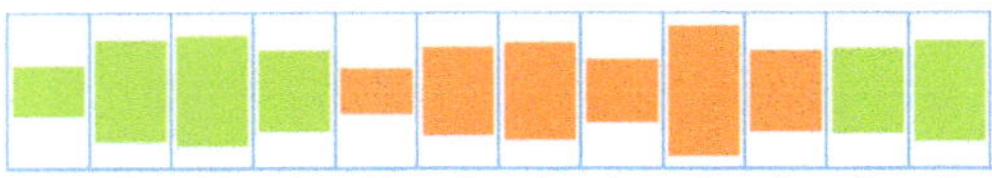

especially at Nightcliff, East Point, between Buffalo Creek and Lee Point, and on some islands in Darwin Harbour. They are occasionally recorded at Palmerston SP on mudflats. There are rarely more than three or four individuals at any time, though there were up to 30 at Camerons Beach in July 1967,[96] and there are records of 10–15 birds irregularly at Lee Point, Nightcliff and East Point.[31] Fluctuations in numbers between roosting[198] sites suggest that the species concentrates on reef habitats during the Dry season, but occurs in larger numbers on sandy beaches during the Wet season. Pied Oystercatchers do not appear to mix with other shorebirds at high-tide roosts on sandbars or shore, instead remaining on reefs until the tide is too high and then moving away from the area. Birds are often observed flying from these sites towards Tree Point[21] where there is a regular roost,[43] although there are no data on numbers at that site.

During surveys in the 1990s[195] there were at least 12 confirmed, or probable, breeding records made of Pied Oystercatchers on Top End coasts - one in each of May, June, August and October, and eight in September. The nearest was on Quail Island, about 45 km west of Darwin, where a nest with one egg, and two nests with two eggs, were recorded, in June and July 1996, and two nests with single eggs in June 1997.[126] The only confirmed breeding within the Darwin region was the observation of a pair with two chicks, estimated at just a few days old, at East Arm Wharf in late September 2010.[199]

Sooty Oystercatcher
Haematopus fuliginosus

RANGE: almost entire coast of Australia. HABITATS: coast; mainly rocky shores and reefs. STATUS: scarce, mainly Dry season visitor. BREEDING: May–Sep (TE).

Northern Australian Sooty Oystercatchers are of the subspecies *opthalmicus*, and differ from those of the nominate race in being smaller and lighter,[200] having a shorter, heavier bill, and having a distinct fleshy yellow to orange eye-ring.[201] It has been suggested that the northern Australian form may warrant specific status,[201] though the subspecies apparently intergrade on the Western Australian coast.[135]

Adrian Boyle

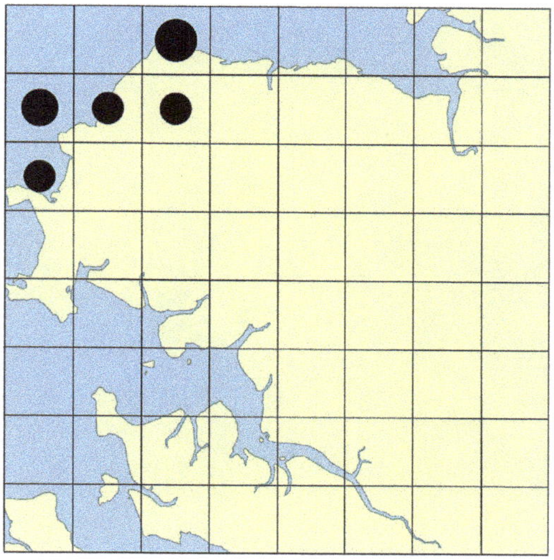

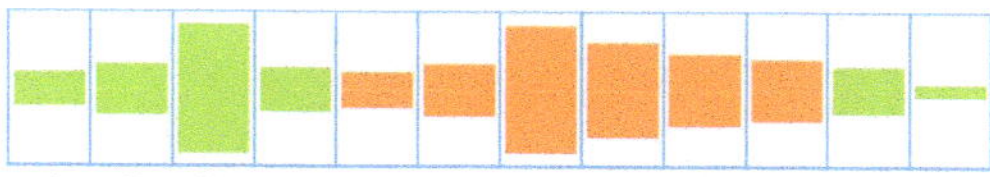

During the 1990s a population of at least 240 Sooty Oystercatchers was estimated for the Top End,[195] with the greatest numbers on the islands in north-east Arnhem Land, and significantly fewer birds further west.[195]

As in the rest of the Top End, the majority of Sooty Oystercatchers in and near the Darwin region are to be found on islands, particularly those with rocks and reefs, and the species appears to be only an occasional visitor to the Darwin mainland, where most records are from the period June to October, at Lee Point, Buffalo Creek, Nightcliff and East Point. Although a flock of 30 birds has been recorded off the coast of Arnhem Land, a maximum group of five birds has been recorded in Darwin, but they are generally seen singly or in pairs.

The nearest confirmed breeding has been found in Fog Bay, west of Darwin, where nests with eggs were recorded on some of the small islands from June to August. Elsewhere in the Top End breeding has been recorded from May to December; however, no data are available on the period of egg-laying.[195]

White-headed Stilt
Himantopus leucocephalus

RANGE: Australasia, NZ, Indonesia, Philippines. HABITATS: wetlands. STATUS: moderately common, mainly Dry season visitor. BREEDING: Nov–Mar (DR), Apr–May (TE).

The population of White-headed Stilts in the Top End was estimated from ground counts during the 1990s to be about 27 600,[27] of which at least 16 000 were found in the Darwin Coastal bioregion, largely from Chambers Bay east to the mouth of the East Alligator River, in Kakadu NP. In the Darwin region, the maximum count of White-headed Stilts was in June 1977, when 1250 were recorded at Leanyer SP.[31] This is exceptional, and maximum counts are not typically in excess of 100 birds.

Niven McCrie

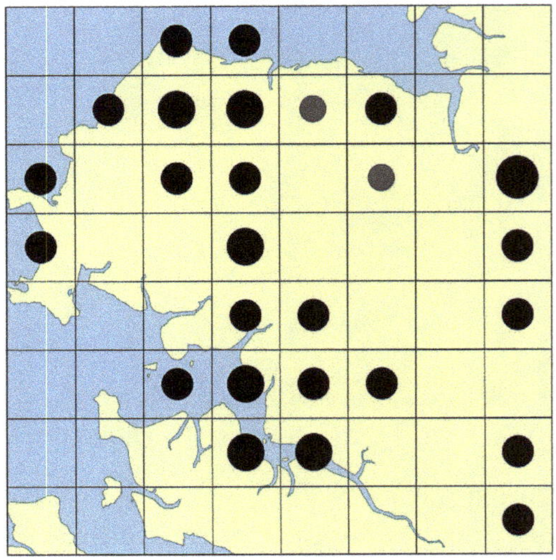

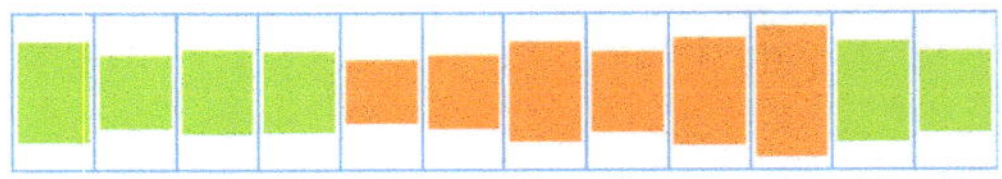

Although there is evidence of movement out of the Top End during the Wet season,[27,28] this is less marked in the Darwin region, and there is considerable variation in seasonal abundances between years. Movement into coastal areas may be due to the drying of wetlands further inland as the Dry season progresses and then, with the onset of the Wet season, the opposite may occur, with wetlands becoming too flooded to provide suitable habitat for the birds. In the Darwin region the majority of Wet season records come from sewage works, where water levels are artificially controlled, but in years when Wet season rains are delayed, there may be large numbers at local wetlands as late as January.

There is an observation of an immature stilt seen swimming in a sewage pond, picking food from the water surface, but the species typically feeds on invertebrates in shallow water.[202] At Leanyer SP they are often seen foraging in the water near the edges of ponds, apparently finding the optimal water depth where the bottom of the ponds slope upwards towards the banks.

Despite the large numbers of White-headed Stilts throughout the Top End, the only observation of colonial nesting comprised five pairs with eggs on the western Gulf of Carpentaria in mid-May 1999.[27] In March and April 1984, over 5000 White-headed Stilts were found breeding in the south-east Gulf of Carpentaria,[203] in Queensland, but this cannot account for the number of birds in the Top End, and their major breeding areas remain unknown. It is probable that the species breeds mostly during the Wet season, when suitable habitat becomes available further inland, and the presence of juvenile birds in the Darwin region largely from March to October, and the few records of local breeding, support this. A nest with three eggs was found in March 1969 at Leanyer Swamp[1] and young birds, about 4-5 days old, were recorded at Leanyer SP in late March 1978.[31] Downy young have also been recorded in the period December to February at Palmerston SP.[31]

Red-necked Avocet
Recurvirostra novaehollandiae

RANGE: largely se. and sw. Australia. HABITATS: terrestrial and estuarine wetlands. STATUS: rare Dry season visitor. BREEDING: Mar–Sep (NT).

Laurie Ross

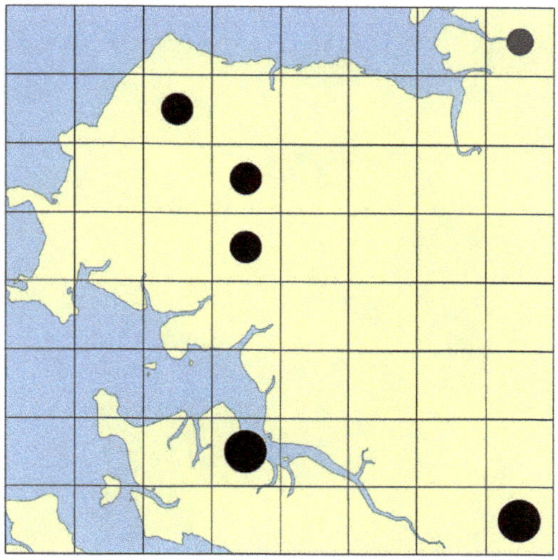

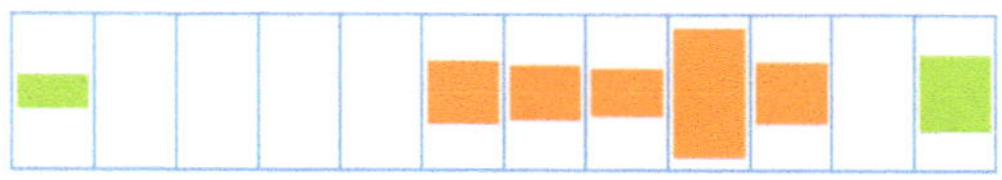

The first report of the Red-necked Avocet in the Darwin region was in September 1981, and there have subsequently been observations of the species in 13 of the years between then and 2013. The majority of sightings have been in the late Dry season, from September to December, probably due to birds in northern inland areas moving towards the coast in response to drying wetlands. Records just outside of the region are more frequent,[27,204] and involve greater numbers, including a single flock of at least 3000 in coastal wetlands along Chambers Bay.[27]

Breeding has been recorded in the Barkly Tableland wetlands,[192] but most occurs in southern Australia,[163] and none has been reported in the Top End. In the Darwin region the only sighting of immature birds was of three in company with five adults at McMinns Lagoon in December 2004.[3]

Masked Lapwing
Vanellus miles

RANGE: e. and n. Australia, NG, NZ. HABITATS: virtually all except mangal and monsoon rainforest. STATUS: very common resident. BREEDING: all months, peaking Sep–Nov (DR).

Niven McCrie

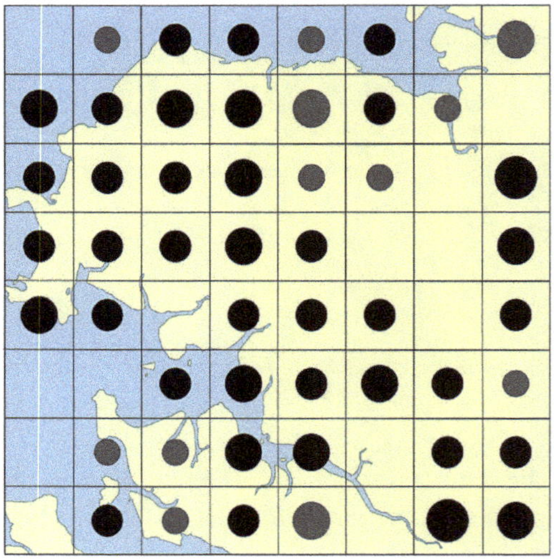

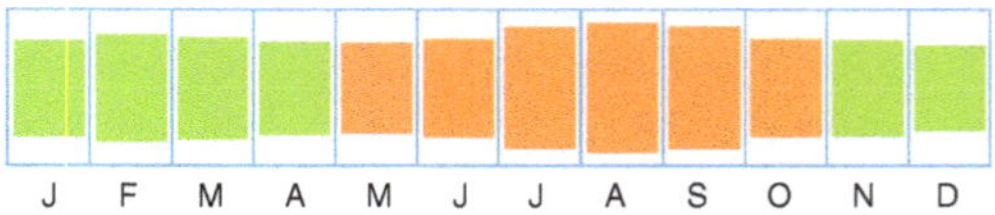

The nominate subspecies of Masked Lapwing occurs across tropical Australia and southern New Guinea. It differs so strikingly from the larger, southern subspecies *novaehollandiae* that some authors consider them separate species.[205]

In the Darwin region, Masked Lapwings occur in a range of habitats including woodland fringes, wetlands, mudflats, mangrove edges, reefs, sewage works, parklands, ovals, road verges and nature strips. While they are most frequently seen in pairs or small loose groups, they will also form dense aggregations. During the hot part of the day, for example, birds sometimes gather into groups of up to 40 or more in the shade of trees, and at high tide those that have been foraging on mudflats concentrate at roosts often comprising 100 or more individuals. There are few data on overnight roosting groups, but on several occasions in the early evening numerous Masked Lapwings have been observed flying from the northern suburban area into Leanyer SP, presumably to roost. On one occasion, in May 1997, over 150 birds were seen to arrive there.[21]

Masked Lapwings are present in the region throughout the year, but counts in excess of 100 birds at individual sites are rarely outside the Dry season months May to September. The fall in concentrations from late September most likely represents birds dispersing to breed, either locally or outside of the region.

Breeding has been recorded throughout the region in a variety of habitats. Nests may be on bitumen, loose gravel, short dry grass, lawns or swamp edges, but the immediate surrounds do not appear to be of much consideration to the birds, with many selecting sites close to foot or vehicular traffic. The majority of nests, however, tend not to be far from water, or are on green nature strips or other grassed areas that receive regular watering. A nest at Tracy Village, on bare gravelly ground in a woodland clearing far from water, appears to be somewhat unusual.

Both eggs and young have been recorded in the Darwin region in every month, but with a distinct peak from September to November, and the fewest breeding events have been reported in June.[21] Birds in the Alligator Rivers region, most frequently nesting on dry banks near water, differ somewhat from Darwin in timing of egg-laying, with most activity from November to March, and only a few as late as June.[28]

Red-kneed Dotterel
Erythrogonys cinctus

RANGE: mainland Australia, excluding arid areas. HABITATS: wetlands. STATUS: uncommon Dry season visitor. BREEDING: May–Oct (TE).

Laurie Ross

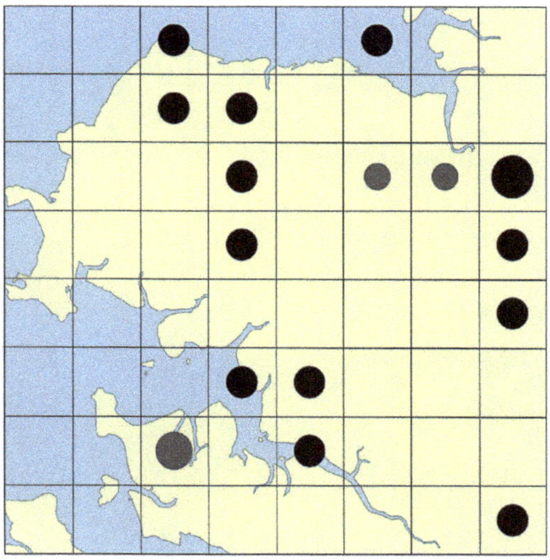

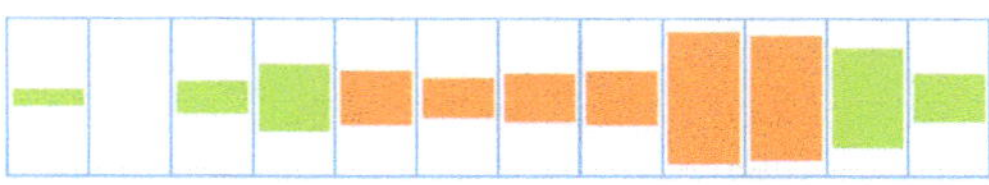

More closely related to Lapwings than to the true plovers,[206] the Red-kneed Dotterel swims well, and while it mainly forages on wet mud, it also wades, sometimes putting its head under water to take prey.[135] In the Darwin region, the species is an irregular visitor to freshwater wetlands, occasionally sewage works, and on rare occasions has been recorded on the shore at Lee Point and the estuarine mouth of Buffalo Creek.

During surveys in the 1990s[195] the Top End population of Red-kneed Dotterels was estimated to be at least 1300 birds, but although the greatest numbers were found in the western Top End, including a single flock of 200 at Chambers Bay 80 km north-east of Darwin, and 50 at Harrison Dam[3] about 50 km east of Darwin, numbers are generally fairly small in the Darwin region. Groups of fewer than about 10 individuals are the norm, although larger numbers occur from time to time, including the maximum counts of 40 birds at Holmes Jungle Swamp in September 1984 and 35 at Leanyer SP in July 1975.[2]

The Red-kneed Dotterel shows a clear peak in reporting rates during the late Dry season, suggesting that birds move towards the coast as inland wetlands dry. The subsequent sharp fall in reporting rates from December and January appears to be directly related to the onset of the Wet season rains, perhaps as flooding and new growth around the swamps reduces foraging areas in the Darwin region, and more suitable wetlands are opened up further inland. Coastal surveys of the entire Top End in the 1990s failed to find any Red-kneed Dotterels during the period January to March, so the movement away from the coast is extensive. Observations at Katherine SP[207] show that the species is only present from November to March, so it may be that birds moving from coastal areas spend the Wet season at that latitude, then move elsewhere, perhaps further inland, to breed from April.

No breeding has been reported in coastal areas of the Top End, but immature birds are frequently seen and, of 40 birds present at Holmes Jungle Swamp in September 1984, only one was in full adult plumage.[34]

Pacific Golden Plover
Pluvialis fulva

RANGE: breeds from Siberia to Alaska; winters from e. Africa to the Pacific islands. HABITATS: coast, grassland, wetlands. STATUS: common late Dry, Wet season migrant. BREEDING: not in Australia.

Ashley Carlson

Morphometric data and band recoveries suggest that Pacific Golden Plovers occurring in Australia originate from separate breeding populations in Alaska and Siberia.[208] During surveys in the 1990s,[195] over 90% of Pacific Golden Plovers located along the Top End coast were in the Darwin Coastal bioregion and, given the vast area of the north-east and eastern coasts where no birds were recorded, it seems likely that Top End birds are from the Siberian breeding population, rather than the Alaskan population which forms the majority of Pacific Golden Plovers on the east coast of Australia.

Non-breeding birds leave their breeding grounds in mid-July, and by late July the Darwin region sees the first migrants

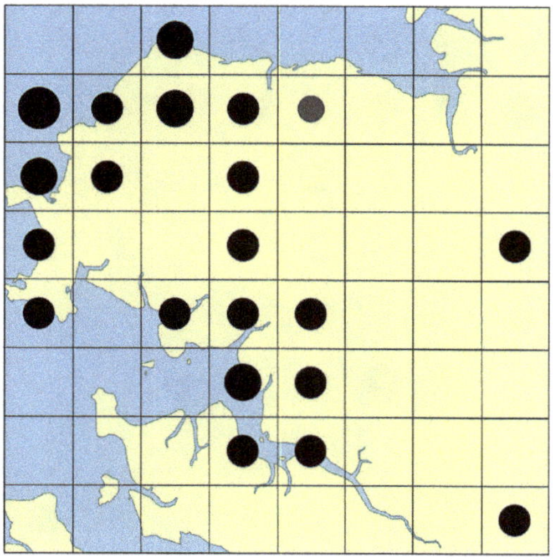

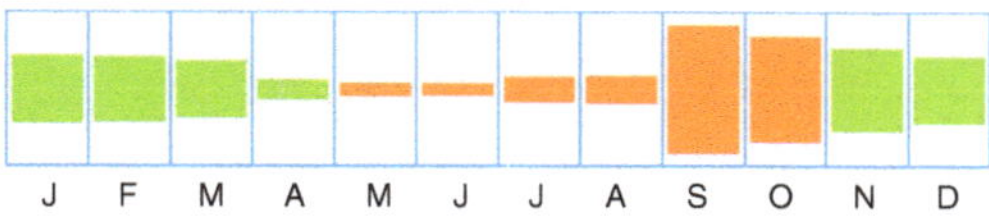

arriving. Breeding adults depart somewhat later, during August, and those birds arriving during September show varying extents of breeding plumage. On arrival, birds tend to remain on the coast, most frequently around Nightcliff and East Point, where the habitat is a mix of reef, muddy shore and mangrove fringe, rather than on beach areas such as Lee Point. During very high tides birds frequently move inland onto ovals, paddocks and other grasslands. Small numbers, generally fewer than about 10 birds, may move into Leanyer SP, but up to 20 or more are often found on the gravel tracks at Palmerston SP, and quite likely comprise birds that feed around the mouth of the Elizabeth River estuary. From October birds begin to disperse, some perhaps moving large distances, others moving to grasslands, croplands and floodplains in the subcoastal area, often loosely associated with Little Curlews which share much of the same habitat.

It has been suggested that most Pacific Golden Plovers fly from southern Australia to south-east Asia without stopping in northern Australia,[209] so it may be that a build-up in the Darwin region in February–March represents a pre-migratory accumulation of relatively local birds.

Grey Plover
Pluvialis squatarola

RANGE: breeds across Arctic Asia and n. America; winters n. and s. America, Europe, Africa, s. and se. Asia, Australasia. HABITATS: tidal mudflats. STATUS: moderately common late Dry, Wet season migrant. BREEDING: not in Australia.

It has been shown that the larger sex of many migratory shorebirds moves the greatest distance from breeding grounds.[210] Grey Plovers exhibit the strongest differential migration known in migratory shorebirds, and over 98% of those reaching Australia are female.[211]

Estimates from aerial and ground surveys in the 1990s indicated a peak of at least 5400 Grey Plovers in the Top End, with birds found all around the coast and on some of the offshore islands.[195] The largest single count, of 750, was from Chambers Bay, 80 km north-east of Darwin. In the

Ashley Carlson

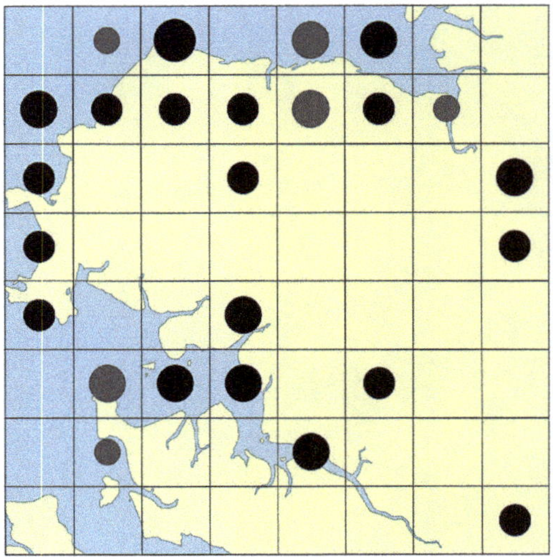

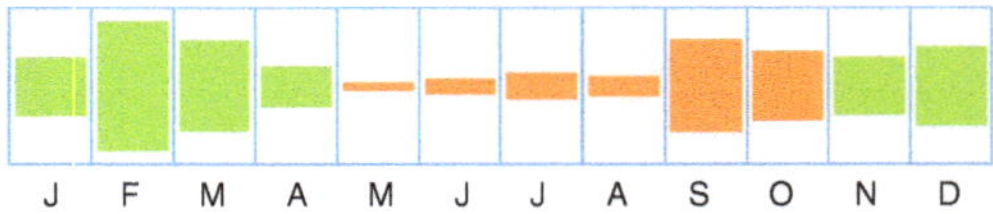

Darwin region, the maximum count has been 160, the only single count of over 100 birds, at Buffalo Creek in February 2011,[3] but roosts are typically much smaller, most holding fewer than about 10 individuals, with up to 30 on rare occasions.

Arrivals have been recorded from late July onwards, though numbers peak in late August to September, then fall sharply in October, suggesting the dispersal of birds shortly after their arrival in the Top End.[209,212] A subsequent rise in numbers the following month, after which numbers remain consistent until March when birds begin their return migration, may represent a second wave of migration, or the dispersal of birds from north-west Australia.[209,212] It has been suggested that the majority of Grey Plovers fly from southern Australia to south-east Asia without stopping in northern Australia,[209] and the absence of an apparent increase in numbers in Darwin or elsewhere in the Top End during the March–May period supports this. The few records of overwintering birds comprise only one or two individuals.

Grey Plovers are typically coastal birds, found on mudflats, beaches and reef areas, although there are records from Leanyer SP, and single birds have been observed at McMinns Lagoon[213] and at Fogg Dam.[21]

Little Ringed Plover
Charadrius dubius

RANGE: breeds n. Africa, Eurasia, China, Japan; winters in Africa, s. and se. Asia, Australia. HABITATS: sand flats, gravel or shingle, wetlands. STATUS: uncommon late Dry, Wet season migrant. BREEDING: not in Australia.

Niven McCrie

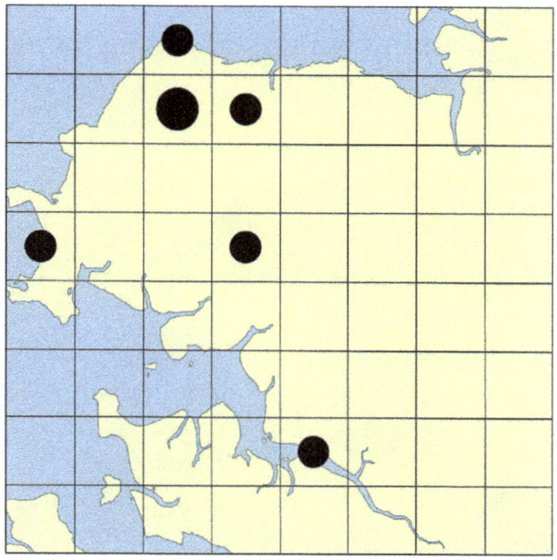

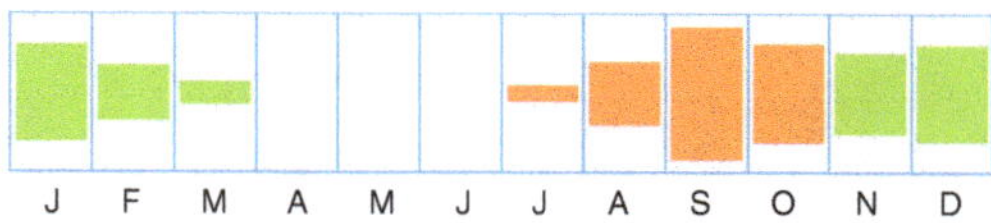

An estimated 2000 Little Ringed Plovers winter in Indonesia,[214] but few continue beyond there into Australia. The first NT record of Little Ringed Plover was of a bird seen along the East Alligator River in Kakadu NP in February 1973,[215] and in September the following year[129] two birds were found at Leanyer SP. Following that observation, Little Ringed Plovers were recorded in eight of the years from 1978 to 1994, then annually from 1998 until 2013. Whether this increase in the regularity of records simply reflects increased site fidelity to Leanyer SP or indicates a change in the pattern of migration to Australia is uncertain. Little Ringed Plovers are known to migrate singly or in small groups, often of less than 10 birds,[201] so in most years those birds present in the Darwin region may be a single migratory group, but in some years numbers increase after the first arrivals, indicating that more than one migratory group is represented. While the majority of records are from Leanyer SP, birds have also been seen at Leanyer Swamp and Knuckey Lagoons, and there has been a single sighting at Vestey's Lake.

The Little Ringed Plover is generally seen singly, though commonly there are up to three birds present at Leanyer SP. There have been occasional sightings of up to five birds together,[21] with a maximum of seven birds at Leanyer SP in February 2001 and November 2008.

One of the earlier migrants, birds normally arrive in August and September, though there is a record as early as 23 July.[21] The species also tends to depart early, and there have been no records after March. Birds in south-east Asia also depart during March,[185] and this fits with the relatively early breeding season of the species, which begins egg-laying as early as late March, probably in eastern Russia, Mongolia, Korea or Japan.[216]

Many birds are still in breeding plumage when they arrive, but lose this quickly. By January, however, they are gaining breeding plumage again, and birds seen in February and March are typically in advanced or full breeding plumage. There have been no observations of juvenile Little Ringed Plover in the Darwin region and, indeed, there appear to have been none anywhere in Australia.[135] Juveniles leave breeding grounds later than adults and perhaps are less likely to travel the distance, particularly as Australia is not a major wintering area for the species.

Kentish Plover

Charadrius alexandrinus

RANGE: breeds temperate n. America, Europe and Asia; winters Africa, S. America, Asia. HABITATS: tidal mudflats, beaches. STATUS: vagrant. BREEDING: not in Australia.

Niven McCrie

A single adult male Kentish Plover, the first to be recorded in Australia, was seen several times at Buffalo Creek during November 1988.[217] The bird was mostly observed feeding at the water's edge or roosting in the sand in proximity with Lesser and Greater Sand Plovers. It was initially considered to be subspecies *dealbatus*; however, recent studies have shown that this name belongs to the distinctive white-faced form,[218] which may be a separate species. It therefore appears that it should be 'eastern' Kentish Plover, subspecies *nihalensis*.

Red-capped Plover

Charadrius ruficapillus

RANGE: throughout Australia, occasionally occurs NZ. HABITATS: coast, terrestrial wetlands. STATUS: moderately common resident. BREEDING: mainly Apr–Jul (DR).

Red-capped Plover is common and widespread throughout the Top End, but survey data from the 1990s indicate that over 75% of the estimated population of 10 000 birds occurs east of the Darwin region.[195] During those surveys Red-capped Plovers were

Nolan Caldwell

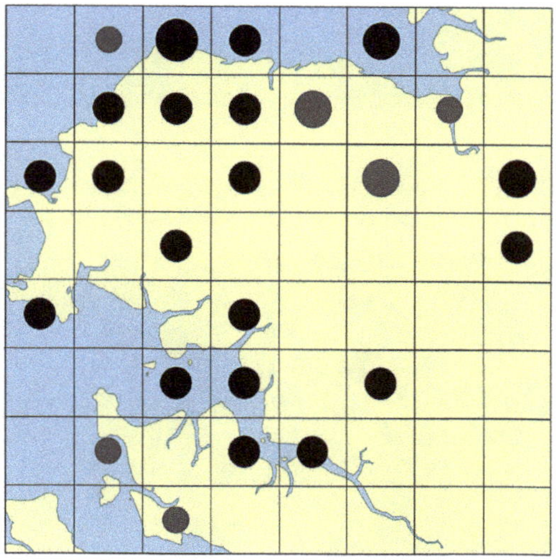

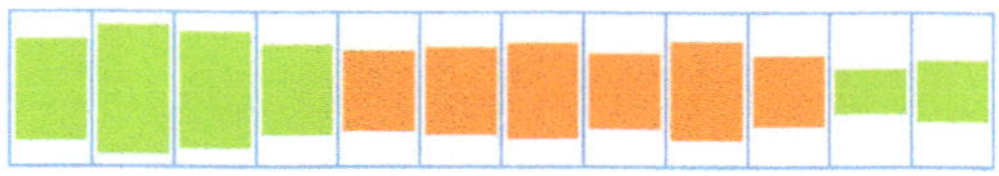

most often recorded in large feeding groups on open saline wetlands and sandy intertidal areas, or as pairs scattered around the coast. These observations, and the scarcity of Red-capped Plovers during surveys of freshwater wetlands in Kakadu NP in the 1980s,[28] suggest that in the north the species is largely confined to coastal areas.

In the Darwin region also, Red-capped Plovers occur most frequently along the coast, particularly on the sandy beach areas from Buffalo Creek to Dripstone Park. Apart from a small number of high counts, for example 70 at Leanyer SP in January 1977, numbers around freshwater swamps and at sewage works rarely exceed about 10 or 20 birds.

Numbers increase during the early Dry season, May through to July, then fall in August before rising sharply again from about late October.[21,198] It may be that the August fall indicates birds dispersing to breed, and the increased numbers in October and November result from the return of breeding birds, along with juveniles.

Red-capped Plovers have been found with eggs either at Lee Point, where nests are placed above the high-tide line, or Leanyer SP, where eggs are laid on the tracks, in all months of the year except August and October, but the majority are in the months April to July. This timing fits with that found in other parts of the Top End,[195] but appears rather at odds with the apparent movement of many birds to breed from August. Given the low numbers involved in coastal breeding, however, it appears that the majority of Red-capped Plovers move inland to breed, and the period from August to October may be optimal for that.

Lesser Sand Plover

Charadrius mongolus

RANGE: breeds Siberia; winters w. Africa, s., e. and se. Asia, Australia. HABITATS: tidal mudflats. STATUS: common late Dry, Wet season migrant. BREEDING: not in Australia.

Adrian Boyle

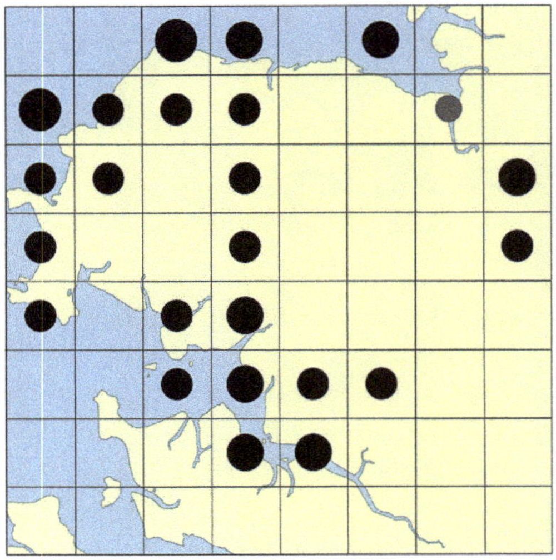

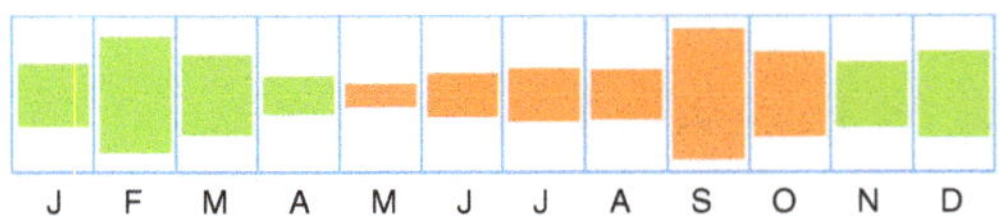

Lesser Sand Plovers occur all around the Top End coast and offshore islands, and occasionally on open saline wetlands close to the coast.[195] Estimates from aerial and ground surveys in the 1990s[195] suggested a population of about 39 000 Lesser Sand Plovers along the Top End coast, with the majority occurring in the eastern Top End. This is not surprising, given the species' more easterly migration path. In north-west Western Australia, Greater Sand Plovers massively outnumber Lesser Sand Plovers, as evidenced by the 10 000 Greater Sand Plovers banded there between 1981 and 2005, compared with only 424 Lesser Sand Plovers.[219] While the proportions of the two species in the Darwin region are far less pronounced, Greater Sand Plovers nevertheless outnumber Lesser Sand Plovers significantly. About 16% of sand plovers caught at Lee Point in October and November 2011 were Lesser Sand Plovers,[220] although counts of sand plovers in the Darwin region typically show this proportion to be only around 10% or fewer. While some may be overlooked among large mixed groups of sand plovers, the Lesser generally tend to roost separately,[3] so underestimates are likely to be insignificant.

Observations of Lesser Sand Plovers overwintering in the region are infrequent, and generally involve fewer than about 10 birds. During July, when Lesser Sand Plovers start arriving in the Darwin region, the greatest numbers are found at Lee Point, but from August it appears that they largely move to the muddy, mangrove fringed shore in the East Point Nightcliff area, as mean counts from August to December show higher numbers there than at Lee Point.[198] From January to March, however, birds appear to move back to Lee Point and, although still quite early, it may be related to subsequent departure northward from that site.

The subspecific status of Lesser Sand Plovers in the Darwin region has not been adequately investigated; however, it appears that most of the birds occurring in north-west Australia, which should include Darwin, are subspecies *mongolus*,[135] while subspecies *stegmanni* is the common form on the east coast. Subspecies *schaeferi*, one of the *atrifrons* group, which according to some taxonomies is a separate species, occurs in Indonesia, so its occurrence in the Darwin region is not unlikely.

Greater Sand Plover
Charadrius leschenaultii

RANGE: breeds from w. China through Mongolia to s. Russia; majority of subsp. *leschenaultii* winter Australasia. HABITATS: tidal mudflats. STATUS: common late Dry, Wet season migrant. BREEDING: not in Australia.

Adrian Boyle

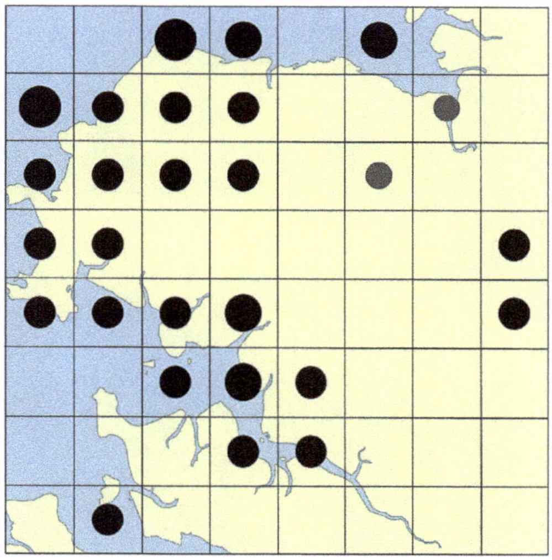

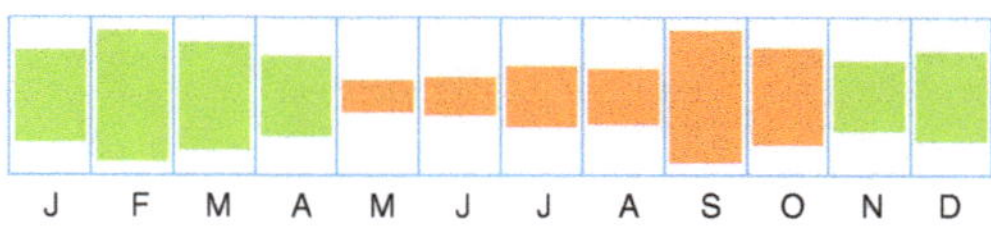

Greater Sand Plovers visiting Australia are of the nominate subspecies, which breeds in the northern parts of the Gobi Desert in Mongolia and in north-western China.[135] Birds begin to form into flocks on their breeding grounds in mid-July, and arrive in north-west Australia as early as late July.[221] It is very likely a similar pattern in the Darwin region, but counts of Greater Sand Plovers overwintering have been as high as 985,[3] and July counts would comprise both new arrivals and overwintering birds, making assessment of newly arrived numbers difficult.

Surveys in the 1990s provided an estimate of at least 40 300 birds for the Top End coast, with the largest single count of about

1800 at the southern end of Fog Bay, west of Darwin, in mid-July 1998.[195] These surveys showed a peak in numbers of Greater Sand Plover through the over-wintering months. While numbers in the Darwin region are also high during the non-breeding period, with maximum monthly counts from May to August all in excess of 500 birds, the highest count, of 2092 birds, was made in February, and maximum counts for the months September to April are higher than those for May to August.

There was a distinct increase in numbers from 2003 compared to prior years. The maximum monthly count during the period 1970–2002,[21,212] of 700 in July 2000, was exceeded in 34 months covering every year from 2003 to 2014,[3,198] but the cause of this sharp increase in abundance is uncertain. It appears likely to be the result of a shift in distribution rather than an increase in numbers migrating. During the 1990s, surveys of the coast to about 60 km either side of, and including, Darwin, showed peak numbers of about 4300 Greater Sand Plovers, and a relatively small shift in distribution may well account for the increase.

Unlike the case with the Lesser Sand Plover, numbers of the Greater Sand Plover are invariably much greater at high-tide roosts on the sandy beach areas around Lee Point, and counts on mangrove fringed mudflats rarely exceed about 15% of those numbers. There are also records from sewage works, and there have been sightings at Knuckey Lagoons[3] and Fogg Dam.

Oriental Plover
Charadrius veredus

RANGE: breeds in n. China and Mongolia, almost the entire population wintering in n. Australia. HABITATS: coast, grassland, floodplains. STATUS: uncommon passage migrant, Sep–Dec. BREEDING: not in Australia.

Danny Rogers

Oriental Plovers start to leave their breeding grounds in July, with males leaving first,[201] but although there are records from August, the species typically arrives during September. Observations of Oriental Plovers in the Barkly Tableland in late August and early September,[21] at about the time the first birds are seen in Darwin, suggest that many birds may continue some distance beyond the coast on their southward migration. On arrival many birds still show some remnants of breeding plumage, but this is quickly lost. Only small numbers of juvenile birds have been observed, but these birds appear to arrive a little later than adults. Numbers in the Darwin region

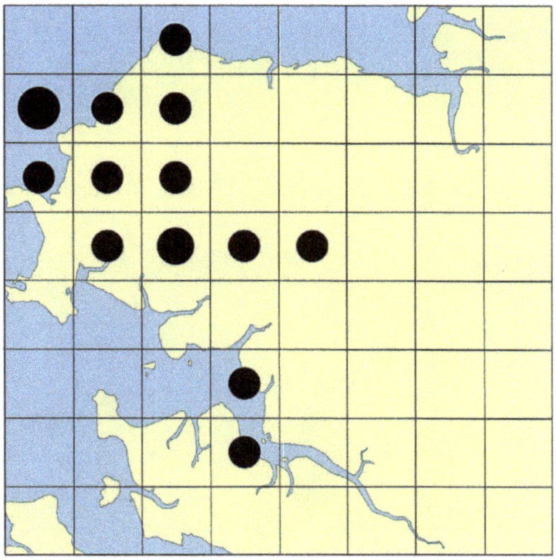

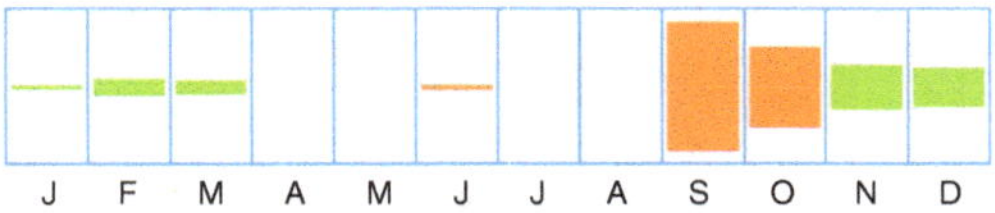

generally build up until October when birds start to disperse, although in some years there are moderate numbers present well into December. Unlike the Little Curlew, the Oriental Plovers' exit from the Darwin region does not typically coincide with the onset of heavy rains.

Oriental Plovers are found in a range of habitats including edges of swamps, dry grassland, reefs, mudflats and occasionally sewage works. During their time in the Darwin region, the species shows some consistent patterns of behaviour. At Nightcliff, where birds are generally present early in their stay, groups fly in, generally from a south-westerly direction, to roost on the reef in the early mornings, predictably between 8 and 9 o'clock. The birds sometimes arrive in a single group but often in two or three groups arriving perhaps 10 minutes or so apart. When they move out, it is normally in a south-easterly direction, with birds likely to be heading to daytime feeding areas away from the coast. It is not certain where birds roost overnight, but in past years, perhaps until about 2000, birds frequently moved to grassland near Lee Point in the late afternoon, presumably to spend the night. What caused the change in habits cannot be ascertained with certainty but it may have coincided with the end of slashing of the grass, after which the growth was too rank to provide suitable habitat.

Numbers in the Darwin region are generally small, 195 at Ludmilla Creek in late September 1970[1] being the maximum recorded. Counts of 150 at Nightcliff in September 1998[21] and October 2002[21] and a similar number in early December 2004[222] are also notably high for the area, the latter unusual in being so late, though it may be a result of the delayed Wet season that year.[124]

No studies of feeding have been undertaken, but it appears that most feeding takes place in grassland, though at Leanyer SP a few birds have been observed foraging in dense floating vegetation.

The generally scarcity of records in the Darwin region during the return migration period agrees with data from north-west Western Australia.[223] It appears that the return migration route differs from the inward, though it is possible that birds fly over without stopping, perhaps at night. Those few records of Oriental Plover in the Darwin region from January onward include two birds at Leanyer Swamp on 6 April 1974,[31] a bird in full breeding plumage collected at Lee Point in mid-June 1971,[1] and a sighting of a bird at Lee Point in late June 2010.[37] The general lack of overwintering records of the species suggests that the latter birds may have been unable to undertake return migration because of injury or sickness.

Black-fronted Dotterel
Elseyornis melanops

RANGE: Australasia, excluding NG. HABITATS: wetlands, occasionally coast. STATUS: uncommon Dry season visitor. BREEDING: Mar–Nov (TE).

Niven McCrie

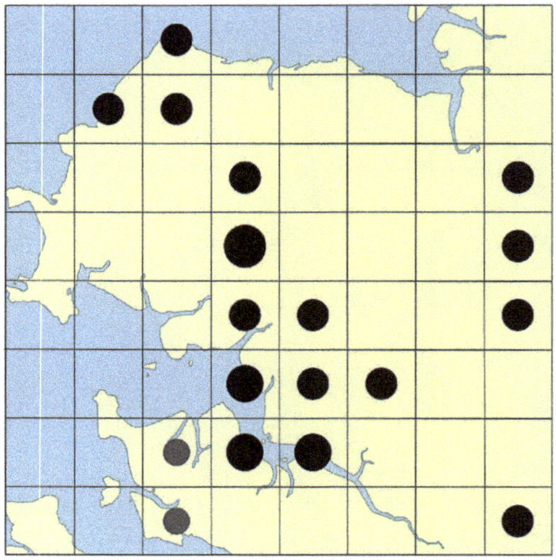

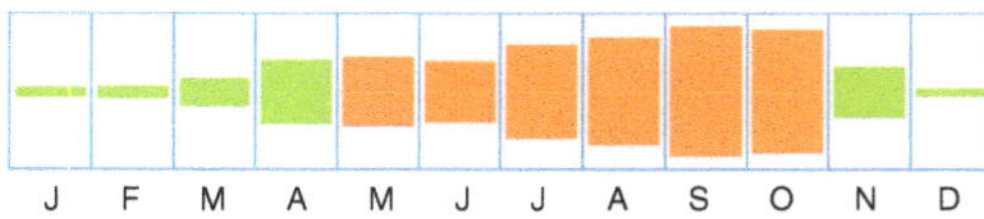

Typically a bird of terrestrial freshwater wetlands,[135] Black-fronted Dotterels are virtually absent from coastal areas of the Top End. Indeed, during surveys of the entire NT coast during the 1990s,[195] the species was found in only two locations outside of the Darwin region and the nearby wetlands.

Numbers of Black-fronted Dotterels in the region are generally very small and, apart from counts of 25 at Palmerston SP in late September 1984 and at Leanyer SP in early July 1986,[31] there are very few records of more than 10 birds. While there is a large variation in the number of records between years, with no sightings in some years, the species shows a clear seasonal pattern, appearing in the region from April and peaking in abundance in the period August to October. There are very few records from the Wet season months December to March.

Breeding records for the region are scant, and it appears that the majority of birds are non-breeding visitors. A Black-fronted Dotterel with a chick was present at Leanyer SP in late March 2013,[224] while downy young were reported at Knuckey Lagoons in November 1996,[21] October 1998[21] and August 2012, and at Leanyer SP in May 2011.[47] Recently fledged young were seen at that site in November 1991, September 1994,[21] October 1996 and in June 1999.

Australian Painted-snipe
Rostratula australis

RANGE: Australia, largely e. and se. regions. HABITATS: wetlands. STATUS: vagrant. BREEDING: May (NT: one record).

Ross Monks

There are few records of Australian Painted-snipe from the NT, and most of those have come from southern and central areas. In May 1993 five birds were found at Tarrabool Lake in the Barkly Tablelands region, and a single, cold egg of the species was also discovered.[225] While this provided the first evidence of attempted breeding for the NT, and is indicative that the species may be more than a vagrant, it nevertheless remains rarely reported. The only record from the Darwin region is the observation of a male bird at McMinns Lagoon in early December 2006.[3]

Comb-crested Jacana
Irediparra gallinacea

RANGE: Australia, NG, Indonesia, Timor-Leste. HABITATS: wetlands with floating vegetation. STATUS: moderately common Dry season visitor, some breeding residents. BREEDING: Nov–May (TE).

Comb-crested Jacana, best known for its extraordinarily long toes that allow it to easily walk on floating lily pads, is common throughout the Top End in swamps and lagoons, particularly those with water lilies. However, it is also a visitor to sewage ponds when there is a suitable cover of vegetation, and is occasionally seen at rainwater puddles. Birds are present in the Darwin region throughout the year but concentrate in greater numbers in the latter part of the Dry season, probably in response to contracting wetlands. As occurs in the wetlands of Kakadu NP,[226] numbers in the Darwin region then fall sharply with the onset of the Wet season, with birds dispersing to the large floodplains as they begin to fill.[27]

In the Alligator Rivers region, the population has been estimated at up to 25 000 birds,[28] although densities in the Darwin region are much lower. The highest counts in the Darwin region have been from Knuckey Lagoons, where 175 birds were present in November 1999,[34] and an estimated 160 birds, mostly immatures, in November 1989. An unusually high concentration occurred at Fogg Dam[27] in

Laurie Ross

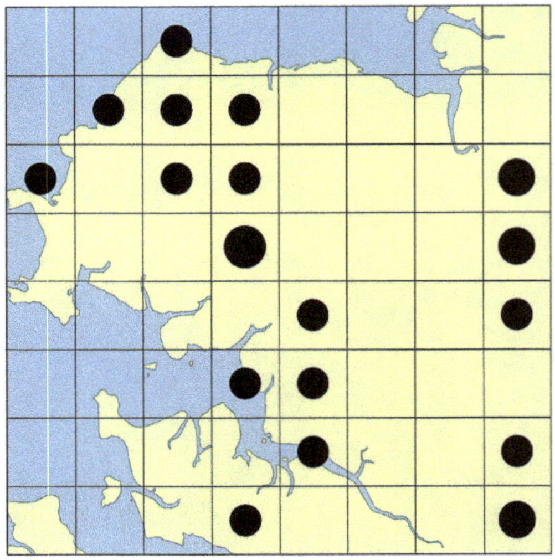

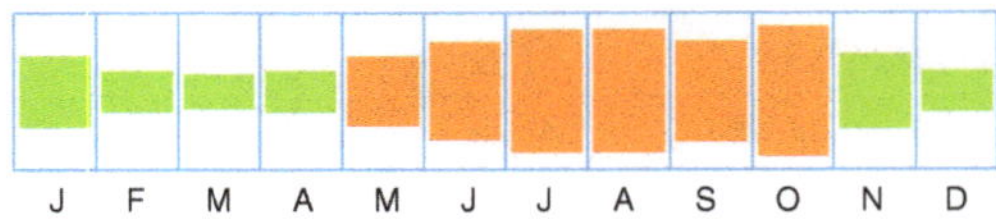

September 1983, when an estimated 2000 Jacanas were seen, spread densely throughout visible areas of the dam.[21,195]

Comb-crested Jacanas forage for insects and seeds in low emergent and aquatic vegetation,[28] with aquatic invertebrates and seeds of aquatic plants forming the greater part of their diet.[226] Although the species becomes vigorously defensive during the breeding season, birds display little aggressive behaviour when foraging, though they will chase other birds some distance from a feeding area.

The species appears to be particularly susceptible to drought conditions, and in 1970 it was estimated that mortality may have been as high as 90% for birds in the Fogg Dam and Darwin regions.[45] Recovery at least from shorter term drought appears rapid, as numbers had increased almost to previous levels by the following year.

During surveys of the Top End coast in the 1990s[27] the species was recorded breeding in each month between March and November, with 18 of the 19 records in or near the Darwin region. No further details are available, and somewhat surprisingly, other than sightings of chicks at Knuckey Lagoons in April and October 1993[26] and in June–July 1997,[21] there appear to be no other direct observations of breeding in the Darwin region. It may be that the majority of Top End Jacanas breed on the large floodplains during the late Wet season and early Dry season, as they are reported to breed as isolated pairs throughout the Magela floodplain in Kakadu NP at that time of year.[28]

Swinhoe's Snipe
Gallinago megala

RANGE: breeds in central and se. Siberia; winters from India to NG and n. Australia. HABITATS: wetlands, floodplains, grassland, cultivated areas. STATUS: uncommon late Dry, Wet season migrant. BREEDING: not in Australia.

Micha V. Jackson

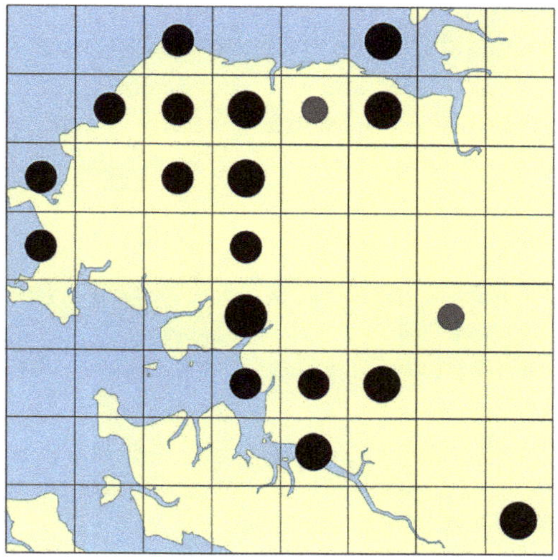

Swinhoe's Snipe, along with Oriental Pratincole, are generally the last of the migrant shorebirds to arrive in the Darwin region. They are rarely present before October, and in some years there are no reports until November. At the end of the migration season, the last of them are gone by late April, and there are no records of birds overwintering.

Numbers fluctuate markedly from year to year, and in recent years, probably due to a reduction in suitable habitat, numbers in the Darwin region appear to have declined. During the 1980s and 1990s, at Knuckey Lagoons, where one section has long been known locally as 'Snipe Swamp', it was not unusual to find 20 or more birds present.[21] Livestock had kept growth around the water edge short, providing open expanses of mud interspersed with areas of short grass, forming ideal habitat for snipe. Birds were frequently seen foraging in water to about belly depth, or on the mud, away from the water, well out in the open. When disturbed at the presence of an observer or possibly a predator, birds would often crouch into the hoofmarks created by the livestock.[21] With the removal of the livestock, those conditions no longer exist and the swamp edges become completely covered in long grass quite early in the Wet season, making the area largely unsuitable for snipe.

The species is not confined to swamps, however, and birds have been recorded in a broad range of habitats and sites covering fresh and brackish water, open woodland, monsoon forest, parklands, sewage ponds, mangrove edges and temporarily flooded roadsides. Two birds were seen sheltering under Beach Hibiscus just above the high-tide line at Buffalo Creek in January 1996, and at Bullocky Point birds have often been seen in the early morning feeding in grass beside a car park. Birds have also been observed feeding by a freshwater section of Rapid Creek at Jingili, within a lightly forested area,[21] and several birds that were flushed at Howard Springs landed in forest.[31]

In addition to Swinhoe's Snipe there is the possibility also that Pin-tailed and Latham's Snipes could occur in the region. These three *Gallinago* snipes present difficulties in identification, and it may perhaps be stated that none can be identified in the field with certainty. However, the only snipes to have been captured, or taken as specimens, have been Swinhoe's, and it is appropriate to assume that any snipe seen in the Darwin region is this species, unless there is clear evidence to the contrary.

Asian Dowitcher
Limnodromus semipalmatus

RANGE: breeds in scattered areas of central Asia, from w. Siberia to ne. China; winters in se. Asia. HABITATS: mainly tidal mudflats; occasionally wetlands. STATUS: rare late Dry, Wet season migrant. BREEDING: not in Australia.

Dr Chan Ah Lak

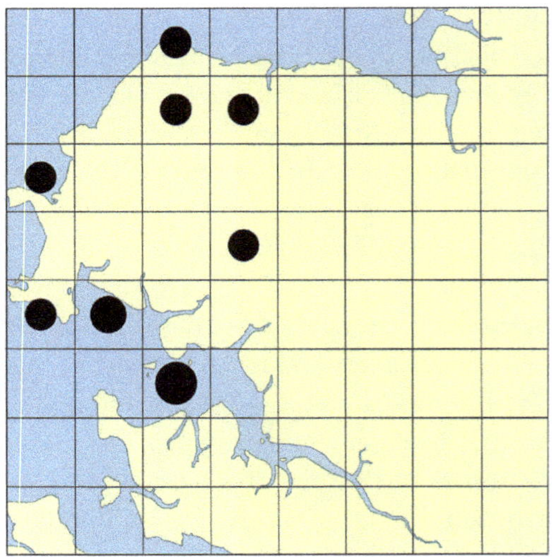

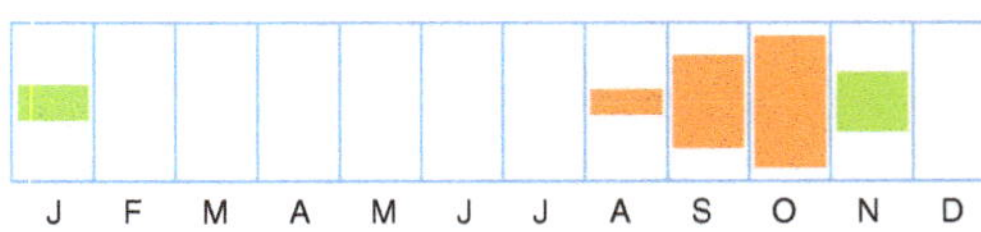

The first Australian record of Asian Dowitcher was a specimen taken from the mouth of Sadgrove Creek near Frances Bay, Darwin Harbour, in mid-October 1971.[227] Between then and 2013 there were records covering only nine years, clearly showing that they are irregular visitors to the Darwin region. During surveys of the Top End coast[195] in the 1990s the largest groups, including one of 70 birds, were found in the eastern Top End, while groups of up to 15 birds, including at four sites within 50 km of Darwin, were the largest in the west. Considering the large numbers that occur in north-west Western Australia,[228] it is perhaps surprising that the distribution of the species along the Top End coast is such.

Birds are also occasionally seen on the Magela floodplain[28] and Mamukala Wetlands,[3] in Kakadu NP. Given that the species' entry into Australia is likely to be from Sumatra and Java,[209] and that there are records from both east and west of Darwin, Asian Dowitchers could reasonably be expected to be more regular in the Darwin region, and it may be that some are overlooked.

Most Darwin records are of single birds, with exceptions including four at both Knuckey Lagoons in October 1993 and Buffalo Creek in September 1999,[21] and three near East Arm wharf in September 2013.[50] Individual birds have been observed at Lee Point, Leanyer SP and East Point.

Records in the Darwin region are from the period September to January, but there are several observations from elsewhere in the Top End in the period May to July,[195] suggesting that at least some birds overwinter.

Black-tailed Godwit
Limosa limosa

RANGE: subspecies *melanuroides*, which occurs in our region, breeds in e. Siberia; winters in se. Asia and Australasia. HABITATS: tidal mudflats, near-coastal wetlands. STATUS: moderately common late Dry, Wet season migrant. BREEDING: not in Australia.

Danny Rogers

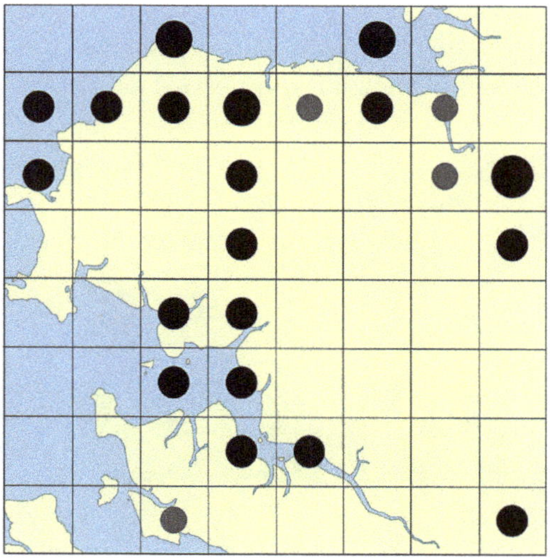

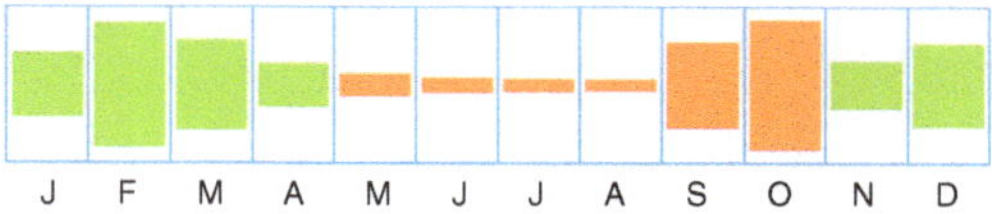

Black-tailed Godwits occur around most of the NT coast, particularly where there are extensive areas of mangroves and intertidal mudflats. From surveys in the 1990s[195] the Top End population of Black-tailed Godwits was estimated to be at least 44 000 birds, with the largest single count, of 5000 birds, at Buckingham Bay in the north-east Top End. The coast of Arnhem Land and the

Gulf of Carpentaria are the most important sites in Australia for this species.[209]

Counts in excess of 2000 Black-tailed Godwits on the Adelaide River floodplain in May and July[195] indicate that many birds overwinter in the western Top End; however, the species is virtually absent from the Darwin region from May to August, and birds do not start returning until September. Even then, numbers remain low until the second half of October and, given that coastal surveys of the Top End[195] found peak numbers south-west of Darwin during August and September, it may be that birds disperse north and eastward from there rather than arriving directly on Darwin's shores. A fall in numbers on the coast from early November suggests that birds largely move to wetlands on or soon after arrival. Coastal numbers peak from late December to March, and almost all coastal counts in excess of 500 birds occur during this period, perhaps indicating movement from inundated wetlands to the coast.

Although numbers build up from February to May in southern Australia, this is not the case in the Darwin region, where numbers fall sharply from late January. The fall is most marked on sandy beaches, but a small concurrent increase in numbers on Darwin's mangrove-fringed mudflats during January and February suggests that the fall is at least partly a result of a shift in habitat, and birds may disperse into similar areas in and outside of the region rather than moving elsewhere en masse.

Intertidal areas are reported to be their main habitat in Australia,[228] and the largest numbers of Black-tailed Godwits in the Darwin region have been recorded at Buffalo Creek with, for example, 1800 in January 1986 and 2000 in February 1987,[21] at Leanyer Swamp where 1600 were observed in March 1983,[34] and near the mouth of Mickett Creek where 700 were seen feeding at the edges of a brackish swamp in January 1996. At McMinns Lagoon, which is not tidally affected, over 1300 were present in early February 2005, and other high counts in excess of 700 have also been recorded there, all from late December through February.[3] Numbers at Holmes Jungle and Knuckey Lagoons are somewhat lower but there have been counts in excess of several hundred.

Bar-tailed Godwit

Limosa lapponica

RANGE: breeds in n. Europe, n. central and e. Russia, Alaska and nw. Asia; winters in NZ, Australia. HABITATS: tidal mudflats. STATUS: common late Dry, Wet season migrant. BREEDING: not in Australia.

Two subspecies of Bar-tailed Godwit migrate to Australia. Those occurring along the east coast are subspecies *baueri*, which breeds in Alaska and the Bering Sea coast of Chukotka, while birds visiting north-west Western Australia have been shown to be subspecies *menzbieri*, which breeds further west.[228] It is likely that most Bar-tailed Godwits in the Darwin region are also subspecies *menzbieri*. There is little information on subspecies in the Darwin region, but 12 of 100 birds scrutinised at

Micha V. Jackson

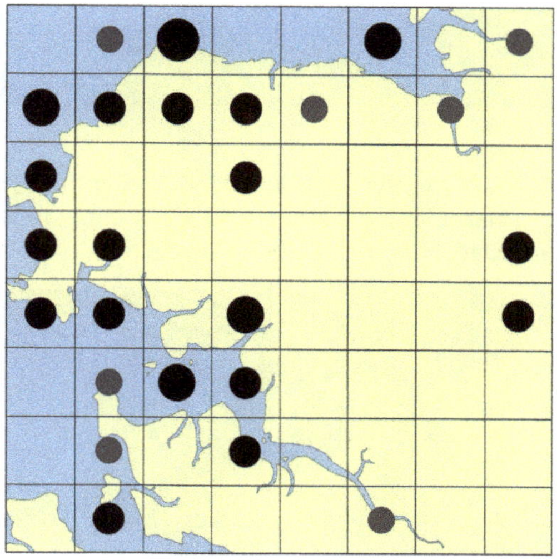

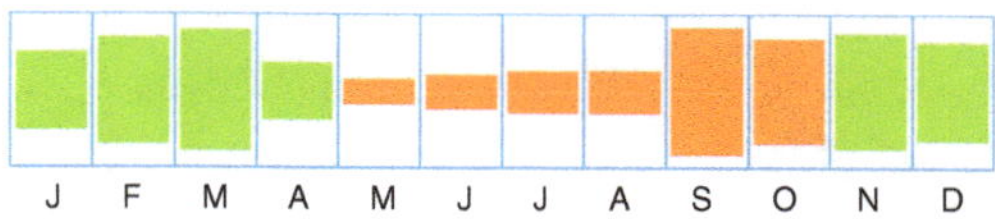

Nightcliff in February 1997[21] showed the darker rumps of subspecies *baueri*, and it may be that some birds move westward from areas in the eastern Top End, or perhaps that this subspecies reaches Australia on a broad front.

An estimated 53 000 Bar-tailed Godwits occur on Top End coasts,[195] making them the second most abundant shorebird, behind the Great Knot. A little over half of the Bar-tailed Godwit population occurs in the Darwin Coastal bioregion. The species is almost entirely confined to coastal sites, and in the Darwin region the greatest numbers occur along the sandy shore in the Lee Point area.

Following a period of low numbers in the Darwin region from May to July, Bar-tailed Godwits arrive during August, but mean counts[198] fall in September, suggesting dispersal, probably relatively local, shortly after arrival. Numbers subsequently build up again and peak in the period November through February. A gradual fall in numbers from late February suggests that the species undertakes return migration early, although this is in contrast with Bar-tailed Godwits in north-west Western Australia, which typically depart in early April.[228]

It has been shown that Bar-tailed Godwits from New Zealand and eastern Australia fly non-stop from their wintering areas to staging grounds in the Yellow Sea region of East Asia[229] on their northward migration. Consequently, a rise in numbers in the Top End during the period before migration most likely comprises local birds augmented by an influx of young birds, which move northward from south-eastern Australia after the adults have departed for their breeding grounds.[209]

Little Curlew
Numenius minutus

RANGE: breeds Siberia; majority of population winter in n. Australia. HABITATS: floodplains, grassland incl. urban grassed areas, croplands. STATUS: moderately common passage migrant, Sep-Dec. BREEDING: not in Australia.

David Webb

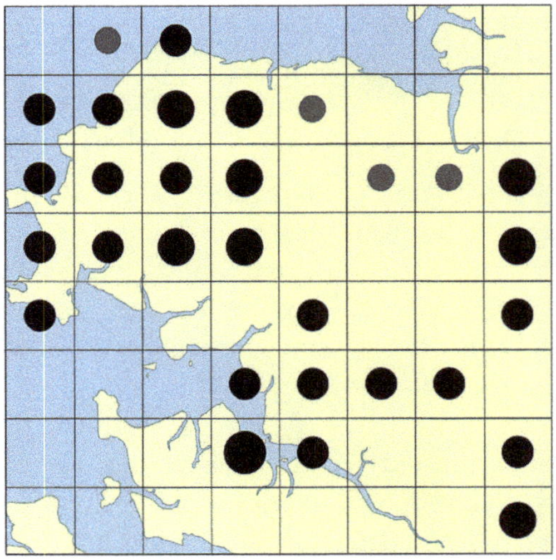

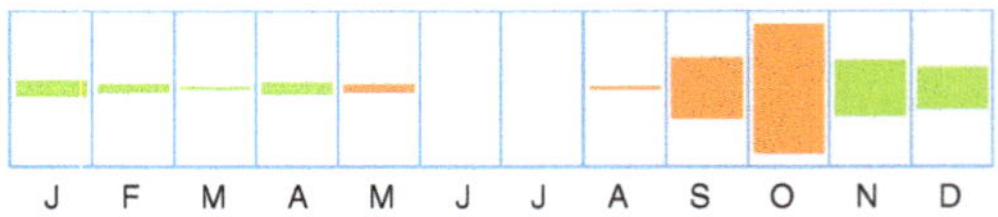

Little Curlews in the Darwin region exemplify the term 'passage migrant', as in most years they are present only from September to November. A single bird was seen at Leanyer SP on 10 August,[21] but typically the first sightings in the region occur from mid-September. After gradually increasing through October, numbers decrease dramatically, almost completely, with the onset of the first heavy Wet season rains. The birds disperse inland, probably to take advantage of newly created feeding habitat. The few January and even fewer February records in the Darwin region have occurred mainly in years of late or poor Wet seasons. Following their departure from the region, Little Curlews do not generally return; however, observations of large numbers in grassland such as near Middle Point after birds have left the Darwin region indicate movements may sometimes be only short distances before dispersal further inland.

Contrasting with the pattern of dispersal at the onset of the Wet season, some birds appear to be present for a very short time, perhaps resting and feeding for only a few days to recover from a non-stop flight from eastern China,[228] before moving on. At McMinns Lagoon in mid-October 1990, an estimated 30 000 birds – the highest number reported in the Darwin region – were present, but three days later only eight birds remained.[97] A similar phenomenon occurred in mid-September 1994 at Lake Finniss, 70 km east of Darwin, when an estimated 4000 birds, clearly recent arrivals, were seen to take flight, rising on thermals until lost from sight.[230]

Return migration of Little Curlews appears to be through north-west Western Australia and the Gulf of Carpentaria,[231] and only on rare occasions are there a few individuals, or minor influxes, into the Darwin region during that period. An influx during the period March to May 1985, for example, may have been a consequence of flooding in the Gulf of Carpentaria and Barkly Tableland due to tropical cyclone Gretel.[232] Similarly, high rainfall inland in early 2001 and 2005 may have caused birds to move to coastal areas. However, in other years for which there are March to May records, there are no obvious indicators of climatic influence.[124]

As evidenced by the scarcity of records of overwintering Little Curlews from anywhere in Australia, the entire population undertakes return migration. The two June and July records[31,106] from Darwin may involve birds unable to make the return migration due to sickness or injury.[233]

In October 1992 and 1993, single day counts covering parks and ovals in the Darwin urban area, along with Knuckey Lagoons, totalled about 2000 Little Curlews. Although similarly coordinated surveys were not conducted in other years, less systematic observations suggest these numbers to be typical, though in some years they appear much more numerous and in other years there may be relatively few.

In Darwin, Little Curlews are mainly recorded around swamps, both dry and cultivated grassland including at Darwin International Airport, golf courses, sports ovals and fields with ground crops. At Knuckey Lagoons, birds come in mainly from about midday to mid-afternoon to drink and rest, occasionally feeding in the surrounding grassland. Many also feed during the morning and later afternoon in the market garden areas around Berrimah, working the soft ground around crops probably for insects. Ninety per cent of the stomach items of birds collected at Darwin International Airport were termites, while the stomach contents of a bird collected in Kakadu NP were entirely seeds of wild rice.[234]

Whimbrel

Numenius phaeopus

RANGE: Australian migrants breed NE Siberia; winter e. India to Taiwan, se. Asia, Indonesia, Australasia. HABITATS: reefs and beaches, mudflats. STATUS: moderately common late Dry, Wet season migrant. BREEDING: not in Australia.

Ashley Carlson

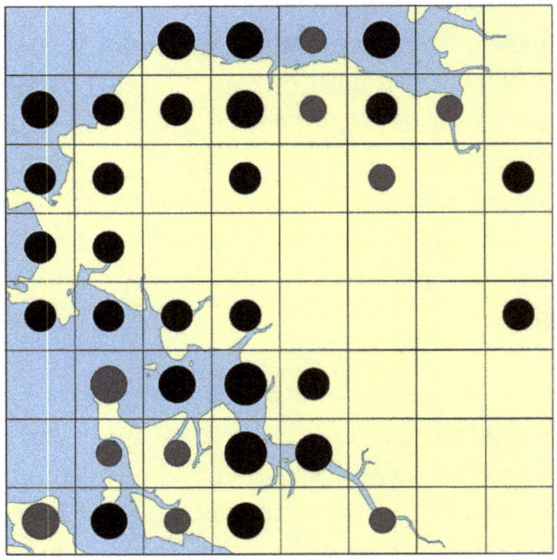

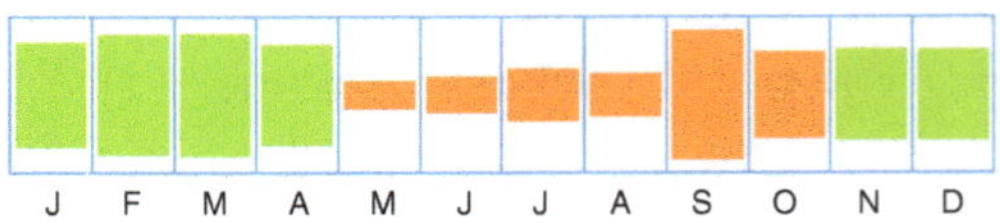

Most Whimbrels visiting Australia are of the subspecies *variegatus*, which breeds in east Siberia.[135] Of an estimated population of 5100 Whimbrels along the NT coast, the greatest numbers occur in the western Top End, including an estimated 1500 birds in Chambers Bay, east of Darwin.[195]

Whimbrels show a preference for sheltered coasts and mudflats with mangroves, and in the Darwin region this is certainly the case. The greatest concentrations have been found at a high-tide roost adjacent to Palmerston SP, where an estimated 300 birds have been observed during January and February.[21] That roost probably comprises birds that feed along the muddy, mangrove fringed edges of the Elizabeth River mouth and associated creeks, which provide the largest expanses of this habitat within the region. Although numbers are much smaller in the East Point and Nightcliff areas, which are also fringed with mangroves, Whimbrels are still found in greater abundances there than along the sandy shores around Lee Point. Indeed, even the relatively low numbers at Lee Point high-tide roosts are more likely drawn from the Buffalo Creek estuary than the open shore. A single bird seen at Knuckey Lagoons[3] is atypical, and appears to be the only observation in the region of Whimbrel away from coastal or estuarine habitats.

Unlike Far Eastern Curlews, Whimbrels show a classic pattern of migration, with numbers increasing from August to October, as birds arrive from the northern hemisphere, then remaining relatively constant until April, when they depart on their return journey. Numbers at Lee Point generally peak at this time,[3,198] suggesting that at least some birds gather to depart from there, but there is little variation in numbers around East Point during the months before departure. A few birds overwinter in the region, but although there is a record of 85 Whimbrels in June 1987,[21] there are few counts of more than 10 birds in the period May through July.[21,198] The consistency of numbers from October to April suggests that birds do not use the Darwin region as a staging point for onward migration, although it may be that some move on, and are replaced by late arrivals, or perhaps by birds moving east or west along the coast.

Movement out of large concentrations is reported to occur during December in much of Australia and it is assumed that birds spread out along the coast in low densities, flocking again before return migration.[209] While there is evidence of this across the Top End as a whole, it does not occur in the Darwin region, and indeed the peak in numbers tends to be in December or January.

Far Eastern Curlew
Numenius madagascariensis

RANGE: breeds in e. Russia and ne. China, with the majority of the population wintering in Australia. HABITATS: tidal mudflats. STATUS: moderately common late Dry, Wet season migrant. BREEDING: not in Australia.

Adrian Boyle

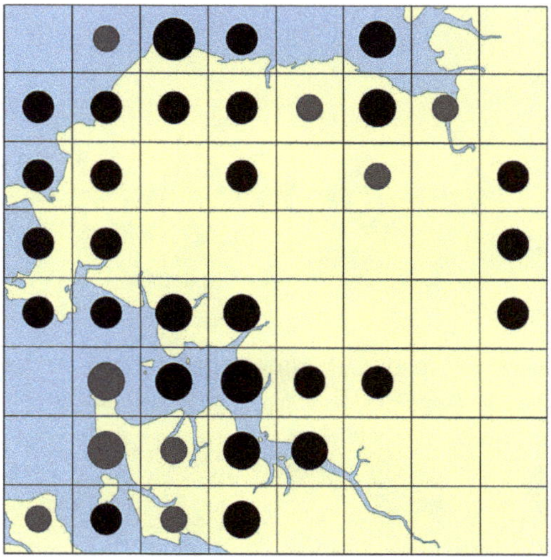

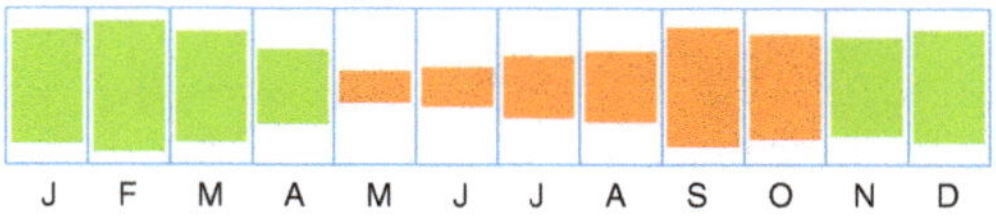

Far Eastern Curlews visiting Australia are reported to be from breeding grounds in south-eastern Siberia and Manchuria.[235] Departure from breeding grounds starts from early June, with most birds departing during July.[235] A sharp increase in numbers in the Darwin region during July and August, followed by a fall in September, suggests birds on passage to southern Australia, as this species, like some other shorebirds, is known to migrate over inland Australia.[235]

A maximum population of 6800 Far Eastern Curlews has been estimated for the coastal Top End,[195] the largest numbers occurring east of Darwin, particularly around north-east Arnhem Land.[195]A roost of 500 birds was found on Chambers Bay, only about 65 km north-east of Darwin,[195] but numbers in the Darwin region are much lower, and counts at high-tide roosts rarely exceed 40 birds. Indeed, an early February count of 70 at East Arm[3] is the highest single count for the Darwin region.

Throughout northern Australia, numbers of Far Eastern Curlews are highest in June and July, as immature birds, which may spend their first three years in Australia, move from southern areas to spend the Austral winter in northern Australia.[236] While that is also the case for the Top End generally, overall numbers in the Darwin region are higher during the Austral summer, although there is an apparent variation between habitats. At high-tide roosts on the sandy beaches in the Lee Point area, greater numbers occur in July and August, suggesting that this is an arrival point, but around East Point, where the shore is fringed by mangroves, numbers are highest during the period October to February.[21,198] While there is a small increase in numbers at mangrove areas in September, there is a sharp drop in numbers at Lee Point, so it appears that the majority of birds continue onward. October sees higher numbers again; these continue until January, after which they fall until May, when the species is virtually absent from the region. An increase in June probably represents an influx of immature birds, and this may continue into July, but the sharp rise in that month can be attributed largely, if not entirely, to new arrivals.

Common Redshank
Tringa totanus

RANGE: breeds w. Europe to central Asia; winters n. Africa, s. and se. Asia. HABITATS: tidal mudflats, wetlands. STATUS: rare late Dry, Wet season migrant. BREEDING: not in Australia.

Adrian Boyle

The first Australian record of Common Redshank was at Buffalo Creek in July 1970, and although 17 birds were recorded there a week later,[237] there appear to be records from only 10 of the years following the first sighting until 2014. Reports have been widespread, however, coming from Lee Point, Leanyer SP, Stokes Hill Wharf, Palmerston SP,[31] McMinns Lagoon,[51] and East Point.[40]

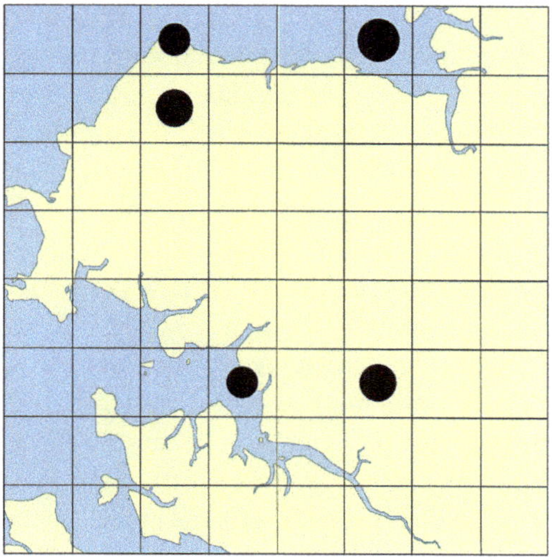

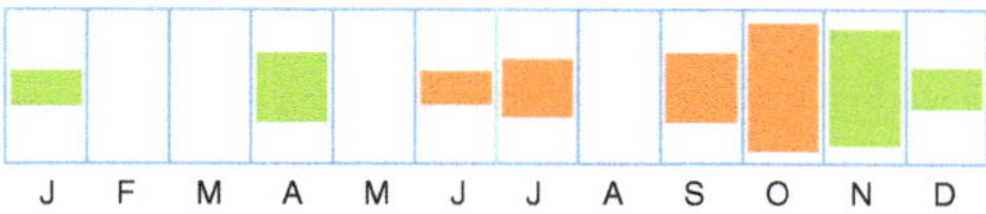

During surveys of the Top End coast in the 1990s Common Redshank were recorded on six occasions, from three areas. There were up to 30 birds recorded during April, September and December, at the western end of Chambers Bay,[195] only about 35 km to the north-east of our region. That area was also the only one where records of multiple birds were obtained. As it appears that the species is an annual visitor to the Kimberley coast,[228] perhaps the only area in Australia where this is the case, it is not surprising that the majority of Top End records are from western areas.

Marsh Sandpiper
Tringa stagnatilis

RANGE: breeds from e. Europe to central Asia; winters from Africa to Australia. HABITATS: wetlands, occasionally tidal mudflats. STATUS: moderately common late Dry, Wet season migrant. BREEDING: not in Australia.

Niven McCrie

Marsh Sandpipers depart breeding grounds from the first half of July to early September, with little difference between the timing of adult and juvenile departures.[228] Birds arrive in Australia from September and apparently move south across the continent from September–December.[209]

Coastal surveys of the Top End during the 1990s found Marsh Sandpipers to be more numerous in western areas, with high counts of 1600 in April 1992 near the mouth of the East Alligator River, 800 near the mouth of the Finniss River and 500 at two sites near the Adelaide River, in May 1993 and July 1998.[195] There were 'many hundreds' at Fogg Dam in October 1986.[21]

The Marsh Sandpiper is a common and widespread visitor to the Darwin region, but reports of more than two or three birds

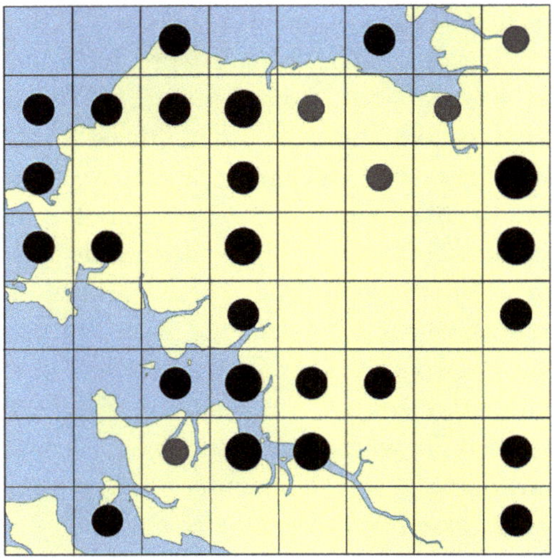

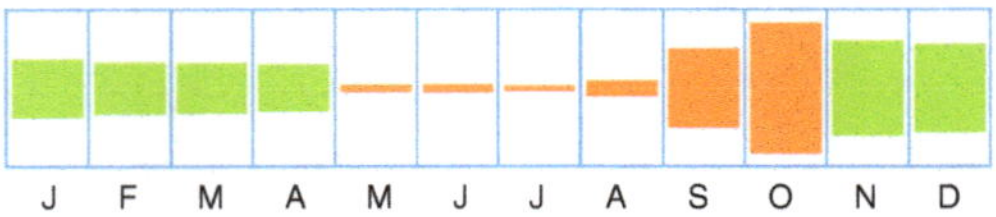

are largely restricted to freshwater wetlands, such as at Knuckey and McMinns Lagoons and Holmes Jungle Swamp. Numbers fluctuate markedly from year to year, with counts of up to 50 or more at various sites in some years, but at comparable times in other years no more than five or 10. The maximum count for the Darwin region was of 75 birds at Leanyer SP in May 1991, where at the time an emptied sewage pond provided ideal conditions.[21] Birds typically arrive from September, and numbers peak in October or November. This is usually followed by a sharp decline in late November or early December, perhaps prompted by the onset of heavy rains that reduce suitable habitat locally. Birds moving even relatively short distances inland would find suitable habitat newly created by the onset of the rains. Those remaining in the region at this time may make use of recently created pools, even small isolated puddles, or move to sites such as sewage ponds. There is a small influx into Darwin during late January and February, perhaps as birds begin to gather for return migration. Numbers drop quickly from March, and by the end of April most birds have departed. Although there are only moderate numbers present in May, there are generally fewer present in June. It may be that some are late for the return migration, but it is as likely that the fall in numbers represents the dispersal of overwintering birds.

Common Greenshank
Tringa nebularia

RANGE: breeds from w. Europe to e. Russia; winters from Africa to Australasia. HABITATS: tidal mudflats, reefs, wetlands. STATUS: moderately common late Dry, Wet season migrant. BREEDING: not in Australia.

The Common Greenshank is a regular visitor to Top End beaches, mudflats, wetlands, mangrove edges, saltpans and sewage works. Estimates of the number visiting the Top End were calculated at 7600 birds, with the highest single flock record of 500 observed in the south-eastern Top End in mid-October 1996.[195] Numbers in the Darwin region fall well short of that, however, and rarely exceed 20 birds at roost sites. A record of 150 birds at Palmerston SP in October 1983[31] is exceptional. Although occurring in a range of habitats, including sandy beaches, reefs, mangrove-fringed

David Webb

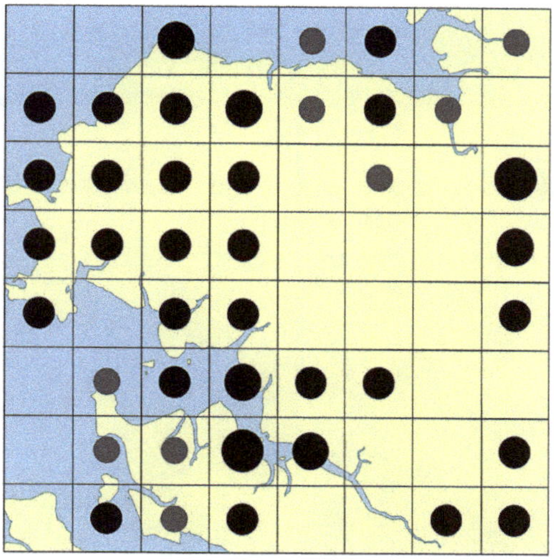

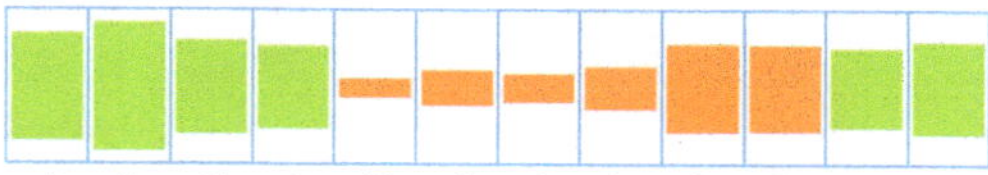

muddy shores and freshwater wetlands, it is in the two latter habitats where the greatest numbers are typically found. At wetlands, numbers peak in September and October, before falling sharply. This fall occurs also in north-west Western Australia, and apparently coincides with an increase in numbers in southern parts of the country.[209] It is also just before the onset of the Wet season, when such habitat becomes inundated and unsuitable for shorebirds. On muddy shores, numbers are lowest from February to May, then increase markedly in June and remain relatively consistent until January, when they fall sharply. While numbers are lower on sandy beaches such as the Lee Point area, there is a tendency for them to peak in October and February, suggesting the arrival of migrants and perhaps, but surprisingly early, departure.

There is, however, no apparent increase in total numbers from January through April to indicate birds gathering in preparation for northward migration. Some birds remain during the breeding season; however, counts in May are typically the lowest for the year in all habitats.

Green Sandpiper
Tringa ochropus

RANGE: breeds w. Europe and Asia; winters Africa to se. Asia. HABITATS: wetlands. STATUS: vagrant. BREEDING: not in Australia.

David Fisher

The Green Sandpiper occurs commonly in south-east Asia and is considered to be an uncommon visitor as far as Indonesia, at least in Sumatra.[238] It is perhaps surprising then that there is only one confirmed record for Australia. This bird was found foraging in a rainwater puddle at the Shoal Bay Waste Management Facility in March 1998.[239] It was observed for a short time, and on flushing returned briefly, but subsequently flushed again and was not relocated.

A Green Sandpiper was reported from Kapalga, Kakadu NP, in September 1979,[240] but the documentation contained insufficient evidence to confirm its identity.[241]

Wood Sandpiper
Tringa glareola

RANGE: breeds from w. Europe to e. Russia; winters in Africa, s. and se. Asia, Australia. HABITATS: edges of wetlands. STATUS: moderately common late Dry, Wet season migrant. BREEDING: not in Australia.

Niven McCrie

Wood Sandpipers may begin arriving in the Darwin region as early as late July, though the major influx typically occurs from late August through to October, with numbers peaking in that month. At this time lagoons in the region are generally shallow with muddy and grassy edges, providing prime habitat. In some years, however, early rains inundate the muddy edges of local wetlands and the subsequent rapid growth of vegetation makes the habitat unsuitable, and in those years numbers may fail to build up, as birds disperse early and later migrants do not stop over. Typically though, the onset of heavy rains occurs from November or

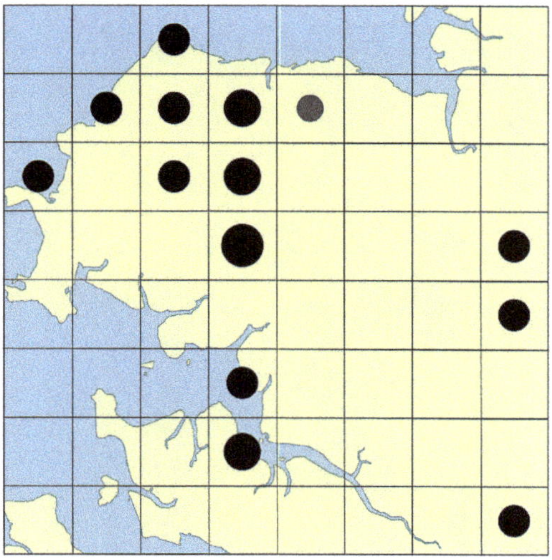

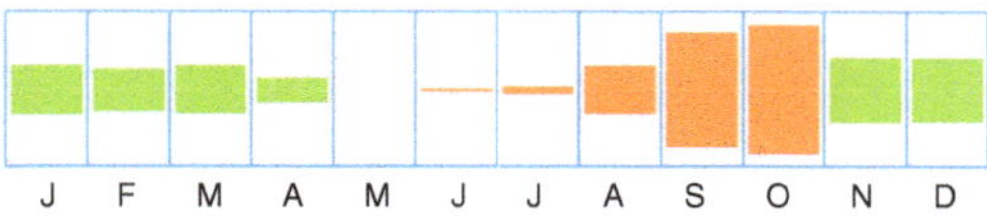

December, and it is then that many birds are forced to move. Some disperse further inland, while those that remain move to newly flooded grassy areas, sewage ponds, or even to roadside rain puddles. In Wet seasons of high rainfall there may be no sightings of Wood Sandpipers from the first major rains until a few are seen on return migration northward.

During February and March, birds that have spent the Wet season in the Darwin region are joined by birds stopping over on their way north, although numbers never reach those of the southward migration. They depart by the end of April and there are almost no records between then and late July. This is perhaps to be expected, as Wood Sandpipers breed in their first year[201] and so the entire population returns to their breeding grounds.

Up to 300 Wood Sandpipers were recorded at Fogg Dam in October 1986, but the highest single counts in the Darwin region are of about 50 birds, recorded around Knuckey Lagoons and Holmes Jungle Swamp. Numbers at Leanyer SP rarely exceed about six individuals, but larger influxes may occur when ponds are drained, increasing the availability of suitable habitat.

Grey-tailed Tattler
Tringa brevipes

RANGE: breeds Siberia; winter se. and se. Asia, Australasia, Fiji, Tuvalu. HABITATS: tidal mudflats, reefs. STATUS: moderately common late Dry, Wet season migrant. BREEDING: not in Australia.

During the 1990s a population of at least 16 000 Grey-tailed Tattlers was estimated for the Top End,[195] with some 70% of these found in the eastern half. The highest counts were in the period May to July, indicating that large numbers overwinter in the Top End. The pattern is similar in the Darwin region, where mean counts show an increase from May to a peak in July,[198] although maximum counts occur somewhat later, from September to November. The July peak may indicate a gathering of overwintering birds, perhaps supplemented by the arrival of early migrants. Lower counts from early August show that some birds disperse, and it may be that birds stay in the region but disperse to mangrove-fringed mudflats that are not included in regular counts.

The highest count in the region was 250, in the Lee Point area in early October 1983,[31] but 400 were present in mid-September 1993[195] on an island off Bynoe Harbour, just

Niven McCrie

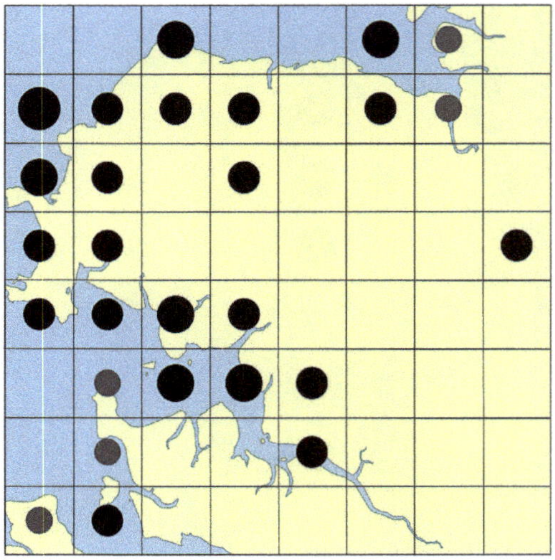

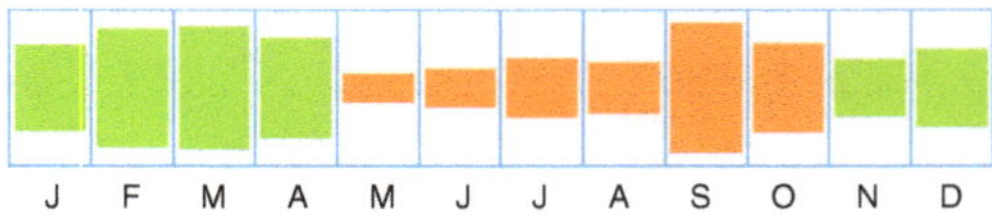

outside of the region. These high counts, when typical maximum numbers in the Darwin region do not exceed 100 birds, may indicate that many Grey-tailed Tattlers stay for only a short time, and go unrecorded, before continuing south or east, or it may be that the species does not regularly use the Darwin region as an arrival point.

Like Whimbrels, Grey-tailed Tattlers are most often seen on mudflats fringed by mangroves, or along reefs. They are often seen individually, but small loose groups of up to about 20 individuals also occur.

At night Grey-tailed Tattlers frequently perch to roost, often in mangroves, and sometimes on boats or marine structures, including an observation of up to 80 birds roosting on the diagonal supports of Stokes Hill Wharf in February 1989.[21] Grey-tailed Tattlers are occasionally seen on beaches late in the evening, and may roost there or move to other roosting sites after dark.

Terek Sandpiper
Xenus cinereus

RANGE: breeds from e. Europe to e. Russia; winters from w. Africa to Australia. HABITATS: tidal mudflats, reefs. STATUS: moderately common late Dry, Wet season migrant. BREEDING: not in Australia.

Terek Sandpipers visiting north-western Australia are most likely from the same population as eastern Indonesian birds, and probably breed farther north than the slightly smaller birds that visit western Indonesia, south-east Asia[209,242] and southern and eastern Australia.[243] Adults leave breeding grounds from the first half of July, females before males, with juveniles mainly leaving from August.

An estimate of at least 15 000 Terek Sandpiper was made from data gathered during surveys of Top End coastal areas during the

Adrian Boyle

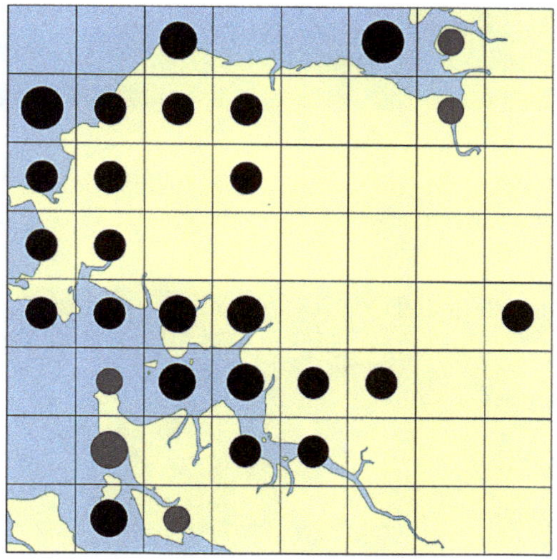

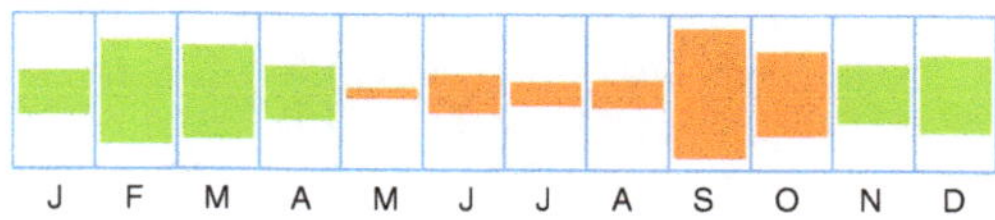

1990s.[195] The largest numbers, including a flock of 1000 birds, were found in western areas. Larger groups tended to be in the period June through August, and eight of 19 single flock counts in excess of 300 birds were in June and July.[195] In the Darwin region also, mean counts are generally highest during those months, and the largest count for the region, 120 birds, occurred in July 1981.[244] There were 50 to 60 birds wintering at Camerons Beach in July 1967.[96]

In mangrove areas such as at Palmerston, high counts are still made during the months December to March and, as well as gatherings at high-tide roosts there, groups of up to 50 birds have been observed resting in the shade of mangroves during the middle of the day.[21] Large numbers, 70 or more, can also be seen around Darwin Harbour, and many roost at night on wharf supports.

Aggressive behaviour, mainly chasing, appears to be most noticeable in February and March, perhaps as a consequence of the need to feed more intensively to build up pre-migration reserves.

Common Sandpiper

Actitis hypoleucos

RANGE: breeds w. Europe to e. Russia; winters Africa, s. and se. Asia, Australasia. HABITATS: coastal and terrestrial wetlands, rivers, creeks. STATUS: common late Dry, Wet season migrant. BREEDING: not in Australia.

Niven McCrie

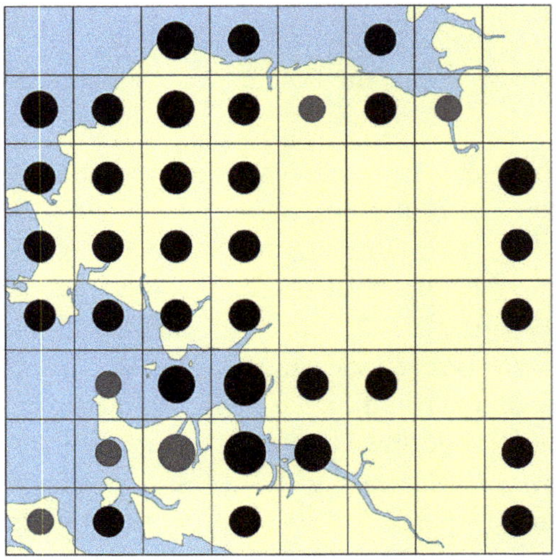

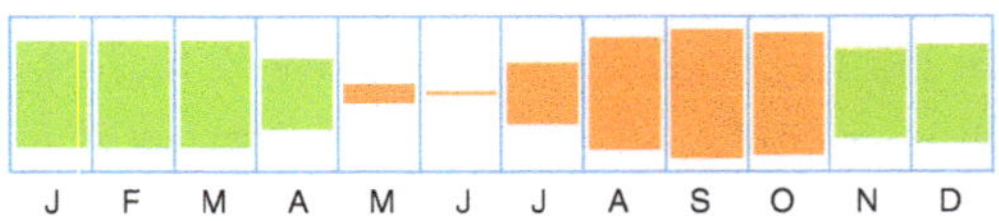

In the Darwin region, unlike the case in southern and eastern Australia, the Common Sandpiper can legitimately be termed common. It is a widespread visitor, seen on reefs, freshwater swamps, drains, sewage ponds, at mangrove edges and along saline or freshwater creeks. At high tide birds may also be seen perched in mangroves or on fences at sewage works, and in Darwin Harbour they also roost at night on boats or on marine structures such as jetty pylons. With the sealing of the tracks in Leanyer SP, numbers of Common Sandpiper seem to have increased in line with a corresponding reduction in other wader species. Total numbers of Common Sandpipers in the Darwin region are difficult to assess, as many are likely to be spread out around the mangrove-lined harbour and creeks. Nevertheless, Darwin has been recognised as a site of national significance for the Common Sandpiper, being ranked third in the country.[245] Kakadu NP holds somewhat more individuals and is considered the most significant national site for the species.[245]

Although an estimated 3000 birds remain in Australia during the non-breeding period,[206] Common Sandpipers are only rarely recorded in the Darwin region in May, and almost never in June, making them one of the few shorebird species for which the entire population departs the region at the time of northward migration. Whether all birds return to breeding grounds or whether some disperse to over-wintering areas is unknown.

On southward migration, birds typically begin arriving in mid to late August, sometimes late July, but numbers do not peak until at least October, and in some years as late as December. The highest counts, some in excess of 100, are mostly from the period August to November. A small but noticeable drop in numbers occurs after December, and it may be that some birds disperse as particular habitats become inundated by Wet season rains. Numbers in January and February are generally consistent, but it is clear that some birds depart during March, and by the end of April they have all but moved out.

During the day Common Sandpipers are usually seen individually, but in the evening birds may roost in larger groups. Gatherings of up to 150 birds have been seen late in the afternoon at Palmerston SP, and night-time surveys have shown that many birds roost on the tracks; many may also roost in the adjacent mangroves. At Buffalo Creek birds have been observed to arrive at dusk, and then take up presumed roosting positions in depressions, such as footprints, in the sand.

At coastal sites where very high tides inundate all of the shore, birds move to high ground, sometimes considerable distances from the shore. In Stuart Park, a Common Sandpiper was seen feeding among a group of Peaceful Doves on the ground under a Banyan Fig at the edge of a sports oval, several hundred metres from shore.[21]

At Nightcliff, Common Sandpipers are frequently seen on grassy areas above the high-tide line, occasionally feeding near rubbish bins, though whether they are scavenging scraps or feeding on invertebrates has not been determined. Birds are often seen resting in shade on hot sunny days, and one at Leanyer SP was observed taking refuge from the heat inside a large concrete pipe.

The Common Sandpiper appears to be one of the few shorebirds that display territorial or agonistic behaviour, beyond simple chasing, during the non-breeding season.[246] Examples of aggression have been observed against several other species, including the Eastern Yellow Wagtail and Long-toed Stint, but most of it, and perhaps the most violent, appears directed at other Common Sandpipers. In two observed instances the aggressive events culminated with the aggressor mounting the other bird, as if to mate, but it is uncertain if that act took place. Aggressive events have been most frequently observed in the period after arrival, and in the period before return migration, suggesting that competition for food, both in replenishing reserves after migration and gaining fat for return migration, is the cause.

Ruddy Turnstone
Arenaria interpres

RANGE: breeds along coasts of Eurasia and n. America; winters on coasts of all continents except Antarctica. HABITATS: coast; mainly rocky shores and reefs. STATUS: moderately common late Dry, Wet season migrant. BREEDING: not in Australia.

Ruddy Turnstones begin arriving in the Darwin region from August, with the sharpest increase in numbers in September. Mean monthly counts in the region[198] show a small fall in numbers in late October and November, indicating continued movement to southern and eastern Australia, where there is an associated increase at that time.

Niven McCrie

Supporting this, sightings of Ruddy Turnstones with leg flags have, with only one exception, been restricted to the period September to December.[3] The majority of these sightings have involved birds flagged in South Australia, Victoria and Tasmania, suggesting that they fly overland to those areas, but bypass or fly over the Darwin region on their return migration.[209] A subsequent increase in numbers during December and January may be the result of local movements, or perhaps late arrivals.

Up to 200 Ruddy Turnstones were recorded in September 1979[34] and 267 were counted during February 1983,[247] but roosts typically peak at fewer than 60 birds. The largest numbers occur around the reef at East Point, with slightly fewer in the Lee Point area and regular but smaller numbers at Nightcliff and around Darwin Harbour. There are occasional records at sewage ponds but these usually comprise only one or two birds.

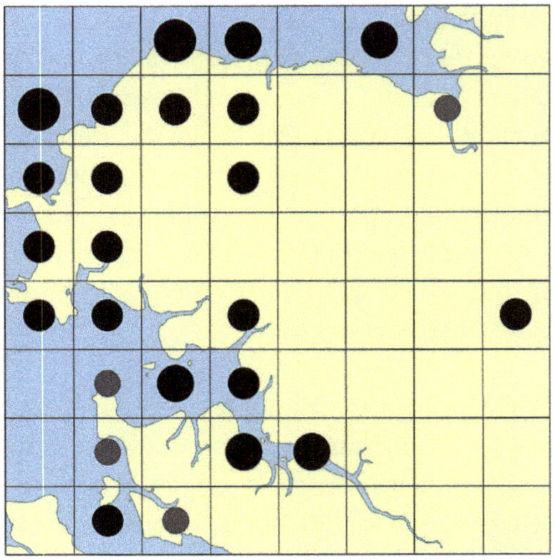

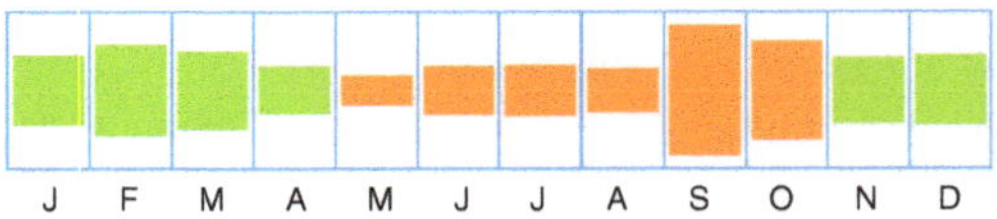

Numbers in the Lee Point area drop sharply by March, while at East Point a similar fall is not seen until late April or May. A small influx in June probably comprises overwintering birds gathering from elsewhere, though it may involve the early return of failed or non-breeding birds.

At high tide the majority of Ruddy Turnstones roost on exposed reef or sand along with other shorebirds, but in Darwin Harbour many birds perch on the wharfs, and sometimes on boats. During very high tides, small numbers may move some distance from the water to feed in grassy areas, and occasionally they are seen scavenging around rubbish bins.

Great Knot
Calidris tenuirostris

RANGE: breeds e. Russia n. of the Kamchatka Peninsula; majority of the population winter in Australia. HABITATS: coast. STATUS: common late Dry, Wet season migrant. BREEDING: not in Australia.

Adrian Boyle

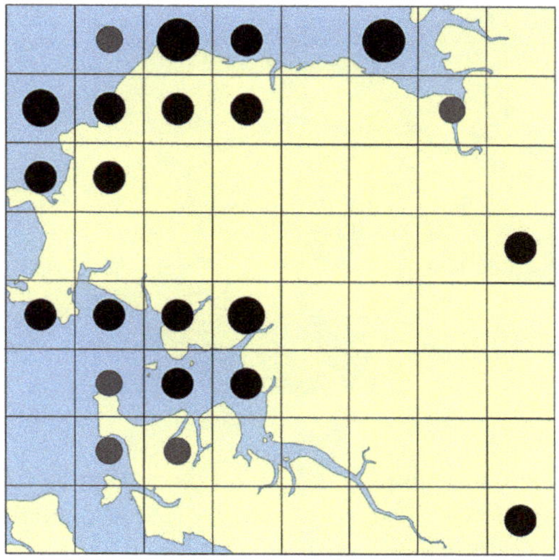

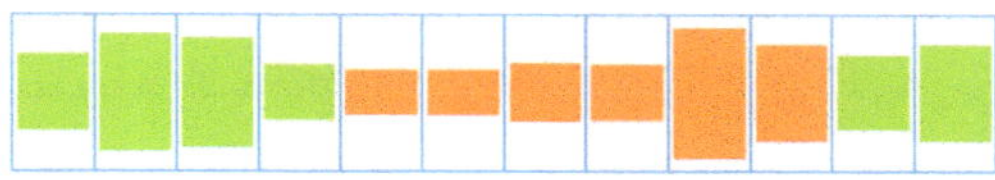

From surveys conducted during the 1990s,[195] the population of Great Knots was estimated to be in excess of 120 000, making it by far the most abundant of the migratory shorebirds that visit the Top End coast. During those surveys, March peaks suggested a pre-departure increase; however, there were no corresponding peaks noted in the August to October period, suggesting that birds may disperse quickly on arrival or perhaps arrive over an extended period of time.

In the Darwin region a sharp rise in numbers occurs during late September and October, indicating this to be the main arrival period, although numbers do not normally peak until December[3,21,198] and sometimes later. The maximum count, of 7800 birds, occurred at Buffalo Creek in February 1990. Great Knots depart the region during the second half of March, and fly non-stop to staging areas in the north Yellow Sea region of China.[248] Those remaining into late April and May are likely to be first-year birds that will stay in the area throughout the northern breeding season. These are sometimes in quite large numbers, with many counts of over 1000 birds in the period May to June.[21] A small increase in numbers of Great Knot during July and August may be failed breeders and perhaps a build-up of overwintering birds.

The largest roosts are on the sandy shores in the Lee Point area. Smaller numbers, rarely in excess of 100 birds, may also be present on the reefs at Nightcliff and East Point, or scattered in a few small pockets around Darwin Harbour, and there is a single sighting from McMinns Lagoon.[3]

Red Knot
Calidris canutus

RANGE: breeds in scattered areas across the Arctic; winters coastal Africa, S. America, Australasia. HABITATS: tidal mudflats. STATUS: moderately common late Dry, Wet season migrant. BREEDING: not in Australia.

Adrian Boyle

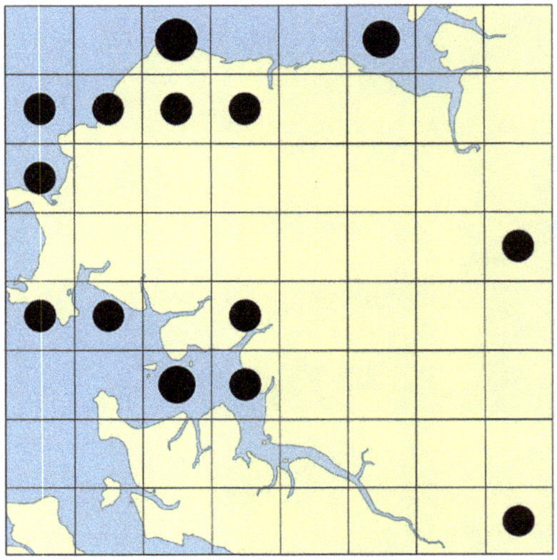

Two subspecies of Red Knot winter in Australia. The subspecies *rogersi* breeds in north-east Siberia, while the more recently described subspecies *piersmai* breeds farther west, on the New Siberian Islands.[249] It appears that individuals of both subspecies occur in the Darwin region, but the majority are likely to be *piersmai*.[250] Identification in the field may only be possible with birds in full, fresh breeding plumage,[249] however, making assessment of proportions of each difficult.

Of an estimated Top End population of 24 200 birds,[195] about one-third occur in the Darwin Coastal bioregion, where the greatest numbers are found between the Daly River and the islands off Bynoe Harbour, particularly the Fog Bay area,[195] south-west of Darwin. The largest concentrations in the Top End, however, occur well east of Darwin, in the western Gulf of Carpentaria. The large gap between concentrations in the eastern and western Top End suggests two populations, perhaps with subspecies *piersmai* dominating in the west and subspecies *rogersi* in the east, although the September capture of a New Zealand-banded Red Knot in Fog Bay suggests there is some overlap.[195]

Across much of the Darwin Coastal bioregion, birds begin to arrive in July, and numbers reach a peak in August and September. In the Darwin region, however, there is no evidence of an August influx, and numbers rise from September until a peak in November. A fall in numbers in December, suggesting local dispersal, is followed by a marked increase in January, and numbers remain high between then and April, with the greatest numbers usually in March.[3,198] Maximum numbers are typically around 200–400 birds, but there were counts of

2000 birds at Lee Point, in April 1983[34] and January 1997.[21]

Red Knots are in the greatest numbers at high-tide roosts on the sandy beach at Lee Point. They also occur in mangrove-fringed mudflats and around brackish swamps, such as at Leanyer Swamp, and in November 1970[1] and October 1986[21] there were birds present at Fogg Dam.

Sanderling
Calidris alba

RANGE: breeds in the high Arctic; winters in the Americas, Africa, Eurasia and Australia. HABITATS: coast; sandy beaches, tidal mudflats. STATUS: uncommon late Dry, Wet season migrant. BREEDING: not in Australia.

Bastiaan Hensen

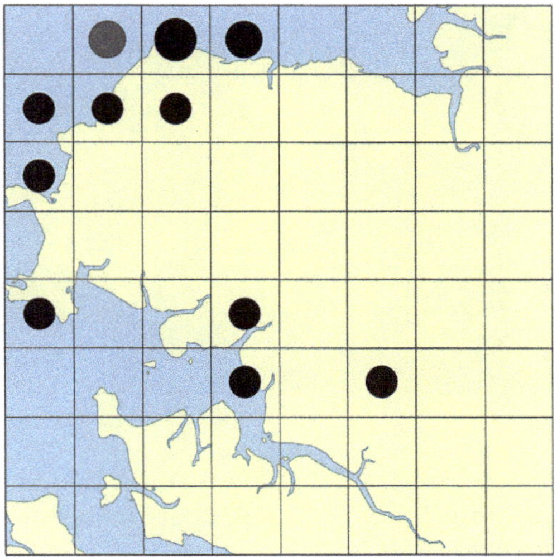

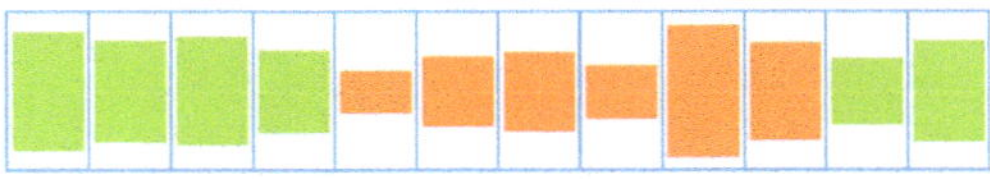

During Top End coastal surveys in the 1990s Sanderlings were recorded in only moderate numbers, with a calculated estimate of 890 birds for the Top End. Most birds were found in the western area, including Darwin.[195]

Sanderlings begin arriving in the Darwin region during August, but an almost immediate fall in numbers suggests that the earliest arrivals are staging before continued southward migration. This is supported by observations in the period August to October of Sanderlings with leg flags, attached in South Australia.[3] Numbers peak in the period late January to March, probably as birds from southern Australia stop over on their northward migration, although there have been no reports of leg-flagged birds at this time.

Sanderlings are almost entirely restricted to the northern beaches from Casuarina Beach to Lee Point and Buffalo Creek, with a few birds present on other sandy beaches, and small numbers occasionally at high-tide roosts on reefs such as at Nightcliff or East Point. Maximum counts are of about 200 birds, at Lee Point in December 1984[34] and early April 2005.[3]

A record of 25 Sanderlings near Buffalo Creek in mid-June 2003[3] suggests that some overwinter in the area but observations during that period are infrequent, and generally comprise fewer than 10 individuals. There were 96 birds recorded in July 1987, but these were likely to have been early returning birds in addition to any that had overwintered.

Red-necked Stint
Calidris ruficollis

RANGE: breeds ne. Siberia, n. and w. Alaska; winters mainly Australasia. HABITATS: mainly coastal mudflats, occasionally terrestrial wetlands. STATUS: moderately common late Dry, Wet season migrant. BREEDING: not in Australia.

David Webb

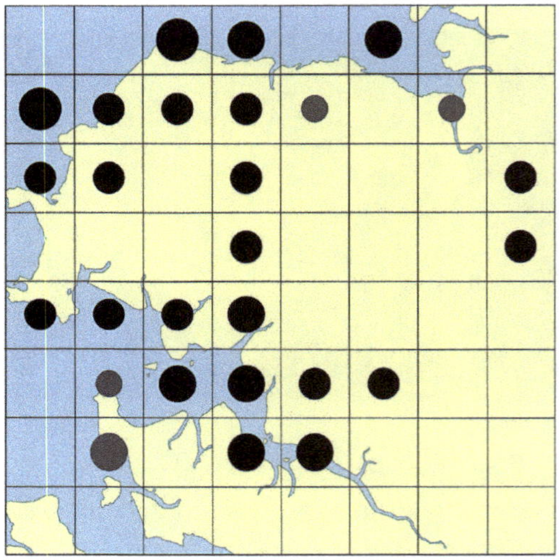

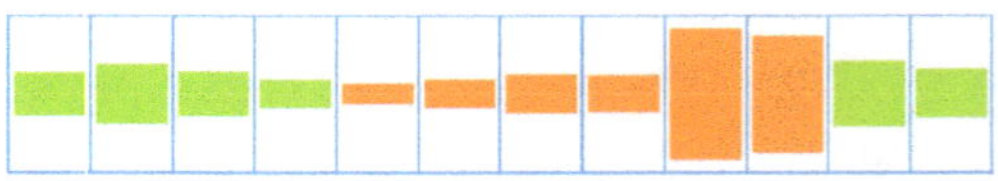

Although Red-necked Stints appear to be the most abundant migratory shorebird visiting Australia,[209,245] their numbers in the Darwin region are very small.

An estimated population of 44 400 Red-necked Stints occurs in the Top End,[195] with about 40% occurring in the Darwin Coastal bioregion, including the Top End's largest single ground count of 1500 birds, in Fog Bay in late December 1992.

In the Darwin region, most concentrations of Red-necked Stints are found on the coast around Lee Point, with small numbers scattered around other coastal areas or at sewage works and lagoons. Maximum counts for the region are 558 in October 2006 and 545 in October 2008.[3]

A sharp rise in numbers in June may comprise a concentration of overwintering birds, although it may be also be failed breeders returning early. New migrants typically arrive in the region from late July and numbers rise gradually until a sharp increase in September, before peaking in October or November. With few years showing an exception, there is a subsequent fall in numbers in December, continuing through to March. Some birds may have continued on southward migration, but by February and March the fall in numbers more probably reflects local dispersion. Although it has been suggested that birds from southern Australia fatten up only enough to reach northern Australia on the first leg of their northward migration,[209] there appears to be only a small increase in numbers in the Darwin region during April. In some years large numbers overwinter in the region. There were 240 birds at Leanyer SP in June 1991,[31] and other counts of up to 150 have been made at Lee Point between late May and the first week of July.[21]

Little Stint
Calidris minuta

RANGE: breeds Eurasia to Siberia; winters Africa, s. Asia. HABITATS: tidal mudflats, wetlands. STATUS: vagrant. BREEDING: not in Australia.

Adrian Boyle

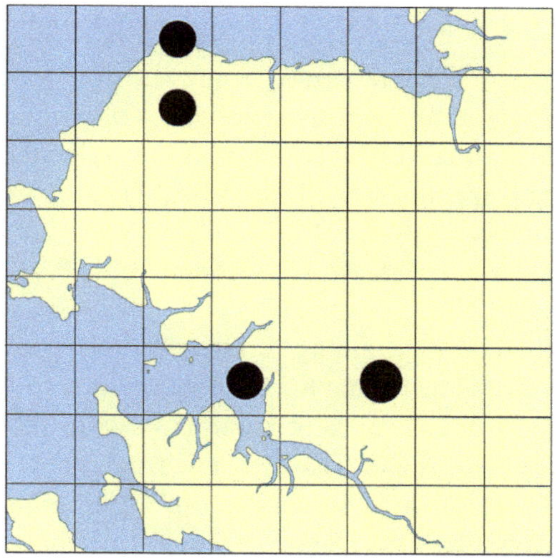

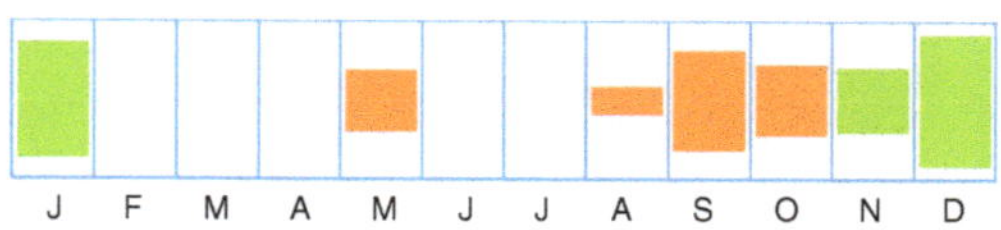

The first Little Stint recorded for the Darwin region was found, and subsequently collected as a specimen, at the edge of a brackish pond near Leanyer SP in November 1979.[251] Six other sightings[228] were reported between then and 1987, although there appear to be no details available to support the birds' identification. In October 2009 a Little Stint was seen by several observers, and photographed, foraging along with Red-necked Stints and Sharp-tailed Sandpipers on an area of floating matted vegetation at Palmerston SP.

Long-toed Stint
Calidris subminuta

RANGE: breeds patchily from central to e. Russia; winters from India to Australia. HABITATS: wetlands. STATUS: rare late Dry, Wet season migrant. BREEDING: not in Australia.

Adrian Boyle

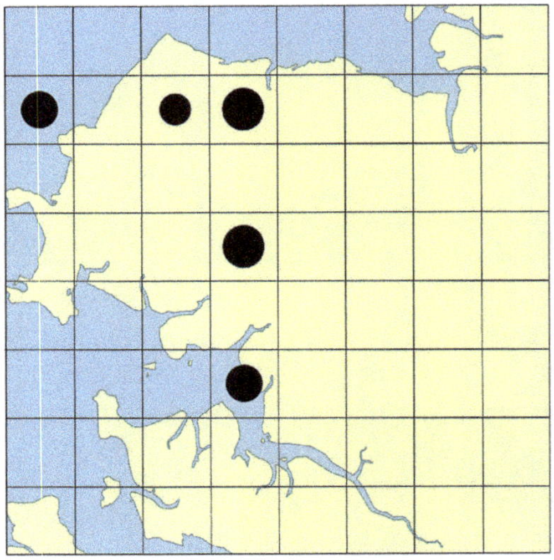

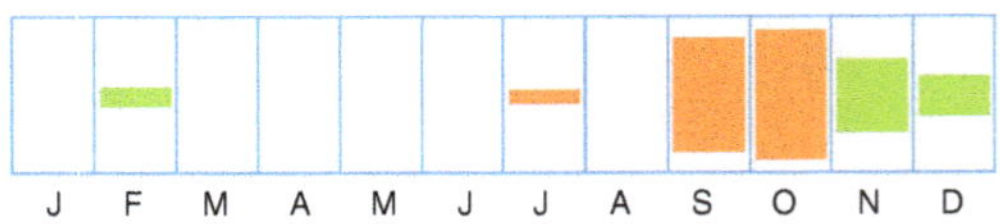

The first NT record of Long-toed Stint was at Harrison Dam in October 1969, with a further record of two birds, one collected as a specimen, at Beatrice Hill lagoons in October 1971.[1] The species was first recorded in Darwin in October 1974.[129]

Prior to its draining, Leanyer Swamp provided ideal habitat for Long-toed Stints, and most occurrences of multiple birds in the Darwin region have been from there. These records included up to 50 birds in December 1983[34] and seven in December 1981.[31]

At Leanyer SP birds have been recorded on dense matted vegetation on the surface of ponds, or in near-empty ponds, when water is shallow and there are extensive areas of mud and short vegetation. In October 2003, one bird was observed being harassed by Common Sandpipers. The bird responded with wing lifting, lowering its breast to ground level and putting its tail straight up. Other records from there include single birds in September 1999, September 2008 and October-November 2011. There is a single record from Palmerston SP, in September-October 2005.

Records from Knuckey Lagoons include two birds in September-October 1994, October-November 1998, October 2003[3] and October 2005, and single birds in November 1992 and November 2004.

With very few exceptions, records of Long-toed Stints are from the period September to November, suggesting that birds disperse following the onset of heavy rains. Birds may move to wetlands that are not yet inundated, or to suitable habitat newly created by the rains.

Baird's Sandpiper
Calidris bairdii

RANGE: breeds Siberia and N. America, winters S. America. HABITATS: terrestrial wetlands, occasionally coast. STATUS: vagrant. BREEDING: not in Australia.

Adrian Boyle

The single confirmed record of Baird's Sandpiper in the NT involved a bird seen feeding among Red-necked Stints in a drained sewage pond at Palmerston in late October 1983.[252]

Pectoral Sandpiper
Calidris melanotos

RANGE: breeds ne. Russia, w. and n. Alaska, n. Canada. HABITATS: wetlands. STATUS: rare late Dry, Wet season migrant. BREEDING: not in Australia.

Mat and Cathy Gilfedder

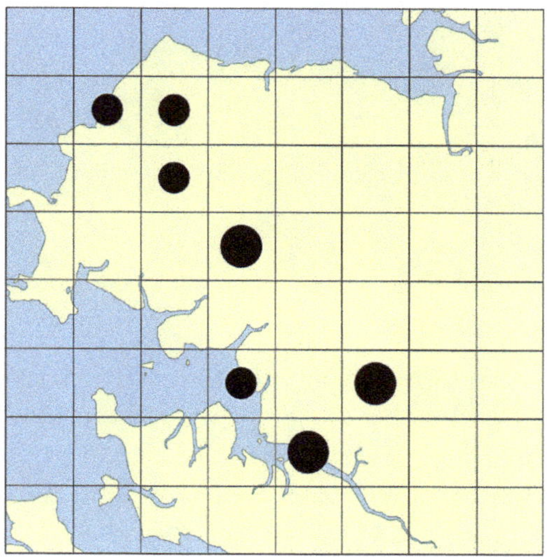

The Pectoral Sandpiper was first recorded in the NT in September 1967, when a bird was found at Beatrice Hill, near the Adelaide River floodplain.[1] The brief description accompanying a subsequent report of five and 14 Pectoral Sandpipers at Lee Point, in October 1969 and January 1970, respectively,[1] is insufficient to preclude Sharp-tailed Sandpiper. There were reportedly 17 records in the Darwin region between 1967 and 1989[228] though details on only a few of these are available. Interestingly, none of the published summaries of uncommon and rare migrants in the Darwin region included a single reference to Pectoral Sandpipers.[2,52,129,251] Consequently, the status of the species is uncertain.

Sightings that can be considered confirmed include two birds at Knuckey Lagoons in October 1993[26] and singles at Knuckey Lagoons in October 1994 and 1997, and November 1998 and 2004;[21] and at Leanyer SP in November 1982, October 1985[24] and September 1999.[21] A record of three birds at Palmerston SP in January 1984[24] is the largest number recorded.

Sharp-tailed Sandpiper
Calidris acuminata

RANGE: breeds ne. and n. Siberia; majority of population winter Australasia. HABITATS: tidal mudflats; wetlands. STATUS: moderately common late Dry, Wet season migrant. BREEDING: not in Australia.

Niven McCrie

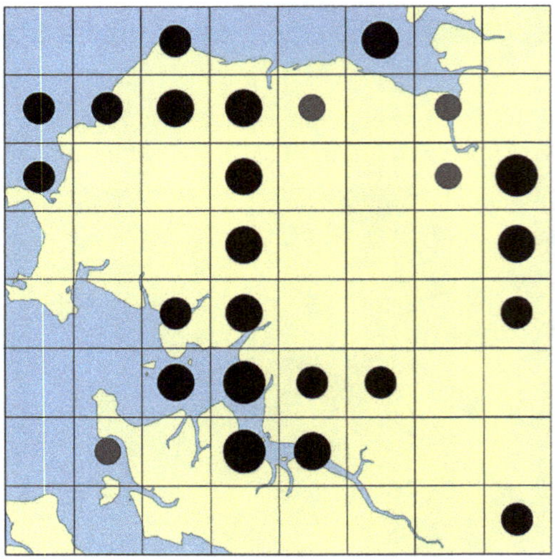

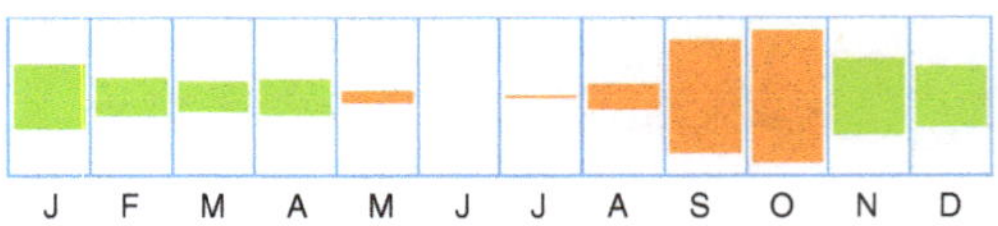

The largest numbers of Sharp-tailed Sandpipers for the Top End occur in the Daly and Finniss River areas to the south-west of Darwin, in the area east of Darwin to Murgenella Creek, and in the Port McArthur area on the Gulf of Carpentaria.[195] From survey data in the 1990s it was estimated that at least 20 100 birds may occur in the Top End.[195] Surveys of Kakadu NP wetlands in the 1980s[28] recorded the species as most abundant from September onwards, and moving out with the first heavy rains.

Birds arrive in Darwin from early to mid-August with inward migration numbers peaking during the period September to early November, after which numbers quickly decrease, so that by late December few remain. Some birds may simply stage briefly in the region and move on, but it is more likely that birds are forced out by the onset of the Wet season. As rains set in and suitable habitat becomes scarce, birds move out, perhaps to areas newly created by the rains, or to the larger wetlands that take longer to flood, or even to southern and eastern Australia. There is a small influx of birds in April, which could be attributed to local birds gathering for migration, as few Sharp-tailed Sandpipers from southern or eastern Australia pass through Darwin on their northward migration.[209] There are no records of birds overwintering.

Despite the high numbers of Sharp-tailed Sandpipers recorded west and east of Darwin, the species occurs only in relatively small numbers in the Darwin region. The two highest counts, of 500 and 700 birds, were from Leanyer Swamp, in November 1978 and October 1983, respectively.[34] Leanyer Swamp has since been drained, and there is little suitable habitat remaining. Outside of that area, up to 300 were present at Holmes Jungle Swamp in September 1977[31] and 398 at McMinns Lagoon on 18 Oct 2009,[3] but numbers are usually significantly lower than this, with rarely more than 100 found at any site. At Leanyer SP, birds are usually restricted to patches of floating vegetation at the edges of the ponds, but when water levels are very low, exposing expanses of mud and emergent vegetation, larger numbers gather, including an undated record of 200.

Curlew Sandpiper
Calidris ferruginea

RANGE: breeds n. Siberia; winters from w. Africa to Australia, NZ. HABITATS: tidal mudflats, wetlands. STATUS: uncommon late Dry, Wet season migrant. BREEDING: not in Australia.

Danny Rogers

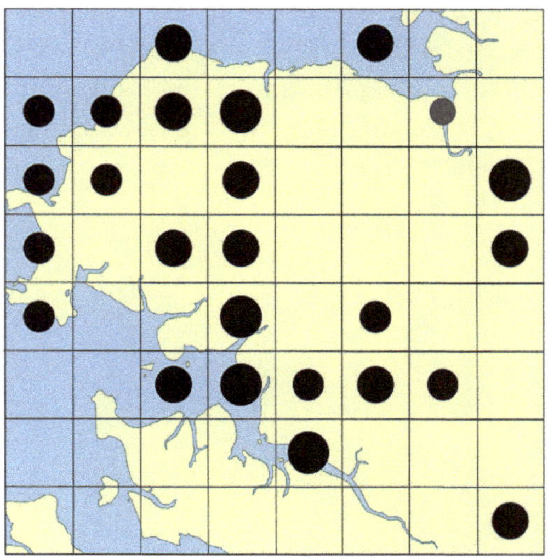

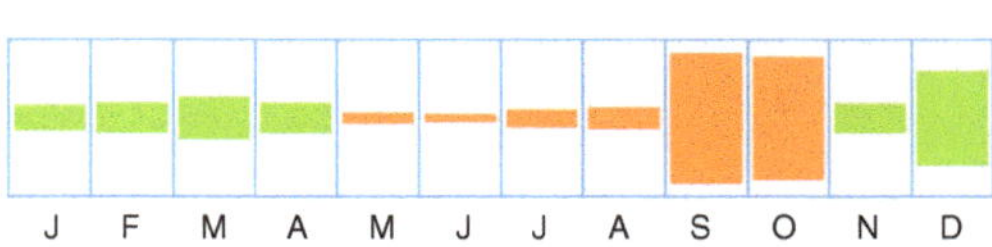

From a Top End population estimated in the 1990s to be about 17 800 Curlew Sandpipers, some 65%, including the largest single count of 1300 in late September 1994, were found west of Cobourg Peninsula.[195] While this suggests that

Curlew Sandpipers enter the Top End largely from the east, the species' southward migration route from their breeding grounds is still uncertain. Many are thought to pass through western Indonesia before entering north-west Australia, and this may include birds occurring in western areas of the Top End. Large numbers stage in north-west Australia before continuing on to south and south-east Australia. A fall in numbers across the Top End during the Wet season also suggests that these birds continue southward.[195] Band recoveries indicate that some Curlew Sandpipers stage in north-west Australia during return migration, but numbers are much lower than on southward migration, and the majority of Curlew Sandpipers are thought to take a more easterly route back to their breeding grounds, moving along the south-east and east coasts of China.[228] There is a sharp increase in numbers in the Top End during April and May,[195] which seems late for continued migration. Of 1400 birds seen during those surveys in May, none showed any breeding plumage, suggesting that they were overwintering birds. There is a significant fall in June, however, indicating that the birds overwinter elsewhere.

The site nearest to the Darwin region that holds large numbers is Chambers Bay, where the maximum count was 750 birds, in mid-August 1992. Numbers of Curlew Sandpipers in the Darwin region are, however, much lower, and counts of 70 birds at Holmes Jungle Swamp in September 1978 and 50 at Leanyer Swamp in October 1983[34] are the highest for the region. During the period December 2008 to May 2014, regular monthly surveys covering a range of coastal sites and habitats resulted in maximum single counts of 22 Curlew Sandpipers in January 2011, and 12 in October 2009, with no other month having in excess of four birds. Indeed, there were no Curlew Sandpipers counted at all during 32 surveys covering six sites in 2012, and only incidental reports of a single bird in each of March and August.[40]

Curlew Sandpiper numbers declined by around 80% at 49 sites across Australia between about 1983 and 2007.[253] There are too few birds in the Darwin region to assess any change in the species' status, but a maximum single count of 15 Curlew Sandpipers in monthly surveys covering five sites between July 1970 and January 1972[254] is not dissimilar to maximum counts in recent years.

Stilt Sandpiper
Calidris himantopus

RANGE: breeds n. Alaska, n. Canada; winters S. America. HABITATS: wetlands; occasionally tidal mudflats. STATUS: vagrant. BREEDING: not in Australia.

David Fisher

The first Australian record of Stilt Sandpiper was of a bird seen and photographed on the edge of a small brackish pond near Leanyer SP, on 30 August and 1 September 1980.[255] A Stilt Sandpiper was also recorded in Alice Springs in October 1991;[256] however, there is no evidence to suggest that the species is other than an accidental vagrant to the NT.

Broad-billed Sandpiper
Limicola falcinellus

RANGE: breeds n. Europe and central Siberia; winters from e. Africa to India. HABITATS: coastal mudflats, occasionally near-coastal wetlands. STATUS: rare late Dry, Wet season migrant. BREEDING: not in Australia.

Adrian Boyle

Shorebird surveys in the 1990s[196] indicated a population of about 2000 Broad-billed Sandpipers for the coast of the Top End. The highest counts, of 200 birds, were south-west of Darwin. Three of the five counts of 100 or more birds during the surveys were in mid to late July.[196] It may be that all were overwintering birds, but it is possible that they were early arrivals, perhaps failed breeders. There were consistent peaks both in numbers and group sizes in the March–April period, suggesting birds gathering in preparation for northward departure.[196]

The Broad-billed Sandpiper is an irregular visitor to the Darwin region, with no reports in some years, but may prove to be more regular if observations are made on

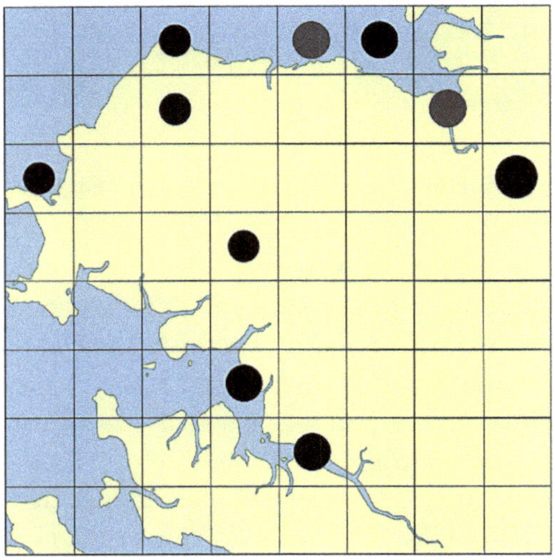

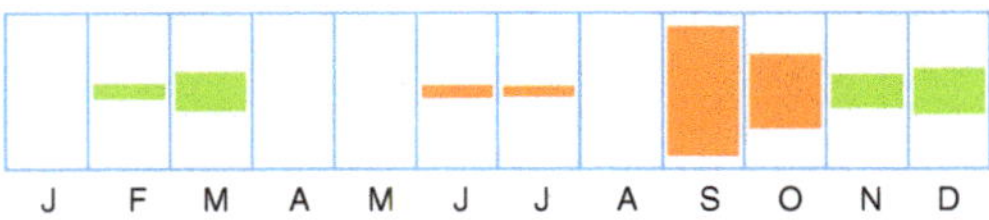

the seaward side of mangroves.[1] While numbers may be underestimated in other parts of Australia due to difficulty of separating them among large flocks of Red-necked Stint,[209] this problem does not occur in Darwin as Red-necked Stints are rarely in large numbers.

The earliest sighting is 30 August 2011, apparently a new arrival, as the bird was in nearly complete breeding plumage. Most records are from the period September–November, and may comprise birds spreading from south-west of Darwin rather than new arrivals. There are no records from December to February, further suggesting the dispersal of birds after arrival. Although there is only one record of a bird in breeding plumage in the pre-migration period, in mid-April 2008,[3] there are a small number of sightings in March and April, following an absence of records for the December–February period, indicating that some birds pass through Darwin on return migration. The only records from the Darwin region of overwintering are single birds seen at Lee Point in June 1977[31] and June 2010.[3]

The majority of Broad-billed Sandpiper reports are from Leanyer SP, with records also from Palmerston SP, Leanyer Swamp, Knuckey Lagoons, Nightcliff reef, Buffalo Creek, Lee Point and Dripstone Beach. Birds are generally seen individually, with sightings of more than three or four birds unusual. Maximum counts are 25 in November 1986 and 21 in October 1981, at Leanyer SP.[34]

Ruff

Philomachus pugnax

RANGE: breeds nw. Europe, n. Russia; winters Mediterranean, Africa, Middle East, s. Asia. HABITATS: mainly wetlands, occasionally estuarine wetlands. STATUS: rare late Dry, Wet season migrant. BREEDING: not in Australia.

Ruff was first recorded in the Darwin region near Holmes Jungle in September 1974.[129] Since then there have been several records, though no indication that the bird is an annual visitor. The maximum count is of six birds, recorded at Howard Springs in 1990;[34] however, there are several records of two birds. In November 1979,[251] two female birds were observed near Leanyer SP, where bird banders had set up a mist-net. One of the birds was caught, banded, photographed and then released. Between

Danny Rogers

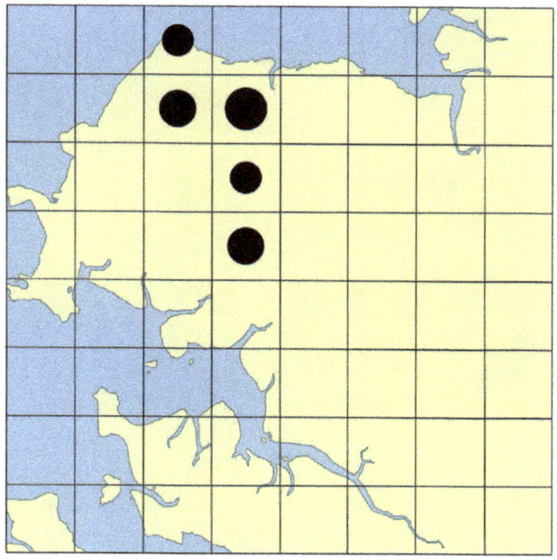

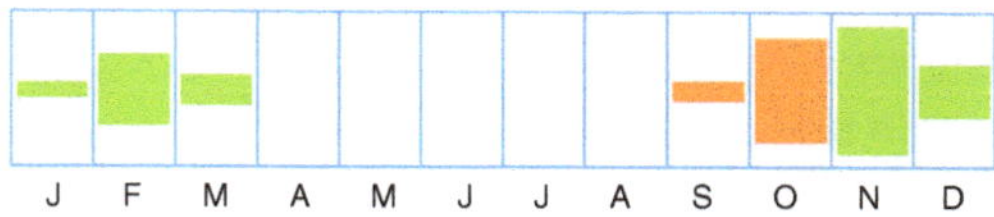

October and December 2004, at least two birds were present in the area, apparently moving irregularly between Knuckey Lagoons and Leanyer SP.[21] Other records of two birds were at Holmes Jungle in November 1981 and at Leanyer SP in November 2008. Single birds have been recorded at Knuckey Lagoons in October 1987, November 1988 and October 1994, and at Leanyer SP in February 2001 and October 2005.[21]

There are records for all months from September to March, excluding January, but most records are from October to December, so it may be that birds disperse as suitable habitat becomes inundated with Wet season rains. While birds may move out of the region, some move only as far as Leanyer SP, where the controlled water levels mean that suitable habitat may be available after other areas have become inundated.

Red-necked Phalarope
Phalaropus lobatus

RANGE: circumpolar breeding distribution; winter largely on oceans between New Britain and the Philippines, n. of NG. HABITATS: pelagic; occasionally inshore and on saline wetlands or sewage ponds. STATUS: vagrant. BREEDING: not in Australia.

The first record of Red-necked Phalarope for the Darwin region occurred in January 1974, when a bird was found at East Point, following a day of incessant storms and rain.[129] A week after this sighting, an exhausted bird was found in the same area. The phalarope subsequently died and was prepared as a specimen. During the following March there were several sightings at Leanyer SP,[129] all involving what was presumed to be the same individual.

There were subsequent records of single birds from Palmerston in February 1987,[257]

Adrian Boyle

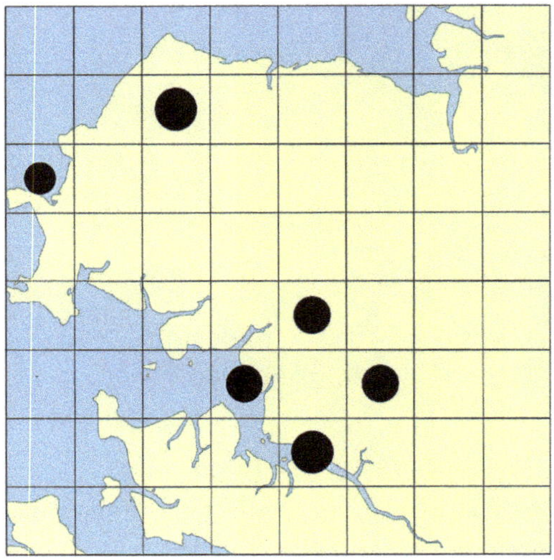

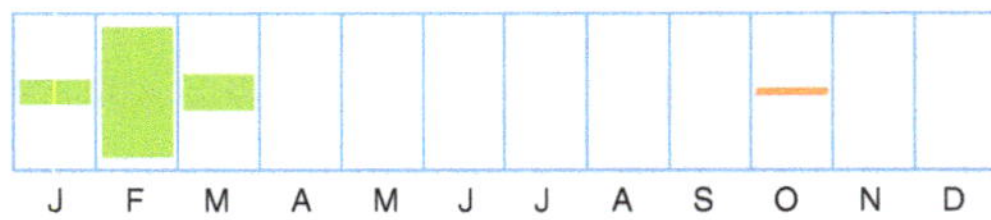

and from Leanyer SP in March 2001 and October 2005. During February 2014, however, there was an unprecedented influx of the species into the region. On 5 February, seven Red-necked Phalaropes were present at Leanyer SP. Two days later the number had risen to 11,[53] and the following day up to 16 were seen.[258]

Large numbers of Red-necked Phalaropes from breeding areas in eastern Siberia and Alaska are found at sea from the southern Philippines, south to Indonesia and at least as far as Timor.[45] The records involving multiple birds can probably be traced to the impact of severe weather within the wintering range.

The records in 1974 refer to at least two individuals, soon after the passage of tropical cyclone Fiona-Gwenda. This cyclone moved west from the Darwin region, and passed north of Port Hedland, where the Red-necked Phalarope is considered to be a regular visitor.[259] It is possible that the Darwin birds were displaced from seas in that area.

The influx in 2014 can be related to the catastrophic cyclone Haiyan, which struck the Philippines in November 2013. Many thousands of Red-necked Phalaropes wintering on the South China Sea would have been displaced, a large number of which would certainly have moved southward. Subsequent severe weather in Indonesia during January and high winds in the Timor Sea may have precipitated further movement, resulting in the occurrence of the birds in the Darwin region.

Red Phalarope
Phalaropus fulicarius

RANGE: breeds circumpolar on coasts of Arctic Ocean; winters mainly w. S. America, w. and sw. Africa. HABITATS: mainly pelagic; occasionally inshore during rough weather. STATUS: vagrant. BREEDING: not in Australia.

Tom McCrie

Red Phalaropes have the most marine wintering distribution of all shorebird species[260] and, given their typical wintering distribution in waters west of South America and Africa, any occurrence in Australia is a rare event.

A single Red Phalarope was found at Leanyer SP on 21 January and 2014[53] and was present for the following three weeks, during which time there was also an influx of Red-necked Phalaropes. The four previous confirmed records of the species in Australia[241] have been on the east and south-east coasts, and suggest an origin from the population wintering off the coast of South America. However, there have also been four records in south-east Asia, one of which was in Indonesia,[261] so it may be that the weather conditions causing the influx of the Red-necked Phalaropes at that time also brought about the occurrence of the Red Phalarope.

Australian Pratincole
Stiltia isabella

RANGE: Australia; occurs outside breeding season also in NG, Timor and Indonesia. HABITATS: wetlands, open grassland, bare areas. STATUS: common Dry season migrant. BREEDING: Oct–Dec (DR: 2 records).

Niven McCrie

The Australian Pratincole is the only Australian breeding wader with a clear pattern of migration, with birds moving from southern to northern Australia, and many continuing northward to Indonesia. A record from Timor of 50 000 birds in November[209] suggests that the numbers involved are significant. At the South Alligator River over several evenings during April 1988 large numbers were seen flying north at a height of several hundred metres, presumably on migration to Indonesia.[262]

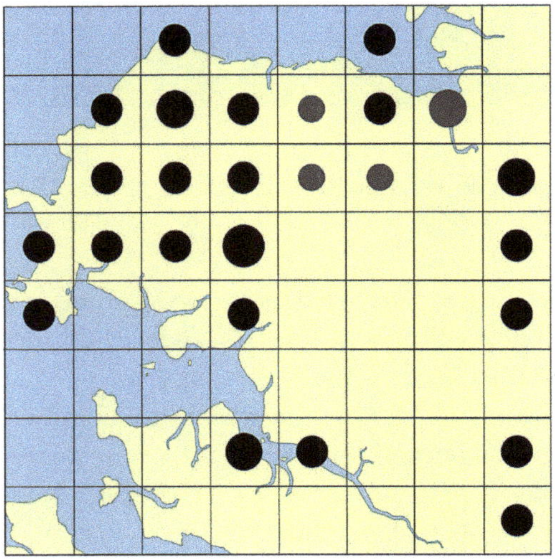

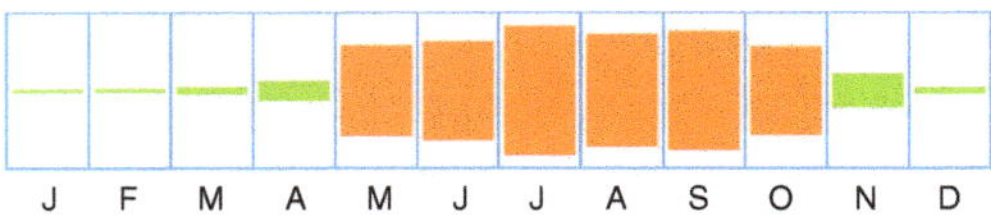

At a melon farm on the rural fringe of the Darwin region, birds appear to arrive in very large numbers in early April, then move out quickly,[263] quite likely continuing to Indonesia, although they may also disperse to other areas in the Top End.

Australian Pratincole is a common and regular Dry season visitor to the Darwin region. Birds occur in a range of habitats including grassland, wetlands, sewage ponds, ovals and roadsides. It is not uncommon to see several hundred birds on the wetland near Holmes Jungle or around Knuckey Lagoons, foraging or resting and drinking in the heat of the day. Birds are also sometimes seen hawking for insects, rather like Oriental Pratincole, although not at such heights as that species.[21]

Arrival in the Darwin region may be as early as the first week in May, rarely before, but birds are generally scarce or absent until June and sometimes into July. While numbers may fluctuate during the length of their stay, they tend to be consistent from about July, possibly because the migrants heading to Indonesia have already passed through by then. Most birds have left the area by the beginning of November, and by December they are typically absent. Data from the Magela floodplain show a similar pattern, although numbers there peak in November, at up to 10 000 birds, before also dropping to almost nothing in December.[28] It may be that return migrants augment the numbers in this region on their way back south.

Numbers vary between years, and in some years, for example 1971 and 1983,[1,34] the species was apparently absent from the Darwin region. In other years they may be in high numbers on any suitable habitat in the Darwin region, and with some thousands present on floodplains such as that surrounding the Adelaide River.

There are only two confirmed reports of Australian Pratincoles breeding in the Darwin region. In mid- November 1991 two downy young and an adult were seen in a paddock near Knuckey Lagoons.[21] At Darwin International Airport, in late October 2004, a single egg was found in a small scrape, and a second nest was located nearby the following day. A newly hatched chick was found at the first nest 22 days after the egg was found. At that time the second nest still held unhatched eggs, but this was not subsequently checked to see whether those eggs had hatched.[71]

Additionally, an observation in December 1978 at Lee Point of a young bird able to fly, but with incomplete primaries, strongly indicates local breeding.

Oriental Pratincole
Glareola maldivarum

RANGE: breeds in ne. Asia; majority of the migratory population winters in n. Australia. HABITATS: grassland, floodplains. STATUS: uncommon passage migrant, Sep-Dec. BREEDING: not in Australia.

David Webb

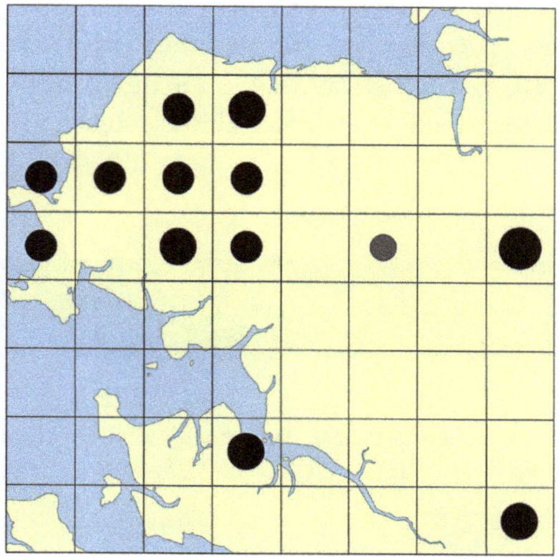

Oriental Pratincole may be considered a regular passage migrant to the Darwin region, though numbers vary widely between years, and there have been some years with no local reports of the species. Birds usually arrive in November, rarely mid-October or before, and their arrival can often be associated with low-pressure climatic conditions or storm fronts, possibly taking advantage of such conditions for feeding. As with the Little Curlew, the species moves out with the onset of heavy rains, and there are few records from January onwards.

Occasionally birds are present in quite large numbers such as 500 at Holmes Jungle Swamp in November 1985[21] and 500 at Leanyer Swamp in January 1983,[34] but generally in the Darwin region they are present only in moderate numbers.

Oriental Pratincoles often feed in the air, hawking for insects, and are sometimes seen in loosely mixed groups with Pacific Swifts.[21] Like the swifts, they may take advantage of particular climatic events in the feeding. They also feed on the ground, around wetland areas such as at Holmes Jungle and Knuckey Lagoons, and particularly during hotter days, birds use these wetlands for drinking and resting. In mid-November 1996 there was an unusually large influx into the northern suburbs of Darwin, with many feeding groups, comprising a hundred or more birds, hawking for insects over the suburbs during successive hot and humid but almost cloud-free afternoons. When the rains began about a week after their arrival the hawking stopped, coinciding with a build-up of birds on the ground around swamps such as Knuckey Lagoons.

Calls of Oriental Pratincole have been heard over Darwin suburbs at night,[31] but no estimates of the number of birds involved has been made, nor has it been determined that these birds continued onward or landed within the region. Very high numbers of Oriental Pratincole occur in the Barkly Tablelands, including an estimated 100 000 birds observed during December 1994.[264] While some of these may have staged in, or passed over, the Darwin region, it seems unlikely that such numbers have gone unnoticed, and passage via north-west Australia seems more likely.

Brown Noddy
Anous stolidus

RANGE: pantropical; subsp. *pileatus* Seychelles, Madagascar, Hawaii, Polynesia, n. Australia. HABITATS: coast, pelagic. STATUS: vagrant. BREEDING: Mar–Jul (TE).

www.wildlifeimages.com.au/Rohan Clarke

Among the rarest of terns in the Top End, the Brown Noddy was first recorded in the NT on Wessel Island, off north-east Arnhem Land in October 1972, and within three months of this discovery two birds were found in Darwin, apparently forced onshore by severe weather. One died after flying into a house in Rapid Creek during north-west gales in early February 1973,[79] and an exhausted bird was also picked up at Fannie Bay at that time.[79] In the following year single birds were sighted at Buffalo Creek in both January and September.[265] Almost two decades later, one was found at Palmerston SP in mid-November 1993.[78] A young bird was seen roosting at Nightcliff reef on 27 February 2008.[3] An exhausted Brown Noddy was found at Charles Darwin University, in Nakara, on 1 February 2014 and rehabilitated.[266] The observation of a single bird foraging among Greater Crested Terns, about 40 km offshore, north-west of Darwin in August 2012 suggests that birds may feed in waters offshore from Darwin. A birdwatching pelagic trip to an area 150 km west of Darwin in November 2013 found none.

In May 1994, 100 Brown Noddies were found nesting along with Roseate and Bridled Terns on Higginson Islet, off Nhulunbuy,[127] about 650 km east of Darwin, and by May 1996, 300 birds were nesting, though all were gone by September. This represents the first and only known breeding colony of the species in the NT. Birds blown onto Darwin shores may have been dispersing from this colony, but it seems more likely that they originated on breeding islands off northern Western Australia, including Ashmore Reef, which holds 54 000 breeding pairs and is the second largest colony of Brown Noddy in Australia.[267] Although this is further from Darwin than the Nhulunbuy colony, it is open sea, whereas from the east birds must enter through Van Dieman Gulf.

Sabine's Gull
Xema sabini

RANGE: breeds n. coastal areas of Greenland, Russia, Alaska, Canada; winters off sw. Africa, nw. South America. HABITATS: pelagic; occasionally inshore. STATUS: vagrant. BREEDING: not in Australia.

Breeding in the Arctic and sub-Arctic regions,[268] the Sabine's Gull normally winters in the cold waters off South America or south-west Africa, making it by far the longest distance migrant of any gull species.[268] The single record for the Darwin region, and the first for Australia, was a bird found near Stokes Hill Wharf in April 1982.[269] Coincidentally, this was only two days after the initial sighting of another vagrant gull, the Black-tailed Gull, in the same general area. The Sabine's Gull was considered to be moulting from first winter to first summer plumage and, as this species does not return to breeding grounds until three years of age,[268] the Darwin bird may have simply been roaming in the Pacific. Cyclonic activity off north-eastern Australia during the previous week may have been a factor, as the bird appeared to have been weak, which perhaps provides some evidence of recent stress.

Silver Gull
Chroicocephalus novaehollandiae

RANGE: Australia, New Caledonia. HABITATS: coast, wetlands, agricultural and urban areas. STATUS: moderately common non-breeding resident. BREEDING: Mar–Nov (TE).

Silver Gull was reported as 'fairly common' on Melville Island during the Wet season of 1911–12,[270] some 70 years before it could be given that status in the Darwin region. Indeed, as recently as the 1970s, Silver Gull was considered to be an 'uncommon visitor' to the Darwin region, and most records comprised only one to three birds.[1] Since then, however, the species has undergone something of a population explosion, probably linked to the growth in Darwin's human population and its attendant rise in available food at rubbish dumps and elsewhere, and perhaps in part due to the development of the fishing industry out of Darwin. By the mid-1980s, counts of

Niven McCrie

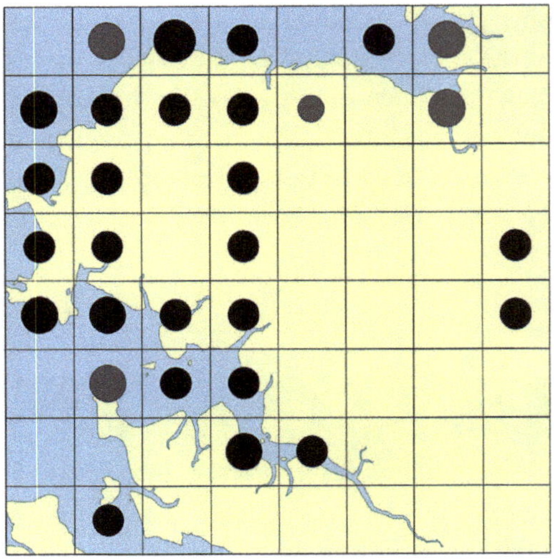

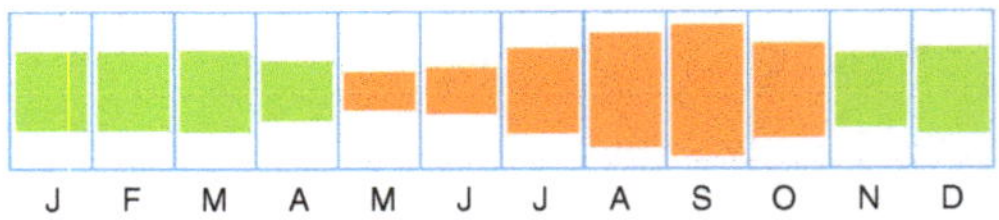

170–180 were made of birds foraging at the Leanyer rubbish dump or roosting near Buffalo Creek.[34] From that period the population appears to have levelled out, and numbers roosting near Buffalo Creek, or foraging over the landfill at the Shoal Bay Waste Management Facility, are rarely above about 200 birds. The largest single count of Silver Gulls in the Top End, outside of breeding colonies, was of 400, 'near Darwin',[127] in October 1995.

Other than at Buffalo Creek, a regular roost of about 50 Silver Gulls occurs along the Nightcliff foreshore, and small numbers congregate at Stokes Hill Wharf to feed on scraps thrown by diners, apparently roosting overnight on the roofs of the buildings. Perhaps surprisingly, there have been few observations of Silver Gulls in large numbers at Leanyer SP, and it may be that there is little food available for them there and that the site is unsuitable as a roost.

The Silver Gull is one of the very few gull species that has tropical breeding populations, and colonies occur extensively along the north and east coasts of the Top End. West of Cobourg Peninsula, only two colonies have been found, both off Melville Island, but one of these is the Top End's largest. Up to 800 birds were present at that site during the 1994 breeding season, although not all were breeding at that time.[127] It seems likely that Darwin birds may be largely or entirely from the Melville Island colonies, as these are about 70 and 150 km from Darwin, whereas the nearest colony eastward is some 200 km distant.

Most Silver Gull breeding in the Top End occurs from March to August,[127] and from May to July the species is scarce in the Darwin region. Extensive aerial counts across the Top End during the 1990s revealed a rapid increase in numbers in July which has been attributed to an influx of young birds into the population after the main breeding period. Birds return to the Darwin region slightly later, from August, and the presence of juvenile birds during August and September, including one bird still being fed regurgitated food by an adult, clearly indicates a return from breeding sites. Numbers continue to build up through September until they peak in the period October to January.

Black-headed Gull
Chroicocephalus ridibundus

RANGE: breeds Greenland, Europe to ne. China; winters Africa, e. and se. Asia. HABITATS: coast; also inland in n. hemisphere. STATUS: vagrant. BREEDING: not in Australia.

Tony Palliser

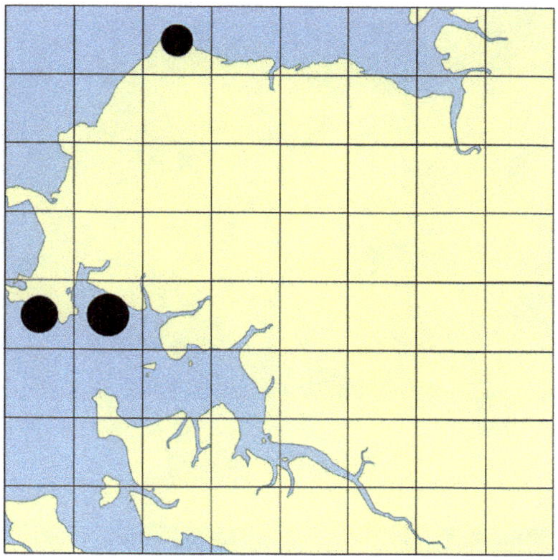

The first record of Black-headed Gull for the Darwin region, the second for Australia, regards a single bird found among a roosting flock of Silver Gulls at Buffalo Creek in early February 1998.[271] Two individuals were observed from mid-December 2002 until March 2003, and in the following Wet season two birds were present between December 2003 and February 2004. Other records, of single birds, were in January 2005, from January to February 2006, in April 2008 and from December 2009 to March 2010. The 2010 bird was considered to be the subspecies *sibiricus*, which on breeding distribution is the one to be expected in the region.[272]

During the day the Black-headed Gulls often foraged along the shore of Darwin Harbour, apparently associating with Silver Gulls, but in the evening they roosted at Buffalo Creek, or joined Silver Gulls at Stokes Hill Wharf to feed on scraps thrown by the al fresco diners, and sometimes slept on the roof of the buildings.

Black-tailed Gull
Larus crassirostris

RANGE: Russia, Japan, Korea, China. HABITATS: coast. STATUS: vagrant. BREEDING: not in Australia.

David Fisher

The single Darwin record of Black-tailed Gull involved a second-year bird found in Darwin Harbour in April 1982,[273] and which was subsequently observed with some regularity at various locations around Darwin, including Dinah Beach, Kulaluk, Lee Point and Leanyer SP, until March 1984.[31] Australia's only previous Black-tailed Gull was found in Port Philip Bay, Victoria, in 1978.[273]

Franklin's Gull
Leucophaeus pipixcan

RANGE: breeds North America; winters s. Mexico to coastal South America. HABITATS: coast, wetlands. STATUS: vagrant. BREEDING: not in Australia.

Niven McCrie

After breeding on marshy lakes in the prairie regions of North America, the Franklin's Gull migrates south to wintering areas, mainly on the Pacific coast of South America. This species has a widespread history of vagrancy, ranging from north-west Europe to the Antarctic Peninsula, including all of the world's continents, and is by far the most frequently reported vagrant gull in Australia. There has, however, been only one case of the species in the Darwin region: an adult in non-breeding plumage, found in late November 2008 at Stokes Hill Wharf.[136] The bird was regularly seen there, and in nearby parts of Darwin Harbour, until about mid-December. At the wharf, usually during the evening, the bird foraged along with Silver Gulls, feeding on scraps of food thrown by diners onto the wharf or over its edge into the water, but during the day, at low tide, it was most frequently seen foraging along the muddy shoreline.

Gull-billed Tern
Gelochelidon nilotica

RANGE: N. and S. America, Africa, Eurasia, Australasia. HABITATS: wetlands, coast. STATUS: moderately common. BREEDING: Sep–Jan (NT).

Niven McCrie

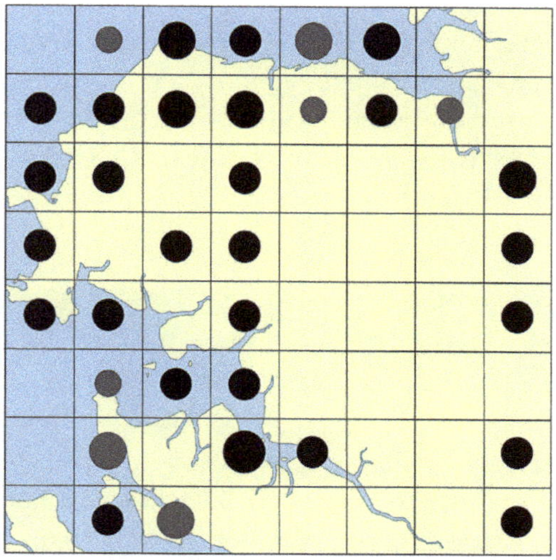

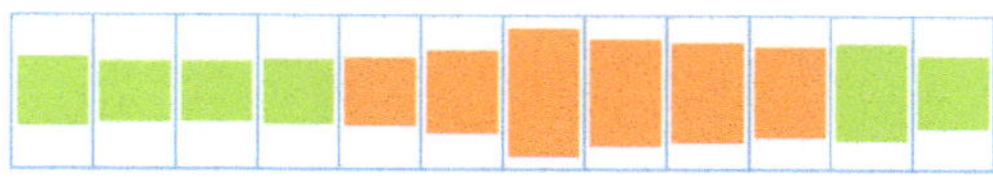

The second most frequently recorded tern species in the Darwin region, the Gull-billed Tern comprises two distinct subspecies, one of which migrates from as far away as China. The Australian-breeding subspecies *macrotarsa* is much larger and paler than the Asian migratory subspecies *affinis*, which was first recorded in Australia with the collection of specimens from the South and East Alligator Rivers in February 1973 and October 1974.[274] A specimen collected at Leanyer SP in December 1989 confirmed its presence in the Darwin region, and it is now known to be a regular visitor to north-west Australia, including the western Top End.

Although records show Gull-billed Tern as present all year round, the seasonality is complicated by the presence of the two subspecies. Adult migrants visit mainly from August or September to April, though some immature birds may remain during the Dry season.[275] Unlike the resident subspecies, which occurs also on subcoastal and inland wetlands,[275] the migratory subspecies is strictly coastal, although birds may be found up to 15 km inland along estuaries, and also occur at sewage ponds. Asian migrants are seen in breeding plumage from October to May, whereas the Australian subspecies shows a more varied timing of non-breeding and breeding plumages,[275] which may be an adaptation to exploit unpredictable breeding opportunities.

Gull-billed Terns were recorded most frequently from the Buffalo Creek–Lee Point area, where counts conducted from 1974 to

1980[276] showed that numbers were relatively low during and just after the Wet season, but increased dramatically in May and built up through the Dry season. During the 1990s, too, extensive aerial and ground surveys along the entire NT coastline showed that Gull-billed Terns were most abundant from July to September. The largest flocks encountered during those surveys held about 350 birds, the majority of these occurring in Chambers Bay, 80 km north-east of Darwin, but the highest count in our region was about 60 birds roosting at Buffalo Creek.

No breeding has been recorded in or near the Darwin region. The largest documented breeding colony of the species in Australia[277] was recorded in 1995 at Lake Sylvester, in the Barkly Tablelands, where at least 1350 pairs were found breeding, and it may be that the majority of birds from the Top End breed in that region.

Caspian Tern
Hydroprogne caspia

RANGE: n. America, Europe, Africa, Asia, Australasia. HABITATS: coast, wetlands. STATUS: moderately common resident (?). BREEDING: Apr–Jun; Sep–Nov (TE).

David Webb

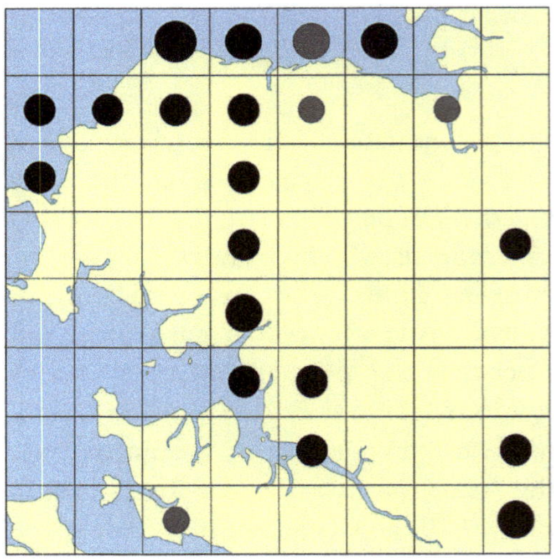

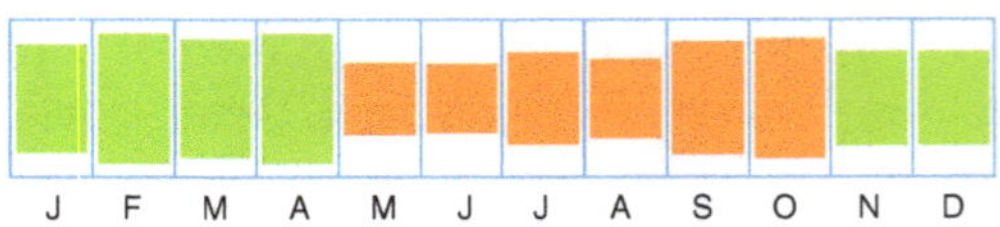

In the Top End, the Caspian Tern is more abundant around the Gulf of Carpentaria,[127] where the largest group, 115 birds, was found in April 1998. The only breeding recorded along the NT coast during surveys in the 1990s was also on the Gulf of Carpentaria, where single nests were present in four widely separated colonies of other terns. Two of these nests had eggs in May while the other two had eggs in October.[127] In contrast, inland colonies can be large, though irregular in occurrence. About 500 breeding pairs were located on Lake Gregory, about 900 km south-west of Darwin, in May 1986,[278] and smaller numbers are reported to breed in the Barkly wetlands, some 800 km south of Darwin. A large gap in distribution of Caspian Tern between the western and eastern Top End may indicate distinct populations, with those in the east moving to the eastern Gulf of Carpentaria to breed, and those to the west breeding inland in north-west Western Australia or inland NT.

Though Caspian Terns are present all year on Darwin's coast, low counts in June, following a spike in numbers in May and a subsequent increase in July, suggest some short-term movement out of the region.[21,276] Counts over the entire NT coast, however, revealed no obvious seasonal movements, possibly the consequence of irregular breeding inland. Caspian Terns normally occur in ones or twos, but larger groups may be encountered, the largest in the Darwin region being 56 birds at Buffalo Creek in July 1999.[21]

Greater Crested Tern
Thalasseus bergii

RANGE: e. and s. Africa, e. to Australia, NZ and w. central Pacific Ocean. HABITATS: coast. STATUS: common non-breeding resident. BREEDING: Mar–Jul (TE).

Niven McCrie

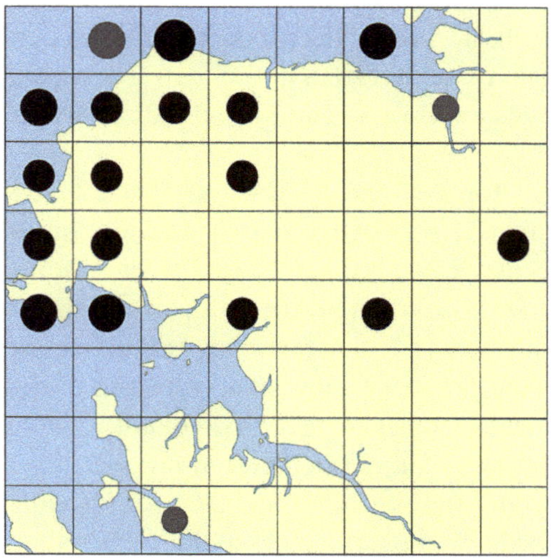

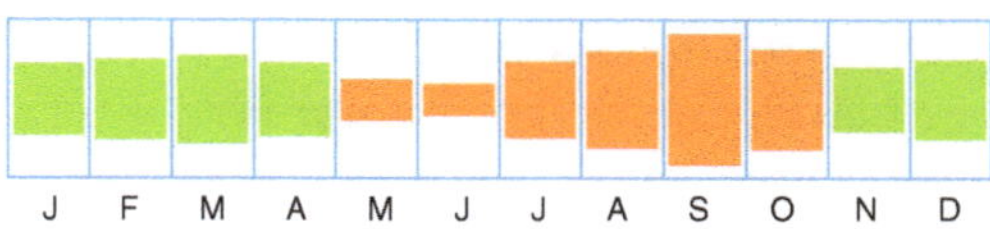

The Top End is home to the largest breeding colonies of Greater Crested Terns in Australia and, indeed, hosts the largest colony of the species in the world, so it is hardly surprising that this species is the most widespread and numerous of all seabirds along the coast and on offshore islands of the NT.[195] The Greater Crested Tern is also the most commonly seen tern species along coasts in the Darwin region, birds often seen plunging into water from several metres to take fish just below the surface, or loafing or roosting on rock platforms and sandy shorelines.

There are no fewer than 20 breeding colonies on islands around the NT coast, most having in excess of 5000 birds and two comprising over 50 000 birds each. Although most of the colonies are on the east and north-east coasts, the largest, and the closest permanent colony to Darwin, is on Seagull Island, north of Melville Island, where at least 25 000 pairs nested from 1998 to 2001.[279] While it seems likely that birds occurring in Darwin will have originated from that colony, it may be that small numbers breed nearer to the region. In May 1996 two small breeding colonies were found in northern Fog Bay,[126] about 50 km west of Darwin. These, however, appear to have been unviable. One of the colonies, on a sand-bank near Bare Sand Island, contained about 200 eggs, laid directly on the sand in positions vulnerable to the tide. The eggs were gone nine days later, most likely having been washed away by spring high tides.[126] The second colony, on a sand-bank at Middle Reef, contained 160 eggs, but three days later, the only remaining eggs were found among high-tide debris.[126]

All Top End Greater Crested Tern colonies are active between March and July; most eggs are laid from late April to early June, and most colonies are finished by July. In the Darwin region numbers start to decline from February, consistent with the movement of birds to breeding colonies, while July sees a rapid build-up in numbers, and from August large numbers of juvenile birds as well as many adults in breeding plumage are present, indicating the return of birds from breeding colonies. Where birds from the Seagull Island colony disperse to during the non-breeding season is uncertain. The small numbers in the Darwin region, which rarely exceed 200 birds, cannot account for them, and while numbers along the north and eastern Top End coast show a tendency for birds to concentrate in larger groups from September to November, there is otherwise only a moderate variation in abundances.[127]

The breeding success of Greater Crested Terns on Seagull Island, the largest Top End colony, was estimated to be less than 7%, with losses largely due to egg predation by Silver Gulls.[280] While Silver Gulls have also been observed pecking at Greater Crested Tern chicks, often inflicting fatal injury, it appears they rarely feed on them.[280] The gulls also steal food from the adult terns, leading to starvation of the chicks. White-bellied Sea Eagles also appear to be significant predators of Greater Crested Terns, their chicks and eggs at some Top End breeding colonies. Sea Eagle feeding perches with hundreds of Greater Crested Tern carcasses at them[127] have been found near Greater Crested Tern colonies, and the species sometimes nests near or even within colonies, to prey upon both incubating adults and young.[127] On Seagull Island, Aboriginal people from the Tiwi Islands harvest about 2000–3000 eggs, about 10% of the total, each year.[280] Nevertheless, Greater Crested Terns may be increasing in number on the Top End coast. An estimate of breeding numbers in 1994 was put at 'over 60 000' but in 1999 two colonies alone contained a combined breeding population in excess of 100 000 birds.[195]

Lesser Crested Tern
Thalasseus bengalensis

RANGE: w. Africa to s. Asia; n. Australia, Indonesia, NG. HABITATS: coast. STATUS: moderately common non-breeding resident. BREEDING: Aug–Nov (TE).

www.wildlifeimages.com.au/Rohan Clarke

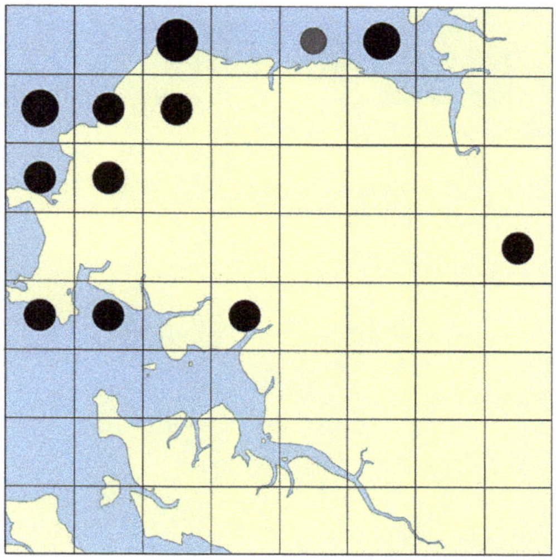

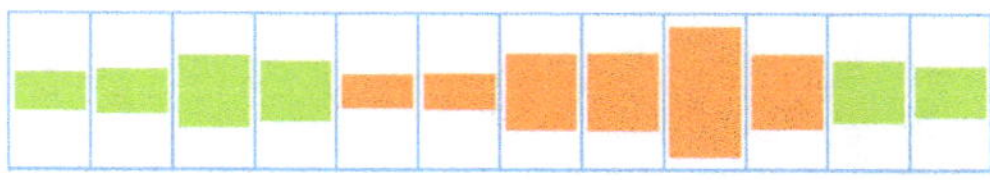

Unlike the widespread Greater Crested Tern, the largely tropical Lesser Crested Tern has a patchy distribution in the Top End, being most abundant in the region from Darwin to the Peron Islands, about 120 km to the south-west.[127]

In the Darwin region Lesser Crested Terns are both less numerous and more restricted in distribution than Greater Crested Terns. While small numbers may be seen along the coast at Nightcliff, East Point and Darwin Harbour, numbers are greatest in the Buffalo Creek–Lee Point area, where there have been maximum counts of 200 and 249 birds in June and July, respectively. Outside of these two months, there is a single September count of 120 birds, but otherwise all are fewer than 100, falling to no more than 50 during the Wet season, from December to March.

The Lesser Crested Tern has a disjunctive breeding range. The only known breeding colony of the species in the NT is on Low Rock, an island in the south-west Gulf of Carpentaria, at least 1000 km from Darwin along the coast. This colony contained around 440 birds, with at least 150 nests, when surveyed in September 1994. Given the distance from Darwin, and the relatively small size of the colony, birds in the Darwin region are more likely to derive from colonies on islands off the Kimberley coast, in north-west Western Australia, which hold a breeding population of several thousand birds.[281,282] The sharp peak in June and July, when many adults in breeding plumage and juvenile birds are present, is consistent with birds returning from these breeding colonies, which are active mainly from March to June. The fall in numbers from August indicates that these birds are largely on passage, some perhaps dispersing around offshore islands near the Darwin region, and others continuing further east. In some years there is an influx during September, again including juvenile and adult birds in breeding plumage, and this may be late breeders on passage. The near absence of the species from the Darwin region during the Wet season is difficult to explain, however, as the exodus appears to be too early for birds to move directly to breeding colonies.

Little Tern
Sternula albifrons

RANGE: Europe, Asia, Australasia. HABITATS: coast. STATUS: moderately common, largely Wet season visitor. BREEDING: Apr–Jan (TE).

Micha V. Jackson

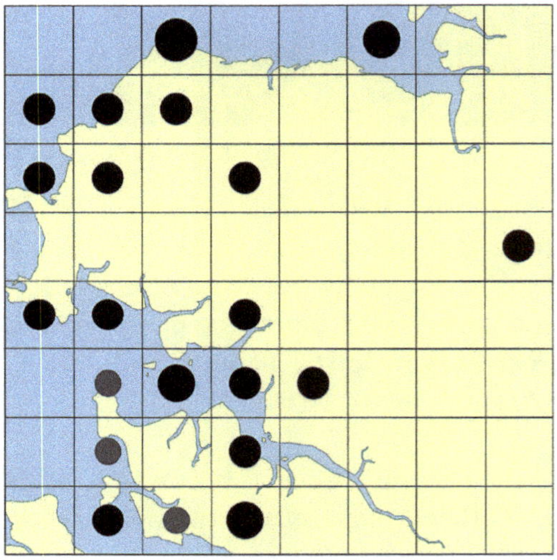

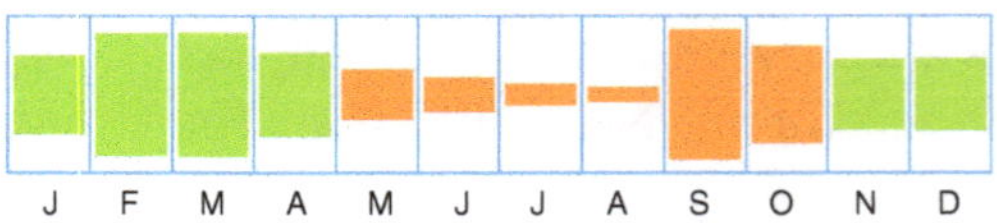

Three populations of Little Tern, all of the subspecies *sinensis*,[228] occur in Australia. Two of these are Australian breeding, one in eastern Australia and one in northern Australia, including the Top End, while the third comprises migrants from Asia, including China, Taiwan, Korea and Japan. The Little Tern is largely absent from the Darwin region during the mid-Dry season, from June to August, but a spike in numbers during September,[3,40] comprising birds in non-breeding plumage, indicates the arrival of migrant birds of Asian origin rather than Top End or eastern Australian breeders dispersing from breeding colonies. A subsequent fall in numbers in October[3] may represent birds dispersing after arrival from Asia, but numbers increase again from January and while there is some fluctuation in numbers during the Wet season, this is the period of greatest abundance for the species in the Darwin region.

The seasonal patterns in the Darwin region follow those for the Top End as a whole, with the exception that there are two distinct peaks in those numbers, in the period March–April and November–December, and numbers drop very sharply in January, rather than in December.

The majority of records and the greatest numbers in the Darwin region are from the Lee Point–Buffalo Creek area, but there are occasional reports of birds at Nightcliff and East Point and around Darwin Harbour. Flocks typically comprise fewer than 60 birds, but 150 were counted in December 1978[34] and 144 in November 2004.[3]

Breeding has been recorded in the Top End in every month except January and August, but the greatest number of colonies are active in May–June and September–October.[127] Numbers around known nesting sites in the Gulf of Carpentaria are similar in winter and summer, suggesting that the population there is resident,[228] but throughout the Top End, including the Darwin region, and in Broome,[283] Western Australia, numbers are substantially lower during the mid-Dry season,[21,276] indicating a significant migration away from the Top End coast at that time of year. Moreover, the breeding population, at around 1000 birds, accounts for far fewer birds than has been estimated for the total Top End population, indicating that the majority of birds found around NT coasts are migrants.

The Little Tern is the only coastal tern species that regularly nests on mainland beaches of the NT, although, like other terns, they also nest on small islands. Little Tern breeding colonies are relatively small, all but one of the 44 found on the NT coast comprising fewer than 100 birds.[127] The majority of breeding colonies are in the eastern Top End, in the Gulf of Carpentaria, with only five colonies occurring in the western Top End, four of which are on the Tiwi Islands. The other, on Bare Sand Island, about 50 km west of, and closest to, Darwin, held up to 40 breeding birds when surveyed in November 2000.

Bridled Tern
Onychoprion anaethetus

RANGE: pantropical; subsp. *anaethetus* s. Japan and Taiwan, through Philippines and Indonesia to Australia. HABITATS: coast. STATUS: rare Wet season visitor. BREEDING: all months (TE).

Micha V. Jackson

With at least 60 000 birds breeding among 43 colonies on small islands off the coast of the NT,[127] records of Bridled Tern in the Darwin region are surprisingly scarce and have almost invariably occurred during or immediately following stormy weather with periods of high winds. Most breeding colonies are situated in the eastern half of the Top End with the most westerly on an island off Cobourg Peninsula, about 180 km north-east of Darwin.[127]

The first documented record for the Darwin region concerned an exhausted bird found in the centre of Darwin in early February 1973, during a period of severe monsoonal weather and north-west gales.[79] The bird was released after 16 days but was later found dead, and was prepared as a study skin. Subsequent sightings have covered only seven of the 40 years from then until 2014. Single birds were reported from Lee Point in February 1975,[2,276] February 1985 and March 2012,[284] from Nightcliff in January 1986[34] and from Darwin Harbour in February 1997.[21] The only observations involving multiple birds were during periods of severe stormy weather in mid-January 2011, when up to 30 or more Bridled Terns were seen at various locations around Darwin Harbour,[285] and in February 2014, when 14 birds were present at Stokes Hill Wharf.[286]

Breeding activity in Top End colonies is almost year-round, but is mainly concentrated into the periods March to June and September to December.

Sooty Tern

Onychoprion fuscatus

RANGE: pantropical; in Australia, largely e. and ne. coasts. HABITATS: coast, pelagic. STATUS: vagrant. BREEDING: not in NT.

Laurie Ross

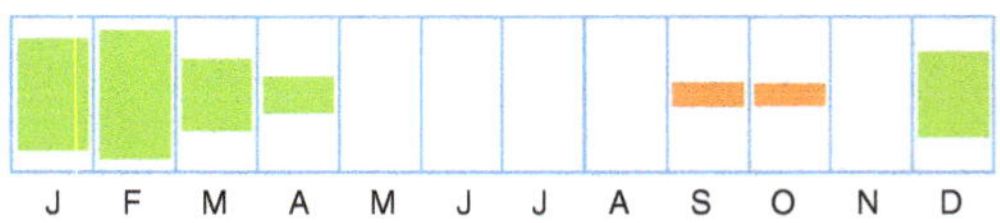

Sooty Tern has not been found breeding in the NT, though like the Brown Noddy very large colonies are found on islands off north-east Queensland, and off the Kimberley coast in north-west Western Australia. Those birds seen in the Darwin region are likely to have originated from breeding colonies on Ashmore Reef, which hold up to 45 000 Sooty Terns.[128]

Like Bridled Terns, the appearance of Sooty Terns in the Darwin region is almost entirely associated with tropical cyclones or severe storms, which have forced these birds inshore. The first verified NT record was of a storm blown bird about 4 km south of Darwin city in February 1973.[79] In January 1974, two birds were also seen inshore,[276] and that month another bird, banded in the Seychelles, was found dead, floating in the Edith River some 230 km south-west of Darwin[287] and almost 9000 km from its banding point. There appear to be no further records until 1982, but from then until 2011 there were reports covering 12 years.

Roseate Tern
Sterna dougallii

RANGE: almost cosmopolitan; Europe, the Americas, Africa, s. and se. Asia, n. and w. Australia. HABITATS: coast. STATUS: rare Dry season visitor. BREEDING: Apr–Jul; Sep–Jan (TE).

www.wildlifeimages.com.au/Rohan Clarke

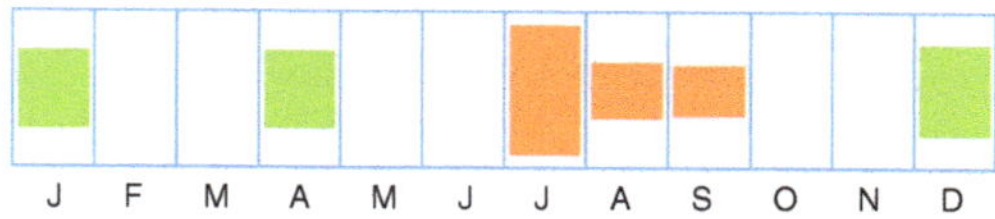

Up to 26 000 Roseate Terns have been found breeding among 38 colonies on small islands off the NT coast, all of them in the eastern half of the Top End. These colonies are the largest in Australia, one containing 7500 birds and four others containing 3000 or more birds.[127] Most breeding occurs between September and January, with a smaller amount between April and July.[127]

Roseate Terns forage at sea, rarely coming inshore and, given that the nearest breeding islands to Darwin are on islands around Cobourg Peninsula, it is not surprising that the species is rarely seen in the Darwin region. Nevertheless, there are records from 10 of the years between 1974 and 2012, making it a more frequent visitor than the Bridled Tern, and unlike that species and the Sooty Tern, sightings have not been restricted to periods of severe weather conditions.

Black-naped Tern
Sterna sumatrana

RANGE: Japan, China, se. Asia, tropical Australia and Pacific islands. HABITATS: coast, pelagic. STATUS: vagrant. BREEDING: Apr–Jun (TE).

Jon Norling

With 72 breeding colonies of Black-naped Tern, one of 1300 birds and several comprising over 500 birds, the Top End has the largest breeding colonies of this species in Australia. Yet despite having more breeding colonies than any other tern species in the Top End and an estimated total of 9000 birds, it is the least frequently recorded of the terns in the Darwin region as, like the Roseate, it forages at sea, and very rarely comes inshore.

The region's first documented sightings relate to two birds at Camerons Beach and several seen from a boat between Darwin and Mica Beach in July 1967.[96] However, no evidence was provided to support the identification in either case. The species was not recorded in monthly surveys of the Lee Point–Buffalo Creek area between 1974 and 1980,[276] but two birds were recorded there in February 1989[34] and one in August 2004.

Though the nearest permanent breeding colonies to Darwin are off the Cobourg Peninsula, 40 birds were found nesting in bomb craters on Bare Sand Island, only 50 km west of Darwin, in mid-December 1989.[288] Most Black-naped Tern colonies were active between September and December, though a few were active in April–May.

Common Tern
Sterna hirundo

RANGE: Race *longipennis* breeds Siberia and nw. China; winters in se. Asia and Australasia. HABITATS: pelagic, coast. STATUS: uncommon late Dry, Wet season migrant. BREEDING: not in Australia.

Niven McCrie

Data from coastal surveys of the Top End during the 1990s suggest that Common Tern is one of the most abundant seabirds occurring along the coast and seas off the NT. At least 20 700 birds were counted during those surveys, and it is likely the total is vastly greater, as most Common Terns feed further out to sea and so were not part of the population surveyed.[27] It is surprising then that the species was first confirmed as a visitor to the Top End coast only as recently as 1973, when three specimens were collected from the mouth of the East Alligator River,[289] and first reported in

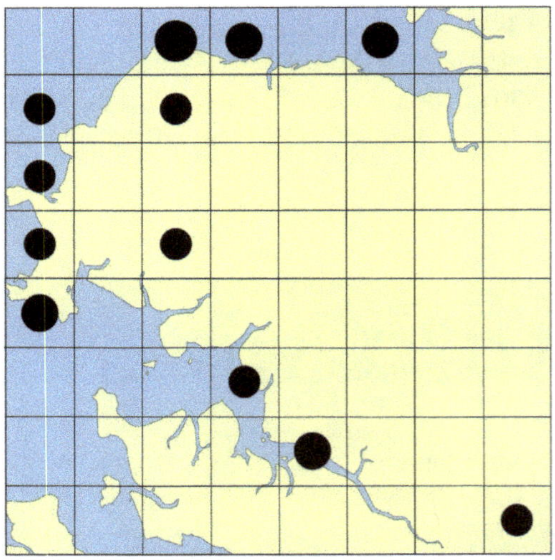

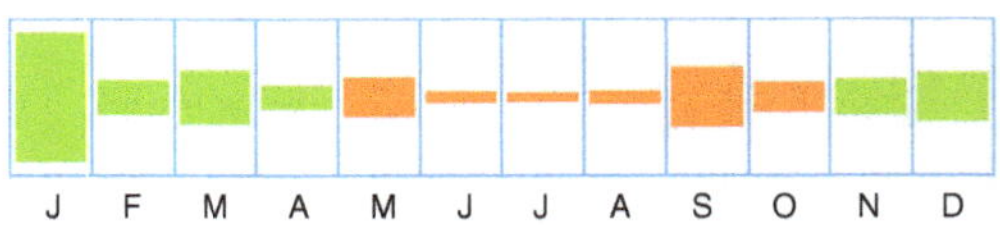

the Darwin region in January 1974.[276] Between 1974 and 1980, Common Terns were irregularly reported in the Darwin region, with a maximum of six birds in December 1980.[276] In January 1986,[34] about 100 birds were present at Leanyer SP and

25 at Knuckey Lagoons, the influx probably caused by tropical cyclone Hector's strong winds forcing birds at sea to seek shelter on the coast or just inland. Since then, the species was reported in 22 of the 27 years until 2013, probably due to more rigorous scanning of flocks of terns by birdwatchers than to an actual increase in regularity of birds visiting the region.

While the Common Tern has been observed in all months, birds typically arrive in the Top End, including the Darwin region, during September. A small, short-lived peak in reporting rates at this time suggests that birds disperse from the Darwin region soon after arrival. Numbers around the Top End coast remain high from September until March,[27] when there is a sharp reduction, indicating return migration. The pattern in the Darwin region differs somewhat, however, in that numbers peak in January and July. Around the Top End,[27] the majority of Common Terns are found at sea, foraging among pelagic fish feeding congregations, so the January peak in reporting rates and abundances may result from increased numbers coming inshore during early Wet season storms.

The majority of Common Tern records, and the largest numbers, are from the Lee Point-Buffalo Creek area. The maximum July count,[21] of 28 birds, was made in mid-July and, along with other records in the June to early August period, probably refers to overwintering birds. Outside of that period groups of 20 or more birds have been seen only in September and January.

Birds in breeding plumage have been recorded from September to November,[21,27] and in May, coinciding with their arrival and departure.

Whiskered Tern
Chlidonias hybrida

RANGE: almost worldwide in temperate and tropical regions. HABITATS: wetlands, irrigated agricultural areas, coast. STATUS: common Dry season migrant. BREEDING: Mar–Sep (NT).

Niven McCrie

The most widespread tern species in the Darwin region, the Whiskered Tern differs from most other tern species in its predilection for freshwater wetlands, although it is also seen off Lee Point and in Darwin Harbour. Numbers of Whiskered Terns peak in the period from late August until early October, sometimes as late as November, then drop off markedly as birds move to breeding areas, so that very few birds remain during the Wet season. Coincident with the fall in numbers of Whiskered Terns, there is an influx of White-winged Terns, and it is an almost smooth transition as numbers of the former drop and numbers of the latter increase. From April onwards the number of Whiskered Terns builds up again, and by mid-May they have all but replaced the White-winged Terns.

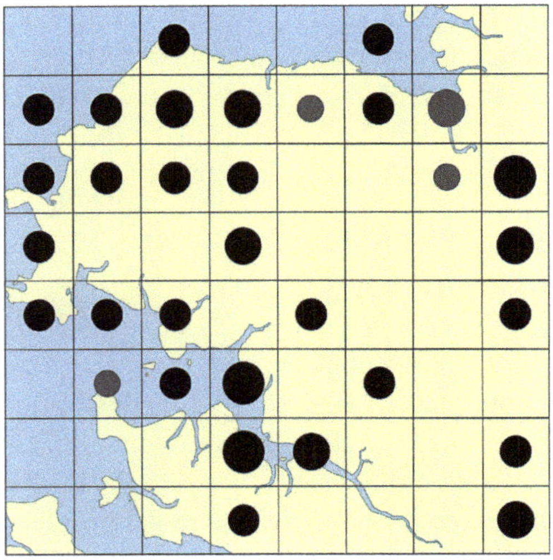

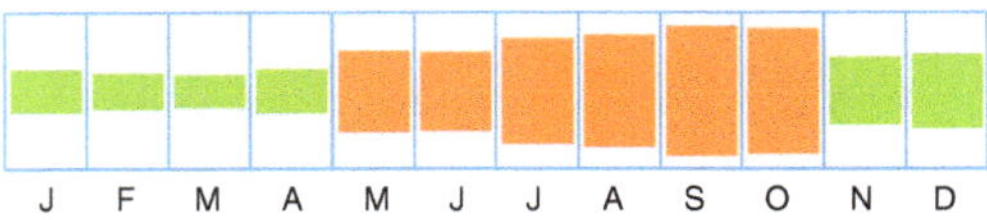

In Australia the Whiskered Tern is migratory and dispersive, breeding mainly in the south-east during late spring and early summer, and from April or May migrating north and north-west to non-breeding areas in northern Australia, as well as New Guinea and Indonesia as far as Borneo. While there is little evidence of northward passage of Whiskered Terns through the Darwin region, the increase in numbers during September and October may include birds on passage from Indonesia[276] to breeding areas in south-east Australia.

Ground surveys of waterbirds across the Top End during the 1990s[27] suggested a total population of around 17 400 Whiskered Terns, but a comparison of ground and aerial counts on the floodplains of the Alligator Rivers region, Kakadu NP during the early 1980s suggested that this region alone supported up to 45 000 birds.[290] In the Darwin region the highest reporting rates for the species are from Palmerston SP, Knuckey Lagoons and Leanyer SP, with the greatest numbers occurring at the latter site, including the region's maximum count of 600 recorded in August 1985.[31]

There are no confirmed breeding records of Whiskered Terns in the Top End, but small numbers were reported breeding in the Barkly Tableland wetlands between March and September 1993,[192] and in May – June 2004 Whiskered Terns 'bred extensively' on the Roebuck Plains in north-west Western Australia.[221] In August 2004 a Whiskered Tern that had been colour flagged on the Roebuck Plains was seen at Katherine SP,[221] indicating that some birds seen in the Top End may breed in north-west Western Australia. While the majority of birds in breeding plumage are seen in the Darwin region during the late Dry season, suggesting that they breed in south-east Australia, birds in breeding plumage have been seen in all months, so it is likely that at least some, possibly the majority of those present during the Wet season, breed during the Dry season in northern Australia.

A study of the diet of Whiskered Terns on the Magela floodplain in Kakadu NP revealed that males consume more fish than do females, while the shorter-billed females ate more invertebrates, such as dragonflies and beetles. Whiskered Terns hunt mainly by fluttering 1–2 m above the water, dipping periodically to take prey apparently on or just below the surface, and at Leanyer SP they are frequently seen in long lines, coursing end to end above the ponds, feeding in this manner. At Knuckey Lagoons, however, where the amount of open water is limited by vegetation, birds are seen to dive for food more frequently than they do on open water such as at the sewage ponds.

White-winged Tern
Chlidonias leucopterus

RANGE: eastern population breed Russia, n. China; winter in se. Asia, Australasia. HABITATS: coast, wetlands. STATUS: moderately common late Dry, Wet season migrant. BREEDING: not in Australia.

Adrian Boyle

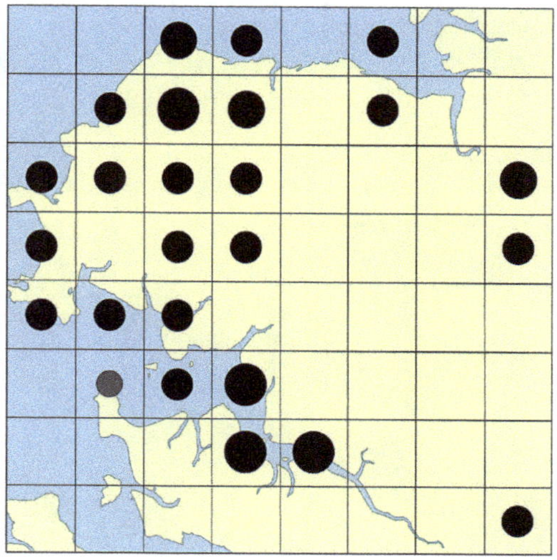

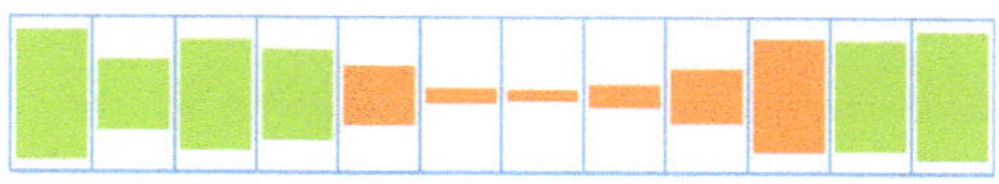

Travelling at least 12 000 km each year, from their breeding grounds in the marshlands of China to Australia and even New Zealand and back again, White-winged Terns typically associate with Whiskered Terns on freshwater wetlands, but they are recorded more commonly on the coast, on islands and at sea than are Whiskered Terns. In Darwin, the arrival of White-winged Terns during late September or October, occasionally earlier, coincides with the departure of Whiskered Terns, so that the period when the two species occur together is brief. While many spend the Wet season in the region, a fall in numbers for January and February indicates that some birds are on passage, although numbers on Top End floodplains peak during February,[27] so birds moving from the Darwin region may be dispersing to those areas. During March and April, numbers build up along the Top End coast,[27] and birds can be seen in large groups, often on beaches, before migration. In the Darwin region numbers increase also from late March, when birds are showing various stages of breeding plumage, and in most years numbers peak in April. However, in some years there is a distinct influx of first-year birds during the period late April to early May, following the departure of the majority of breeding-plumaged birds.[21] In some years also, a few birds may remain through the Dry season, but as none develop breeding plumage, they are also likely to be in their first year of life.

Ground surveys across the Top End during the 1990s suggested that at least 29 000 birds winter in the area, the maximum single count being over 15 000 birds on North Peron Island in mid-March 1992. In the Darwin region, most records, and the greatest numbers, come from Leanyer

SP, where a maximum of 1200 birds were present in December 1985. At Palmerston SP, the maximum count was 800 birds in November 1984,[34] while numbers at other coastal or wetland areas are typically small with few counts in excess of 20 birds.

Pomarine Skua
Stercorarius pomarinus

RANGE: breeds Russia and N. America n. of Arctic Circle; winters largely in seas off se. S. America, w. Africa, se. Asia, Australia. HABITATS: coast, pelagic. STATUS: vagrant. BREEDING: not in Australia.

www.wildlifeimages.com.au/Rohan Clarke

Three species of the piratic jaeger and skua group migrate south from their breeding grounds in the high latitudes of the northern hemisphere to spend the Austral summer in the southern hemisphere, including Australian waters. Of these, the Pomarine Skua is the only one to be found regularly during the non-breeding season in the tropics and, despite being typically pelagic, is the one most frequently recorded around the Top End coast, although there are few records for the Darwin region. Two dark morph birds were observed near Lee Point in October 1982,[34] and one was seen at Stokes Hill Wharf in November 1995. More recent records are from East Point in February 2007, and from Lee Point in January–February 2009 and March 2010.[291]

Parasitic Jaeger
Stercorarius parasiticus

RANGE: breeds in circumpolar region of n. hemisphere; winters in coastal seas of S. America, S. Africa, Australia, NZ. HABITATS: mainly coast. STATUS: vagrant. BREEDING: not in Australia.

Jon Norling

During the non-breeding season, the Parasitic Jaeger relies almost exclusively on kleptoparasitism to obtain its food, but while it also uses this foraging method to feed its chicks,[292] the species is the most significant predator of birds and their eggs in its breeding areas. It appears that pale-morph birds are less common than dark-morph birds in Australian waters, and studies have shown that dark-morph outnumbers pale-morph by three to one in those breeding areas from which it is assumed Australian birds derive. The sole Darwin record of the Parasitic Jaeger was of a single pale-morph bird seen harassing Little Terns roosting at Lee Point, in February 1978.[293] Parasitic Jaeger is the most likely of the skua-jaeger group to be seen inshore, so the scarcity of records for the region suggests that birds only rarely occur in Top End waters.

Long-tailed Jaeger
Stercorarius longicaudus

RANGE: breeds circumpolar, n. hemisphere; winters mainly in S. Atlantic seas, S. Africa, S. America; se. Australia, NZ. HABITATS: pelagic, coast. STATUS: vagrant. BREEDING: not in Australia.

Long-tailed Jaegers visiting the Australasian region probably originate from breeding grounds in east Russia and Alaska.[228] In Australian waters, the species occurs regularly only in the south-east, and is uncommon, rare or vagrant to more northern and western parts of the country. Long-tailed Jaeger is the most pelagic of the skuas and jaegers in wintering areas, so it is not surprising that the single record from the Darwin region involved a bird blown in during a period of stormy weather. It was

Adrian Boyle

found exhausted in the northern suburb of Malak in March 1983.[34] Another live bird, probably forced inland by tropical cyclone George, was found and photographed on the Stuart Highway about 450 km south of Darwin in March 2007.[294]

African Collared Dove
Streptopelia roseogrisea

RANGE: native, Africa; introduced to several cities, Australia. HABITATS: settled areas. STATUS: known from aviary escapees only. BREEDING: Feb, Nov (DR).

The taxonomic treatment of African Collared Dove is somewhat uncertain, with some authors using the name Barbary Dove for the domesticated form. The species is kept widely in aviaries in Australia, and there have been numerous instances of escapees breeding in the wild.[228] In Alice Springs, African Collared Doves became established ferally during the early 1990s. By the time control measures were put in place, around 2001, the feral population was estimated at about 4000 birds.[295] The species appears to have been successfully eradicated, and there have been no further reports since about 2006. In the Darwin region, feral breeding[34] of African Collared Doves was reported in Stuart Park and Nakara, during February 1975 and November 1990, respectively. There have been sporadic sightings of birds since, but no further reports of breeding.

Rock Dove
Columba livia

RANGE: cosmopolitan. HABITATS: urban areas and rural properties. STATUS: scarce localised, introduced resident. BREEDING: Oct (DR).

Niven McCrie

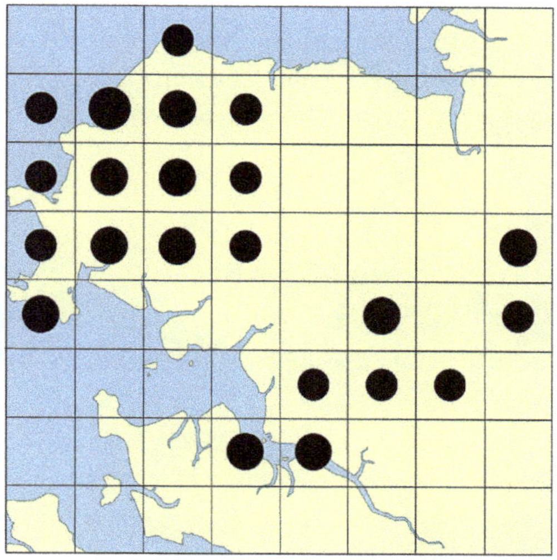

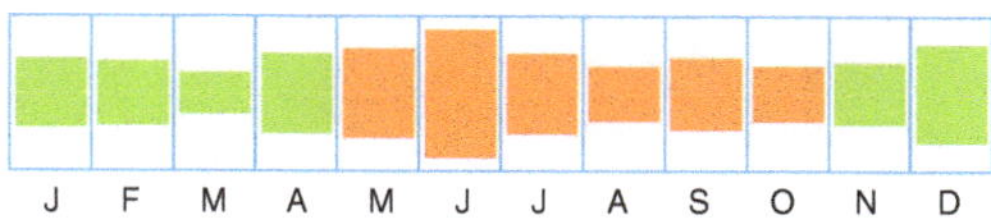

Among the most familiar and most studied bird species in the world, the Rock Dove or Feral Pigeon has the dubious honour of being the only non-indigenous bird species that has established long-term populations in the Darwin region. While there were no published reports of its presence before 1975,[2] large numbers came to the attention of the Parks and Wildlife Commission of the Northern Territory (PWCNT) after tropical cyclone Tracy (1974). During the Millner bird survey from 1988 to 1993, this was among the 10 most abundant species, though the number of birds rarely exceeded 20 and mean abundance fluctuated between years.[296] However, it was not until October 1997 that the Commission undertook systematic surveys to determine their distribution and abundance and, as required under PWCNT policy, instigate an eradication programme.[297] In that month, pigeons were found at 35 sites extending from Tiwi and Jingili in the northern suburbs, along McMillans Road and the Stuart Highway, to Coolalinga, where the greatest concentration was found. The Darwin Bakery at Winnellie and the Police Horse Stables at Berrimah had the highest numbers, both exceeding 100 birds, but most sites averaged fewer than 30 birds. The food source of large flocks was the grain around horse stables, while smaller flocks relied on seed provided at schools, sporting arenas and shopping centres.

Although the counts suggested a total regional population of around 970 birds, a media campaign resulted in the discovery of 786 birds using a two-storey, purpose-built coop on one property in Howard Springs, just 200 m from another property where three surveys had failed to detect the species![297] With the cooperation of the owner,

PWCNT staff successfully caught and euthanised all birds. From 1999 to 2004 around 5500 birds were found in Palmerston and nearby rural areas, and all were eradicated.[298] Since then small flocks of six or seven birds have been found at Casuarina and Fannie Bay, and dispatched with low velocity firearms. At the time of writing there were two pigeon racing clubs in Darwin, as well as a company that releases white doves at funerals and weddings.[298] Occasionally such birds go astray but are not always reported. Clearly birdwatchers and other members of the public can play an important role in reporting sightings of this species to PWCNT. Up until 2001 Rock Doves were recorded in 34% of cells, 68% of records emanating from the northern suburbs. A nest of a feral pair, containing eggs, was found in a Tiwi home on 7 October 1999.

Pacific Emerald Dove
Chalcophaps longirostris

RANGE: tropical and subtropical e. Australia, NG and e. Indonesia. HABITATS: monsoon rainforest. STATUS: moderately common resident. BREEDING: Jan–Mar (TE).

Mat and Cathy Gilfedder

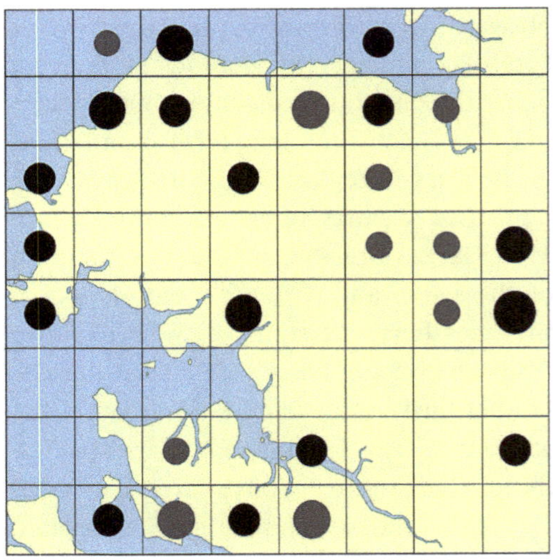

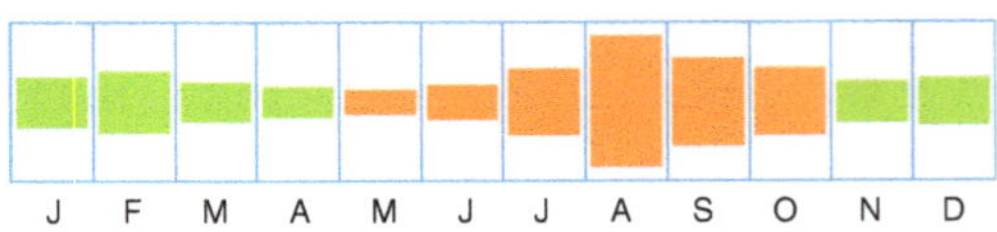

Unlike the loud accelerating call of the Rose-crowned Fruit Dove, the main call of the Pacific Emerald Dove is a series of drawn-out mournful notes that increase in volume but is generally soft and easily missed. This species is moderately common in monsoon rainforests and their edges, as well as in riparian forest, where it feeds mostly on seeds and fallen fruit on the ground, rather than in the canopy. Although recorded in 41% of cells, one-third of all records came from Casuarina

CR, and another third was shared between Howard Springs NR and Buffalo Creek.

Reporting rates suggest that the species is seasonal in our region, with a pronounced peak in August, followed by a steady decline during the late Dry season and Wet season to a trough in the early Dry. While there is no evidence of seasonal movements in eastern Australia,[228] the population on Melville Island was thought to leave the island in December, returning before June.[270] Thus the latter population was reputedly absent over the period when Darwin birds are least often recorded. However, these seasonal trends probably reflect variation in conspicuousness, possibly associated with breeding activity or local movements between rainforest patches.

A nest containing an egg and a very small chick, found near Channel Island on 15 March 1987,[43] appears to be the only nest record from the Darwin region, though an adult was seen carrying sticks, presumably to a nest, near Buffalo Creek on 10 March 1998.[47] However, a bird survey on the Cobourg Peninsula during 1965–66 yielded six nests with eggs or young from 30 February to 3 March,[141] when it was also noted that males were calling continually. In other parts of the Top End nests have been reported in June, and October through January,[228,299] but it appears from the above that most egg-laying occurs during the late Wet season.

Birds of the Top End and Kimberley race (*longirostris*) are markedly larger than those of the east coast race (*rogersi*).[228] Since the latter occurs mostly south of the range of the former, this size disparity is the reverse of the usual trend for tropical birds to be smaller than their temperate counterparts. The local race also has a stronger wash of blue and mauve on the head and underparts, respectively, and a larger patch of white on the shoulder, than does the eastern form.

Common Bronzewing
Phaps chalcoptera

RANGE: whole of Australia. HABITATS: open woodland. STATUS: vagrant and possible aviary escapee. BREEDING: reputedly Jul–Sep (NT).

Though widespread and abundant in many parts of southern Australia, the Common Bronzewing is generally scarce in the Top End, and rare along the north coast.[96,103,110,143,164] The species is regularly reported from the Pine Creek region,[300] and occasionally from Litchfield NP.[93] The first known record for Darwin concerns two birds seen at Leanyer SP in April 1974.[93] In 1980 a wild bird was seen in the grounds of a caravan park in Howard Springs,[43] and during the 1980s and 1990s a group of up to four birds resided at Larrakeyah Barracks, near present-day Cullen Bay.[93]

Laurie Ross

Although it is possible that the Larrakeyah birds were escapees from an aviary, the species has been reported from several localities on Cox Peninsula, and around Bynoe Harbour, to the west of Darwin since 1978.[43] Indeed, in December 2001 a Common Bronzewing was observed on a vacant bush block within the township of Wagait Beach,[301] near Mandorah, on the west side of Darwin Harbour and only 6 km from Cullen Bay. Thus it is feasible that the Larrakeyah birds immigrated from Cox Peninsula.

Flock Bronzewing
Phaps histrionica

RANGE: inland Australia, mainly tropical. HABITATS: grassland. STATUS: rare Dry season visitor. BREEDING: not in TE.

Adrian Boyle

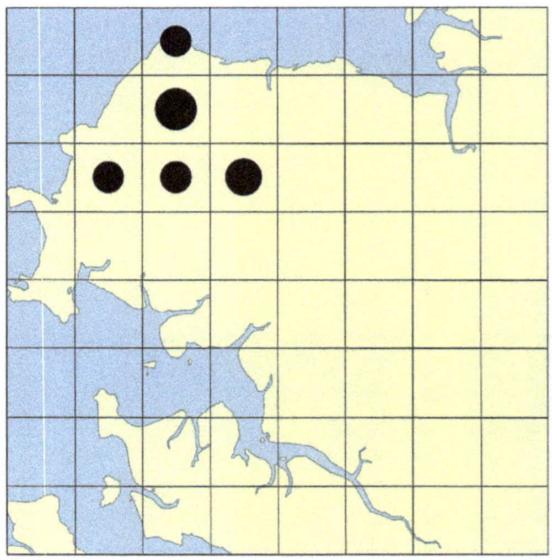

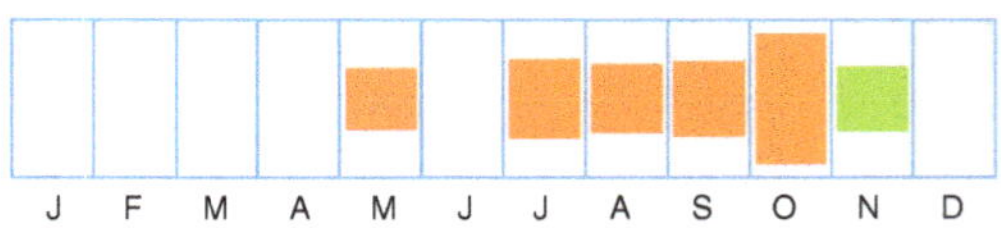

An occasional visitor to Darwin, the Flock Bronzewing has been mostly reported from the grassland and swamp adjacent to Holmes Jungle, and along tracks at Leanyer SP. At the sewage ponds sightings are generally of one or two birds nestled among the large, sometimes very densely packed, gatherings of waterfowl, and it may be that these dense packs of birds are what attracts this gregarious species to the site. An unusual but somewhat similarly suggestive observation in November 1996 concerns a bird that was seen flying among a group of Bar-tailed Godwits at Buffalo Creek, and later sighted standing among the godwits in seawater.[146]

The largest groups have been seen at Holmes Jungle, where 26 birds were observed in September 1974.[2] During September and October 2011, up to 19 birds flew in to drink at the swamp during the mornings and late afternoons on most days. On one occasion the birds appeared to alight momentarily on the surface of the water before flying off again. This, and a similar observation of a bird alighting for several seconds on the surface of water at Leanyer SP, may have been birds drinking. The presence of the observer in the Holmes Jungle observation may have forced the birds to drink more quickly, and at Leanyer SP the pond banks were too high to allow the Bronzewing to drink while standing. While most records are from July to November, there have been several observations of birds in May, two of which were made at Darwin International Airport,[47,71] the largest expanse of grassland in the region.

Crested Pigeon
Ocyphaps lophotes

RANGE: whole of Australian mainland. HABITATS: woodland, rural and urban areas. STATUS: rare visitor or resident, possibly including aviary escapees. BREEDING: no records in TE.

Laurie Ross

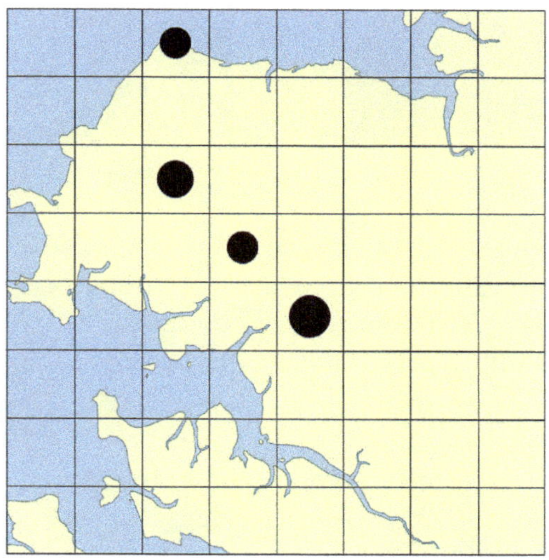

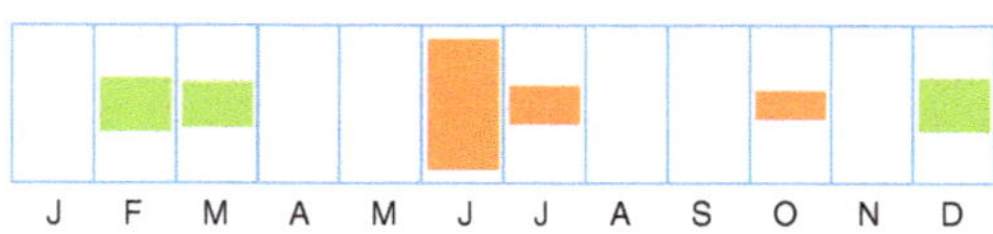

Once restricted within the NT to the arid zone in the south,[41] the Crested Pigeon expanded its range northward as European settlements arose during the late 1800s, reaching Larrimah by the mid-1940s.[164] It is now common in Pine Creek,[300] but only occasionally recorded further north. The first record in the Darwin–Humpty Doo region is of a bird sighted on 16 August 1969 at the old Marrakai Homestead, 39 km from the Darwin region.[302] Since early 1975 and until at least the mid-1980s, small populations of this species became established in Darwin city, and around Berrimah and Marrara.[21,60,228] As these populations apparently did not spread and may now be extinct, it is possible that they were descended from aviary stock that escaped after the destruction caused by Cyclone Tracy in December 1974.[228] Yet after a gap of 13 years, the species was reported in 2000 from both the Marrara and Knuckey Lagoons areas, and in January 2007 two birds were sighted at Fogg Dam,[136] 23 km from the Darwin region. These records, along with the 1969 one from Marrakai, suggest that wild birds occasionally disperse towards coastal areas.

Partridge Pigeon
Geophaps smithii

RANGE: endemic to monsoon tropical Australia. HABITATS: eucalypt woodland. STATUS: rare resident near s. edge of region. BREEDING: Mar–Jul (TE).

Micha V. Jackson

Considered a threatened species because of the extinction of populations in south-east parts of its former range, the Partridge Pigeon is still patchily common in the northern Top End.[303,304,305] The only record from the Darwin region is of two birds observed on the verge of the road from East Arm to Palmerston on 20 March 2001.[306] The adjoining habitat was swamp grassland with scattered pandanus. However, the species is known to occur just south of the Darwin region, in the Berry Springs area, where it has been recorded sporadically since 1971,[1] as well as in the Marrakai region.[3,21] It is quite possible that small numbers persist in the extreme south-west of our region. These birds breed during the late Wet and early Dry seasons, nesting on the ground, where their eggs and young are vulnerable to both predators and early Dry season fires.[47,305]

Diamond Dove
Geopelia cuneata

RANGE: whole of Australian mainland, mostly inland. HABITATS: floodplains, lightly wooded grassland and woodland. STATUS: scarce Dry season migrant. BREEDING: reputedly Jan–Aug (NT).

The diminutive Diamond Dove is a Dry season migrant to the Top End,[4,228] and while it is a regular visitor to subcoastal areas such as the Marrakai region[3,21,47] and Coomalie Farm,[47] it rarely reaches as far north as Darwin.[228] In the Humpty Doo-Marrakai region this species was recorded during 1969 and 1970, when inland parts

Bastiaan Hensen

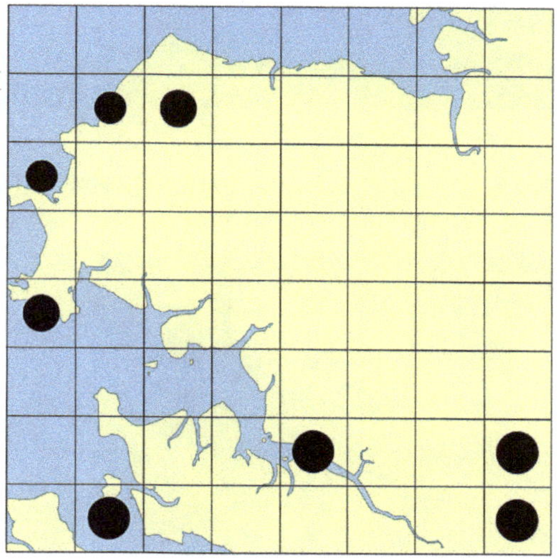

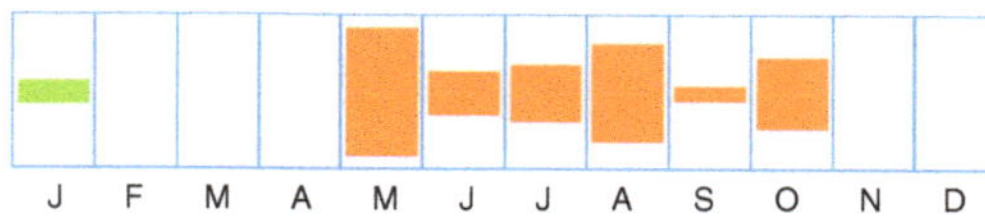

of the NT and Queensland experienced drought, but not in the preceding two years or 1971, when the interior had good rains.[1,124] In the Darwin region, it was reported in only seven of the 36 years since 1977. The majority of records occurred in 1990 and 2008, when there was widespread drought in the Top End and the southern NT, respectively.[124] Thus its appearance in our region, and probably other coastal areas, tends to coincide with times of droughts in other parts of the NT. It was reported from only 13% of cells, and almost a third of all records pertain to the Leanyer SP.

In the Humpty Doo-Marrakai region birds arrived around 12 April in 1969 and 1 June in 1970, but departed in early October in both years.[1] Reporting rates in the Darwin region are consistent, with the first birds arriving in May and the last departing in October, although there is one record for January. This migration is said to be the result of birds moving away from the more arid parts of the NT during the Dry season,[41] some traversing 300 km of woodlands to subcoastal areas.[299] Yet there are few breeding records from the NT.[4,5] In the Murray–Darling region of south-eastern Australia the species is a breeding migrant from November to March,[228] and while this corresponds to the period over which it is absent from our region, there is no evidence of a northward migration. Although it reputedly breeds from January to March in southern parts of the Top End,[41] at Victoria River Downs, one bird was found brooding its single egg on 3 October 1976.[98] Like the Peaceful and Bar-shouldered Doves, this species is gregarious at times and has been noted in flocks of more than 60 birds in the Marrakai region,[1] and in western Arnhem Land even 100 birds.[110]

Peaceful Dove
Geopelia placida

RANGE: tropical and e. Australia, NG. HABITATS: eucalypt forest, woodland and suburbia. STATUS: very common resident. BREEDING: year-round, but mostly Mar–May and Aug–Sep.

Niven McCrie

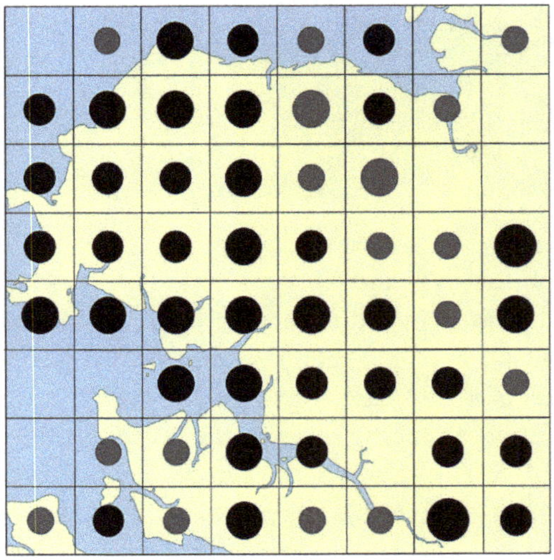

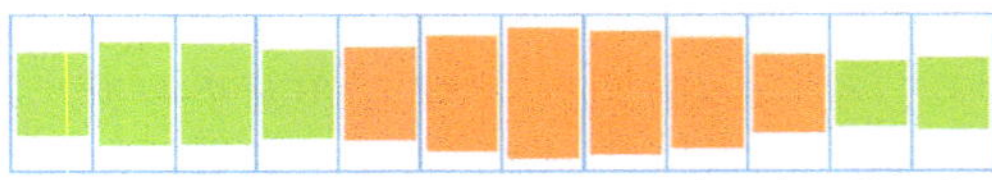

The smaller and less colourful of the two ground-foraging doves frequenting Darwin suburbs, the Peaceful Dove ranks second only to the Bar-shouldered Dove among the most often recorded birds of the region. It is exceptionally tame in urban areas, and may be seen foraging for scraps around tables at outdoor cafes both in the city and suburbs. While it was recorded 20% less frequently than the Bar-shouldered Dove, the Peaceful Dove was found in more cells (86%). In Palmerston, it is far less common than the Bar-shouldered Dove, recorded on only 11% of transect counts during the 2007 Dry season, but like the latter species was most often encountered in older suburbs, built during the 1980s and 1990s, and rarely seen in the surrounding bushland.[307] During the Howard's Peninsula survey the Peaceful Dove was recorded on half as many occasions and at half as many sites as the Bar-shouldered.[308] It occupied 11 of the 20 sampled habitats, but was most abundant in open paperbark woodland, tall paperbark swamp and tall dense grassland.

In the 1970s it was suggested that this species was more numerous in the Darwin-Humpty Doo regions during the Dry season than the Wet, due to an influx of birds from inland areas.[1] However, during five years of censuses between 1988 and 1993, numbers of this species around Millner were highest in January-February and lowest in June-July,[296] while the highest numbers at Darwin Golf Course during 2006/07 were recorded in October-November.[309] Reporting rates were slightly higher in the mid-Dry season than at other times, but show a very similar pattern to those of the Bar-shouldered Dove, which is generally considered sedentary. They were relatively low during the early Wet season, possibly indicating local movement associated with food availability. Along with several finch species, Top End

Peaceful Doves suffer a food shortage during the early Wet season, due to the scarcity of ripe or unspoilt grass seed, forcing them to switch to the seeds of sedges.[310] Coincidentally, at Yinberrie Hills, about 200 km inland from the Darwin region, the species was more abundant in the three months from October to December, when reporting rates were lowest in Darwin, than in all other months of the year except June.[311] Nests have been found in all months in our region, but breeding records show that egg-laying is biannual, with the major peak during September and a minor one in March–April.[47] Fewest nests were recorded in February and June.

Bar-shouldered Dove
Geopelia humeralis

RANGE: tropical and e. subtropical Australia, and NG. HABITATS: mainly monsoon rainforest, mangal and suburbia. STATUS: very common resident. BREEDING: year-round, but peak in Feb–Mar and Aug–Sep.

Niven McCrie

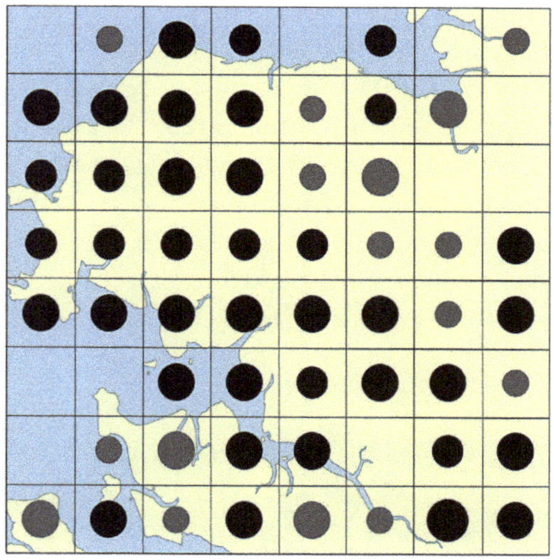

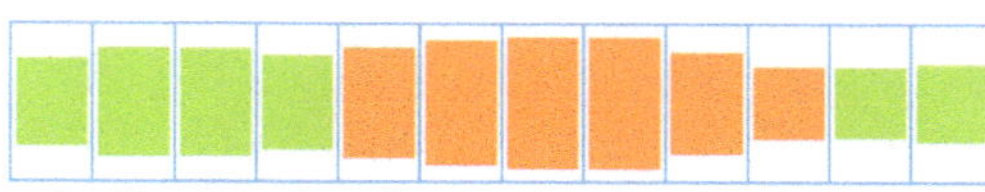

Indisputably one of the commonest birds in our region, the Bar-shouldered Dove was recorded more frequently, and from more cells (84%), than any other bird species. In Palmerston, it was both the most frequently recorded and most numerous species during surveys conducted from May to September 2007.[307] Its frequency and abundance were demonstrably higher in the older suburbs, built during the 1980s and 1990s, than in suburbs built after 2000 or in the surrounding bushland. Generally more timid than the much smaller Peaceful Dove, it is nevertheless highly urbanised, feeding and nesting in backyard gardens and perching on power lines beside busy roads, especially in the early morning and late afternoon. Indeed, along one well-traversed 2 km stretch of road in Nightcliff, the number of doves perched on power lines averaged 55, or one bird per 36 m, though up to 60 were seen perched within 20 m of each other.[47]

During the Howard's Peninsula survey this species was the third most frequently recorded species, being found in 43% of sites and in 16 out of the 20 sampled habitats.[308] It was most abundant in monsoon vine thickets and rainforest, Black Wattle forest and mangal. Like the Peaceful Dove, this species often rests and nests in mangals, but does not forage in this habitat.[187,312] Indeed, in a bird survey of 13 mangal patches around Darwin it was one of the commonest species in two of the seven mangal zones.[313] The average density of Bar-shouldered Doves at the 13 sites was twice as high as that of the Peaceful Dove, and maximum density three times as high. During weekly censuses from 1988 to 1993 around Millner, numbers were low during the early and mid-Dry season, then rose over the late Dry to a peak in November, dropping suddenly after March.[296] A similar trend was evident at Darwin Golf Course during 2006–07, though numbers peaked a month earlier and dropped gradually thereafter.[309] Reporting rates for the region do not reflect these trends in abundance, however, and may relate more to the timing of breeding and moult than to movements.

Males are often seen courting females on suburban ovals and roads. In the aerial display, the male advertises his sexual activity by flying 20 or 30 m vertically on loud clapping wings, then gliding back to his perch on outstretched wings and tail.[299] In the bowing display, the male courts the female by walking behind or around her, and at intervals bowing suddenly while raising his fanned tail to almost 45 degrees, revealing the white tips to the tail feathers. Local birds have been recorded breeding in all months of the year, but egg-laying peaks in February–March and, to a lesser extent, August–September.[47] Few clutches are laid during the early Wet, when reporting rates are also relatively low. One pair nested repeatedly in a hanging basket little more than a metre from the front door of a Tiwi home,[21] and up to four pairs have been noted nesting concurrently in densely vegetated yards in Nightcliff and Rapid Creek.[93]

Rose-crowned Fruit Dove
Ptilinopus regina

RANGE: tropical and e. subtropical Australia, and e. Indonesia. HABITATS: monsoon rainforest, mangal. STATUS: common resident. BREEDING: Aug–Feb.

While the eastern Australian race (*regina*) of the Rose-crowned Fruit Dove migrates north for the winter, the Top End and Kimberley race (*ewingii*) is resident,[228] occurring throughout the year in all rainforest areas in the Darwin region. Camouflaged in green, this species could be easily overlooked were it not for its frequent calls, the most distinctive of which is an accelerating and diminishing series of deep 'coos'. In the Top End, it is known to eat the fleshy fruits of over 30 species of rainforest trees,[314,315] which in turn depend on the Rose-crowned Fruit Dove and other fruit-eaters to disperse their seeds. It has been recorded in 41% of cells in the Darwin region, though 83% of all records came from the five cells encompassing Casuarina CR, Howard Springs NR, East Point, Holmes Jungle and Buffalo Creek. Reporting rates

Mat and Cathy Gilfedder

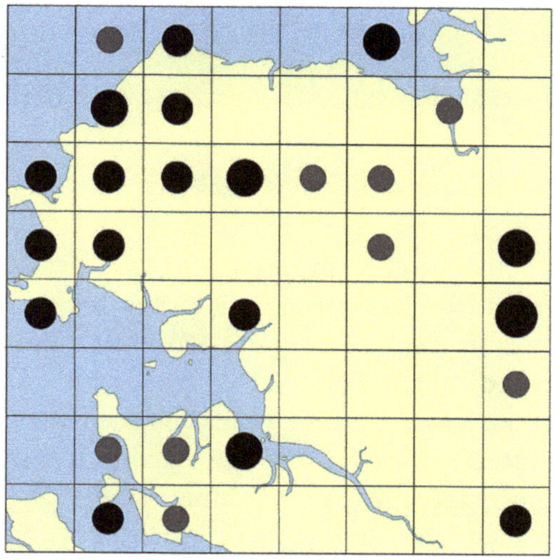

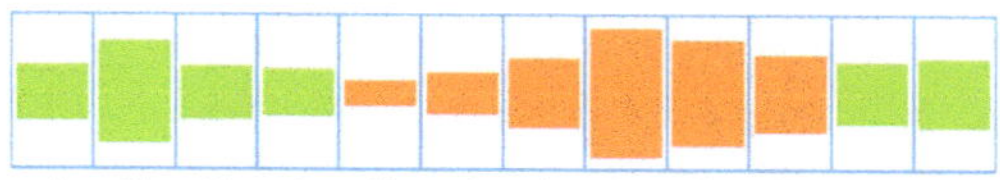

fluctuated markedly over the year, being lowest during the late Wet–early Dry season transition period, and highest in the latter half of the Dry, with a minor spike in February.

A study of fruit-eating birds in the Gunn Point-Black Jungle region, just north-east of the Darwin region, showed that Rose-crowned Fruit Doves were most abundant in wet, usually spring-fed, rainforests, especially those with a high density of Carpentaria palms.[316] However, they moved between different rainforest types depending on the availability of fruits. Numbers were highest in wet rainforests during the Wet season, when fruit abundance was also highest in this habitat, while during the early Dry season they were more abundant in dry rainforests, where the abundance of fruit showed less seasonal variation.[316,317] One radio-tracked Fruit Dove moved 5.2 km between rainforest patches,[315] though this species moved less frequently between rainforest patches than Figbirds or Torresian Imperial Pigeons.

The nest is a flimsy platform of sticks, through which the single egg can often be seen from below. Records of seven nests in Darwin suggest that eggs are usually laid from August to February,[47] though on Cobourg Peninsula seven nests containing eggs were found in February and March.[143] Thus the breeding season starts well before the Wet season peak in abundance of rainforest fruits,[317] possibly to ensure that young birds have an ample food supply.

Elegant Imperial Pigeon
Ducula concinna

RANGE: islands off e. Indonesia and nw. NG. HABITATS: forest, forest edges, cultivated areas with trees. STATUS: vagrant. BREEDING: not in Australia.

The occurrence in Darwin of the Elegant Imperial Pigeon is a somewhat unusual story in birdwatching. The bird was first noted, but not identified, by its call, in about May 1991, then heard and seen intermittently for over two years before it was finally identified in November 1993.[318] The pigeon was quite regularly seen in Nightcliff, and on occasion in Jingili and Leanyer, where it was last seen, in December 1994. In November 1993 the bird was reportedly moving twigs in apparent nest-building behaviour, but there were no further observations indicating a breeding attempt. A report of a second individual was not substantiated.[241]

The Elegant Imperial Pigeon inhabits the smaller islands in eastern Indonesia, where it is common in flocks of up to 40 birds.[86] Its range overlaps that of the Torresian Imperial Pigeon and it is possible that the Elegant Imperial Pigeon travelled to Darwin with migrants of that species.

Torresian Imperial Pigeon
Ducula spilorrhoa

RANGE: Tropical Australia, NG. HABITATS: rainforest, mangal, urban areas. STATUS: common Wet season migrant and resident. BREEDING: mainly Aug–Feb, peaking Sep–Nov.

Niven McCrie

A familiar sight to many people in the suburbs of Darwin and Palmerston,[319] the fruit-eating Torresian Imperial Pigeon is one of the few Australian-breeding birds that annually migrates to New Guinea.[228] Indeed, research conducted just outside the Darwin region provided evidence for this migration among local birds. Captured in rainforest 16 km east of Howard Springs NR, and fitted with a radio-transmitter on its back, one bird was tracked as it travelled eastward along the north coast to Copeland Island, off Cobourg Peninsula. Starting on 21 March 1996, the journey of 220 km took six days and included one flight of 80 km.[315] Although its transmitter was ominously recovered below the nest of a White-bellied Sea Eagle, the bird's trajectory suggested that it was headed for Dolak

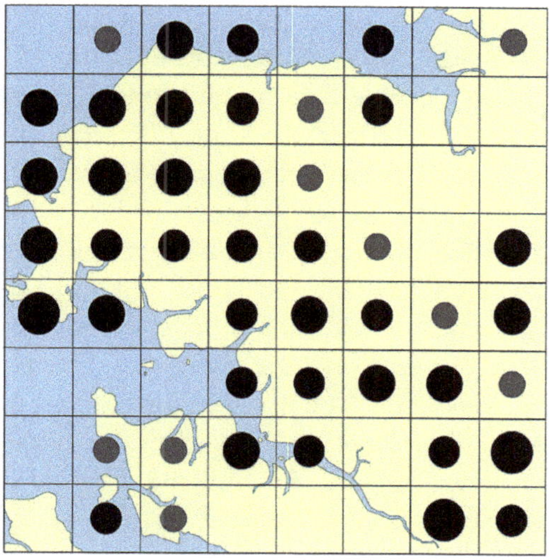

Island, adjacent to south-eastern Papua (Irian Jaya), involving a flight of another 630 km, or if it continued in a more northerly direction, the Aru Islands, 500 km away.

During the late 1960s and 1970s, the species was regarded as a strict migrant in the Darwin–Humpty Doo region, with most birds arriving in the third week of August and departing before the second week of February; in the early 1980s, too, it was described as being largely absent from April to August.[1,60] Indeed, reporting rates since 1983 show a gradual increase from June to September, and a sharp decline after February, yet they also reveal that some birds are present throughout the intervening months. Although early Dry season records exist from as early as 1967,[1,98] there were few until 1990, when 17% of the year's records for the species were from May through to July. From 1990 to 2013, the percentage of records from these three months averaged 15%, but in one year (1998) reached 25%. Almost daily bird counts around Millner from 1988 to 1993 show that there were no records of the species from June or July until 1992, when 10 birds were observed through these two months.[296]

The present extent of overwintering behaviour was unknown, however, until a communal roost tree was discovered in the mangal at Dinah Beach, just north of the city, in 2010. In the following year, monthly counts of birds flying to this roost site at dusk revealed 1325 individuals in early June and 1208 in July. This was followed by a steep decline until only around 70 birds occupied the site over October–November, when presumably many birds were nesting.[3,320] During 2012, however, the maximum number counted at this site over the Dry season was 408 in July. Resident birds may be restricted to Darwin. In Palmerston, the species was barely recorded during the 2007 Dry season until the third week of August, when small numbers arrived.[307] Populations in Kakadu NP[103] and Arnhem Land[110,143] are largely migratory, whereas those in the Kimberley region of Western Australia are resident.[321]

In the Darwin region, the species is widespread, occupying 73% of cells, with significantly more records coming from urban than rural or forested cells. An abundant supply of cultivated fruit-bearing plants,[322] including the native Carpentaria Palm, Weeping Fig, Banyan and Umbrella Tree, has undoubtedly facilitated overwintering behaviour or long-term residence in Darwin. Over half of the seeds collected below a day time roost in Millner during September 1991 belonged to Carpentaria palms. Individual birds have been observed swallowing up to 43 Carpentaria berries over short bouts of continuous feeding,[21] causing obvious distension of the crop. Outside Darwin, the species is largely restricted to monsoon rainforests and vine thickets, but it also visits isolated fruiting trees, such as figs, in eucalypt woodland up to 5 km from rainforest patches.[316] The aforementioned radio-tracking study revealed that these birds flew an average of 4.5 km between rainforest patches every three days in search of food sources.[315]

Nesting has been observed locally in all months except March and April, but egg-laying peaks in the three months from September to November, which account for almost three-quarters of all records.[47] Unlike the very large populations in eastern Australia, local birds do not breed in colonies,[228] though single pairs frequently share their nest tree with other bird species such as the Figbird, Magpie-lark and friarbirds.[47] In Darwin, pairs often build their flimsy stick nests in exotic tree species, including African Mahogany, mango and Milkwood. The single egg is incubated by the male and female on alternate days, not alternately by day and night as with some other pigeon species.[299]

Pheasant Coucal
Centropus phasianinus

RANGE: tropical and subtropical e. Australia, NG and Timor. HABITATS: grassy eucalypt forest and woodland, wetlands. STATUS: common resident. BREEDING: Nov–Mar.

Nolan Caldwell

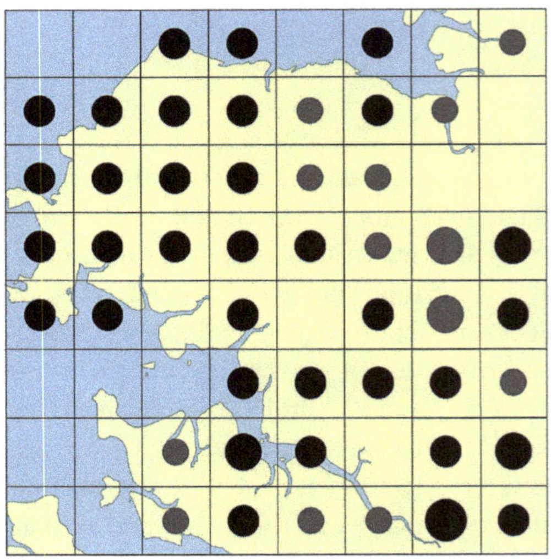

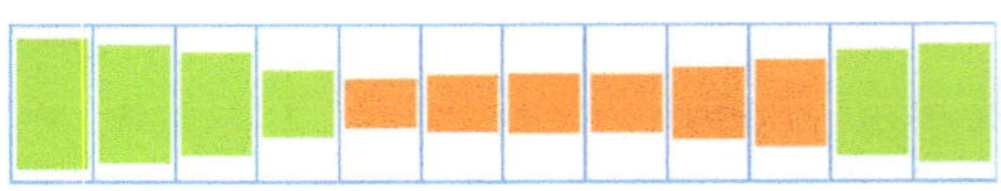

Although over half of the cuckoo species in the world, including all coucals, build their own nests and care for their own young, the Pheasant Coucal is exceptional in that the male not only builds the nest and incubates the three or four eggs, but also provides 80% of the food to the nestlings.[323] Yet some of the young that they care for are not their own. In a study at Howard Springs Hunting Reserve, only 2 km outside of our region, about 20% of young were not sired by their putative father. Females are almost 50% heavier and slightly larger than males, and while they are socially monogamous, staying with one male throughout the breeding season, they seek matings from males in neighbouring territories while their mates are incubating their first clutch.[323,324] In the above study each male territory overlapped only one female territory, but was 4–5 ha larger, preventing females from monopolising more than one male.[324]

The familiar, resonating calls of the Pheasant Coucal can be heard from distances of over 1 km. Of the two territorial calls, the 'descending whoop' is the more frequent, consisting of 8–29 evenly spaced 'oop' notes, steadily descending in pitch.[325] The much less frequent 'scale call' consists of 19–39 'oops' that initially descend in pitch while accelerating, then ascend while decelerating towards the end. In both call types, the female version is lower-pitched and shorter in duration than that of the male. Males call more frequently than females, and in the above study they produced about 80% of the calls during the dawn chorus. Pairs often performed duets, most of which comprised a single scale call by each partner, but mixed duets, in which the sexes gave different calls, were also given, and each sex started duets with equal frequency.[325]

Records for the Darwin region cover 73% of all cells, significantly more records coming from rural areas than urban or natural areas. Reporting rates rise gradually during the late Dry season and early Wet, peaking in January, then decline gradually until May. This seasonal pattern reflects the frequency of their vocalisations, which is highest from November to March,[326,327] coincident with the local breeding season.[324] The number of calls given during the morning chorus falls through the season.[325] Coucals call much less frequently during the Dry season, when they moult into their all-brown non-breeding plumage and probably wander over an area larger than their breeding territory, especially after fires remove the grass layer in which they forage.[1] This is usually the period when they are seen skulking in suburban gardens. Although the diet mainly comprises large insects, as well as frogs and reptiles,[323] Coucals are also known to take birds' eggs and nestlings.[327] One horrified Darwin resident watched an adult Coucal in his backyard as it systematically raided four nests of Bar-shouldered Doves in quick succession, eating and dropping eggs as it went.[62]

Pacific Koel
Eudynamys orientalis

RANGE: tropical and e. Australia, NG and e. Indonesia. HABITATS: rainforest and riparian vegetation. STATUS: common Wet season migrant. BREEDING: Dec–Feb (TE).

Laurie Ross

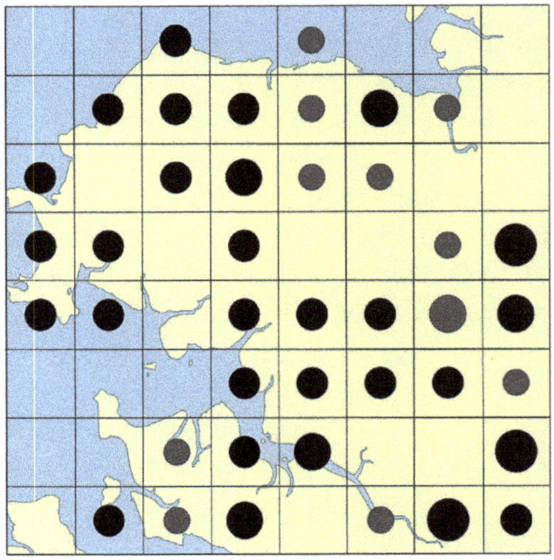

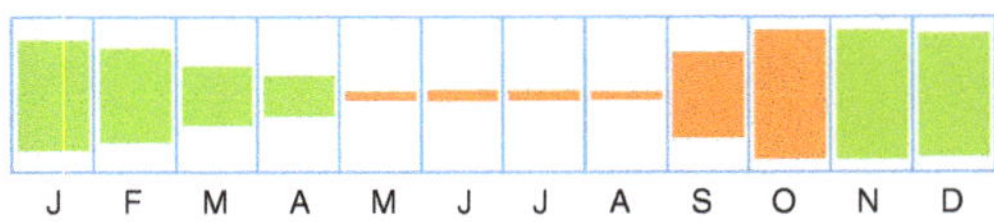

Probably due to its loud, unmistakable calls, the Pacific Koel has been recorded as frequently as the resident Little Bronze Cuckoo, despite being largely absent for four months of the year. Unlike the other migratory cuckoos, this species almost completely forsakes Australia during the winter, but while populations breeding in the south-east are said to migrate to eastern Indonesia, those from the tropics, including our region, are thought to migrate to New Guinea.[327] Interestingly, birds begin to arrive in the south-east and Darwin region around the same time, in mid-September, and the former often arrive as early as August, despite having to travel much further.[327]

Darwin's Koels seem remarkably punctual. In 16 of the 20 years between 1984 and 2011 the first arrivals were recorded in the last two weeks of September, the earliest record being 4 September, and the median date being 25 September. Thereafter the reporting rate appears to increase until December. Although censuses in the Leanyer SP area showed a peak in numbers in November, this was at least partly due to a large number of observations in November 1974, just before tropical cyclone Tracy. The Darwin population apparently plummeted in the cyclone's wake, owing to the destruction of local rainforest patches.[326] Reporting rates drop gradually from February through April, undoubtedly reflecting a decline in the frequency of calls and other conspicuous behaviour after breeding. In at least three years the last recorded date was in the last week of April, though stragglers in May have been reported[1,328] and a few birds apparently stay throughout the Dry season.

One of only two fruit-eating cuckoos in Australia, the Koel often ventures into suburbs where there are trees producing fleshy fruits, such as Banyans, Weeping Figs and Carpentaria Palms. It was recorded in roughly two-thirds of all cells in the Darwin region, though significantly more often in urban and rural areas than in natural areas. In a Millner home garden, beside a moderately busy street, one male Koel was frequently observed resting only 2.5 m above the ground in a small Sandpaper Fig tree over eight consecutive days in November.[328] During a broad-scale survey of fruit-eating birds of rainforest patches across the Top End, Koels were found at only 17 of 78 patches, being absent from the smallest (less than 1.5 ha) and very rare in patches isolated by more than 3 km from the nearest patch.[329]

Surprisingly, there are no confirmed breeding records of this species for our region but, based on 18 records for the Top End, local female Koels lay their eggs from December to February.[330] A courtship display, in which the male repeatedly bows at the female, was witnessed near Brinkin in mid-November.[326] The species' major hosts in tropical Australia are friarbirds and, although most records in the Top End refer to the Little Friarbird,[330] two observations suggest that the larger friarbird species are also parasitised locally. A full-grown juvenile was attended by Helmeted Friarbirds near Buffalo Creek on 15 January 1978,[326] while on 8 February 1987 a fledgling was fed by Silver-crowned Friarbirds at the Berrimah CSIRO laboratories.[43] It is curious that most species that are parasitised by the Koel are primarily insectivorous, indicating that the diet of the Koel chick differs markedly from that of adults.

Channel-billed Cuckoo
Scythrops novaehollandiae

RANGE: tropical and e. Australia, NG and e. Indonesia. HABITATS: monsoon rainforest and riparian vegetation. STATUS: scarce visitor or passage migrant, mainly late Wet season. BREEDING: May–Jun and Sep–Oct (TE).

Remarkably little is known of the breeding biology of this, the largest, species of Australian cuckoo, though its major hosts are predictably large, comprising crows and Pied Currawongs. Like the Pacific Koel, the Channel-billed Cuckoo eats fruit, so while the fruit-eating currawong seems an apt choice of host in eastern Australia, the carrion-eating crow seems less so. However, tame young Channel-bills have reputedly been reared on meat.[330] This species also migrates from breeding grounds in eastern Australia to New Guinea and eastern Indonesia for the austral autumn and winter,[327] but as it is apparently less common in the monsoon tropics, its status and movements in the Top End are poorly understood. Its general scarcity in the

Mat and Cathy Gilfedder

Darwin region may be partly related to the relative scarcity of its sole host in the monsoon tropics, the Torresian Crow.

Reporting rates suggest that small numbers visit Darwin throughout the year, but a noticeable influx occurs in April–May, with a possible secondary influx in January. At Banjan Point, Djukbinj NP, 37 km outside our region, they were recorded mainly in the Wet season, with 13 records between mid-October and late April but only one in August, from 1969 through 1971.[1] Moreover, in the South Alligator region of Kakadu NP, records between 1980 and 1995 indicate a marked influx in November in many years, while there was no evidence of an influx during April–May.[43] At East Point, one was seen flying in from the sea on 27 March 1971,[1] but this bird may have flown in from the Tiwi Islands rather than Timor or New Guinea. It has been reported from 30% of cells in the Darwin region, significantly more in urban areas than in rural or natural areas.

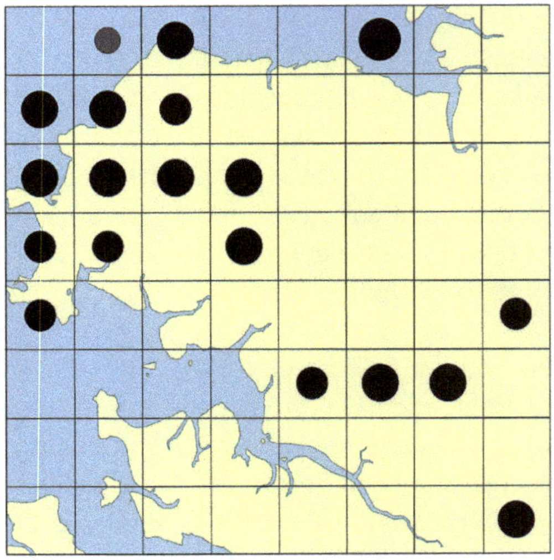

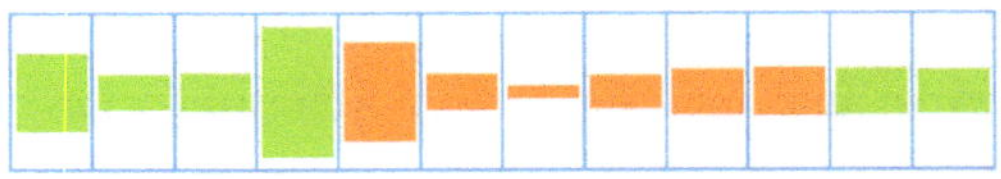

The only probable breeding record in Darwin concerns a fledgling being attended by a pair of Torresian Crows at Shoal Bay Waste Management Facility on 5 November 2011.[93] This fledgling presumably came from an egg laid in September, although local Crows nest mainly from January to May. Four other records of crows attending young Channel-billed Cuckoos from Mary River NP[43] and Kakadu NP suggest egg-laying in May–June and September–October.[47] This cuckoo is unusual in often laying more than one egg per host nest, more so in those of crows than currawongs. More than a half of 102 nests in eastern Australia contained two or more Channel-billed Cuckoo eggs or young,[330] yet most breeding records in the Top End concern only one fledgling or juvenile.

Horsfield's Bronze Cuckoo
Chrysococcyx basalis

RANGE: whole of Australia, s. NG, Indonesia and peninsular Malaysia. HABITATS: woodland and forest edges. STATUS: scarce visitor in both seasons. BREEDING: Dec–Apr (TE).

Bastiaan Hensen

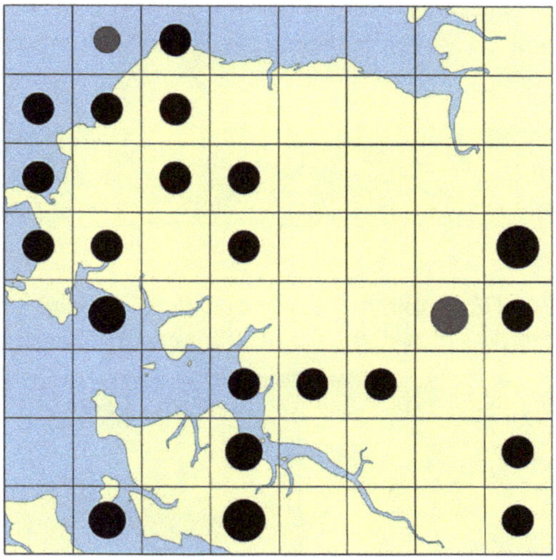

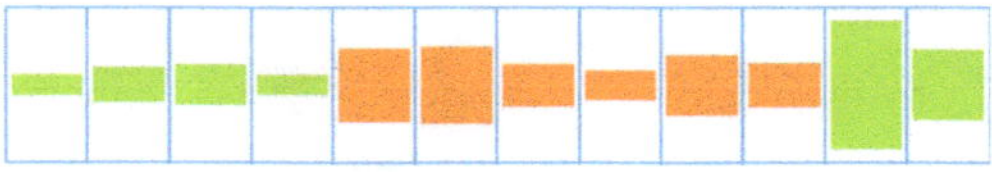

Parasitising the broods of more bird species than any other Australian cuckoo apart from the Pallid Cuckoo, the Horsfield's Bronze Cuckoo nevertheless prefers to exploit fairywrens throughout its wide range.[331] Not only does the female lay eggs that mimic those of its hosts, but its nestlings also mimic the appearance and begging calls of the hosts' nestlings.[332] In northern Australia its major host is the Red-backed Fairywren.[330] In the southern half of the continent, this species is largely migratory, breeding mostly during the spring and moving north during summer or autumn[327] but, like the Pallid Cuckoo and a few passerine species, eastern birds appear to take a more inland route on their southward journey than on their northward journey.[157]

The status of the Horsfield's Bronze Cuckoo in the Top End is unclear. Although generally considered a non-breeding visitor during the late Dry and early to mid-Wet seasons,[41,103] there are a handful of breeding records from southern parts of the Top End,[330] and scattered reports of the species outside this period. The timing and number of birds visiting the Darwin region appear to vary from year to year. From 1967 to 1971, the species was recorded 23 times in the Darwin–Marrakai regions, mainly during the late Dry and early Wet seasons, but in the first four years the number of months represented varied from two to nine, and in the fifth it was not recorded at all.[1] During surveys of the Leanyer SP area between 1974 and 1980, it was recorded in all months except February through April, but there was a dramatic spike in numbers during November of 1980.[326] In the whole Darwin region it was recorded in 20 of the 28 years (71%) from 1986 to 2013, and November was represented in more years (8) than any other month, while September,

October and May were each represented in five years. Reporting rates show the same spike in November, but also show a small peak in May–June. The latter peak may represent an influx of birds migrating from inland parts of the Top End, where they have been recorded breeding from December to April.[330] However, the November spike may represent a different population of birds, possibly passage migrants returning to Australia from Indonesia, where the species is a non-breeding visitor from March to October.[86] In support of this suggestion is a record from West Island, Ashmore Reef, off the Western Australian coast, on 9 November.[333]

Black-eared Cuckoo
Chrysococcyx osculans

RANGE: whole Australian mainland, vagrant to s. NG and e. Indonesia. HABITATS: woodland and margins of wetlands. STATUS: rare Dry season visitor. BREEDING: none in TE.

Mat and Cathy Gilfedder

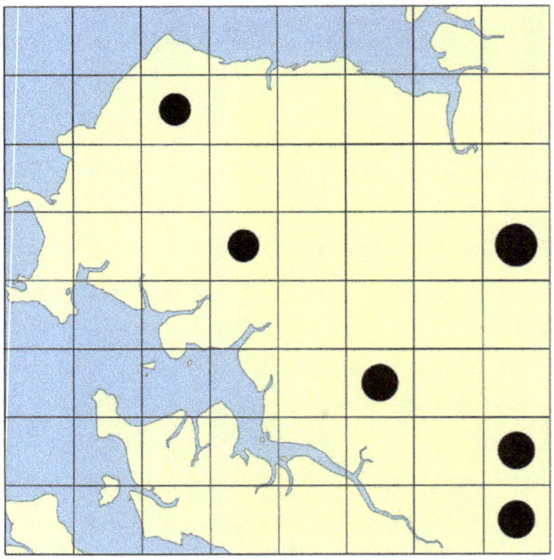

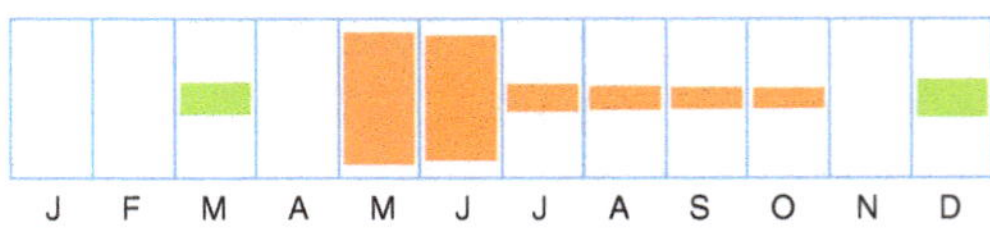

Predominantly a bird of the open woodland and shrublands of arid and semi-arid Australia, the Black-eared Cuckoo is a rare visitor to the Top End. Indeed, it was unknown in the Darwin region until a bird was sighted at Knuckey Lagoons on 26 July 1974.[326] However, the second record for the species came only three years later, on 15 June 1977, at Leanyer SP, where the species was subsequently reported in three non-consecutive years (1979, 2005, 2008). Contrary to some of the literature,[4,327] the species is largely a Dry season visitor to the Darwin region, with only two records in the Wet and half of all records referring to the months of May and June.[326] Of the 10 years in which it has been reported in the Darwin region, eight followed droughts in some part of the country, including the Barkly Tablelands in three cases.[124] Moreover, the year with the highest number of records (2005) followed widespread drought during summer and autumn in central Australia, affecting areas as far north as Elliott.[124] The effect of these dry conditions on their insect prey may have forced more birds than usual to the better watered regions of the north. All records refer to single birds.

The species is migratory in southern Australia, most populations leaving in summer and returning in winter.[327] Analysis of bird atlas records indicated a north–south annual shift of some type, but the pattern of movements was confused.[157] Given the total lack of breeding records for the Top End and most of the tropics,[5,330] it seems certain that visitors to Darwin are migrants from southern Australia. Occasionally such migrants fly beyond Australia to islands of Maluku in eastern Indonesia, where they have been recorded between July and November.[86]

Little Bronze Cuckoo
Chrysococcyx minutillus

RANGE: tropical and subtropical e. Australia, NG, Indonesia, Malaysia, Indo-China. HABITATS: mangal and monsoon rainforest. STATUS: common resident. BREEDING: biannual, Feb–Apr and Jul–Nov.

Niven McCrie

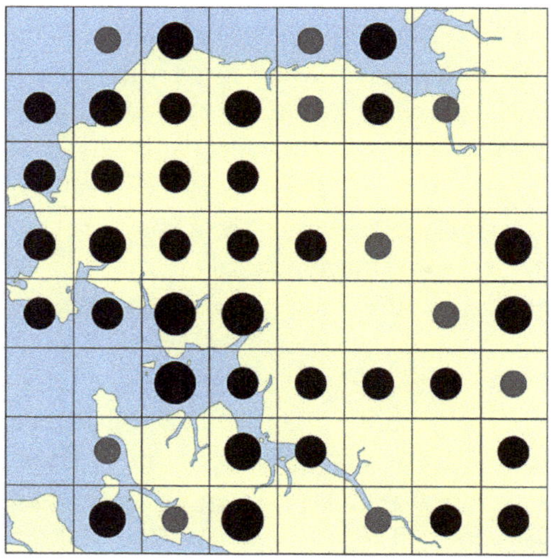

Darwin has earned a special place in the worldwide study of the co-evolutionary 'arms race' between brood parasites and their hosts, for it was in the mangals of Ludmilla Creek that research on the commonest local cuckoo revealed a previously unknown behavioural 'weapon' among its hosts. The smallest and most host-specialised of Australian brood-parasitic cuckoos after the Black-eared Cuckoo, the Little Bronze Cuckoo is the scourge of gerygone warblers from eastern Australia to Vietnam. This is particularly so in Darwin where parasitism rates are among the highest known in the world.[334] Here its major hosts are the Large-billed Gerygone and Mangrove Gerygone,[326,330] and along Ludmilla Creek more than a third of all nests found of both species contained cuckoo eggs.[334] The female cuckoo lays one egg in each gerygone nest that she parasitises, after removing one of the eggs from the gerygone's clutch. In Darwin mangals, however, it is common to find two or even three cuckoo eggs in a gerygone nest, presumably laid by different females. Yet as the first cuckoo egg to be laid in the nest will be the first to hatch, and this hatchling will eject all other eggs in the nest, the additional eggs are destined to fail. This multiple parasitism suggests not only a high density of cuckoos, but an inability among females to recognise the eggs of their own species when removing an egg from the nest.

Unlike other cuckoos, the Little Bronze Cuckoo and the closely related Shining Bronze Cuckoo of eastern Australia produce eggs that do not mimic the rufous-speckled white eggs of their hosts, instead laying a uniform dark olive-green or bronze-coloured egg. However, inside the poorly illuminated domed nests of their hosts these dark cuckoo eggs are effectively hidden.[335,336] Thus when the female cuckoo arrives at a nest she is more likely to see, and remove, an egg of a gerygone than one laid by a rival female. Competition among female cuckoos for host nests may have been even more important in the evolution of egg coloration in this species than the response of their hosts.[337]

While the eggs of this cuckoo do not mimic those of their hosts, their chicks are a different matter. Newly hatched Little Bronze Cuckoo nestlings look identical to those of the Large-billed Gerygone, being black-skinned, with a sparse covering of white down, unique among cuckoos.[334,338] This evolution of mimicry in cuckoo chicks suggests that their hosts discriminate against them. So it was a surprise when, in 2006, video footage showed a female Gerygone pulling a cuckoo chick out of its nest – the first case of cuckoo chick ejection by a host.[339] Many hours of filming in the mangal at Ludmilla Creek revealed that several individual Large-billed and Mangrove Gerygones engaged in this retaliatory behaviour. However, the gerygones are still learning to discriminate these imposters as they sometimes make the mistake of ejecting their own young! Moreover, a study of the Horsfield's Bronze Cuckoo in Canberra showed that their chicks can mimic the food-begging calls of their hosts soon after hatching.[332] This may explain why gerygones fail to pick out the interloper if ejection takes place one or two days after hatching. For the young cuckoo that evades ejection, however, its greatest challenge may well be fledging without drowning in the water often present below the nests of their hosts. In north Queensland young cuckoos have solved this problem by creating an exit hole above the nest entrance.[340]

In an extensive survey of mangals around Darwin, Little Bronze Cuckoos were found in all 13 patches surveyed. The species has been reported from about two-thirds of all cells in the Darwin region and, not surprisingly given their ubiquity in mangals, significantly more records are from coastal than subcoastal cells. However, it is also moderately common in vine thickets, rainforests, paperbark forests and other riparian habitats.[308] In the latter habitats it often parasitises nests of the Green-backed Gerygone.[47] Egg-laying has been recorded in all months, but in mangals breeding is biannual, with peaks in the late Wet and late Dry seasons, and few records in December,

January or June.[334] This pattern corresponds to the biannual breeding seasons of the two main hosts, and both gerygone species are equally frequently parasitised during both peaks.[47,334] The higher reporting rate during the late Wet season may reflect conspicuous behaviour before the start of the first breeding interval. Prior to breeding, up to six males sometimes congregate in one or two trees, calling, displaying and chasing one another as if competing for territories or females.[47] In addition, during the months of peak egg-laying, males court females with food offerings, invariably caterpillars.

Pallid Cuckoo
Cacomantis pallidus

RANGE: whole of Australia, vagrant to s. NG and Timor. HABITATS: eucalypt woodland and floodplains. STATUS: scarce visitor, mostly Wet season. BREEDING: May and Oct (TE).

Adrian Boyle

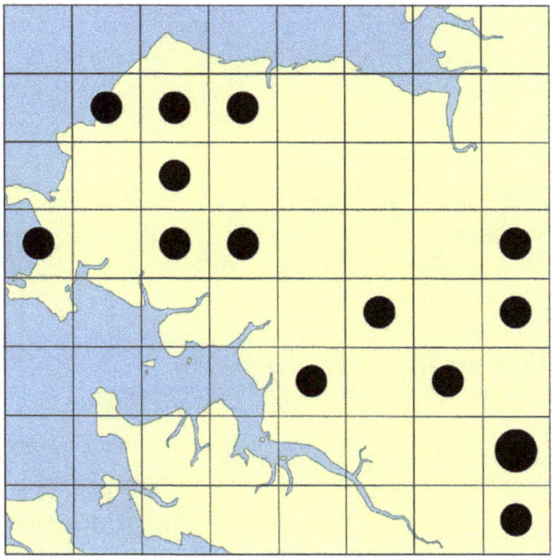

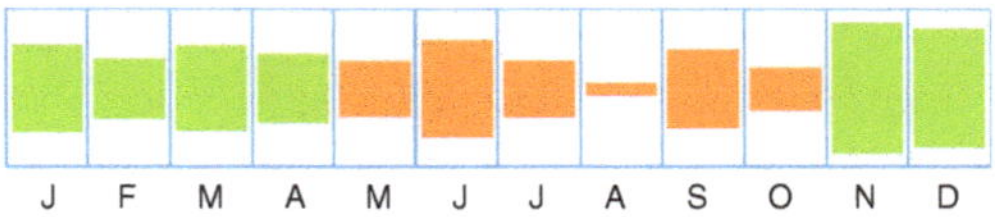

Generally scarcer in the tropics than in temperate Australia, the Pallid Cuckoo appears to be a highly irregular visitor to the Darwin region. In the Humpty Doo–Marrakai region, over the five years from 1967 through 1971, it was recorded once in each of six months, comprising the four months from October through January, plus May and June.[1] During monthly bird surveys around Leanyer SP between 1974 and 1980[326] it was recorded in all but one month, but was noticeably more abundant in November. Since then, until 2013, it was reported in the Darwin region in 12 years (35%), and reporting rates showed a similar peak during November–December. During the Kakadu survey, however, the species was recorded almost exclusively during the Dry season,[103]

although in 1990 one or two birds were sighted repeatedly during November and early December in burnt areas of the South Alligator region of the park.[43]

In southern Australia the Pallid Cuckoo is migratory, breeding during the late winter and spring, then moving north between October and May, and returning from June to September via an inland route.[157,327] The timing of this northward passage corresponds to the tropical Wet season when reporting rates in our region are highest. In the Kimberley region of Western Australia, too, the species is considered a mainly Wet season visitor, as well as a passage migrant,[321] moving northward in October and southward in April. The destination of the latter birds is unknown, but it is noteworthy that the species has been recorded on Timor in October, as well as in August, and occasionally reaches islands in the North Moluccas.[86]

Although the Pallid Cuckoo parasitises the broods of more host species than any other Australian cuckoo, its most frequent hosts are honeyeaters.[330] Indeed, the only local probable breeding records involved fledglings being fed by White-gaped Honeyeaters in Darwin suburban gardens on 18 November 1980 and 11 November 2004.[47,326] In addition, courtship feeding was observed in the Marrakai area on 19 November 1970.[1] In Kakadu NP, however, a fledgling was observed being fed by a Bar-breasted Honeyeater on 10 June 1979.[326] These records suggest that egg-laying takes place in the Top End during both the early and late Dry seasons.

Brush Cuckoo
Cacomantis variolosus

RANGE: tropical and e. Australia, NG and e. Indonesia. HABITATS: forest and woodland. STATUS: common resident. BREEDING: Oct–Apr.

Intriguingly, the choice of host species by the Brush Cuckoo varies geographically, and possibly seasonally.[326] In south-east Australia, where it is migratory, the Brush Cuckoo mostly parasitises the broods of cup-nesting bird species,[330] such as the Grey Fantail and Leaden Flycatcher. In contrast, the tropical race *dumetorum*, which is at least partly resident, appears to specialise in dome-nesting hosts. The major host in the Top End and Kimberley regions is the dome-nesting Bar-breasted Honeyeater,[330] although half of the 18 known breeding records for the Darwin region involve five other species (Northern Fantail, White-throated and Rufous-banded Honeyeaters, and Broad-billed and Paperbark Flycatchers), all of which build cup-shaped nests.[47,326] All but two of these breeding records indicate eggs being laid in the Wet season. Four of the five records involving fantails and flycatchers occurred from October to December, whereas all 13 records of honeyeater parasitism were made from January to July.

While Brush Cuckoos of south-east Australia migrate to New Guinea, leaving their breeding grounds from January to April, and returning around October,[327] Top End birds are assumed to be at least partly resident. Reporting rates in the

Vince Bugeja

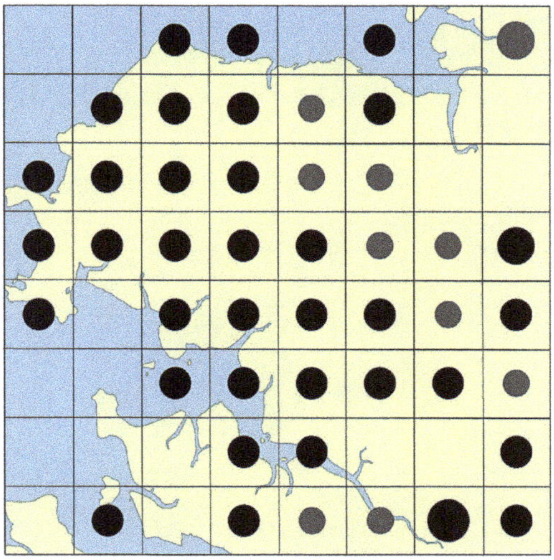

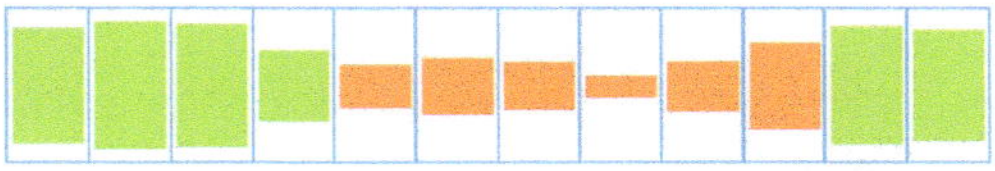

Darwin region suggest that local birds are seasonal in their occurrence, with relatively few records during the Dry season, but this may relate, at least partly, to their detectability, as birds may call less during the non-breeding season. Bird censuses in the Leanyer SP area from 1974 to 1980 showed a dramatic spike in abundance of this species during November, as well as a smaller peak in March.[326] One possible explanation for these spikes in abundance is passage migration to and from eastern Indonesia. Australian-breeding Brush Cuckoos have been recorded on islands of the Moluccas as far north as Moratai in May and September,[86] two months after and before the Darwin peaks, respectively, so it is feasible that some of these birds pass through the Darwin region. Resident birds may undertake seasonal movements between habitats. In both the Darwin and Kakadu regions the species appears to be largely confined to wet habitats, such as rainforests and paperbark swamps during the Dry season, but spreads into open forest and woodland during the Wet season.[1,103,308]

Oriental Cuckoo
Cuculus optatus

RANGE: tropical and subtropical e. Australia, e. and se. Asia. HABITATS: forests, woodland, floodplain margins. STATUS: scarce Wet season migrant. BREEDING: not in Australia.

Kirri Hardy

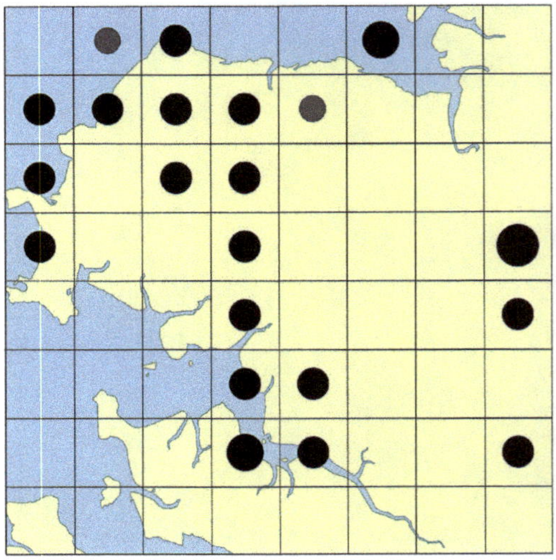

The monsoon rainforests, paperbark forest and eucalypt woodland of Australia may seem unlikely habitats for a species that breeds in the coniferous and birch-aspen forests of Russia, northern China and Japan.[327] Nevertheless, this is exactly the case for the Oriental Cuckoo, the only arboreal land bird species that migrates to Australia from the northern hemisphere. Every year, these birds leave their breeding grounds around July, and begin a journey of at least 7000 km to Indonesia, New Guinea and Australia. They arrive in small numbers in Darwin during October or November, or occasionally as early as September, and presumably wander locally until they depart, which in most years is by early April.[1,326] At Kapalga, near the South Alligator River in Kakadu NP, arrival dates for the species over six consecutive seasons (1984–89) varied by only 12 days, from 9 to 20 November.[43] Bird censuses in the Leanyer SP area from 1974 to 1980 showed that their abundance was highest in November and March, suggesting a passage movement through Darwin at either end of the Wet season.[326] Reporting rates support this proposition. Moreover, at Kapalga 75% of observations over eight years fell in these two months.[43] In 1986, at least 12 individuals were sighted in this area in the first week of March.

Oriental Cuckoos were recorded in a third of all cells in the Darwin region, equally frequently in coastal and subcoastal areas. There is a possibility that the Himalayan Cuckoo also visits Australia,[134,341] but the two species are identical in plumage. However, the Himalayan is slightly larger than the Oriental and on its breeding grounds gives a call consisting of three or four 'hoop' notes, higher-pitched than the two-note call of the latter species.[341,342] Unfortunately these cuckoos rarely call on their 'wintering' grounds, including Australia. Only 600 km from Darwin, Timor hosts both species, as well as the very similar resident Sunda Cuckoo.[86,342]

Eastern Barn Owl
Tyto delicatula

RANGE: whole of Australia, Indonesia, NG. HABITATS: Open country, cultivated areas, woodland, urban areas. STATUS: rare resident. BREEDING: Apr–Jun (NT).

David Webb

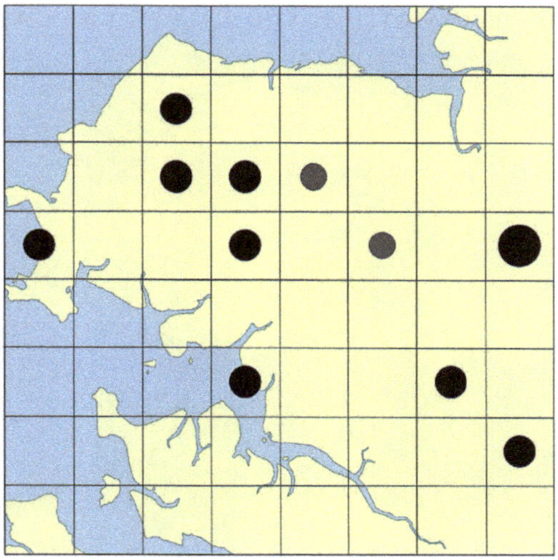

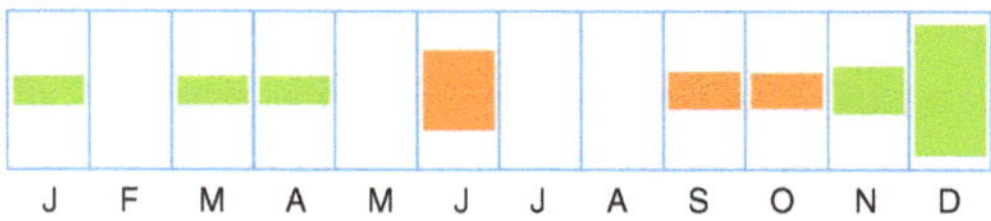

A familiar resident in many parts of Australia, the Eastern Barn Owl is sporadic or localised in most parts of the Top End. In the Middle Point area, near Fogg Dam, however, up to 20 Eastern Barn Owls have been seen along 3 km of road.[21] The birds are most frequently seen perched on power lines, fences or fence posts, but sometimes standing on the ground in the adjacent grassy paddocks.

There are relatively few records of Eastern Barn Owl in the Darwin region, but these are widespread and cover most months with a peak in the early Wet season. It may be that after the onset of Wet season rains, when hunting ground-dwelling prey becomes difficult due to the thick cover of tall speargrass, the birds concentrate in cleared or open areas, such as at Middle Point, where the highest numbers occur from January to April.[1]

The species' diet in the Top End has not been studied, but in the arid southern NT they are reported to feed predominantly on the House Mouse, and in the Barkly Tableland examination of Eastern Barn Owl pellets indicates that the Stripe-faced Dunnart is their main prey.[343]

An undated report of an adult and two nestlings along Oolloo Road,[34] in the Douglas–Daly region, about 150 km south-west of Darwin, is the only record of breeding in the Top End.

Eastern Grass Owl
Tyto longimembris

RANGE: India to se. Asia, NG, n. and e. Australia, New Caledonia, Fiji. HABITATS: tall grassland in dry or wetland areas. STATUS: rare irruptive visitor. BREEDING: Jul (TE).

Ross Monks

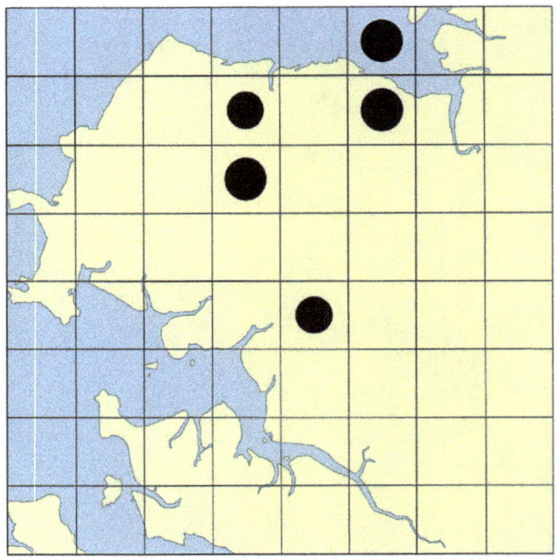

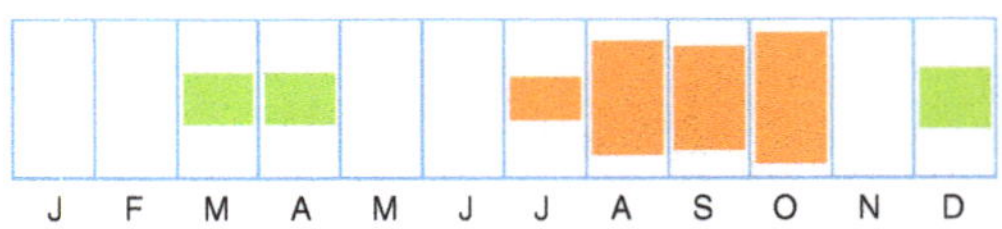

Though very similar to the Eastern Barn Owl in appearance, the Eastern Grass Owl differs markedly from that species in its ecology, hunting from the air instead of from perches, and roosting and nesting on the ground rather than in tree hollows. Perhaps ironically, due to its rarity, more is known about this species in the Top End than the more common Eastern Barn Owl.

Like the Letter-winged Kite, the Eastern Grass Owl is well known for its irregular irruptions, when birds appear in areas from which they have been absent for years, or even move into areas where they have not previously been seen. During the mid-1970s such an event occurred in the Darwin region, and birds were found in several suitably grassed areas, from March 1975 until 1979.[344] Sightings were most frequent, and numbers highest, on the black-soil floodplain at Holmes Jungle, where up to 11 birds were seen during August 1977. The birds were reported to hunt until about two hours after sunrise, and from around 50 minutes before sunset. Two owls were seen to capture rats, but no other observations of prey were made. Other sightings during that period were at Leanyer Swamp,[34] the '11-mile peg' on the Stuart Highway, now part of urban Palmerston,[344] and Kings Swamp, near Howard Springs.[43] Three recently deceased birds were found at Holmes Jungle in August 1977,[344] and this may indicate that death, as much as dispersal, is the reason that irruptions are short-lived.

There were subsequent observations of single Grass Owls at Camerons Beach[34] during 1985 and at Holmes Jungle in December 1997.[345] Following a 15-year absence of sightings, there was a small influx during September and October 2012, when up to four birds were seen, also at Holmes Jungle.[179] The sighting of a bird near Katherine in October 2012[53] may indicate that this irruption was more widespread, but there were no reports from elsewhere.

An Eastern Grass Owl was flushed from long grass at Middle Point during February 1968,[1] and there were further sightings there in May and December 1973.[34,344] It is not unlikely, however, that some Eastern Barn Owls, which are common in the area and which will stand in short grass, have been misidentified as Grass Owls.

In July 1980,[34] a nest containing three downy chicks was found in a small Grey Mangrove beside a tidal channel of the South Alligator River.[346] The nest was barely above high water level. The adults fed the chicks on Dusky Rats, an abundant floodplain-specialised rodent. Coming at the end of the 1970s irruption, it is possible that other birds remained and bred in that area. Nevertheless, those floodplains provide the most consistent source of reports of Grass Owls in the Top End, and may hold a small resident population.

Rufous Owl
Ninox rufa

RANGE: tropical Australia, NG and Aru Islands. HABITATS: monsoon rainforest, paperbark forest, riverine woodland. STATUS: scarce resident. BREEDING: Feb–Jul (DR).

The Rufous Owl has been found in at least 25 locations around the western Top End[347] and is likely to occur almost anywhere there is suitable habitat. Records are widespread in the Darwin region also, but most sightings are from the Botanic Gardens, where for many years the resident breeding pair attracted the attention of birdwatchers from around the world. Rufous Owls have also been seen at Casuarina CR, Holmes Jungle,[26,348] Palmerston, Howard Springs[21] and Coconut Grove,[349] and some of these may hold resident pairs, but no breeding has been confirmed.

In the Darwin Botanic Gardens, the Rufous Owl has been reported since 1996, and was first confirmed breeding there in 1999. In most years adult birds were observed near the nest hollow, situated about 8 m from the ground in an Earpod tree from about February, suggesting that this was the beginning of the breeding season. Young birds, either one or two, still

Niven McCrie

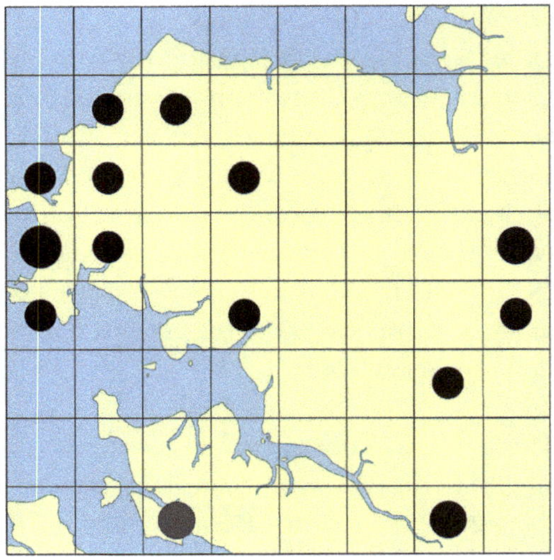

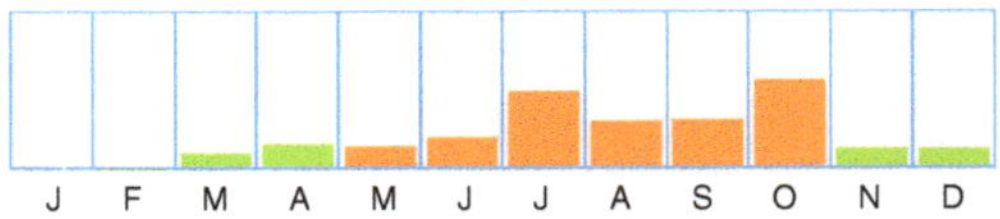

bearing downy plumage, have been seen near the nest from late May to early October, but usually in July or August. Given an incubation period of about 37 days and a further 35 days for the young to fledge,[327] this suggests that egg-laying occurs from March to July.

Each year, after the fledging of young, the owls moved about 300 m from the nest tree up to their non-breeding roost area, in the denser rainforest section of the gardens, an area of approximately 2 ha. The adjacent area of a little over a square kilometre of semi-open habitat was perhaps the birds' main hunting ground. The move to the non-breeding roost area generally took up to two weeks, with the adults and young spending time at staging roosts along the way. Immature birds were typically seen with their parents until about November or December, when they moved out of their natal territory. Individual birds seen occasionally in Darwin suburbs may be immature birds dispersing from their natal territory. A bird reported from the Palmerston suburb of Driver, in May 2004, for example, may have come from Howard Springs, and another seen in Ludmilla in November–December 1998 and November 1999[67] appears likely to have come from the Botanic Gardens.

In Kakadu NP the diet was found to comprise mainly small- and medium-sized mammals and few birds,[347] whereas in north Queensland, although small mammals predominated, avian prey were a major food source.[350] In the Darwin region, the Black Flying Fox has been the most frequently reported prey item, but the owls have also been seen eating Northern Brush-tailed Possum, Blue-tongued Skink,[67] Bar-shouldered Dove,[32] Torresian Imperial Pigeon and Red-collared Lorikeet. As flying foxes and possums are readily available within the Botanic Gardens, the resident Owls may not need to hunt far from where they roost. Night-time observations suggest they may also hunt around the nearby golf course,[67] and Owls have been seen hunting Red-collared Lorikeets at a city roost, about 2 km from the Botanic Gardens.

Rufous Owls roost in a range of tree types, heights and foliage densities, although they apparently prefer to roost on horizontal limbs of trees that have dense foliage above, and little foliage below. They generally perch close to the trunk, at heights from little more than 3 m above the ground to 10 m or more, with most perches about 5–8 m high.[21,351]

Carcasses of the species were found at the Botanic Gardens in July 1998, September 2000[21] and September 2007.[21] The cause of those deaths is unknown, but a dead Rufous Owl found in November 2007 at Noonamah, about 35 km south-east of Darwin, had its death attributed to Cane Toad poisoning.[352] The nest tree in the Botanic Gardens was lopped in December 2006, removing the nest hollow. There were irregular reports of the species from then until July 2011, then none until October 2013,[88] when a single bird was seen, with subsequent observations of two adults and a juvenile from late the following month.[353] While the latter indicates breeding relatively nearby, a new nest site has not yet been found, but a Rufous Owl seen in a suburban yard in Stuart Park during June 2012[354] may have been one of the pair from the Botanic Gardens.

Barking Owl
Ninox connivens

RANGE: tropical, e. and sw. Australia, s. NG and ne. Moluccas. HABITATS: monsoon rainforest and edges, eucalypt forest, paperbark forest, urban areas. STATUS: moderately common resident. BREEDING: Jul–Sep.

The 'woof-woof' calls of the Barking Owl, especially those of the male, could easily be mistaken for the yapping of a domestic dog. The female's shorter and higher-pitched 'wok, wok' is recognisable when pairs perform their frequent duets.[327] The commonest owl in the Darwin region, this species was recorded in 28% of cells, with significantly more records in urban-dominated cells than in rural or forested areas. Indeed, about one-third of all records emanate from the cell encompassing Royal Darwin Hospital and Charles Darwin University Casuarina campus, sites where these birds are regularly seen or heard. In April 1997, three pairs were heard calling between the hospital and Tracy Village.[21]

Barking owls are more active at dusk and dawn than other Australian owls.[327] On six of 12 nights, mostly during August–September 1999, birds in the hospital area started calling at least 58 minutes, and up to 98 minutes, before sunset.[21] During the breeding season this species preys on bird species as large as the Tawny Frogmouth, Sulphur-crested Cockatoo, Blue-winged Kookaburra, Pheasant Coucal and Blue-faced Honeyeater, as well as mammals such as the Sugar Glider.[327] When not breeding, however, insects are its staple diet. One bird was watched foraging for insects around the spotlights on the rooftop car park of Casuarina shopping centre.[47]

Reporting rates were higher during the Dry season than the Wet, dropping sharply after October, and increasing gradually after the mid-Wet. This suggests that these birds are most vocal in the Dry, as they are

Nolan Caldwell

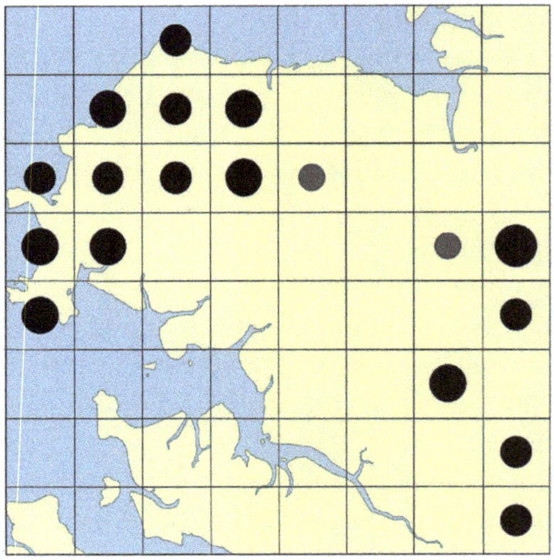

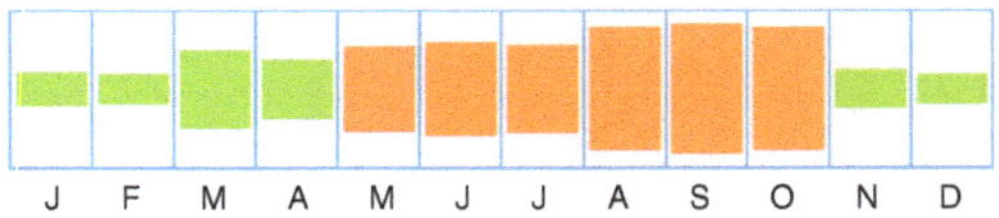

typically detected by their calls. Four birds were heard calling simultaneously between 22:00 and 22:30 hours on 27 July 2006, apparently advertising their territories.[21] Observations at six local nests suggest that egg-laying occurs from late June or early July to September.[21,93] One pair, though not necessarily the same individuals, nested in a hollow in a Bloodwood tree at Nakara Primary School for at least three seasons (1997, 1999 and 2003).[21] The incubating female was sometimes heard giving calls from inside the nest, including a low rumbling growl 'wuwuwuwuw' as well as the typical bark, in response to its mate's calls. A month later, scratchy grating noises, presumably given by chicks, emanated from the nest. The owners of this nest succeeded in raising three young to fledgling in at least one year.

Southern Boobook
Ninox boobook

RANGE: whole of Australia, s. NG and e. Indonesia. HABITATS: eucalypt forest and woodland, paperbark forest, monsoon rainforest and edges. STATUS: scarce resident. BREEDING: Sep and possibly Jan.

The hooting call of the Southern Boobook, aptly described by its vernacular name 'Morepork', is less urgent and penetrating than that of the Barking Owl and, compared to the latter species, the Southern Boobook rarely performs vocal duets, females calling less frequently than males.[327] The diet of the two species is broadly similar, but the much

Will Riddell

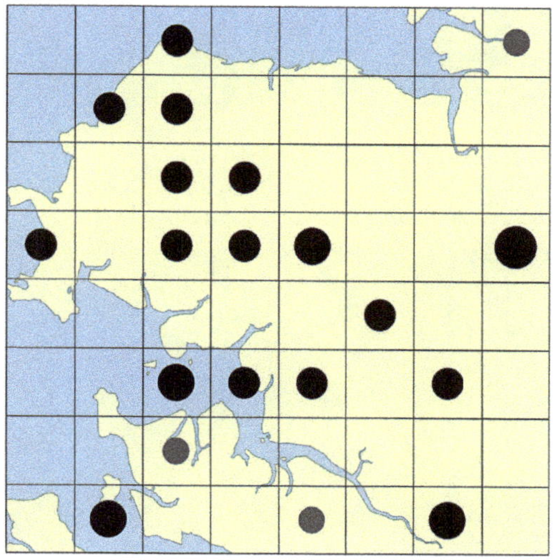

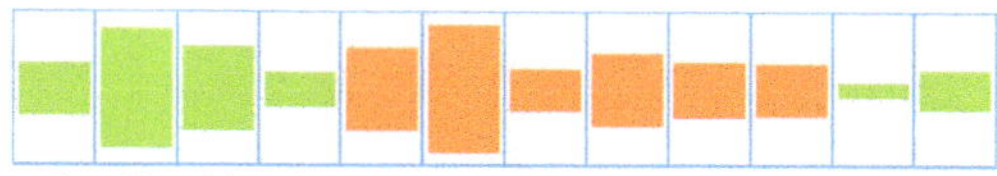

smaller Southern Boobook is unable to tackle such large prey items as adult rabbits or cockatoos, preferring to hunt smaller mammals and birds, such as mice and passerine birds.[327]

Reported from 31% of cells in our region, the Southern Boobook appears to be as widespread as the Barking Owl, yet scarcer, with about one-third the number of records of the former species. The reporting rate shows two spikes, one in February and the other in June, but given the small number of records for most months these fluctuations may not represent any real seasonal pattern. In Kakadu NP, the species appears to make seasonal movements between habitats. During the Wet season it is most abundant in monsoon rainforest,[103] but is very scarce in this habitat during the Dry,[355] when it is common in eucalypt forest and woodland. The dense cover of Annual Speargrass and other grasses in the latter habitats during the Wet season probably makes hunting too difficult to be profitable.

Almost nothing is known of the Boobook's annual cycle in the Top End. The only known breeding record for the Top End is of three food-begging fledglings seen near the South Alligator Inn, Kakadu NP, on 11 December 1992.[43] As the nest cycle, from egg-laying to the chicks fledging, takes almost 10 weeks in this species, at least in southern Australia, the South Alligator clutch was probably laid in September. Indeed, based on historical records, the breeding season in the NT includes both September and January.[41]

Tawny Frogmouth
Podargus strigoides

RANGE: whole of Australia. HABITATS: forests, woodland, urban areas. STATUS: moderately common resident. BREEDING: Aug–Dec.

Niven McCrie

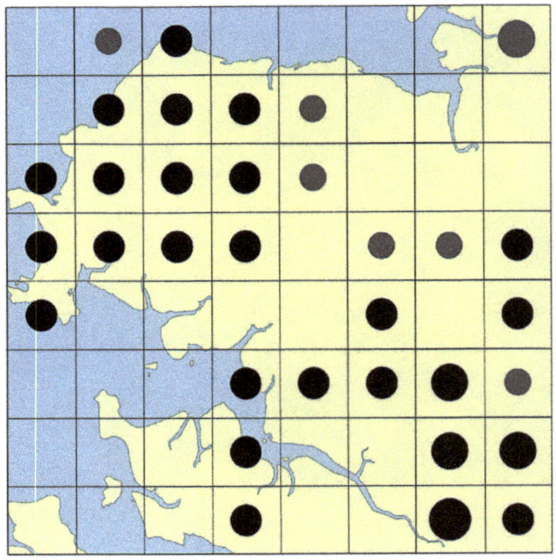

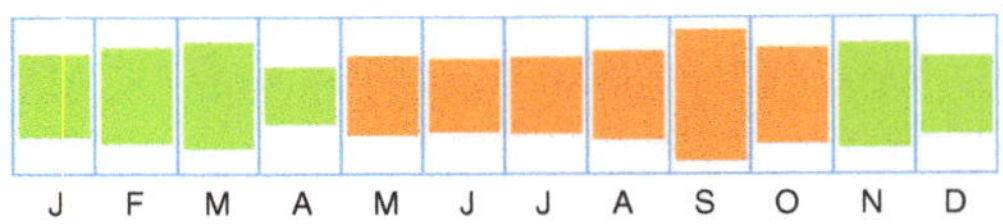

The commonest nocturnal bird in the Darwin region, with the possible exception of the Bush Stone-Curlew, the Tawny Frogmouth is still occasionally referred to as the 'frogmouth owl', despite the facts that frogmouths are totally unrelated to owls and are largely insectivorous rather than carnivorous. Local birds belong to the tropical race *phalaenoides*, which is only about 80% of the length, and as little as 60% of the weight, of the race (*strigoides*) in south-eastern Australia.[327] The general plumage coloration of males of all races is always grey, but that of females may be either grey or red-brown, birds of the latter colour being known as rufous or chestnut morphs.

As plumage colour and patterning have evidently evolved in these birds to provide them with camouflage against their day roost sites,[356] we might expect rufous females to roost in different sites to those chosen by grey birds. This may partly explain the lack of reports of rufous females since these birds are often spotted when on the nest, and only the grey male incubates the eggs by day, though females often roost close by.[327] As soon as they detect an intruder, Tawny Frogmouths adopt a camouflage posture, which consists of stretching the head and body upwards and aligning them in the direction of the branch, rather than at right angles to it, as well as sleeking down all feathers and closing the eyes to a thin slit.[356] They then remain motionless until danger has passed.

Frogmouths were reported in 53% of cells, significantly more in subcoastal than coastal cells, and less frequently in forest than in urban and rural areas. Despite considerable fluctuations in the reporting rates, local birds are probably sedentary, as they are elsewhere in Australia. Around Darwin Golf Course, up to three birds were seen in 46% of all weeks from July 2006 to June 2007, the longest period of absence being nine consecutive weeks,[309] suggesting that these individuals were long-term residents. In Kakadu NP the species was recorded more in open forest and monsoon rainforest than in woodland.[103] They feed primarily on nocturnal insects and spiders, mostly captured on the ground after pouncing from a low perch.[356] They often seize insects from roads where, sadly, many are hit by passing cars.

There is some evidence that the breeding season of Tawny Frogmouths in tropical Australia is longer than that of their temperate counterparts, allowing two broods to be raised in a season.[327] Indeed, at Darwin Golf Course, one pair with a dependent youngster began incubating a second clutch in October 2006, and by February 2007 had another fledgling,[309] suggesting that local birds are double-brooded. Nevertheless, breeding records in the Darwin region show that most (92%) clutches are laid from August through October, with few as late as December.[47]

Spotted Nightjar
Eurostopodus argus

RANGE: whole of Australian mainland and far e. Indonesia. HABITATS: woodland, floodplains. STATUS: scarce Dry season migrant. BREEDING: Oct (TE).

Adrian Boyle

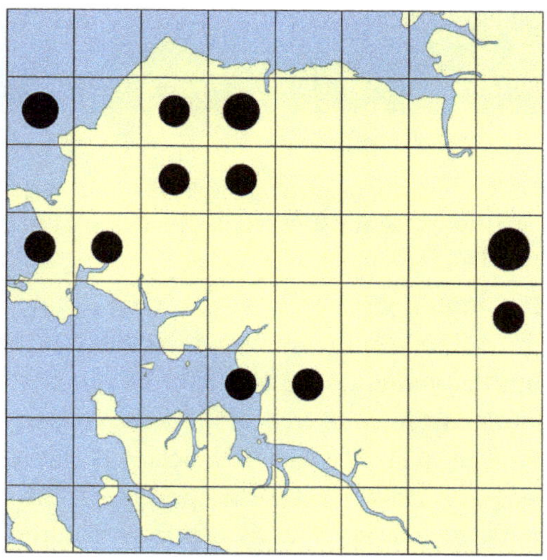

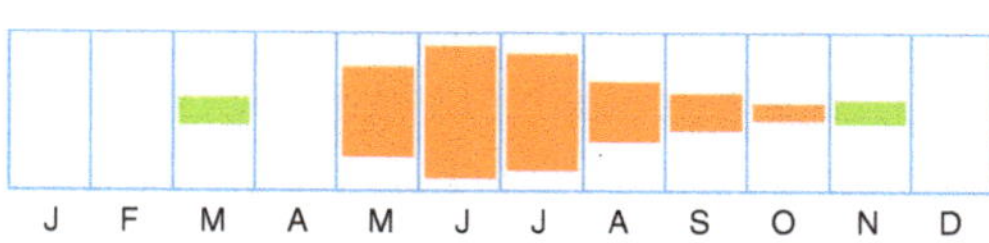

Although sometimes considered to be resident in the tropics,[327] the Spotted Nightjar is largely absent from the Top End during the Wet season.[5] Moreover, analyses of atlas data revealed that this species, like the Black-faced Cuckooshrike, White-winged Triller and White-breasted Woodswallow, migrates in a north-westerly direction from its breeding grounds in inland south-eastern Australia, a trajectory that takes birds towards the Gulf of Carpentaria and Top End.[157] From 1967 through 1971, it was considered locally common in the Humpty Doo–Marrakai areas from early April to early September in both 1969 and 1970, but was scarce or absent in the remaining three years.[1] However, at Coomalie Farm, 45 km south of the Darwin region, it was recorded in May–June in all five sampled years during the 1990s.[47] In the Darwin region, reporting rates peak in June and July, suggesting that many of these birds pass through the region on their northward migration, but the steady decline afterwards and the lack of any peak later in the year suggest that they return south by a different route. Some birds may be destined for small islands to the east of Timor, where they have been recorded from July to September,[86] or even the Aru Islands, but they also visit the Tiwi Islands at this time.[5,357]

Since 1977 Spotted Nightjars have been reported from 19% of cells in the Darwin region, significantly more often in subcoastal cells than coastal ones. The species roosts during the day on bare ground, often on stony ridges.[358] At Victoria River Downs 11 birds were flushed from an area of only 5 m^2 on a sparsely vegetated ridge strewn with large rocks.[98] A similar 'communal' daytime roost comprising seven birds was encountered on a stony ridge at Coomalie Farm in June 2002.[47] Spotted Nightjars often sit on roads at night, where they are very approachable, and, sadly, many are hit by vehicles. The only record of breeding in the Top End is of one egg found under a small bush in the Adelaide River District in either 1942 or 1943.[177] This record is undated, but elsewhere in the NT it reputedly breeds in October, December and April.[41] In the Kimberley region of Western Australia the species is known to breed in June and July.[359]

Large-tailed Nightjar
Caprimulgus macrurus

RANGE: coastal tropical Australia, se. Asia and NG. HABITATS: mainly mangal, monsoon rainforest, vine thickets and paperbark swamps. STATUS: moderately common resident. BREEDING: probably Aug–Oct (TE).

The monotonously repeated 'chop, chop, …' call of the Large-tailed Nightjar is a familiar sound of the night across much of south-east Asia, as well as along the coast of tropical Australia. Yet the species is rarely seen as it usually calls from perches within dense vegetation, and by day roosts on the ground, where its patterned plumage provides excellent camouflage against the litter-strewn forest floor. Sightings typically involve birds observed at dusk as they 'screen' the air for flying insects while circling above a clearing, or by day when flushed from their day roost or nest sites.

In our region they have been recorded in 30% of cells, over half of all records coming from the monsoon rainforests at Buffalo Creek, Lee Point, East Point and Howard Springs Nature Park. As early as 1948 the species was noted calling nightly in Nightcliff,[110] when a large patch of rainforest was still present in this suburb. Nowadays it can still be heard calling from the mangals

David Webb

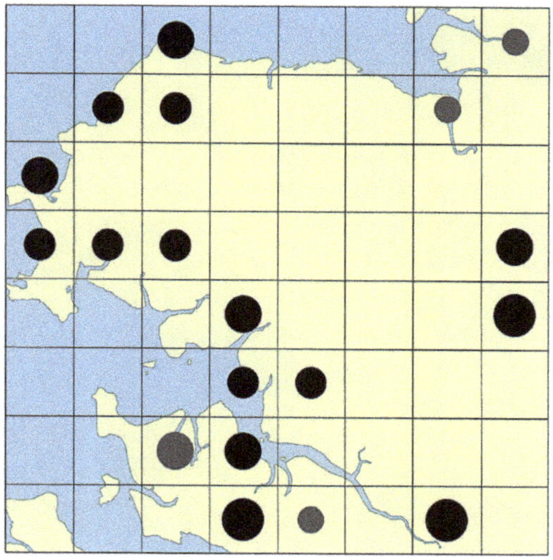

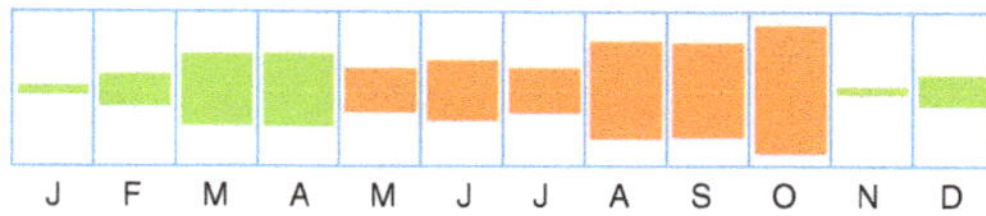

just south of Nightcliff. Reporting rates were lowest during the early and mid-Wet season, and highest in the late Dry. Given that the species is sedentary in both Australia[327] and eastern Indonesia,[86] and that the majority of records are likely to be auditory, these fluctuations probably reflect calling rates. Thus calls may increase in frequency during the late Dry season, when the species appears to breed in the Top End.[41] The sudden drop in records in November suggests that birds are quiet at the time when they might be expected to have chicks, which are highly vulnerable to predation until they are able to fly. Nightjars do not build a nest, laying their eggs on the ground instead, usually among leaf litter, but occasionally along roads and tracks.[327] The only breeding record in our region relates to a bird incubating two eggs that had been laid on bitumen, just a few centimetres from the fence surrounding the Palmerston SP.[21] The sitting bird was partly shaded by tiny shrubs growing up against the fence.

Australian Owlet-nightjar
Aegotheles cristatus

RANGE: whole of Australia and s. NG. HABITATS: eucalypt forest and woodland, monsoon rainforest edges. STATUS: moderately common resident. BREEDING: Sep, Oct and Jan (NT).

Unlike true nightjars, but similar to some owls, the diminutive Australian Owlet-nightjar both roosts and nests in tree hollows. This species is unusual among nocturnal birds in its habit of calling from its roost-hole during the day.[327] These calls, once known, are the best means of detecting this bird, and may show that it is more widespread than currently known in the Darwin region. During the late 1960s and

Vince Bugeja

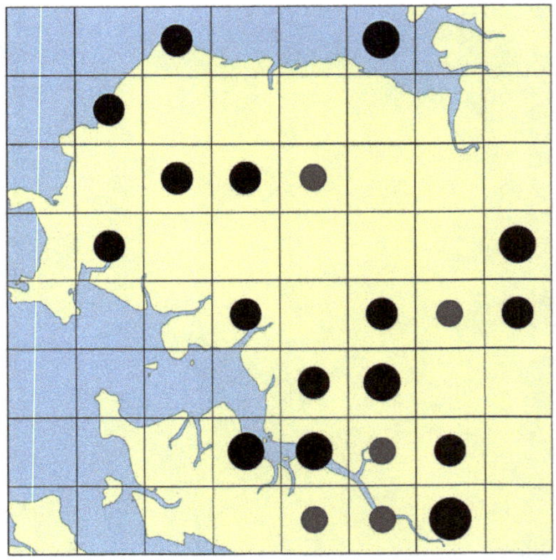

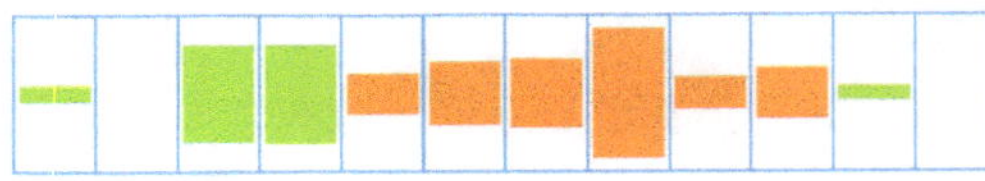

early 1970s it was recorded only twice, both records of road-killed birds well outside Darwin.[1] Since then it has been reported from 33% of cells in the region, both coastal and subcoastal, and including suburbs such as Tiwi and Fannie Bay.[60] Reporting rates fluctuated considerably over the year, but were lowest during the early and mid-Wet season. As these fluctuations probably reflect variation in calling frequency or other conspicuous behaviours rather than seasonal movements,[327] the species may be most vocal or active during the late Wet and mid-Dry seasons in the region. In the NT the species purportedly breeds mainly in the late Dry season, based on four century-old clutches, two of which were collected in the McArthur River district, one in October 1913, the other in January 1914,[360] and the remaining two, of unknown provenance, in September and October.[29,361] Of the four colour types across Australia, the rufous-grey morph, with a buff or rufous wash across the head and breast, apparently predominates in the tropics.[327,362]

White-throated Needletail
Hirundapus caudacutus

RANGE: breeds c. Siberia to Japan, s. Asia to sc. China; winters mainly e. Australia. HABITATS: aerial. STATUS: vagrant. BREEDING: not in Australia.

Scott McGregor

A common spring-summer visitor to eastern Australia, White-throated Needletail is a vagrant anywhere west of the Gulf of Carpentaria, including the Top End. Like the Pacific Swift, it breeds in Siberia, Mongolia and northern Japan, but its migration route to Australia is more easterly, with most birds entering Australia via Torres Strait.[86,327,363] Also like Pacific Swift, its appearance often coincides with frontal weather change or falling barometric pressure.

There are three records of White-throated Needletails in the Darwin region, all from Lee Point: one was sighted in December 1969,[1] 16 birds in November 1980,[43] and an unknown number in April 1985. At Kapalga, in Kakadu NP, small numbers were recorded in April 1979, February 1981, November 1992 and April 1994.[34,43,327]

Pacific Swift
Apus pacificus

RANGE: nominate subspecies breeds s. Siberia, n. Mongolia, n. China, Japan; winters Indonesia, Australia, NG. HABITATS: aerial. STATUS: moderately common Wet season migrant. BREEDING: not in Australia.

In an annual migration paralleling that of the migratory shorebirds visiting Australia, the Pacific Swift leaves its breeding grounds in Siberia, Mongolia and Japan during August–September, migrates southward to Indonesia and Australia, and returns in April–May.[327] Passing through the Lesser Sundas mainly from September to November,[86] these birds usually start arriving in northern Australia during October. Individuals and small groups of up to six birds have been observed flying south-east across the Timor Sea in the first weeks of October and November.[333] The appearance of this species, in flocks of up to 3000 birds,[327] often coincides with low pressure weather systems, though this association may be exaggerated because it appears to call more often, and is therefore more noticeable, during thunderstorms.[327]

The earliest arrival date recorded during 1967–71 was 15 October, and the latest return passage date 30 April.[1] Of 15 years

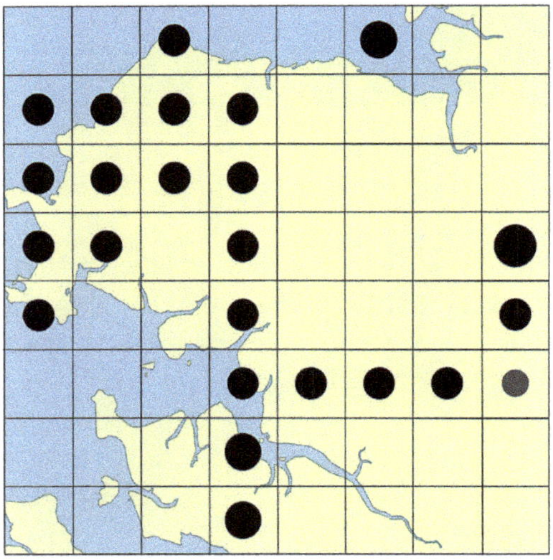

between 1979 and 2008 in which the first arrivals were noted, the earliest was 10 October and latest 18 November, with a median date of 21 October.[21,43,47] The latest recorded return dates between 1985 and 2010 varied from 20 March to 2 May, with a median date of 10 April. There are few records of the species in the wettest months of January and February, when birds presumably move inland. In late

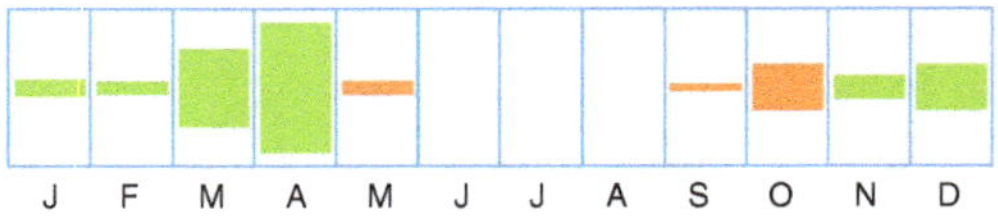

January 1945, however, up to 20 widely spaced birds were seen moving southward singly or in pairs at Batchelor, some 50 km south of the Darwin region, and in 1943 several were sighted in late January and late February at Larrimah, 400 km south-east of our region.[164]

Reporting rates were much higher in March–April than in October–November, suggesting that the northward passage occurs across a narrower front than the southward one. Yet larger numbers tend to be seen during the southward passage, with many records of 40–80 birds in Darwin[21,47] and flocks of more than 200 in two years (1982, 1985) at Kapalga,[43] 140 km east of our region, in October and November. The largest numbers reported during the northward passage were around 60, in March 1986[43] and March 2010.[47]

Oriental Dollarbird

Eurystomus orientalis

RANGE: tropical and e. Australia, NG, e. and se. Asia. HABITATS: woodland, forest edges and clearings, including parks. STATUS: common late Dry and Wet season migrant. BREEDING: Oct–Nov.

Leaving their breeding grounds in mid-February or March, and migrating north by night and day, south-east Australian populations of the Oriental Dollarbird arrive in New Guinea as early as mid-March.[327] They do not return to the south-east again until October, after a period of absence of some seven months. In striking contrast, Dollarbirds of Darwin are absent for half that time, and their wintering grounds are as yet unknown. Arrival and departure dates show that local birds are quite punctual. In the Darwin–Marrakai region, arrival dates over the five years from 1967 to 1971 ranged from 22 August to 23 September, and departure dates from 3 April to 25 May,[1] with median dates of 7 September and 29 April, respectively. Between 1989 and 2007

Kirri Hardy

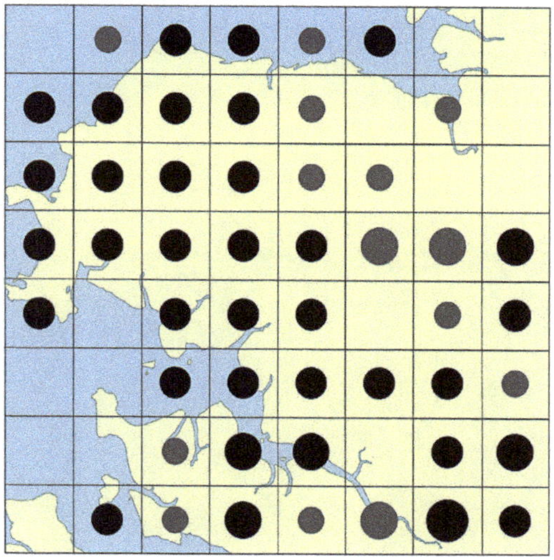

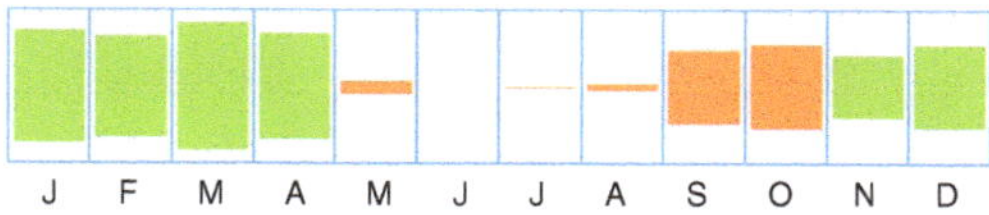

reported arrival dates in Millner[296] and other Darwin suburbs were surprisingly similar, ranging from 27 August to 20 September over 12 sample years, while departure dates ranged from 11 April to 22 May over 16 sample years; median dates are 3 September and 18 April. These data and reporting rates show that the species is absent from Darwin for only three to four months, mainly from late May to late August.

Counts around Leanyer SP during 1974–82 showed two distinct peaks in abundance, the higher one being in March (and, to a lesser extent, April) and the lower one in November (and, to a lesser extent, December).[364] The first peak corresponds with the main northward passage of southern birds and precedes the departure of local birds by a month. The second peak, on the other hand, seems too late to represent the passage of southern birds, though it is interesting that Australian birds winter in both New Guinea and the Moluccas from April until November.[86] However, if large numbers arrive in Australia during November their destination is a mystery, as their breeding season starts one month earlier, in October, over much of the continent.

Oriental Dollarbirds have been recorded in about three-quarters of the cells in our region. In open areas, such as floodplains and forest clearings these birds use tall, often dead, trees as launch pads for their aerial feeding forays and swooping display flights,[327] while in metropolitan and rural areas power lines are used to the same effect.[365] Like kingfishers, they nest in tree hollows, usually quite high, and nest trees have been found in parks and car parks in several suburbs, such as at Wanguri shopping centre, Nakara school and Royal Darwin Hospital. Egg-laying probably occurs mainly in October and November.[47] Adults are said to migrate soon after their young fledge while the young depart later.[327] In the Darwin region, both adults and juveniles are present in January, but by March immature birds, with their blackish bills, seem to predominate.[47] These birds are normally seen in pairs or family parties, but they sometimes congregate into loose flocks of 15 or more before migrating.

Blue-winged Kookaburra
Dacelo leachii

RANGE: tropical and subtropical Australia, and s. NG. HABITATS: mainly eucalypt forest and woodland. STATUS: common resident. BREEDING: Aug–Oct (TE).

Laurie Ross

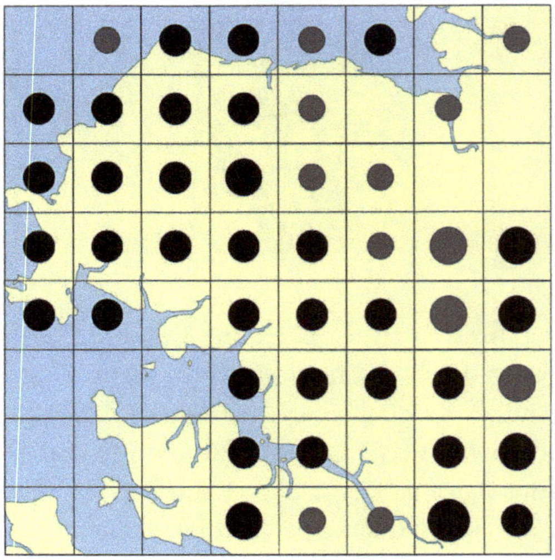

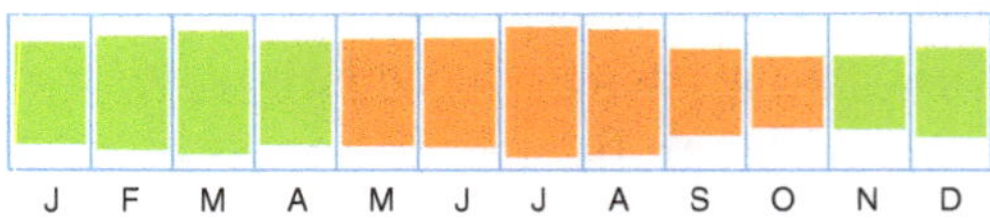

Compared to the familiar, often brash Laughing Kookaburra of eastern Australia, the Blue-winged Kookaburra is generally a timid and unobtrusive bird that has not adapted well to the urban landscape.[60,327] Despite being abundant in the open forests and woodlands surrounding the outer suburbs of Darwin, it is rarely seen in suburban parks and gardens. Nevertheless in several parts of the Top End the species has been observed taking food provided by humans,[47,60] and in the Humpty Doo region was noted foraging through the night under street lights, pouncing on large insects and frogs from power lines or other vantage points.[1,366] Recorded in 73% of the Darwin region cells, its occurrence in many suburbs of Darwin probably relates to the presence of forest remnants, such as around East Point Reserve and Kulaluk, and corridors of bushland, such as along Rapid Creek and around the airport. During the Howard's Peninsula survey in April–May 1986, it was recorded on 23% of plots, representing 12 out of 20 sampled habitats, but was most abundant in forests of Black Wattle, Swamp Box, Paperbarks, and mixed Darwin Stringybark-Woollybutt.[308]

Bird counts around Leanyer SP from 1974 to 1982 suggested that numbers of Kookaburras increased gradually through the Dry Season until November, then suddenly dropped and stayed low over the mid-Wet season.[364] The peak in November coincides with the end of the breeding season, when the addition of young birds might be expected to swell the local population, while the sudden decline thereafter may indicate that adults are moulting and generally less vocal.[364] This seasonal pattern is not, however, mirrored in the reporting rates for the whole of the Darwin region, which are fairly uniform apart from a drop during the late Dry season, when the species breeds. The latter may be the result of the aggregation of birds around nest sites, making them less widely dispersed. In Kakadu NP territories averaged 40 ha in area.[327]

Like the Laughing Kookaburra, Blue-wings are social.[327,367] Although groups of eight have been reported in the Top End, in Kakadu NP, where the species was studied from 1987 to 1991, the commonest group size was three.[368] These groups comprised a monogamous pair of breeders and one or more 'helpers' that contributed to the care of eggs and young, and the defence of territories. 'Helpers' were offspring of the breeding pair from previous breeding seasons, and in 85% of cases were males. As in the Laughing Kookaburra, female Blue-wings are larger and up to 20% heavier than males, and before laying are frequently provisioned by the breeding male. While the breeding female broods the eggs and young by night, the breeding male performs these roles by day, with some assistance from male helpers.[327]

From the moment they hatch, Kookaburra chicks aggressively compete with each other for food, and use the sharp hook on the tip of the bill to inflict wounds on their siblings, often resulting in the death of the youngest bird when food is in short supply.[367] Such siblicide is probably more common in the Laughing Kookaburra than in the Blue-winged, as the latter usually lay two rather than three eggs. Siblicide tends to occur if the first and second chicks hatch close together in time, and when a male chick hatches before a female. However, the hatching sequence and even the sex of these chicks are under the control of the breeding female, and depend partly on whether the group has a male helper or not. Females with three male helpers are likely to produce a brood of only female chicks, whereas one with a female helper is likely to produce more males.[367]

The breeding season of the species in the Top End is relatively short, based on the Kakadu study.[368] Over four consecutive years, egg-laying always started during the last week of August and ended in the third week of October, despite considerable differences in the timing of the first rains. Over three-quarters of nest hollows in Kakadu were in Darwin Woollybutts, and two-thirds of all nest sites were re-used in subsequent years.[368]

Forest Kingfisher
Todiramphus macleayii

RANGE: tropical and subtropical e. Australia, and s. NG. HABITATS: paperbark and eucalypt forest, and wetland fringes. STATUS: common resident. BREEDING: Sep–Dec, mostly Nov (TE).

While Blue-winged Kookaburras and Sacred Kingfishers occasionally use arboreal termite mounds as nest sites, only the Forest Kingfisher uses them almost exclusively. Yet its predilection for these hard, globular structures has been known to have disastrous consequences, as excavation of the nest tunnels requires flying straight at the termite mound, and striking it with such force that birds have been

Niven McCrie

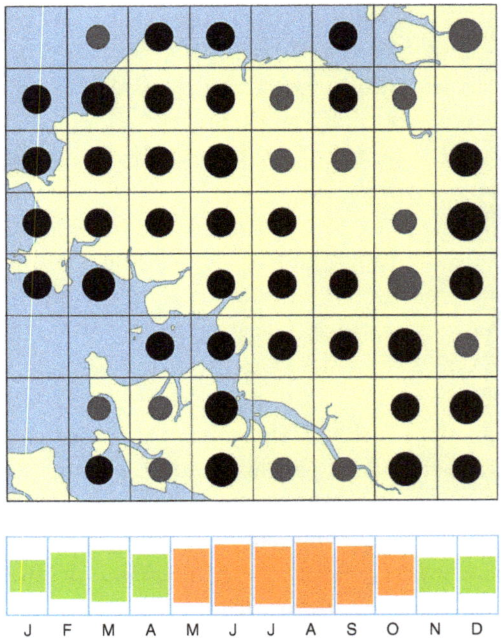

known to be fatally injured.[368] This species is also the only Australian kingfisher other than the kookaburras that is known to engage in cooperative breeding, in which young birds sometimes remain in the natal territory, helping their parents to raise more offspring. A detailed study in Kakadu NP from 1987 to 1993 revealed that while most territories were occupied by pairs, several had three birds and occasionally four or more.[368] These extra birds helped their parents with nest excavation, provisioning chicks and possibly even incubating. The study also suggested that younger chicks are sometimes killed by their older siblings, as is common in kookaburras.

In the Darwin region, this species was recorded in 80% of cells, but significantly more often from urban areas than forest and rural areas. Though rarely seen in suburban backyards,[60] it is sometimes observed on the edges of ovals and, like the other medium-sized kingfishers, often perches on power lines. During the Howard's Peninsula survey, it was recorded on 18% of plots, which represented nine out of 20 habitats, but it was most abundant in riparian forest dominated by Swamp Box and Ghost Gum, followed by paperbark swamps and tall grassland.[308] In Kakadu NP, it was most abundant around billabongs, along creek lines and in seasonally inundated grassland, where territories shared a common border. Territories in eucalypt forest, on the other hand, were not touching, suggesting that habitats associated with wetlands are preferred.[368]

Counts of Forest Kingfishers around Leanyer SP from 1974 to 1982 showed a spike in August–September, followed by lower numbers over the Wet season.[364] Similarly in Kakadu NP, the species was more abundant and widespread in all habitats during the Dry season than during the Wet.[103] Although some populations of the eastern Australian race (*incinctus*) migrate to eastern Indonesia, there is no evidence of long-distance migration among Top End birds (race *macleayii*).[364] Thus it is more likely that the late Dry season spike reflects higher calling rates associated with courtship and breeding than population changes. Reporting rates were also lower during the first half of the Wet season than during the remainder of the year, but they tail off gradually after the peak in August, possibly as birds concentrate around nest sites, and are therefore less dispersed. In Kakadu NP clutches were laid from September to December, most (88%) in late October and November.[368] Indeed, most breeding records in the Darwin region also fell in November.[47] Nests have been found in the Botanic Gardens, at Royal Darwin Hospital, and on the edges of mangals at Charles Darwin University Casuarina campus and near Palmerston SP.[21,47]

Collared Kingfisher
Todiramphus chloris

RANGE: coastal tropical and subtropical Australia, NG, Indonesia and Indo-China. HABITATS: mangal. STATUS: locally common resident. BREEDING: Sep–Dec.

Niven McCrie

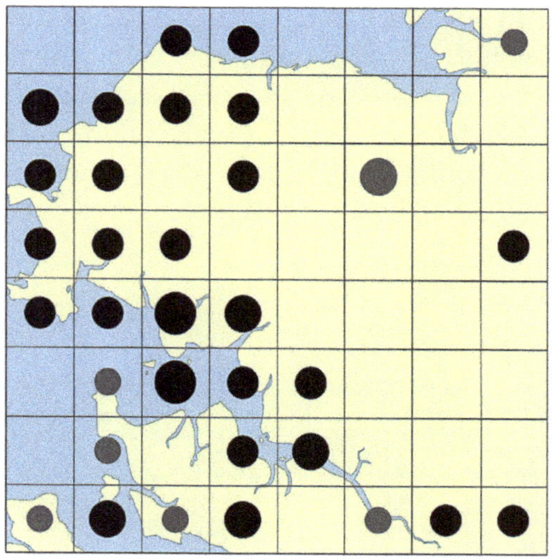

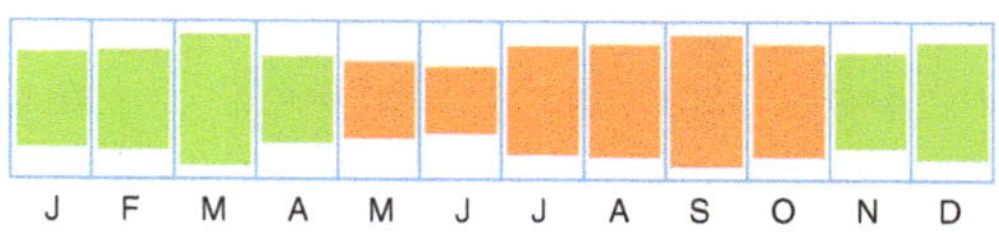

Apart from its voice there is little about the mangal-specialised Collared Kingfisher in Australia that is reminiscent of this species in south-east Asia, where it is common and widespread from coastal cities to village gardens at 2000 m above sea level, even in neighbouring Timor.[86] Not only are Australian birds confined to mangals but, in the Top End at least, are largely restricted to the seafront and tidal river fringes, rarely penetrating far into the surrounding forest.

Local birds feed mainly on crabs,[187] but also eat mudskippers obtained from the muddy ground. At low tide they often forage on exposed reefs or bare mudflats up to 100 m from the seaward edge of the mangal.[47] Consequently it is a mystery how these birds obtain food when the mudflats are inundated by high tides, sometimes for several hours. It is possible that they cache food when foraging during low tides, as fresh crab remains have been found in hollow spouts of old mangrove trees.[47]

Four pairs occupy adjacent territories, each about 200 m long, in the thin strip of mangal fringing the southern foreshore of Nightcliff, where five birds were captured and banded between 1996 and 2009.[47] One bird was recaptured seven years after it had been banded in the same location. Here and right across the Top End, these birds are typically associated with Star Mangroves,[313] oak-like trees that form a distinctive zone along the seaward front of many mangals and which are flooded on all high tides. Nevertheless, at least two pairs hold territories along the upper reaches of Ludmilla Creek, 2 km from the sea, where tall Grey Mangroves line the narrow channel, which fills on most high tides.[47] In stark contrast, this species was rarely seen, and never captured, during a nine-year study conducted in mangal dominated by Stilt and Spurred Mangroves near Palmerston SP, where 21 individuals of the much smaller Sacred Kingfisher were banded.[312]

Nests are sometimes built in arboreal termite nests, but more typically this species nests in holes in old, often dead, mangrove trees.[327] However, one nest was situated inside the dead trunk of a tree in a suburban backyard, about 50 m from the high water limit, and almost 200 m from the nearest mangrove trees.[47] Observations of nests at Nightcliff and Ludmilla Creek suggest that clutches are laid from September through November. Counts of Collared Kingfishers around Leanyer SP from 1974 to 1982 showed a sharp spike in October and a smaller one in August,[364] possibly due to higher calling rates during courtship and the breeding season. Reporting rates over the whole Darwin region, however, show relatively minor fluctuations, with fewest records in the early Dry season.

Sacred Kingfisher
Todiramphus sanctus

RANGE: whole of Australia, NG, NZ and Indonesia. HABITATS: all wooded habitats. STATUS: common Dry season visitor and passage migrant. BREEDING: Oct–Jan (TE).

The longest-distance migrant among Australian members of its family, the Sacred Kingfisher flies north from its breeding grounds around the continent to New Guinea and Indonesia, as far as Borneo and Sumatra.[327] While the species is familiar to many in southern Australia as a noisy inhabitant of eucalypt forests and woodland from September to March, on the islands of Timor or Bali it is a quiet denizen of mangals and nearby cultivation over the remaining months of the year.[86] Consistent with birds passing through the Darwin region to and from their winter quarters in Indonesia, observations during the late 1960s suggested two peaks in abundance: the major peak in April and May, and a smaller one in August and September.[1] Counts in the Leanyer SP area from 1974 to 1982 confirmed the first (autumn) passage,

Niven McCrie

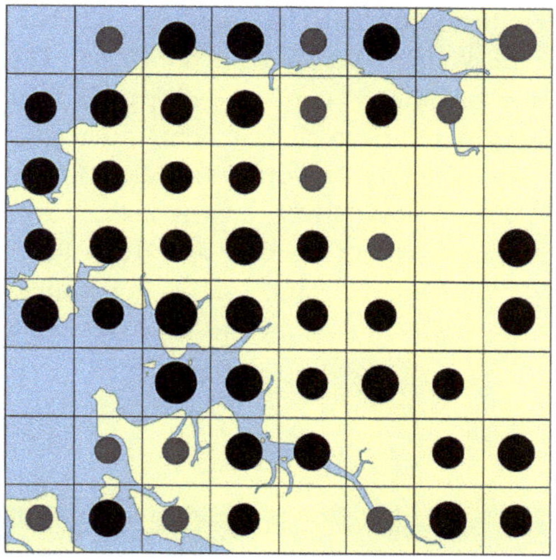

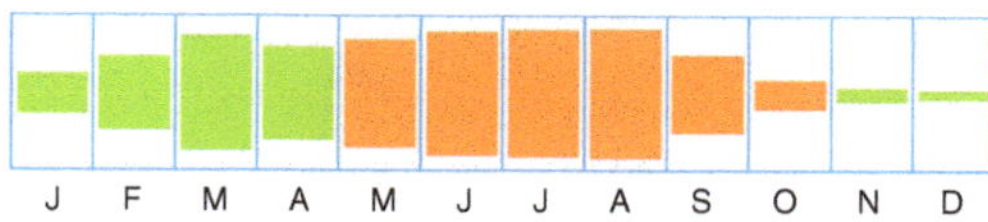

albeit slightly earlier, with a dramatic influx of birds in March–April, but showed no evidence of a return passage, numbers during the Dry season remaining relatively low until November, when they sank even lower.[364] Thus the return (spring) passage presumably follows a different route.

Reporting rates for the whole Darwin region confirm that Sacred Kingfishers are scarce from October to January. This coincides with the breeding season in both northern and southern Australia, but the records suggest two possibly separate influxes. The first probably represents the previously mentioned March passage, while the second, more gradual rise in records from April to June may flag the arrival of birds that winter in Darwin, as suggested by the plateau in reporting rates. Surprisingly, during the Howard's Peninsula survey in April–May 1986 the species was recorded on only six out of over 1000 censuses, mainly in mangals and paperbark swamps.[308] Yet in mid-July 2012, 21 birds were counted along 2 km of Buffalo Creek, with up to three perched in the same tree.[47] Moreover, loose aggregations of up to six individuals have been seen on reefs at East Point and Nightcliff in June and July in several years. During a bird survey of mangals in 2008–09, the species was recorded in all 13 patches sampled, the highest density being at the smallest patch, Mindil Beach, where they were twice as abundant as at other sites.[313] The species has been reported from 78% of cells, most frequently from urban areas, where it sometimes perches on clothes hoists, as well as power lines.[60]

Some populations of Sacred Kingfishers breed in the Top End and, indeed, much of our understanding of the species' breeding biology derives from an intensive study of a population in the eucalypt woodland of Kakadu NP between 1987 and 1991.[368] In each of the four study years, pairs established their territories in late August or September, and the females started laying eggs in their nests in tree hollows in late October. The last clutches were laid in mid-January and the last chicks fledged in late February.[368] Adult birds abandoned their territories and dispersed soon after their last brood became independent, whereas some young birds remained behind. The only breeding record of the species near Darwin refers to a nest containing eggs east of Humpty Doo on 29 October 1978.[364] In the Victoria River Downs area,[98] over 350 km south of Darwin, and Keep River NP[369] to the west, the species is resident, with some influx before the breeding season. Given the time it presumably takes breeders from south-east Australia to reach northern Australia, it is feasible that those wintering in the Darwin region emanate from such distant sources,[157] while those passing through may belong to Top End breeding populations. An evaluation of this hypothesis, however, awaits genetic or satellite-telemetry studies.

Red-backed Kingfisher
Todiramphus pyrrhopygius

RANGE: whole of Australian mainland. HABITATS: woodland and floodplain edges. STATUS: moderately common, largely Dry season visitor. BREEDING: Oct (TE).

Laurie Ross

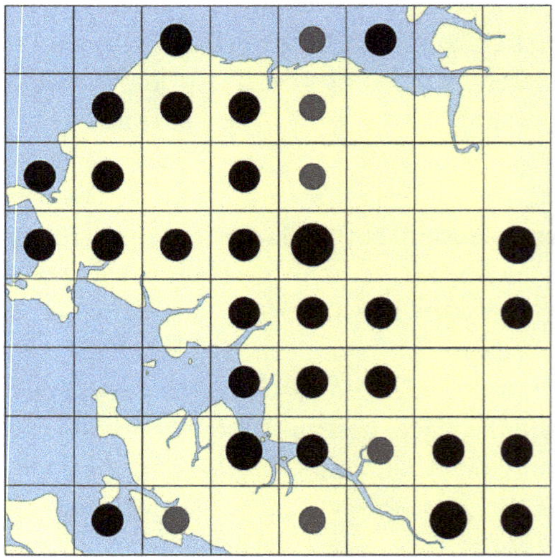

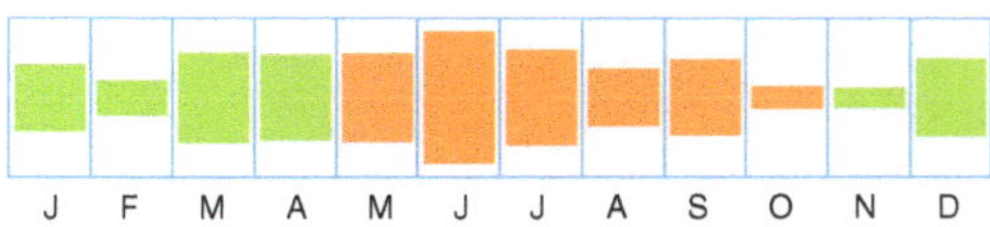

More characteristic of arid and semi-arid environments than the humid tropics, the Red-backed Kingfisher differs from other medium-sized Australian kingfishers in preferring to nest in tunnels in vertical river banks or terrestrial termite mounds, particularly those of the Magnetic Termite.[368] Populations in south-east Australia are migratory, leaving their breeding grounds in February and March, and flying northward possibly as far as the coast of Queensland,[157] where they spend the winter before returning in late August to October.[327] Though recorded in all months in the Darwin region, the species appears to be more abundant and widespread during the Dry season both here and in Kakadu NP,[103] suggesting the possibility that local populations are augmented at this time by birds from southern climes. Bird counts in the Leanyer SP area from 1974 to 1982 suggested a gradual increase in the numbers of Red-backed Kingfishers from May through to August, followed by a precipitous fall in September to zero in October and low numbers throughout the Wet season.[364] Reporting rates since 1977 for the whole of the Darwin region do not, however, show any obvious evidence of an influx of migrants, though a peak is evident in June, and a possible decline in October–November.

During the late 1960s and early 1970s the species was reported to be moderately common in the Darwin region from June to October, but present in the subcoastal Humpty Doo–Marrakai region only during the Wet season.[1] At Coomalie Farm, 45 km south of the Darwin region, the species was recorded in all months of the Wet season but rarely during the Dry season.[47] These scant observations support the suggestion that birds from near-inland localities move towards the coast during the Dry season, returning before the first rains of the Wet.[58,327] The few breeding records for the Top End[98,327] suggest that egg-laying occurs mainly in October, when the species seems scarce in Darwin. Nevertheless, in November 1973 two fledglings were found in the Coonawarra area,[364] indicating that breeding occasionally take place in the Darwin region. The species has been reported from 53% of cells in the region, with significantly more records in sub-coastal and rural areas than on the coast and in urban or forest areas, respectively.

Azure Kingfisher
Ceyx azureus

RANGE: tropical and e. Australia, NG, Timor and Moluccas. HABITATS: mangal, wet monsoon rainforests and freshwater wetlands. STATUS: common resident. BREEDING: Dec–Apr (TE).

Laurie Ross

Although both Australian river kingfishers occur in Darwin, the widespread Azure Kingfisher is commoner than the tropical-endemic Little Kingfisher. Both species occur along tidal and non-tidal reaches of rivers and creeks,[364] as well as around billabongs, but the Azure Kingfisher seems more catholic in its habitat choice than its diminutive relative, often using open areas with little cover. Both species eat small fish, crustaceans and insects, which are captured by making shallow plunges into the water from low perches.

Having a bill that is 50% larger than that of the Little Kingfisher,[47] and being twice its weight (29 g), the Azure is able to feed on larger prey. Assuming that larger fish require deeper water, the Azure may be expected to

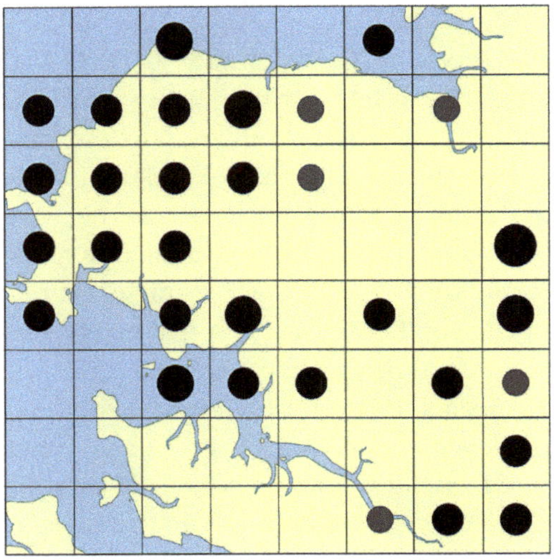

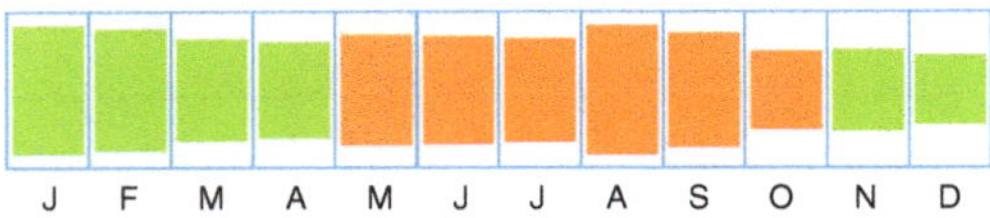

utilise deeper water-bodies more often than the Little Kingfisher does, and to limit its use of tidal channels mainly to times of high tides. Indeed, in dense landward mangal near Palmerston SP, where there were no tidal channels of any depth and few openings in the canopy, Azure Kingfishers were much less abundant than Little Kingfishers, being observed only twice during 54 censuses.[312] Similarly, on the upper reaches of Ludmilla Creek, about 1.5 km from its mouth, the Azure was captured less often than the Little, even though the 5 m-wide creek channel flooded to a depth of up to 1 m on high tides. On the other hand, in short mangroves near the mouth of Rapid Creek, where the creek was 20–60 m wide, Azure Kingfishers were captured three times over 28 banding sessions, whereas Littles were never recorded.[47] Similarly in the tall Star Mangrove-dominated mangal beside the sea at Nightcliff, Azures were seen frequently, and trapped three times over 16 banding sessions, while the Little was rarely sighted and never trapped. Azures were recorded from 50% of cells, almost a quarter of records coming from the Buffalo Creek area.

Azure Kingfishers invariably nest in burrows tunnelled into the banks of tidal rivers or freshwater creeks.[327] The few breeding records in our region suggest eggs are laid during the Wet season, from December to April.[47] However, the reason for the relatively low reporting rates from October to December is unknown. Banding studies show that the species is sedentary and lives for up to 10 years or more, at least in eastern Australia.[327] The local race *ruficollaris* differs from the south-eastern form in being slightly smaller and having more violet-blue on the flanks, in some cases forming an incomplete band across the lower breast.

Little Kingfisher
Ceyx pusillus

RANGE: tropical Australia, NG and Moluccas. HABITATS: mangal, wet monsoon rainforest and freshwater wetlands. STATUS: scarce resident. BREEDING: Jan–Mar (TE).

Laurie Ross

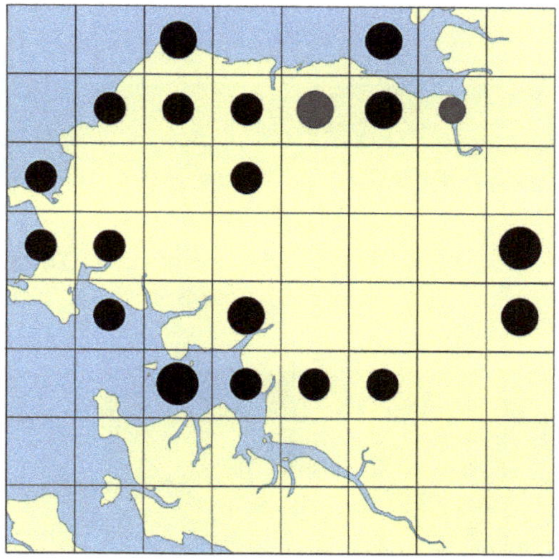

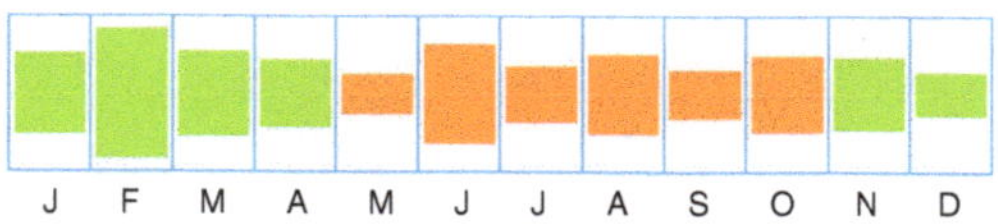

With a bill that is only 3 cm long and a body weighing a mere 14 g,[47,327] the tropical-endemic Little Kingfisher is limited in the size of fish and crabs it can capture and swallow. While the Azure Kingfisher may be found around watercourses and billabongs of all sizes, often with sparse fringing vegetation, the Little tends to confine itself to the upper reaches of tidal and freshwater creeks with dense cover.[312] Consequently the species is generally much less common across the Top End than the Azure,[1,364] and does not penetrate far inland[300] or even reach the Western Australian border.[327] Moreover, there is a gap of 600 km between the NT population and that of western Cape York.[82]

This species was recorded from 31% of all cells in the Darwin region, with significantly more records from coastal than subcoastal cells. In dense mangal with few openings near Palmerston SP, it was recorded in 25 of 54 months sampled from 1985 to 1994, equally frequently during the Wet and Dry seasons.[312] During 44 banding sessions from 2000 to 2009 along the upper reaches of Ludmilla Creek, about 1.5 km from its mouth, it was captured seven times in nets erected beside the 5 m-wide creek channel. In stark contrast it was never captured in mangal near the mouth of Rapid Creek, where the creek was 20–60 m wide, nor in the mangal at Nightcliff, where it is occasionally observed fishing in the rock pools beside the sea.[47] Oddly, in a Nightcliff backyard, about 150 m from the mangal, one over-optimistic bird was observed perching on the handle of a watering can for about 5 minutes before giving up on its search for aquatic life and chasing an insect under the eaves of a nearby veranda.[203]

The February peak in reporting rates coincides with its breeding season. The few local breeding records suggest eggs are laid mainly in January and February. Although this species is known to nest in tunnels on the banks of creeks, as does the Azure Kingfisher, all nests observed in the Darwin–Fogg Dam regions were inside tree hollows, mainly in mangroves, but once in a paperbark.[47]

Rainbow Bee-eater
Merops ornatus

RANGE: whole of Australian mainland, NG and e. Indonesia. HABITATS: airspace above all habitats. STATUS: very common Dry season visitor and passage migrant, some birds resident. BREEDING: Aug–Oct.

Niven McCrie

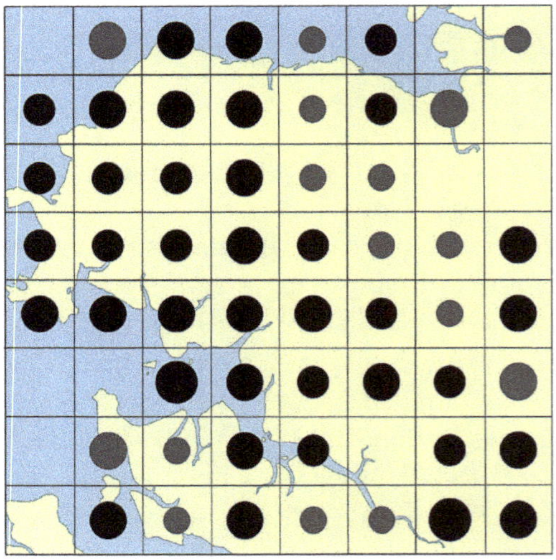

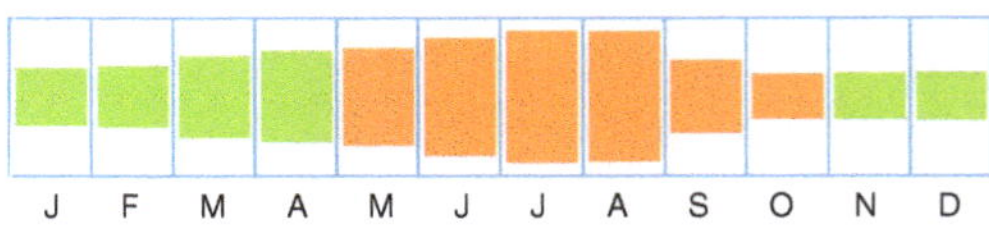

Four weeks of growing up underground seems an incongruous start for a bird that forages in the air, but soon after emerging from its nest burrow the young Rainbow Bee-eater is catching bees and other aerial insects, and within a month is ready for migration.[327] In southern Australian populations, families aggregate into large flocks a week or two before migrating, building up in numbers daily until they set off north. They travel by day, and probably at night, sometimes flying at considerable altitude.[327] These populations depart mainly in February and March, and by early April some birds are crossing the central mountain range of New Guinea at 3000 m above sea level. The species also overwinters on many islands of eastern Indonesia, mainly from late March to late September[86] but, unlike the Sacred Kingfisher, rarely crosses Wallace's Line, and is only a vagrant to Bali. Regardless of their wintering quarters, most birds arrive back in southern Australia between mid-September and mid-October.[327]

In contrast to southern Australia, the Darwin region hosts Rainbow Bee-eaters of one population or another throughout the year. A familiar sight throughout the suburbs of Darwin and Palmerston[307,364] they frequently exploit power lines, clothes hoists and fences as launching pads for their assaults on airborne insects.[365] Reporting rates show that the species is recorded more frequently during the early to mid-Dry season than in the late Dry and early Wet seasons. Around Millner, arrival dates varied from 28 March to 19 April over the five years from 1989 to 1993.[296] Counts around Leanyer SP from 1974 to 1982 showed that numbers started to build up in late March, reaching a peak in May and staying high until the end of August. After this, numbers plummeted and remained very low from September through February.[364] The higher numbers from March to August are consistent with an influx of migrants from southern Australia, but the peak in May suggests that some birds may 'leap frog' those overwintering in Darwin and continue north to Indonesia.[364]

The species roosts in dense colonies[364] and, during the Dry seasons of 1999–2001, 13 roosting colonies were found across Darwin, the average distance between neighbouring colonies being 1.7 km.[370] Almost all colonies roosted in Black Wattles or African Mahoganies situated in public parks, school grounds or roadsides, and most were used in at least two of the sampled years. During the third week of August 2000, the number of birds per roost varied from 31 to 310, and averaged 92.[370] The largest colony was in Millner, where numbers dropped from a peak of 310 birds on 16 August to less than 100 by 26 August, then declined gradually to 14 birds by 7 September; by 18 September the roost was abandoned. Similarly, the roost at Nightcliff, where 160 birds had been counted in May, persisted until at least 25 September, when only five presumably resident birds were counted.[370] A similar pattern emerged from transect surveys of 10 Palmerston suburbs, where Rainbow Bee-eaters were the twelfth most frequently recorded species during the 2007 Dry season.[307] Recorded in 19 of the 20 weeks sampled, the average number of birds was 14, but there were three 'spikes' in abundance between late June and late August, when the highest number was 41. After 6 September total numbers stabilised at close to their pre-June levels.

From late August through October, pairs of Bee-eaters may be seen excavating or visiting nest burrows in coastal sand dunes at localities such as Casuarina CR and Nightcliff.[47,364] Whether such breeding pairs are sedentary or not is unknown. At Darwin Golf Course, the species was recorded in 55 out of 59 consecutive weeks during 2006–07 and, while fewer than 12 birds were normally present, 20–30 birds were recorded from early January to early April,[309] suggesting an aggregation of local or Top End breeders and their young.

In summary, the Bee-eaters of Darwin appear to comprise several different populations, whose origin and movements remain unclear. They include both breeding residents and migrants from southern Australia, a proportion of which may pass through Darwin en route to Indonesia. There is no evidence of a return passage through Darwin, which suggests that birds move south over a broader front.

Nankeen Kestrel
Falco cenchroides

RANGE: Australia, NG, e. Indonesia, s. Pacific. HABITATS: grassland, woodland, cultivated and urban areas. STATUS: uncommon; mainly Dry season migrant. BREEDING: July (one record, DR).

Aiden Webb

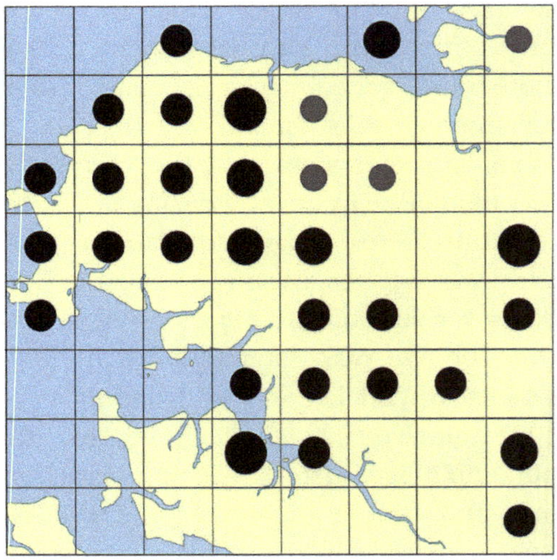

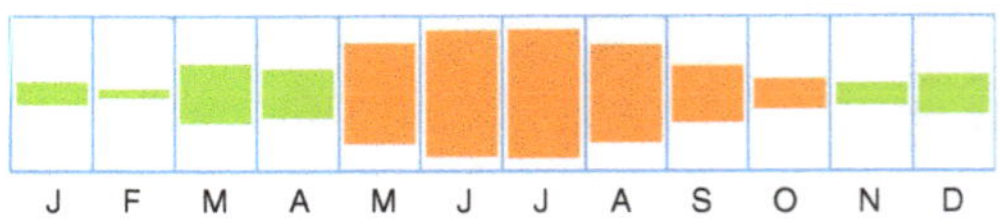

Records of Nankeen Kestrel are widespread, but this is expected of a species that is largely seen in flight, and may cover a considerable area when hunting. Most sightings are of one or two birds, with few reports of multiple birds, and a count of nine at Darwin International Airport in June 2005[371] is a high density for the region. Nankeen Kestrels are most frequently observed hunting in open areas such as at Lee Point, Berrimah, Darwin International Airport and Howard Springs, but they are also seen occasionally in urban areas, in association with tall buildings in the city, at Darwin Hospital and at Marrara Sporting Complex. At the latter site, they often hunt moths or other large insects attracted to stadium lights at night-time sporting events, but on one occasion a bird was seen pursuing an Australasian Grebe under the lights, making repeated strikes but failing to capture it before the grebe passed into the darkness.[146]

Nankeen Kestrels have been recorded in all months, but there is a sharp fall in reporting rates during the Wet season. Heavy rain has a direct and immediate effect on hunting success, but it is also likely that as ground cover, particularly Annual Speargrass, becomes taller and denser during the Wet season, ground prey becomes more difficult to capture. This is probably the case for most of the ground-hunting species, as many move out of the region during the Wet season, but small numbers of some, including the Nankeen Kestrel, can still be seen where suitable habitat remains. The greatest abundances of

Nankeen Kestrels in northern parts of the NT have been found in the Barkly region,[371] and it may be that many from the Top End move there during the Wet season. Numbers of Nankeen Kestrels in the Darwin region are small, however, and relatively local dispersal may fully account for the species' Wet season movements.

There was an influx of Nankeen Kestrels into the Darwin region during May 2012, although it was more marked between Humpty Doo and the Mary River, and along the Daly River.[372,373] This event coincided with influxes during April and May to some islands off north-west Australia,[374] so may have resulted from a population explosion in south-western or inland areas of the country. Other apparent influxes occurred during the Dry seasons of 1975[34] and 1999.

Outside of the region the species becomes increasingly abundant, particularly around floodplains, which provide vast areas of open grassland for hunting. Nankeen Kestrels may also gather at Dry season fires in grassland or woodland along with other birds of prey.[31]

The only report of Nankeen Kestrel breeding in the region was during July 1990, when a pair was found nesting at Royal Darwin Hospital,[34] but there are no further details available.

Australian Hobby
Falco longipennis

RANGE: whole of Australia, NG, e. Indonesia. HABITATS: woodland, wetlands, urban areas. STATUS: uncommon resident. BREEDING: Jun–Sep (TE).

Adrian Boyle

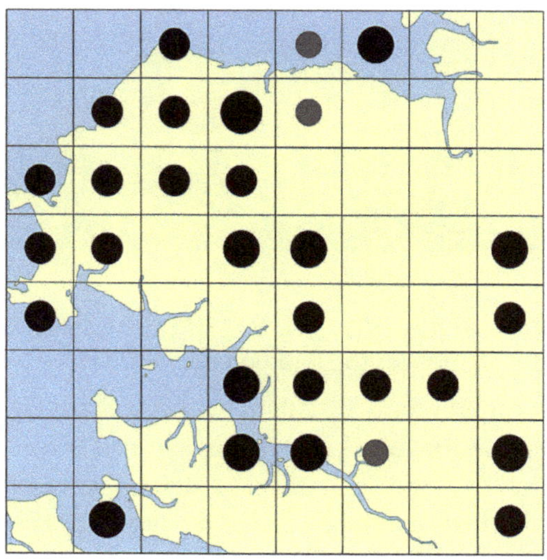

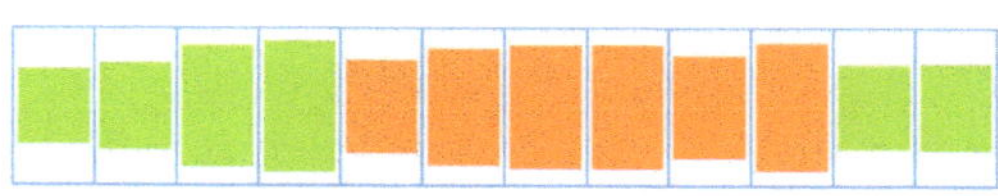

Australian Hobbies are a regular sight around Darwin suburbs as they hunt in the late afternoon, perhaps choosing that time to take advantage of flocks or individual birds heading to roosts. They will hunt at almost any time, however, and have been observed taking Great Knot,[21] Peaceful Dove and a small bat,[43] and pursuing, though not capturing, lorikeets, Little Curlews and Dollarbirds. While most prey is captured in flight, they have been seen to take nestlings of Magpie-larks and Yellow-throated Miners, and a pair was seen attempting, unsuccessfully, to take a Little Friarbird from the dense foliage of a small tree.[43] In Kakadu NP one Hobby was observed chasing a Peaceful Dove until the latter dropped into a shallow waterhole, where it remained motionless for 30 minutes, while the Hobby flew off after 10 minutes of waiting.[375] Insects also form part of the Hobby's diet and, in addition to sightings of Hobbies taking dragonflies from above the water at Knuckey Lagoons, there are frequent observations of them at night, hunting insects attracted to the stadium lights at Marrara Sporting Complex.

Records suggest that the species is resident in the Darwin region, though the only record of local breeding was in early August 2012,[40] when a pair was seen at a nest with unknown contents, on a tower in Leanyer. Breeding is widespread across the Top End. At Darwin River Dam,[44] a pair with two young was observed near a nest on an electricity pylon during November-December 2002. Pairs were attending their nests at Mary River Park and Swim Creek, near Wildman River, on 30 July 2000 and 21 September 2010, respectively.[47] Breeding in Kakadu NP has been recorded somewhat more frequently. At Kapalga, nests with small chicks were found on 19 and 28 October 1988. Another nest, only 50 m from an active Brown Falcon nest, contained two small chicks on 20 July 1993 and there was also a nest with small chicks there on 11 October 1993. Birds nested atop a television satellite tower at Cooinda.[150] In Katherine, a Hobby nest containing two eggs was found on a windmill in October 2003,[146] and west of there, at Victoria River Downs, nesting was reported in July. These records suggest eggs are laid in the Top End from June to September.

Brown Falcon
Falco berigora

RANGE: Australia, NG. HABITATS: woodland, grassland, cultivated areas, floodplains. STATUS: uncommon visitor, rare breeder. BREEDING: May–Jul; Sep (TE).

Although one of Australia's most widely distributed birds, and broadly spread across the Top End, the Brown Falcon is one of the less common birds of prey in the Darwin region.

There is strong evidence of north-south movements[91,157] by Brown Falcons within Australia, and although there is a resident breeding population in the Top End, reporting rates in the Darwin region indicate that the species is largely a Dry season migrant.

The broad distribution of Brown Falcon records in the region may exaggerate its abundance, as records from different localities are likely in many cases to refer to the

Nolan Caldwell

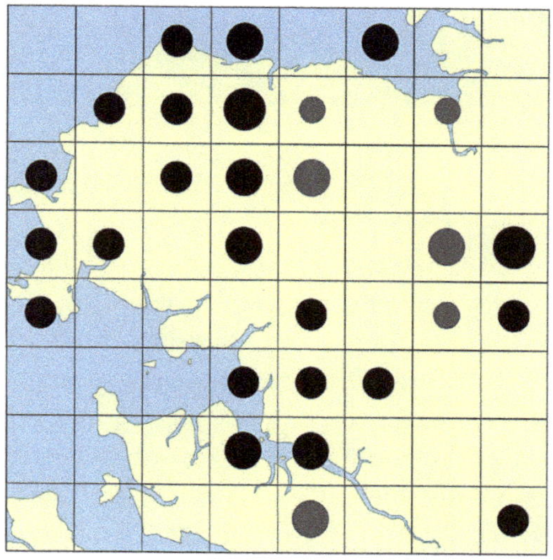

same birds. There are sporadic sightings from in and near the city and suburbs but, as expected, most records are from semi-rural and rural areas, where birds are often conspicuously perched on power lines by the roadside.

Feeding observations are scant. Incidental observations outside of the region include one attempting to catch a Brown Quail on the road, one catching a snake and one hunting insects, probably dragonflies, over water together with Whistling Kites and 'hundreds' of Pacific Swifts.[3] During the Dry season Brown Falcons are frequently observed together with Black Kites and other birds of prey soaring above fires, hunting prey that is flushed by the flames. More so than Black Kites, Brown Falcons often remain to continue hunting or scavenging after the fire has gone.[95]

The single confirmed record of Brown Falcon breeding in the region is of a nest containing two chicks, found on a communication tower at the edge of suburban Leanyer in early August 2011.[40] Near to the region, there is a report of Brown Falcon on a nest at Beatrice Hill in mid-July 1997,[43] and another in the same area in August 2011,[40] but the contents of the nests were not seen. Nests found at Kapalga,[43] in Kakadu NP, contained eggs in mid-May 1989, chicks in mid-July 1993, and two fully fledged young in early November 1987, suggesting that egg-laying occurs in May–July and September.

Grey Falcon
Falco hypoleucos

RANGE: inland Australia. HABITATS: open woodland, shrublands, inland wetlands. STATUS: vagrant. BREEDING: Jun–Nov (NT) TSP.

The Grey Falcon has the most specialised dietary and climatic requirements of the falcons occurring in Australia.[376] It is Australia's rarest raptor and, largely due to the remoteness of its preferred habitat, also one of the country's least known and least frequently seen. Difficulties in identification of the Grey Falcon can come from confusion with the pale plumaged Brown Falcon and Brown Goshawk, as well as the Peregrine Falcon and other species,[377] making reports difficult to assess.

There are breeding records from south of the Top End, including the Barkly Tableland[376] and near the Victoria River, at a similar latitude,[378] but it is only a sporadic visitor to more northerly regions.

Although recorded only once in the Darwin region, with a sighting at Howard Springs in April 1990,[34] reports from near Fogg Dam in October 1994[21] and along the Marrakai road[21,47] in November 2002 indicate that the species occasionally wanders to the far north. It may be significant that there were sightings in Litchfield NP[44] in May 2007 and March 2008,[3] and near Pine Creek in June 2006[3] and July 2008,[376] as these areas are only 150 km apart. The June 2006 observation concerned two birds that retreated to electricity pylons after being harassed by White-breasted Woodswallows. It appears to be the only record of two Grey Falcons seen together in the northern Top End, and its occurrence at the beginning of the species' reported breeding season and near pylons is perhaps some indication of northerly breeding, as 19 of 27 Grey Falcon nests found between 2003 and 2011 were situated on radio repeater towers.[376]

Black Falcon
Falco subniger

RANGE: whole of Australia, but rarely in WA and coastal regions. HABITATS: woodland, grassland, wetlands. STATUS: uncommon Dry season visitor. BREEDING: late Dry season (TE).

The separation of the Black Falcon from the dark morph Brown Falcon has long been an issue for many birdwatchers and in the Top End, where the proportion of dark plumaged Brown Falcons is higher than in eastern and southern parts of the country,[135] this is even more so. Prior to 1973, there were few reports of Black Falcons in the Top End, but between that year and 1977 there were a total of 13 sightings, all occurring in the months April to November. Nine of those records were in the Darwin region,

Will Riddell

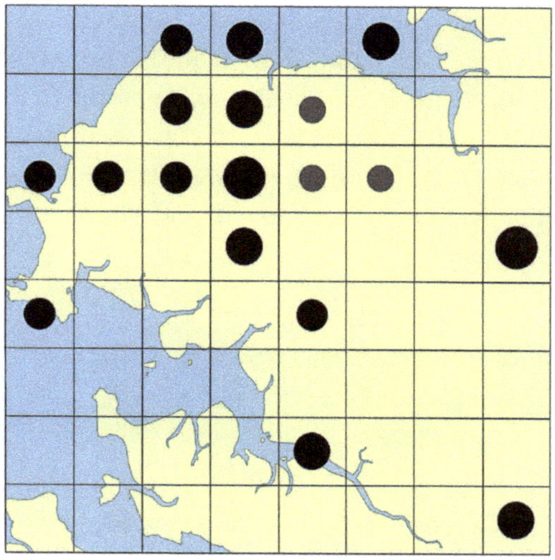

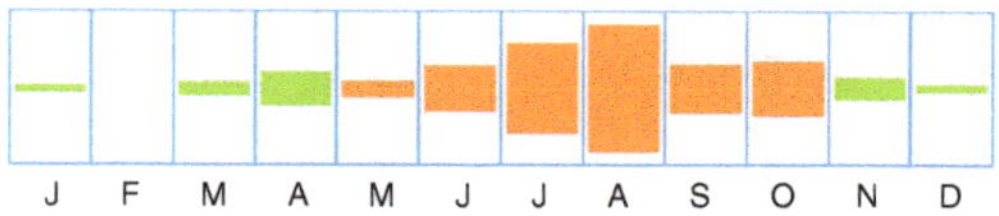

and the species was recorded in the region in 24 of the subsequent 36 years until 2013.

The Black Falcon is virtually absent from the Top End during the Wet season and, unlike Brown Falcon, there is no indication of a resident population. There is evidence of widespread movement of Black Falcon, but it appears predominantly dispersive rather than migratory,[135,157] and this is reflected in the sporadic nature of records in the region.

The highest reporting rates are from Holmes Jungle and Knuckey Lagoons. The only city record of the Black Falcon was in July 2011, when a bird was seen flying over Mitchell Street in the middle of the day.

Live prey items identified include Tunney's and Dusky Rats,[379] Tree Martins, and the chick of a Masked Lapwing or Black-fronted Dotterel. Black Falcons will also eat carrion or steal prey from other birds. One was seen feeding on a road-killed brown bandicoot on the gravel edge of the Arnhem Highway near Humpty Doo and, despite heavy traffic passing within metres of the bird, it did not flush during more than 10 minutes of observation.[21] Also near Humpty Doo, a Black Falcon was seen to attack an Eastern Grass Owl, then steal a rat the owl had just caught.[380] At Fiddler's Lane Swamp a Black Falcon attempted to take a bird, possibly a honeyeater, from an Australian Hobby, but in that instance was unsuccessful, as the Hobby, aided by another, chased it off. Although not habituated to doing so, as are Black Kites, Black Falcons will attend fires, along with other raptors, apparently preying on vulnerable animals after fires rather than feeding at the edges of flames.[95]

There has been no evidence of breeding in the Darwin region. The first NT breeding record was in mid-November 1986,[43] when a nest containing small chicks was found in a Ghost Gum at Kapalga, Kakadu NP. An adult with recently fledged young was also seen in the same area in mid-September 1991,[43] and in 2011, also in September, a pair with two fledglings was seen near Fogg Dam. Based on incubation and fledging periods of around 34 and 42 days,[135] respectively, it appears that clutches were laid in June or July and October. South of the Top End, Black Falcons have been reported nesting in the Barkly region, but otherwise there are very few records from anywhere in the NT.

Peregrine Falcon
Falco peregrinus

RANGE: cosmopolitan, including whole of Australia. HABITATS: coast, woodland, wetlands, urban areas. STATUS: rare visitor. BREEDING: May–Sep (TE).

David Webb

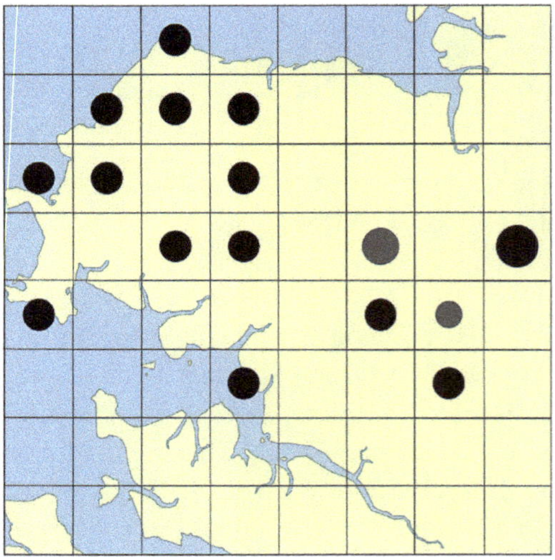

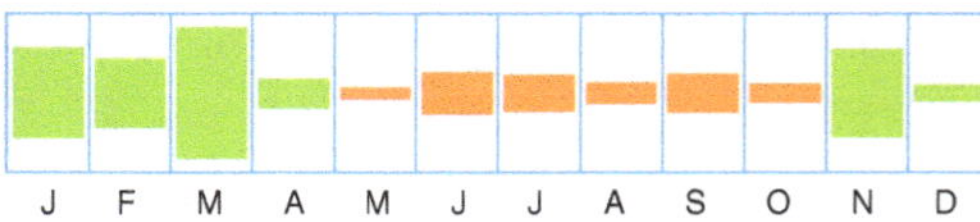

The most broadly distributed of the world's falcons, there are up to 19 subspecies of Peregrine Falcon recognised, of which two, *macropus* and *submelanogenys*, are endemic to Australia, the former throughout much of the country and the latter restricted to the south-west of Western Australia. Recent observations and analysis of photographs, however, have indicated that a third subspecies may occur as a migrant to the Top End.[263]

While it appears that there is some post-breeding dispersal, there is no evidence of regular migration of the species in Australia,[135] so movements in the Top End are likely to be relatively local, and perhaps largely restricted to juvenile birds. Although an erratic visitor to the Darwin region, Peregrine Falcon has been more frequently recorded during the Wet season than the Dry, in contrast to both Black and Brown Falcons, and it may be that the Top End population is augmented by a small number of northern migrants, probably, based on photographs, subspecies *japonicas*. The Peregrine differs from Black and Brown Falcons in not being dependent on ground-dwelling prey, so is not directly disadvantaged by inundation of wetland areas or the increased height and density of ground covers at that time of year. In addition, the reporting rates may relate simply to conspicuousness, as many of the Wet season records are from coastal areas, and it may be that the flocks of migratory waders are an attractive source of prey.

Peregrine Falcon was recorded at Humpty Doo, on the edge of the region, in four of the five years from 1967 to 1971, with three records in the Dry season and two in the Wet.[1] From 1977 to 2013 it was reported in 20 of 36 years in the region but, despite there being records for all months, the maximum in any year was five months, in 1977, and in 12 of those years, only one month was represented. Most reports are of birds in flight, with the few sightings of perched birds all on tall, artificial structures such as communication masts and towers;[40,53] one was seen perched by the nest of an Eastern Osprey, situated on a television tower in Stuart Park. Perhaps surprisingly, there are yet no reports of birds perching on buildings, a common practice of the species in other cities, although it very likely occurs. Peregrine Falcons have been seen stooping at waders at high-tide roosts and around wetlands, chasing terns over the water off Stokes Hill Wharf, and pursuing Tree Martins at Leanyer SP. However, the only report of a successful catch is a wader, probably a Red Knot,[3] taken at Buffalo Creek.

Breeding in the Top End has been reported only from Kakadu NP. In early September 1986[43] two fledglings were seen at a nest, previously used by Wedge-tailed Eagles, at Freezing Gorge, near Jim Jim Creek. Breeding was also reported at Mount Brockman in May 1975,[34] though no details are available. Birds are frequently seen in escarpment areas of Kakadu NP, and it may be that breeding is more widespread and that a resident population occurs there.

Red-tailed Black Cockatoo
Calyptorhynchus banksii

RANGE: tropical Australia, and parts of s. Aus. HABITATS: eucalypt forest and woodland. STATUS: common Dry season visitor, and scarce resident. BREEDING: Apr–Sep.

Unlike most populations of the Red-tailed Black Cockatoo in southern Australia, the Top End population is common, and in the Darwin region is regularly encountered in metropolitan areas. During the late 1960s flocks of 500 or more were recorded in the Humpty Doo region, where the species was declared a pest to agriculture because of the damage it caused to rice crops.[1] The Top End population is at least partly migratory, many birds moving from subcoastal and inland areas to coastal areas in the

Kirri Hardy

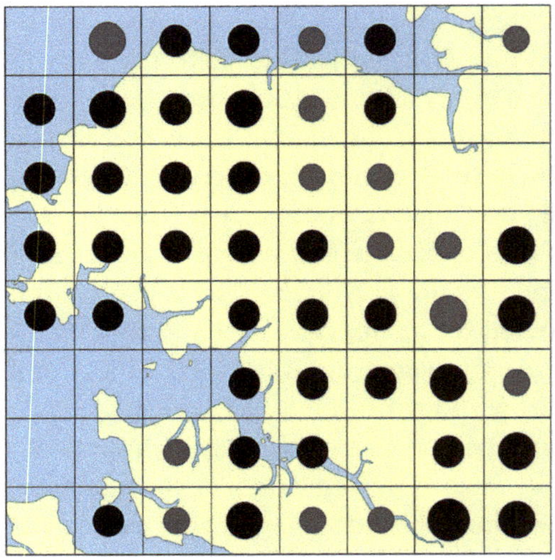

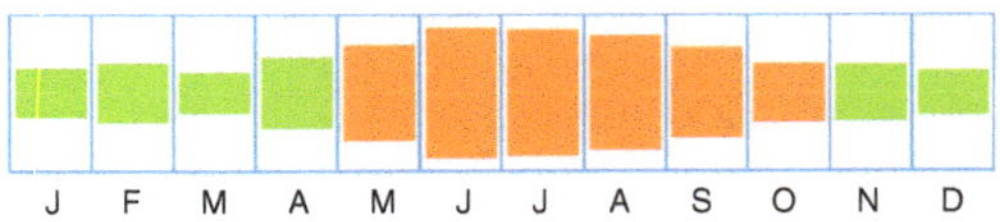

early Dry season, and returning south during the late Dry and early Wet seasons.[1,103,143] In the far south of its range in the NT, such as around Tennant Creek, the species appears only in January–February,[41,327] corresponding with the wettest time in the north.

In Darwin Red-tailed Black Cockatoos are present in all months, but reporting rates increase gradually over the early Dry season, consistent with an influx of birds from inland areas. This movement does not appear to represent migration between breeding and non-breeding areas as it coincides with the breeding season in the Top End, and thus presumably mainly involves non-breeding birds. Indeed, sexual maturity in this species is not reached until at least four years of age.[327] Eucalypt forests and floodplain edges are the main habitats of the species,[103,308] which was recorded in 81% of cells. Flocks of 100–250 birds are regularly seen in the mid-Dry season.[21,381] A study conducted in the Darwin region in 1998–99 suggested that a population of around 30 birds was resident in the Lee Point area.[381] Around Darwin Golf Course, birds were recorded in all months from June 2006 to July 2007, with a maximum number of 34 in the third week of May.[309] Surveys conducted throughout the 2007 Dry season in Palmerston revealed small numbers, mostly in suburbs built in the 1990s, but also on the urban fringes in areas that have since been developed.[307]

The species is strongly attracted to areas that have been recently burnt,[95] where fires have removed the thick grass cover, increasing the detectability and accessibility of seeds, which form an important part of its diet.[327,381] At Munmarlary, Kakadu NP, it was often recorded in plots that had been experimentally burnt during the early Dry season, but was absent from unburnt plots or those burnt in the late Dry.[382] During the Dry season grass seeds, especially those of Annual Speargrass, make up most of the diet of local birds, but the unripe fruits of eucalypts are also eaten.[381] Flocks foraged in the same open areas for two to four weeks after the fire as seed numbers became depleted.[381] The diet of local birds in the Wet season is less well known, but for at least two weeks in January 2006, a flock of over 50 birds was regularly seen feeding on the seed 'cones' high in the canopy of Beach She-oaks in Casuarina CR.[47] Nests are placed inside hollow trunks of eucalypt trees, mostly Long-fruited Bloodwoods and Darwin Woollybutts.[381] Among 16 nests found within 100 km of Darwin, the average depth of the nest hollow was 78 cm and the height of the nest entrance 9.2 m. Eggs are thought to be laid from April to September,[41,43] though birds in captivity appear to lay earlier, from March to May.[383] The single egg takes four weeks to hatch, and the chick remains in the nest for at least 10 weeks.[327]

Galah
Eolophus roseicapilla

RANGE: whole of Australia. HABITATS: floodplains, eucalypt forest and suburbia. STATUS: patchily common resident. BREEDING: Jan–Feb (DR), reputedly Feb–Jun (NT).

Niven McCrie

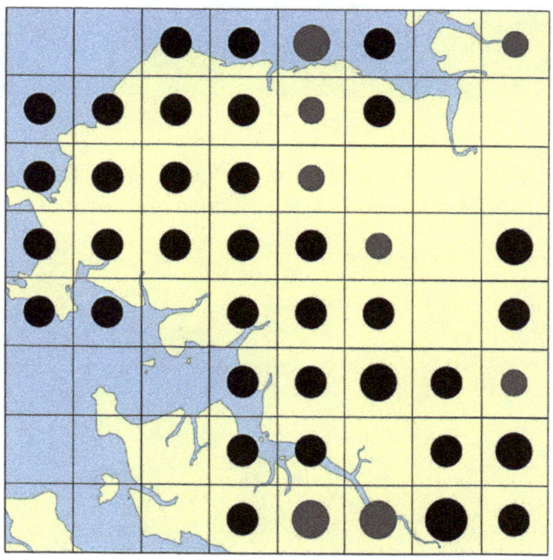

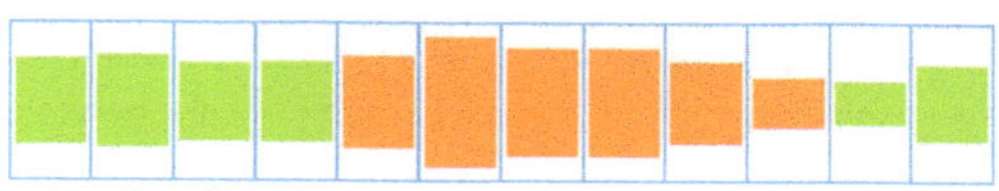

The number of Galahs in Darwin appeared to increase after 1974, when tropical cyclone Tracy thinned out the trees in the open forests surrounding the outer suburbs.[365] In the 1980s they were observed mostly in the northern suburbs of Darwin, where they were suspected of nesting.[60] Since then they have been recorded in 67% of cells, significantly more often from subcoastal than coastal cells, and more from rural and urban than forest areas. In Palmerston, the Galah was the eighth most frequently recorded, and second most abundant, of all birds recorded during a five-month survey over the 2007 Dry season.[307] It was most numerous in the relatively new suburbs of Gunn, Rosebery and Farrar, and least numerous in the oldest suburbs. The species is regularly seen perched on power lines along major roads in both Palmerston and Darwin.

As early as 1948, the Galah was described as a Dry season visitor to coastal parts of the Top End, largely vacating the region in the Wet.[110] During the 1960s, when it was deemed a pest of rice crops in the Humpty

Doo-Fogg Dam area, flocks of up to 200 birds were commonly seen from May to October, but in all other months it was scarce.[1] Bird surveys in Kakadu NP during the early 1980s, however, showed no evidence of movements out of the park during the Wet season.[103] Indeed, in the Darwin region, average reporting rates in Wet season months are not significantly lower than those of Dry season months. Reporting rates peak in June but then gradually decline over the late Dry season. Whether this seasonal pattern of occurrence reflects abundance is unclear. At Palmerston, numbers increased from May to August, with a maximum of 90 birds.[307] Around Darwin Golf Course, on the other hand, numbers fluctuated from week to week over 14 consecutive months in 2006–07, the maximum number of 70 birds being recorded in January and February.[309] There are few breeding records from the Top End,[4,103,360] but in the arid parts of the NT the species reputedly breeds from February to June, and possibly from August to November.[41,327]

Long-billed Corella
Cacatua tenuirostris

RANGE: se. Australia, introduced to se. Qld and Tas. HABITATS: woodland, grassland, urban areas. STATUS: aviary escapee. BREEDING: not in TE.

Adrian Boyle

A small colony of Long-billed Corellas was present in the Nightcliff–Rapid Creek area for several years from the late 1980s until about 1995. More recent reports, including up to four birds in the Palmerston area[384] and a single bird in Rapid Creek,[3] are likely to represent recently escaped aviary birds, rather than remnants of the Nightcliff colony.

Little Corella
Cacatua sanguinea

RANGE: whole of Australia and s. NG. HABITATS: grassland, floodplains, woodland, rainforest and wetlands. STATUS: common visitor. BREEDING: Jun–Sep (NT).

Niven McCrie

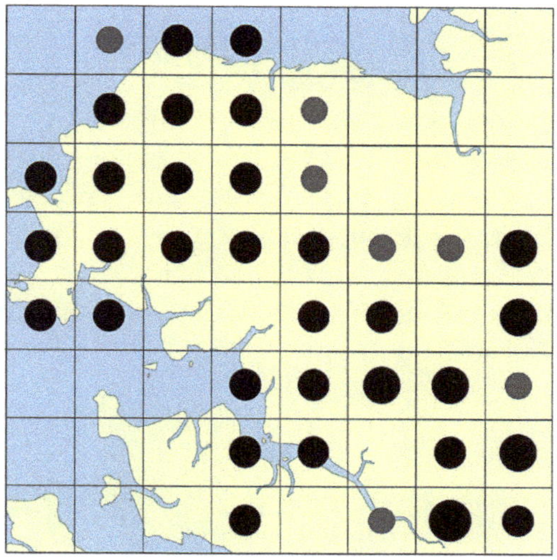

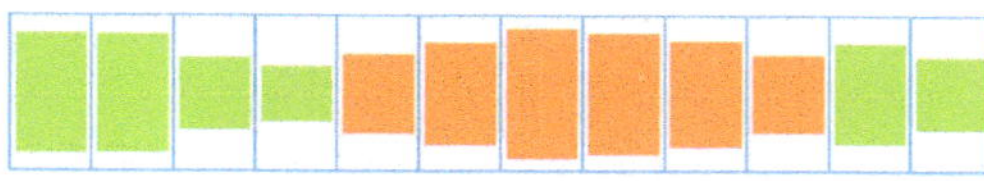

Like the Red-tailed Black Cockatoo and Galah, the Little Corella achieved some notoriety as an agricultural pest during the 1950s and 1960s, when rice-growing was tried in the Humpty Doo–Fogg Dam region.[1] It was the most abundant of the cockatoos, with flocks of more than 2000 birds forming during the mid-Dry season.[1] Yet apart from the Knuckey Lagoons area, where it has been recorded since at least 1964,[96,105] the species was only an occasional visitor to Darwin before the 1980s, mainly to the fringes to the northern suburbs.[60] Since then, it has been recorded in 59% of cells in the Darwin region, significantly more often from subcoastal than coastal cells, and more from rural and urban than forest areas. The largest flock sizes reported were 400 and 430 in June 1996[47] and October 2005,[21] respectively. During the Palmerston Dry season bird survey in 2007, the Little Corella was the eighth most abundant species overall, the largest flocks numbering 48 birds.[307] It was recorded most frequently in the recently developed suburb of Rosebery, but was rarely seen in suburbs built before the 1990s.

Though they often rest and roost on forest edges, Little Corellas favour open plains or fields for foraging. While there is no evidence of large-scale migration within Australia,[157] some southern populations appear to undertake seasonal movements of up to 250 km.[327] In the Fogg Dam region during the late 1960s and early 1970s, the species was present in large flocks from June to August, yet from January to March it was scarce.[1] In Kakadu NP, too, it was more abundant in the Dry season than during the Wet, when flocks were thought to move southward.[103] Reporting rates in the Darwin region also suggest an influx during the mid-Dry season, but another peak in records occurs in the mid-Wet season. Indeed, at Darwin International Airport, more than 400 surveys over 2.5 years (2003–2005) revealed that the species visited mainly from the mid to late Wet season, but was totally absent from April to September.[385] It is possible that these Wet season visitors emanate from regions to the east rather than the south. A large population of many thousands was reported from Gunbalanya, western Arnhem Land, 50 km from the coast, in September–October 1948.[110] Other local populations may be largely sedentary. Around Darwin Golf Course, up to 50 birds were recorded in 11 consecutive months during 2006–2007.[309] There are few confirmed breeding records from the NT,[4,5,327] but in Keep River NP, near the NT-WA border, eggs were laid in August and September,[369] and elsewhere in the NT, as early as June.[41,327] A study of a large population of over 30 000 birds in the Ord River area of the Kimberley region, Western Australia, during the early 1970s suggested that only 20% of birds were breeders, the remainder comprising young and unpaired birds, and non-breeding pairs.[386]

Sulphur-crested Cockatoo
Cacatua galerita

RANGE: tropical and e. Australia, and NG. HABITATS: eucalypt woodland, paperbark forest and monsoon rainforest. STATUS: common resident. BREEDING: May–Aug (TE).

Niven McCrie

While the Sulphur-crested Cockatoo is the most frequently recorded of the four cockatoo species in the Darwin region, in strong contrast to its counterparts in south-eastern Australia, it is the least gregarious.[98,110,143,144] During the late 1960s it was regarded as the least common of the cockatoos in the Humpty Doo–Fogg Dam region, where the largest flocks numbered only 30 birds.[1] As these birds did not visit rice fields they were not considered pests of rice crops, unlike the other three species.[1] In a more recent survey of bat and bird pests of agriculture in the Top End, however, this species emerged as a serious bird pest, second only to the Red-collared Lorikeet in the amount of damage it caused to fruit crops, such as mango, rambutan and pawpaw.[387]

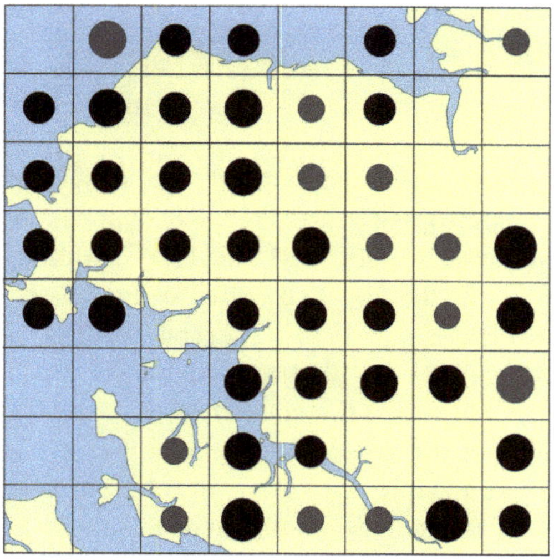

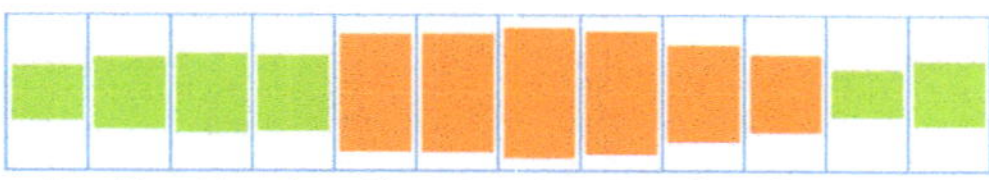

It has been suggested that, like the Galah, the Sulphur-crested Cockatoo withdraws from the Darwin region during the Wet season,[41] and while there is no published evidence of large-scale movements,[157] reporting rates indicate that it is recorded more frequently in the early Dry season than during the early and mid-Wet. This early Dry season peak coincides with the apparent breeding season of Top End birds (from May to August),[29,93,98,110,366] suggesting that the higher reporting rates may be partly due to conspicuous behaviour around nests.

The species was recorded in 73% of cells, significantly more records coming from urban than forested areas. In Palmerston, it was not as frequent or numerous as the Little Corella, the largest group size being seven birds.[307] Around Darwin Golf Course it was recorded in 12 consecutive months during 2006–07, with the maximum number of 20 birds in August.[309] The only cockatoo recorded in more than 5% of censuses during the Howard's Peninsula survey, it was recorded in five of the 20 habitats, most frequently in tall paperbark swamps, but also in open eucalypt woodland and monsoon rainforests.[308] A study of birds inhabiting and utilising Common Pandanus in Kakadu NP showed that the Sulphur-crested Cockatoo was the only species that ate their fruits and dispersed their seeds.[388] While seeds and fruits form the staple diet of the species, one bird was observed feeding on the nectar of the Fern-leaved Grevillea at Limmen NP in the Gulf Country of the NT.[389]

Cockatiel
Nymphicus hollandicus

RANGE: whole of Australian mainland. HABITATS: woodland and grassland. STATUS: rare Dry season visitor. BREEDING: Jun–Jul, Sep–Oct (TE).

While often considered highly nomadic,[41,103,327] the Cockatiel is nevertheless known to migrate north and inland from breeding grounds in south-eastern Australia to wintering quarters in the tropics.[157] It is a moderately common Dry season visitor to southern parts of the Top End[311] and the semi-arid region to the south,[41] but largely vacates the NT in the Wet season[5] and there are few NT breeding records. In most years small numbers visit the Marrakai area, about 40 km south-east of the Darwin

Brian Thistleton

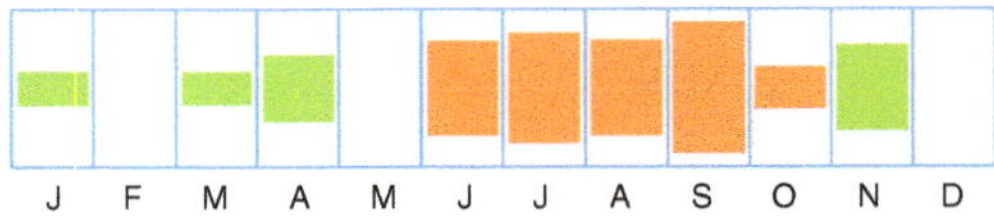

region, and Coomalie Farm, about 45 km to the south,[47,177] during the Dry season, though it rarely reaches Darwin itself. From 1968 to 1971 it was recorded annually in the Humpty Doo–Mt Bundy region, but was common only in 1969 and 1970, after poor Wet seasons both locally and in the interior.[1] In the latter two years Cockatiels arrived in April or early May and stayed until mid or late October, whereas in the previous and following years (1968 and 1971), after good Wet season rains, they arrived later, in June or July, and left earlier. Flocks of up to 30 birds were noted.

Until 1985 there were few records of Cockatiels visiting Darwin but, during this and the following year, they were frequently sighted in the suburbs and rural fringes of the metropolitan area.[60] Perhaps not coincidentally, in both 1985 and 1986, southern parts of the Top End suffered their worst drought since 1970.[124] Several records of the species in 1990 followed widespread drought across the northern half of the NT but not further inland. On the other hand, a presumably wild flock of 15 birds was seen flying over woodland at Lee Point in July 1997,[21] which followed a good Wet season inland.[124] Over the next decade, the species was recorded in five years, only two of which experienced drought in the NT, although some of these records almost certainly refer to aviary escapees, such as an individual that resided at Leanyer SP for over a month[21] and another in Nightcliff for several months.[47]

Since 2000 there have been only two reports of the species: one in 2006 when drought afflicted much of south-eastern Australia and the other in 2013, when there were severe rainfall deficiencies in south-east NT.[124] At Yinberrie Hills, about 50 km north-west of Katherine, the Cockatiel was among the most abundant bird species from April to June, but was scarce or absent in other months.[311] Although the breeding season in the southern NT is reputedly autumn,[41,327] nestlings were found at Victoria River Downs in July 1976,[98] and breeding was noted at Yinberrie Hills in October 1986.[311]

Red-collared Lorikeet
Trichoglossus rubritorquis

RANGE: endemic to monsoon-tropical Australia. HABITATS: forests, woodland, suburbs. STATUS: very common resident. BREEDING: Jan–Oct (TE).

Niven McCrie

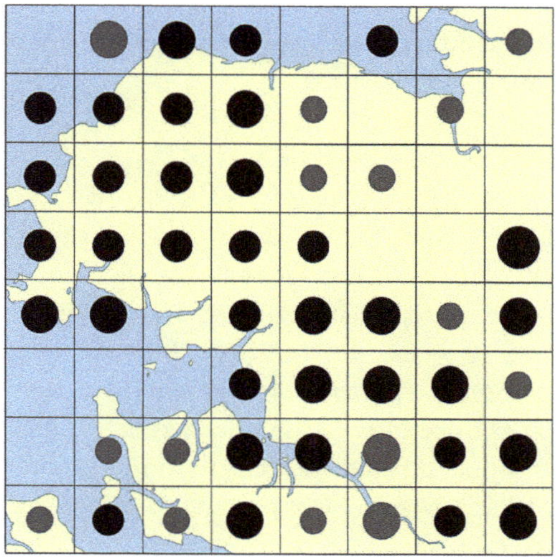

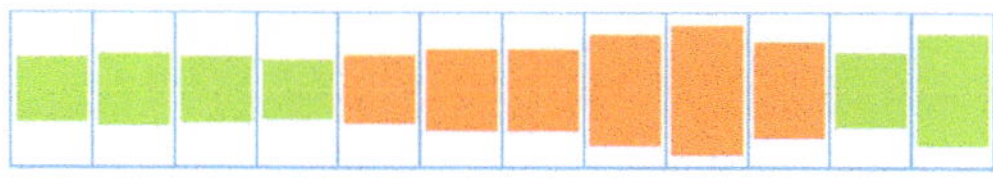

With its adaptability to anthropogenic environments, the gaudily coloured Red-collared Lorikeet is arguably the most familiar, if not the most popular, of parrots in the Darwin region. In rural areas of the Darwin and Humpty Doo regions, the species is regarded as a major pest of mango crops, second only to Black Flying-foxes in the extent of damage it causes.[387] It has also become a nuisance in some popular public spaces of urban areas, where these birds often roost in one or a few isolated trees and in such large numbers that their nocturnal chattering and screeching can be almost deafening, while their droppings potentially pose a health risk to humans.[390] In 1996, one such communal roost site adjacent to an outdoor café in Nightcliff was estimated to support 1500 to 2000 birds until it was eventually disrupted.[390] Up to 30 minutes before dusk many waves of small flocks could be seen converging on their roost site from many directions, suggesting that they were widely dispersed during the day. However, a study at the Territory Wildlife Park, 13 km south of the Darwin region, revealed that the local population was relatively sedentary during the Dry season, staying within an area of only about 100 ha.[391]

The Red-collared Lorikeet was the seventh most frequently recorded bird species in the Darwin region. During the Palmerston Dry season survey of 2007 it was the fifth most frequent and seventh most abundant species, occurring in suburbs of all ages.[307] Average numbers rose steadily from May to September. It was also the most abundant of all species recorded around the suburb of Millner over the five years from 1988 to 1993.[296] Numbers rose sharply in July or August in four of these five years, with a maximum of 338 birds in August 1992. It was among the 10 most frequent and abundant species around Darwin Golf Course over 52 consecutive weeks during 2006–07.[309] Finally, it was among the 10 most frequently recorded species during the Howard's Peninsula survey, when it was encountered in 28% of all plots and in nine of the 20 habitats, though it was most abundant in tall eucalypt forest and paperbark forest.[308]

The species was recorded in 78% of cells in the Darwin region, with significantly more records from urban areas than rural or forested areas. Counts of Lorikeets along a 20 km loop between Nightcliff and Lee Point over 18 weeks during 2002–2003 yielded a maximum estimate of over 7000 birds in early December.[390] However, this number dropped through December and January, probably due to birds moving to paperbark swamps, such as Marrara Swamp, to exploit the nectar of Broad-leaved Paperbarks, which flower profusely at this time.[392,393] This seasonal movement may partly explain the pronounced peak in reporting rates in December, which was evident in most years. Indeed, the Marrara, East Point and Palmerston cells showed a higher proportion of December records than expected based on total proportions over the year. Like other parrots, Red-collared Lorikeets nest in tree holes, the contents of which are difficult for observers to ascertain. The breeding season in the Top End is therefore ill-defined, but data suggest that clutches are laid from January to October.[47,70]

During the Dry season, Red-collared Lorikeets apparently track the floral nectar of trees such as Darwin Woollybutt and Fern-leaved Grevillea, which flower at this time of the year, showing a distinct preference for individual trees with large blossoms.[391] However, during the early Wet season, when the nectar supply in Top End savannas is low, Lorikeets in Darwin have been found to feed mostly on the seeds of Coastal She-oaks and nectar or pollen from almost 30 tree species, the most important of which were cultivated and naturally occurring native eucalypts.[390] Lerp, the sugary secretions of tiny bugs found on the leaves of eucalypts,[394] comprised 14% of the diet. Despite the availability of fruits of many plant species in local monsoon rainforests, the only fruit on which birds were seen feeding was mangoes. The attraction of lorikeets towards the flowers of Umbrella Trees[60] is well known, and indeed they were the second most abundant visitors, after White-gaped Honeyeaters, to one flowering Umbrella Tree in Nightcliff that was monitored over six weeks in 2001.[395]

Varied Lorikeet
Psitteuteles versicolor

RANGE: endemic to monsoon-tropical Australia. HABITATS: eucalypt forests and woodland. STATUS: moderately common visitor. BREEDING: Apr–July (TE).

Laurie Ross

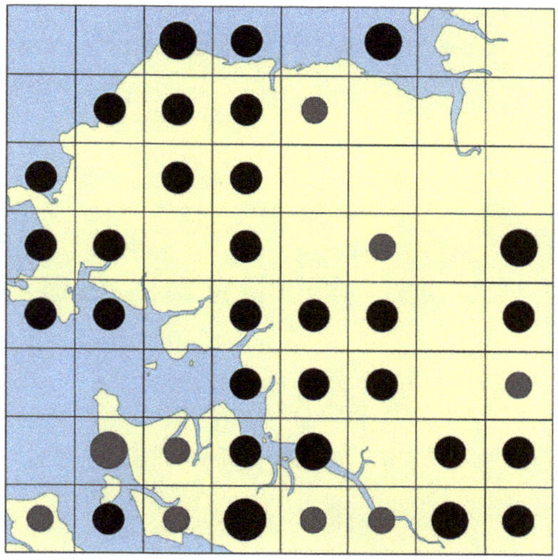

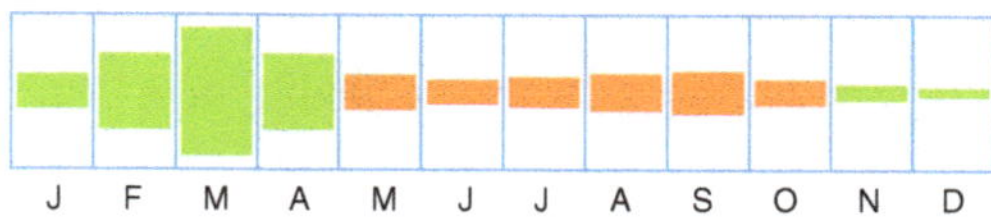

Among all the nectar-feeding birds of the Top End, the Varied Lorikeet, along with the Banded Honeyeater, is the most fastidious in its choice of nectar sources, having a particular penchant for eucalypt blossoms.[392] Though both species are generally deemed to be highly nomadic,[41,327] their appearance in many parts of the Top End is sufficiently regular to suggest migration.[47] In the late 1960s, the Varied Lorikeet was considered an irregular visitor to the Darwin–Humpty Doo region, but was most likely to be seen during the seasonal transition periods of April–May and September–October.[1] In the Darwin region it was recorded in 25 (76%) of the 34 years between 1977 and 2010, but six of the nine years in which it was *not* recorded were also among the seven 'worst' years for the Red-collared Lorikeet, with six or fewer records. Reporting rates show a clear spike from February to April, corresponding to the latter half of the Wet season, but there was no indication of a similar spike during the late Dry season, as noted in the 1960s.

Unlike the Red-collared Lorikeet, the Varied is rarely seen foraging outside eucalypt woodland or paperbark forests,[103,308] consistent with its more specialised diet. At Yinberrie Hills, 50 km north-west of Katherine, this species and the equally specialised Banded Honeyeater were by far the two most abundant bird species during January 1987 and the preceding month, but scarce or absent in the remaining months of the year.[311] In the same area, almost a decade later, numbers were small in August 1995, but from December increased until tens of thousands of birds were present in late January and early February 1996, when Round-leaved Bloodwoods were producing masses of flowers, on which the birds fed.[396] Soon afterwards Lorikeet numbers declined until May, when none were present. Their prevalence in the Katherine region during the early Wet season[96,311] and appearance in coastal and subcoastal areas over the Dry season[47,397] indicate a northward migration that probably continues as far as the Tiwi Islands.[270] Indeed, small flocks are often seen flying in a northerly direction, high over Darwin suburbs, and on 12 March 2001 some 300 birds were counted at Lee Point as they flew due north towards Melville Island.[21]

Remarkably little is known of the breeding behaviour of this species in the wild, and there are few published breeding records. However, observations of nests in the Top End suggest egg-laying occurs from April or May to July.[43,70,366]

Northern Rosella
Platycercus venustus

RANGE: endemic to monsoon-tropical Australia. HABITATS: eucalypt woodland and forest. STATUS: scarce resident. BREEDING: Mar–Jul, possibly also Jan (TE).

Laurie Ross

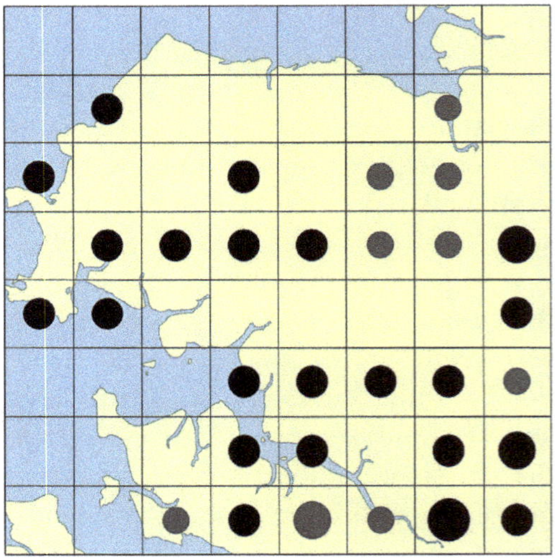

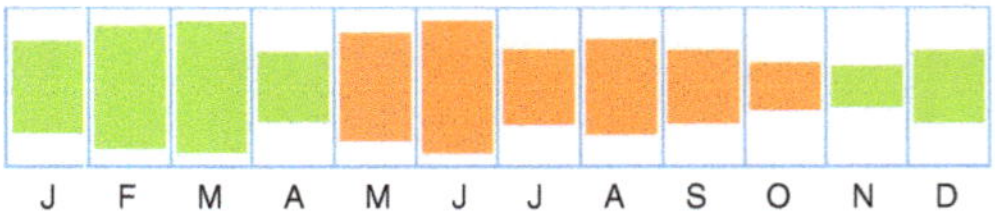

The least known of all rosellas, the Northern Rosella appears to be scarcer in the Darwin region than in the Pine Creek-Katherine region to the south[300] and Kakadu NP to the east.[103] In the late 1960s and early 1970s it was described as locally common in the rural areas of Berrimah and Humpty Doo, but uncommon around Shoal Bay, Marrakai and south of Noonamah.[1] Since 1977 it has been seen in 48% of the cells in the Darwin region, with significantly more records from subcoastal and rural areas than coastal, urban or forest areas. In the mid-1980s it was considered a scarce visitor to Darwin's fringes, but still quite common around Palmerston.[60] During a Dry season bird survey of Palmerston in 2007, however, it was encountered on only 11 (3%) of 320 transect counts, most sightings involving a pair of birds in the young suburb of Rosebery.[307] In 1948, a small band of rosellas was regularly seen near Nightcliff when the original monsoon rainforest was being cleared,[110] but nowadays the closest site to Darwin's metropolitan area where the species is regularly seen is Charles Darwin NP.[21,47]

Northern Rosellas are often flushed from the sides of roads[1] where they feed on the seeds of grasses and herbs. However, they also eat the flower buds and seeds, and occasionally take nectar,[398] of eucalypts and Grevilleas.[327] Unlike the Red-winged Parrot they rarely form flocks of any size, eight being the maximum recorded in the Top End.[1,47,143] Their relatively low density and patchy distribution may relate to fire regimes. In the Woollybutt-dominated forest around Solar Village, just south of Humpty Doo, Rosellas were significantly more abundant in annually burnt plots than in long-unburnt plots that supported many rainforest trees and shrubs.[399]

Almost nothing is known about the breeding biology of this species in the wild. The clutch reputedly consists of two to four eggs only, while all other rosella species lay up to seven or nine eggs per clutch. The few breeding records of Northern Rosellas in the Top End suggest that egg-laying occurs from March to July, and possibly January.[143,311,327,400]

Budgerigar
Melopsittacus undulatus

RANGE: Australian endemic; whole of Australia, but rarely near coast. HABITATS: open country. STATUS: vagrant. BREEDING: not in TE.

Mike Reed

Numerous reports of Budgerigars in the Darwin region clearly relate to escaped or released aviary stock. However, the occurrence of one in April 1952, during a prolonged period of severe, Australia-wide drought,[401] most likely represents a naturally occurring wild bird, as the species is well known to move towards coastal areas during such conditions.

Red-winged Parrot
Aprosmictus erythropterus

RANGE: tropical and subtropical Australia, and s. NG. HABITATS: mainly eucalypt forest and woodland, but also rainforest and suburbia. STATUS: common resident. BREEDING: Jan–Jul (TE).

Laurie Ross

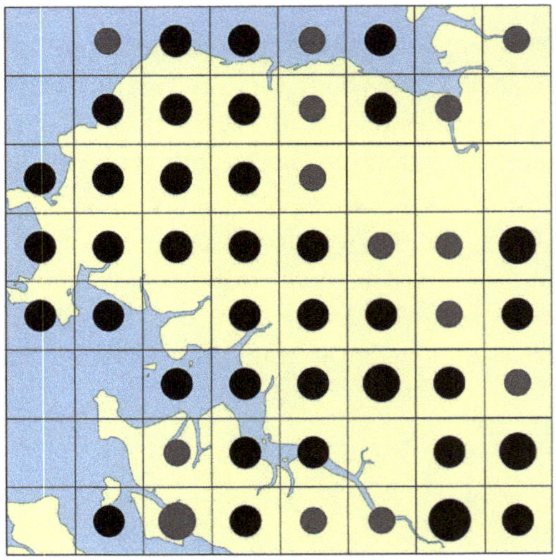

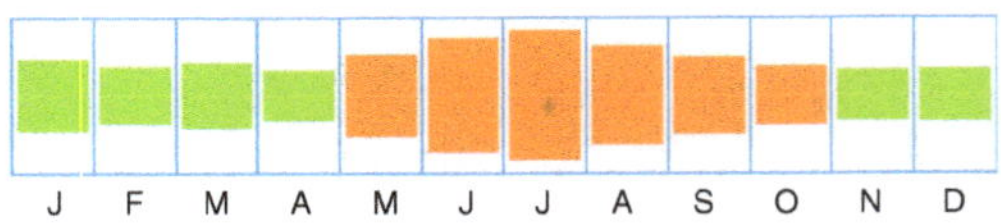

Instantly recognisable at a distance by its characteristically slow, floppy flight, the Red-winged Parrot is the only common seed-eating parrot in the Darwin region. The species is widespread, being recorded in 80% of cells, equally frequently in coastal and subcoastal areas, but more often in urban and rural than forested cells. During the Howard's Peninsula survey it was encountered at 12% of the sites sampled, and in seven habitat types, though the majority of records came from eucalypt forest dominated by Darwin Stringybark and Woollybutt, including areas damaged by tropical cyclone Tracy.[308] During the Palmerston Dry season bird survey in 2007, the species was seen in all five sampled months, and in suburbs of all ages, but most frequently in the newer suburbs built in the 2000s, particularly Rosebery.[307]

Reporting rates show a steady increase during the early Dry season that might indicate an influx into the region but, as there is no evidence of large-scale seasonal movements in the species, this peak is most likely the result of post-breeding dispersal, possibly into more open habitats. Red-winged Parrots are mostly seen in small groups, though they often form flocks of 40 to 60 birds, especially during the late Dry months from August through October.[47] Such flocks are often composed of mostly immature birds,[47] distinguishable from adult females by their paler bills and darker eyes.[327] Around Darwin Golf Course, groups of up to 15 birds were seen in all but nine weeks over 12 consecutive months during 2006–07.[309]

Though its diet consists primarily of seeds derived from Acacias and other shrubs and vines,[327] the species often feeds on the fleshy fruits of a variety of plant species, including figs,[314] mistletoes[164] and (occasionally) mangoes.[387] During the Dry season, it sometimes feeds on the nectar, and chews the flowers, of Fern-leaved Grevilleas and Darwin Woollybutts.[398] It has also been observed feeding on the green pods of cultivated Soya Beans,[1] and the seeds of Hyptis, an introduced aromatic weed that forms dense swathes in disturbed areas.[60]

Despite the prevalence of this widespread species, few nests have been reported, and information about its breeding biology is mostly based on observations of captive birds. Sightings of food-begging juveniles in the Darwin and Kakadu regions, however, suggest that local Red-winged Parrots lay their eggs mainly from January to July.[43,47,358,361]

Rainbow Pitta
Pitta iris

RANGE: endemic to monsoon-tropical Australia. HABITATS: monsoon rainforests. STATUS: common resident. BREEDING: Oct–Feb.

Niven McCrie

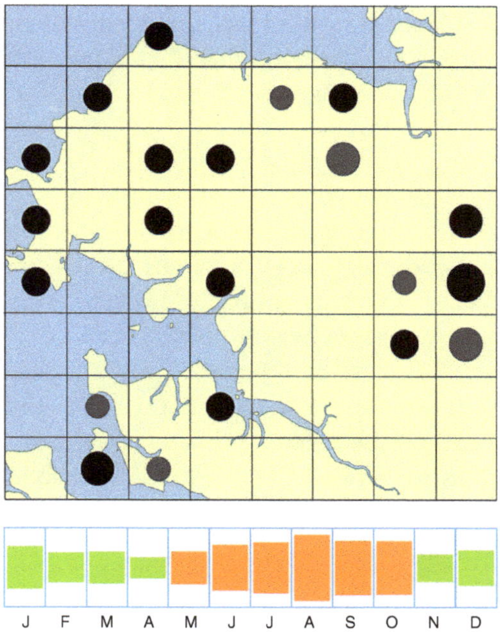

It is arguable that the Rainbow Pitta would not have evolved were it not for the seasonal abundance of earthworms in the Top End, for without a supply of these invertebrates to feed its young it is doubtful that this handsome bird could reproduce successfully.[402] Indeed, its very existence is remarkable given its dependence on rainforests, which in the monsoon tropics are restricted to relatively small areas, increasingly fragmented by the advance of fire-promoting savannas and degraded by feral animals and weeds.[355] Even more astonishing is their occurrence in the Kimberley Division of Western Australia, where rainfall is much lower, and rainforests even more patchy, than in the Top End.

Despite some loss of local coastal rainforests in recent times, the Rainbow Pitta can be found relatively easily at many sites both within and around metropolitan Darwin. Indeed, about one-third of all records since 1970 came from Buffalo Creek and Casuarina CR, while another third came from Howard Springs NR.

Rainbow Pittas are sedentary and territorial, although the boundaries and sometimes the occupants of territories change from year to year. In the first detailed study of any pitta species in the world, 48 Rainbow Pittas were colour-banded at Fogg Dam, and another six at East Point and three at Holmes Jungle. The average size of territories of nine banded pairs at Fogg Dam was about half that of three pairs at East Point (1.6 vs. 3.1 ha). Counts of calling birds also suggested that the population density was highest at Fogg Dam and lowest at Holmes Jungle.[402]

The distinctive four-note whistle of the Pitta is given most frequently towards the end of the Dry season and during the early Wet,[402] and this probably explains why reporting rates peak from August to October. Eggs are laid between mid-October and late February, though the start and length of the breeding season varies annually with rainfall.[403] The Pitta's major food item during this period is earthworms, and nesting activity is correlated with the abundance of these invertebrates, which itself correlates with rainfall.

The nests of pittas are domed structures with a side entrance, but while that of the Noisy Pitta of eastern Australia is usually placed on the ground, the Rainbow Pitta prefers to build its nest at the top of a tree stump or in the fork of a tree trunk, at an average height of 5 m above the ground. The placing of nests off the ground may be related to the risk of flooding, as some rainforests become inundated during the Wet season. A peculiar habit of some individuals of this species is the placement of wallaby dung on the landing platform at the entrance to nests.[404] This may have an anti-predator function, by providing an odour 'camouflage' that may prevent predators from smelling eggs or nestlings. Nevertheless, in the above study about three-quarters of nests were robbed by predators, including tree-dwelling snakes and rats. Fledglings remain in their natal territory for a relatively short period, then apparently disperse to other forest patches. Of 70 nestlings banded at the three study sites, only one was recorded more than one month after fledging. Banded at East Point in mid-March 1995, this bird was caught four months later in a building at Charles Darwin University, 6.5 km north-east of its place of birth.[47]

Great Bowerbird
Chlamydera nuchalis

RANGE: endemic to tropical Australia. HABITATS: eucalypt forests, woodland and rainforests. STATUS: moderately common resident. BREEDING: Sep–Jan (TE).

Laurie Ross

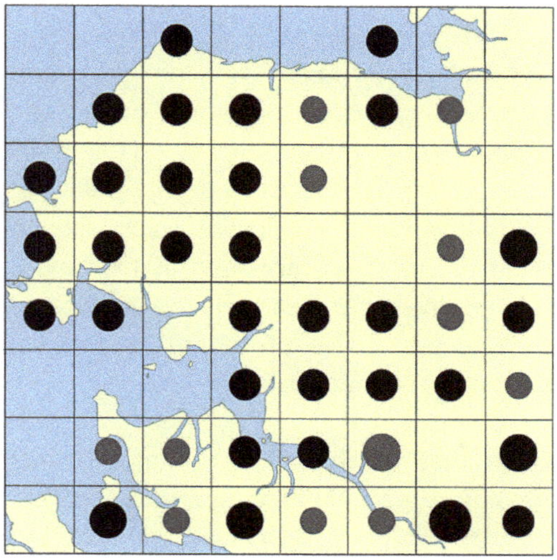

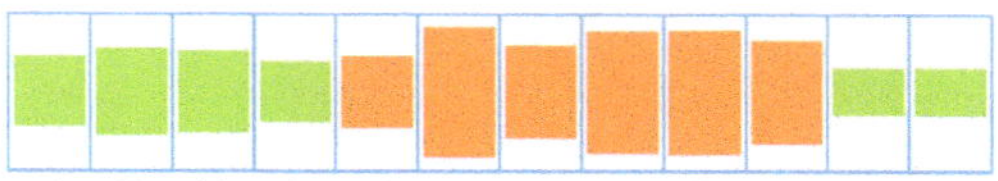

What the male Great Bowerbird lacks in colourful plumage, it makes up for in body size and bower-building prowess. The largest of its family, this species makes the largest of the avenue type of bowers. The arching avenue walls comprise close to 2000 vertical sticks, while an average of 600 decorations are placed on the platforms at both ends of the avenue.[405] But bigger is better in this species. Video footage and measurements of many bowers at Coomalie Farm, 45 km south of the Darwin region, showed that the males whose bowers had the longest avenue and thickest walls also had the highest mating success.[406]

But size isn't everything, as the male must perform a two-stage audiovisual courtship display to entice the female into the avenue for mating purposes. He begins with a stationary display at the side of the avenue entrance, during which he jerkily bobs his head while giving ticking noises, and eventually exposes the normally hidden lilac crest on his nape, which he presents to the female by turning his head.[407] During this display he often picks up a decoration at his feet, then drops it before starting the peripheral display, in which he prances around the bower with wings drooped and tail partly cocked, until he resumes the stationary display and repeats the process. Acoustic analyses of ticking calls recorded during videos of over 100 displays revealed that the males giving the loudest and longest ticks attracted the most females. Thus females based their choice of mates on both bower size and ticking calls.[407]

The most abundant decorations on platforms in Coomalie bowers were grey stones, followed by whitish snail shells, white stones and pieces of transparent and green glass, but invariably included some green fruits, mainly those of cycads.[408]

Although both grey and white objects were widely spread over the entire platform, the former tended to be placed near the avenue and the latter towards the outer platform. Green objects, however, were consistently placed on both sides of the entrances to the avenue, even after researchers experimentally displaced them.[409] This position is practical as the male often picks up and holds a green object in his bill during the stationary display. Red objects, on the other hand, are an anathema to this population; males consistently rejected them during experiments, even carrying them far away. This contrasts with eastern populations of the species, males of which favour red objects for decorations.

Given that much of the savanna in our region burns annually or biennially, bowers are at risk of being destroyed, thus ruining a male's chance of attracting females. But fortunately, whether by design or an accidental spin-off of sexual selection, bowers are fairly fire-proof.[410] First, they are usually built at sites with sparse grass cover, such as beneath the canopies of Turkey Bush or sand palms. Second, the flammable material around the bower, comprising fallen leaves and twigs, is typically removed or covered with decorations by the males. The lack of fuel around the bower, therefore, creates an effective 'fire break'. Indeed, fires came within 50 m of nine bowers at Coomalie in 2006, burning 95% of the area within a radius of 50 m, yet only three were damaged. The compass orientation of the bower is also adaptive. Bowers of this species are invariably oriented north–south, presumably to maximise illumination of the platforms by the sun or to minimise shadows cast by the walls during the peak courtship period.[405]

While fruit is an important part of the diet of Great Bowerbirds, the species occurs only irregularly in monsoon rainforests,[329] and is commonest in eucalypt woodland and riparian forests dominated by paperbarks. It was recorded in 71% of cells in our region, and almost half of all records came from the area stretching from the Casuarina campus of Charles Darwin University (CDU) to Buffalo Creek. In Palmerston it was most abundant in newer suburbs, such as Rosebery.[307] Active bowers have been found at many sites around Darwin, including the Royal Darwin Hospital and CDU.[47] The local breeding season appears to be from September to January.[47,411] Nests are shallow bowls of sticks, built entirely by females, and she alone incubates the eggs for three weeks, and feeds the chicks over a similar period.[405] At CDU and the hospital, nests were found up to 9 m from the ground in trees within 20 m of tall buildings.[47]

Black-tailed Treecreeper
Climacteris melanurus

RANGE: endemic to monsoon-tropical Australia. HABITATS: eucalypt forest and woodland. STATUS: very rare. BREEDING: May–Nov (TE).

Mat and Cathy Gilfedder

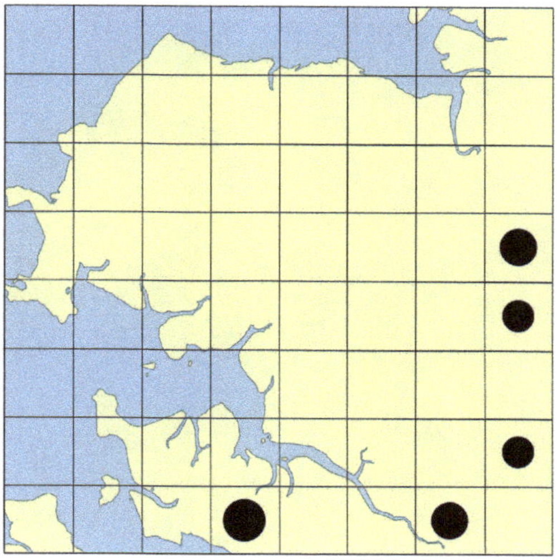

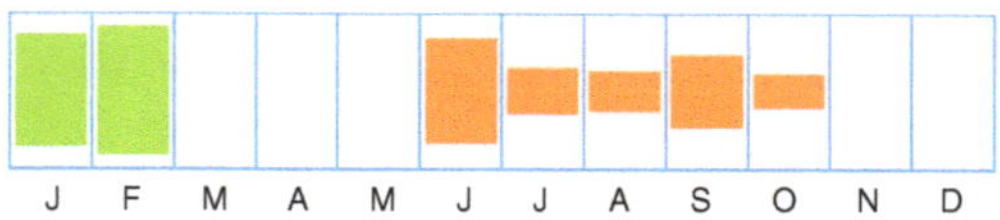

Though a common species of eucalypt forest and woodland in the Top End generally,[103,143,382] the Black-tailed Treecreeper is surprisingly scarce in the Darwin region, with only a handful of records, all from cells on the southern and eastern fringes of the area. One pair was known to occupy a small home range in stringybark woodland on the Channel Island Road from 1978, and was last seen at the site in 2001. Since then the only reports of the species are from cells around Howard Springs NR and McMinns Lagoon. It was not recorded during the 1986 Howard's Peninsula survey,[308] but was considered moderately common in the Humpty Doo region during the late 1960s and early 1970s.[1]

This is the largest, yet least known, of the Australian treecreepers. Observations in four territories at Coomalie Farm, 45 km south of the Darwin region, revealed that, like the Brown Treecreepers of eastern Australia, breeding pairs often have one or two helpers that assist them with the raising of young.[47] Although the literature suggests that Top End clutches have been found only in September and October, observations at Coomalie Farm indicate that egg-laying occurs from late May to early November, possibly making it the longest breeding season among treecreepers.

Red-backed Fairywren

Malurus melanocephalus

RANGE: tropical and subtropical e. Australia. HABITATS: eucalypt forest and woodland. STATUS: very common resident. BREEDING: mostly Jan–Apr (TE).

Vince Bugeja

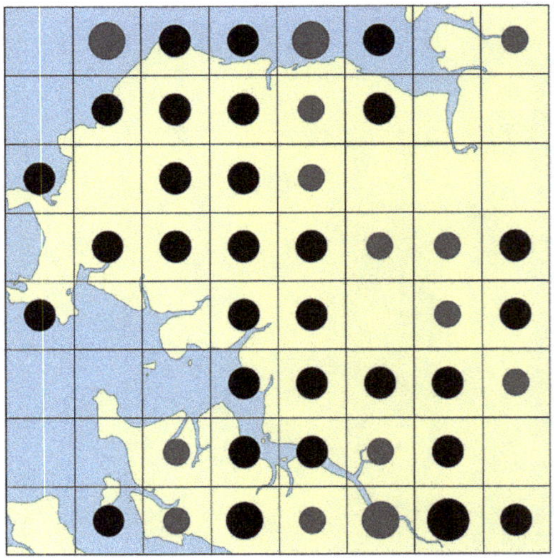

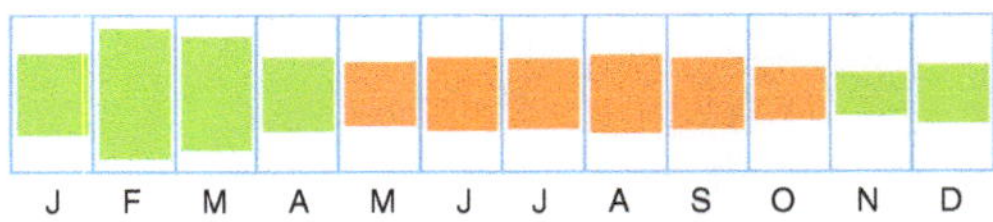

The only fairywren in the Top End, apart from the sandstone-specialised Variegated (Lavender-flanked) Fairywren, the Red-backed Fairywren is common in most habitats with an understorey of grass, including open forest and woodland dominated by eucalypts, paperbark and pandanus, as well as treeless grassland.[1] Indeed, during the Howard's Peninsula survey, it was the fourth most frequently recorded species, encountered on about 40% of sites, and in 12 out of 20 habitats sampled.[308] Unlike the familiar Superb Fairywren of south-east Australia, the Red-backed has not adapted to the urban environment although it can often be found on the fringes of suburbs adjacent to bushland or grassland.[60] It was recorded in 70% of cells, significantly more records coming from subcoastal cells than coastal cells.

Although the species is sedentary, the chart shows a lower reporting rate during the Dry season than during the Wet. This apparent seasonality probably reflects the conspicuousness of birds, which relates to several aspects of their behaviour and social organisation, such as the size and mobility of groups and singing frequency. The breeding season coincides with the Wet season, though relatively few clutches are laid before January.[412,413]

During this period, these birds, like other fairywrens, form pairs that *appear* monogamous. Yet polygamous mating is common, resulting in many offspring that are not fathered by their mother's partner. To confuse matters further, non-breeding birds often help the breeding pair to raise offspring, though such 'cooperative breeding' is less frequent in this species than in other fairywrens.[414] As with other fairywrens, this species shows age-related sexual dimorphism. Females always have brown plumage, but males take up to four years to acquire or retain their bright red and black breeding plumage.[414] In general, young males have dull female-like plumage, and even three-year-old males in breeding plumage will moult back into a brown plumage with black spots during the non-breeding season. A six-month study at Holmes Jungle showed that both bright and dull-plumaged males form pairs with females, but helpers were absent.[415] However, during a three-year study at Coomalie Farm, 70 km inland, most dull males became helpers.[413]

Several surveys have shown that Red-backed Fairywrens prefer unburnt areas to burnt areas.[382] However, at Solar Village, just outside our region, it was one of the most abundant bird species in eucalypt forest that was annually burnt during the early Dry season, and was entirely absent from forest from which fire had been excluded for 23 years, due to the lack of a grassy understorey.[399] Studies in the Kimberley region of Western Australia revealed that fires did not directly kill the birds, but their breeding success was lower in burnt sites than in unburnt sites, possibly due to the lack of understorey vegetation in which to conceal their nests.[416] At Coomalie Farm, a fire forced birds to leave their territories and move to surrounding unburnt areas.[413] However, as unburnt areas were already occupied, displaced birds aggregated in large groups, which wandered locally until the burnt areas recovered. At the start of the Wet season these groups split up to form breeding units.

Dusky Myzomela
Myzomela obscura

RANGE: Tropical coastal and subcoastal Australia; also NG and e. Indonesia. HABITATS: all forests and woodland. STATUS: common resident. BREEDING: all year but mainly Apr–May.

David Webb

With a squeaky song remarkably reminiscent of a 'rubber ducky', the Dusky Myzomela is easily distinguished from other small honeyeaters by its uniform grey-brown coloration, lacking any markings or other colour on the wing or head. Among local honeyeater species, it has the longest bill relative to its bodyweight,[392,417] reflecting the importance of nectar in its diet,[418] and indeed it appears to feed its young largely on nectar.[419] However, like most other honeyeater species, it also takes invertebrates, some taken in the air, to supply its protein needs.

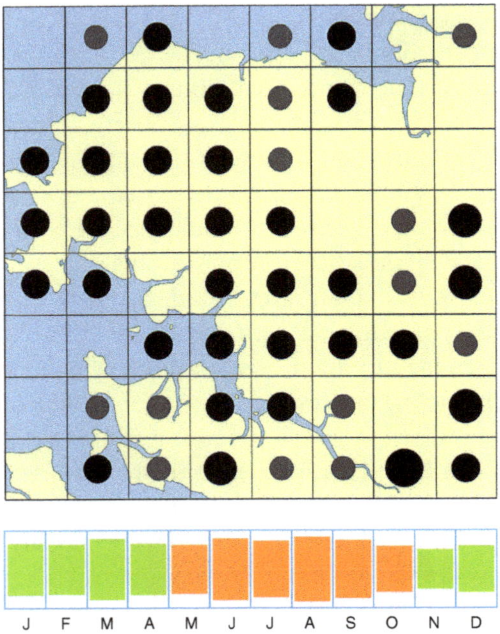

Dusky Myzomelas occur in a wide variety of wooded habitats, but they were most abundant in eucalypt forest with a dense understorey and in rainforest during the Howard's Peninsula survey.[308] Research in mixed eucalypt and paperbark woodland close to a spring-fed rainforest at the Territory Wildlife Park indicated that the species was sedentary within large home ranges.[420] The recapture rate of banded birds over three years was slightly lower than that for the much more numerous Brown Honeyeater, but Dusky Myzomelas were more likely to be recaptured at distances of over 200 m from their original banding site. The maximum recorded movement was 1.2 km, and the maximum time between captures 2.4 years. However, away from the coast the species may move over longer distances. In Kakadu NP, it was most abundant in monsoon rainforests, especially those on the escarpment, but repeated counts suggested considerable seasonal movement between rainforest types, and even among patches of the same type.[355]

Generally scarce in Darwin's suburbs[60] the Dusky Myzomela was recorded in only 27% of weekly surveys around Darwin Golf Course over one year,[309] and rarely encountered at all over five years of surveys around Millner, only 4 km away.[296] In Palmerston, however, it was among the 20 most frequently recorded species, most abundant in the oldest suburbs of Driver and Moulden, as well as in the woodland surrounding the newer suburbs of Bellamack and Mitchell Creek. Indeed, the species was recorded twice as frequently as the Rufous-banded Honeyeater in Palmerston, where the Brown Honeyeater is by far the most abundant small honeyeater, whereas in Darwin the Rufous-banded is overwhelmingly numerically dominant to both species. Perhaps the Rufous-banded is less tolerant of the Dusky than the Brown. Unlike the latter, the Dusky Myzomela seldom visits mangals. Near the Palmerston Sewage Ponds, they were encountered on only 8% of censuses, and of some 1400 birds banded during 77 mornings over 10 years, only 10 were of this species.[312] Similarly, of some 1120 birds caught and banded in mangals around Darwin, only eight were Dusky Myzomelas.[47]

Little was known of the breeding biology of this species until a recent study in Darwin. Breeding was recorded in all months, but just over 50% of estimated egg-laying dates were in April and May, corresponding with the transition from the Wet to the Dry season.[419] In Darwin, nests were placed up to 9 m above the ground in a variety of trees or tall shrubs, including exotics. The rate at which parents fed the nestling was exceptionally low compared with other honeyeater species, possibly because the food contained a high proportion of energy-rich nectar.

Red-headed Myzomela
Myzomela erythrocephala

RANGE: coastal tropical Australia and NG. HABITATS: mostly mangal. STATUS: very common resident. BREEDING: Apr–Oct.

Nolan Caldwell

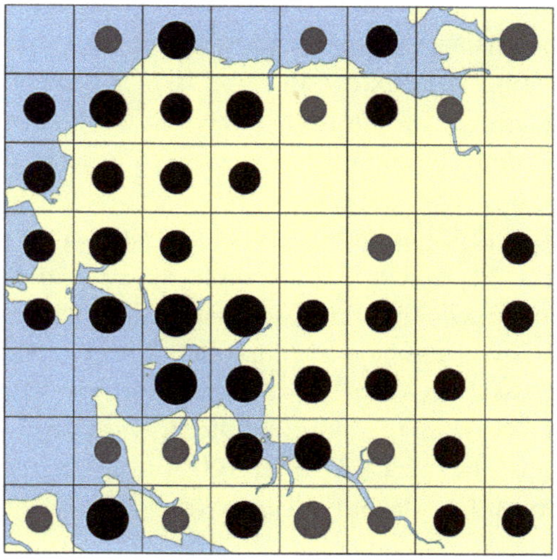

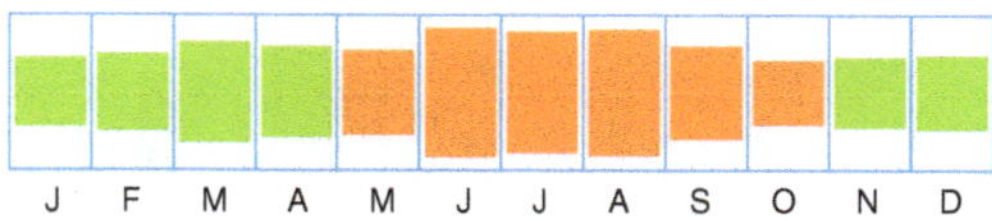

The smallest honeyeater in Australia, the Red-headed Myzomela would probably not occur in Darwin were it not for the Rib-fruited Mangrove, as it is the nectar of this species that provides the energy required for its breeding activities. The dependence is mutual, as flowers of this plant are primarily pollinated by small nectar-feeding birds, which in Darwin comprises the Red-headed Myzomela, and to a lesser extent, the Brown Honeyeater.[421] Male Red-headeds set up tiny feeding territories around two or more Rib-fruited Mangrove trees in May and June as these plants begin to flower, and defend them vigorously against all birds other than their mates, including the slightly larger Brown Honeyeater.[312] With assured access to this energy source, the females are able to build their nests and incubate their eggs, and pairs may raise two broods between April and October, which corresponds with the long flowering season of this mangrove.[412] The symbiotic relationship between the Honeyeater and Mangrove is even exploited by a tiny nectar-feeding mite that hitch-hikes on the bird to travel from one flower to the next.[421] As many as 50 mites can be found in the feathers around the bill, and even in the nostrils, of a single bird![422]

Although the Red-headed Myzomela is the most abundant mangal-specialised bird species overall,[312,313] its density varies according to the availability of its nectar sources. Dense patches of flowering Rib-fruited Mangroves can support more than 50 birds per hectare during the Dry season. At the end of the Dry season, adult and recently fledged birds exploit the nectar of several smaller-flowered mangrove species, but even these sources finish flowering before the end of the Wet.[312] At this time the only mangrove that may be flowering is the Star Mangrove, which is restricted to the seaward fringe, and is exploited by many honeyeater species, all larger than the Red-headed.[421] This forces many Red-headeds into surrounding habitats, especially paperbark swamps where males again set up temporary feeding territories around blossoming branches, up to six males per paperbark tree. Other birds appear in suburban gardens, visiting flowers of exotic plants such as Bauhinia, Pawpaw, Umbrella Tree and Coconut.[60] Many birds also travel up tidal rivers to subcoastal sites, such as Howard Springs NR, 11 km from the mouth of the Howard River, and Marrakai Crossing, 100 km upstream from the Adelaide River mouth.[1]

Despite this ability to disperse, banding studies have shown that adults are sedentary, at least within large tracts of mangal. Of 489 birds colour-banded between 1985 and 1994 at a site adjacent to Palmerston SP, 30% were recaptured or re-sighted over a month after banding, and seven of these were at least five years old when last seen. At the mouth of Rapid Creek, however, one banded bird lived for at least six years and three months.[47] The nest is tiny, about 5 cm wide and 4 cm deep, and so thin that its contents can often be seen from below. Nests may be placed as low as 1.2 m, but more typically 1.5 to 3 m, above the ground in mangroves.[47] Despite their diminutive size, these birds boldly defend their young from intruders, and feign a broken wing as a distraction.

Banded Honeyeater
Cissomela pectoralis

RANGE: endemic to monsoon-tropical Australia. HABITATS: eucalypt and paperbark forests, woodland. STATUS: irregular visitor, mostly Dry season. BREEDING: Mar–Jun (TE).

Laurie Ross

A so-called 'blossom nomad', the Banded Honeyeater is an irregular Dry season visitor to our region, materialising every two years on average, though it was widespread through Darwin during both 2009, when it was particularly numerous, and 2010. In some years it arrives in March or April, in others not until August, but it usually disappears in mid-November before the Wet season rains start in earnest. In the Humpty Doo–Marrakai region it was recorded each year from 1968 through 1971, but only one to seven times in three of those years and 34 times in the other (1970).[1] In the former case, small numbers visited the region late in the Dry only, while in 1970, flocks of up to 50 birds were observed in several areas from March to

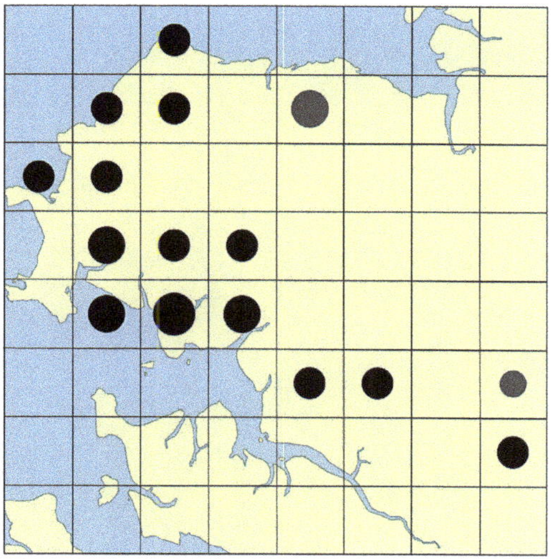

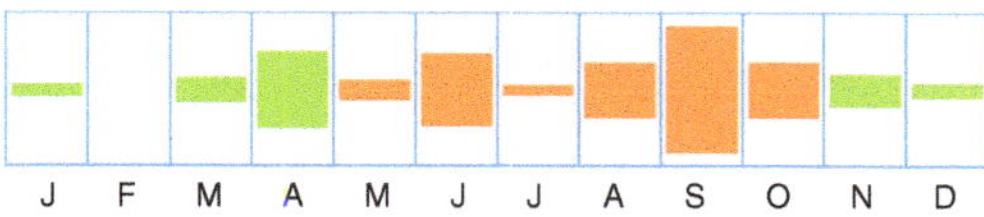

October. The latter irruption followed a poor Wet season inland, with well below-average rainfall falling in the Katherine region, whereas in the other three years this region received average or above-average rainfall during the preceding Wet season.[1]

However, the movements of Banded Honeyeaters may be more predictable than is often presumed. At Yinberrie Hills, 50 km north-west of Katherine, numbers of this species fluctuated spectacularly. Between November 1986 and February 1987, coincident with the Wet season, it was by far the most abundant species, with a density of up to 18 birds per hectare.[311] During this period, up to eight tree species blossomed, chief among them the Sandstone Penda, Darwin Box and Broad-leaved Carbeen, the nectar of which was shared by large numbers of that other 'blossom nomad', the Varied Lorikeet. By May, however, when these and most other nectar-bearing plants had finished flowering, both birds had all but vanished. It seems likely that the population from this area, and perhaps other parts of the southern Top End, migrate annually towards the wetter coast, after the Wet season, but the distance they travel varies from year to year, depending on the intensity of flowering of nectar sources, such as Darwin Woollybutt and Darwin Stringybark, which also varies enormously annually.[393] For example, at Coomalie Farm, 45 km south of the Darwin region, it was abundant over 3–6 months during both 2002 and 2005, but in only one month during 2003,[47] whereas in Darwin it was recorded over three months in 2005, one month in 2002 and not at all in 2003.

Few nests of this species have been found since the early 1900s, and certainly none near the NT coast until 2007, when one was found with nestlings on the Marrakai Road, about 40 km from our region, in late May.[423] Recently fledged birds were found at Bird Billabong, Mary River NP, in July 2011, suggesting eggs were laid in June. At Yinberrie Hills, breeding was reported in March and April 1987,[311] long after the major nectar resources had dried up. Recent observations at a site near Katherine revealed that this species nests in small colonies of at least six pairs,[366] males staking out territories or attracting females with a song flight, behaviour reminiscent of the White-winged Triller, another black-and-white colonial breeder. Clutches here were laid in late March and April, consistent with the above records and suggesting that the main breeding season in the Top End, contrary to the literature,[41] is the early Dry season.

Brown Honeyeater
Lichmera indistincta

RANGE: whole of Australia except south-east, NG and e. Indonesia. HABITATS: all wooded habitats. STATUS: very common resident. BREEDING: Apr–Aug.

Niven McCrie

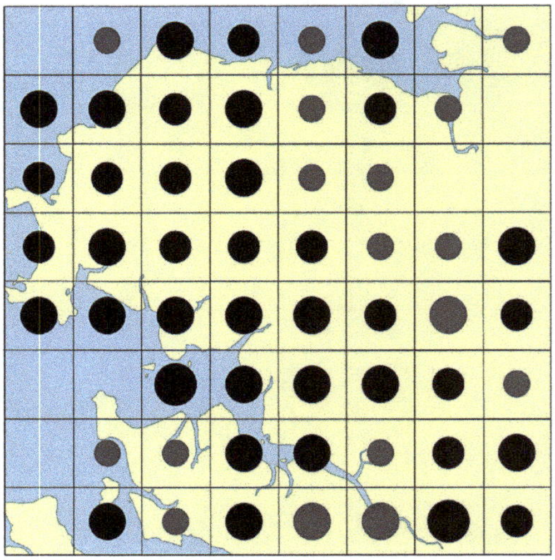

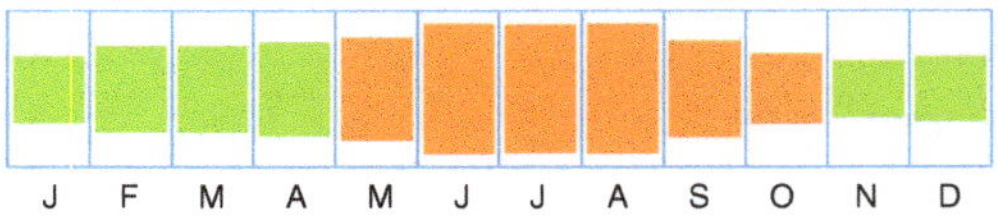

Like dragonflies, Brown Honeyeaters are harbingers of the Dry season in the Top End, their strident songs announcing both their breeding season and the arrival of cooler weather. Incessant singing during the Dry season is the likely reason for the peak in reporting rates at this time as the species is resident but less conspicuous in the Wet. But the singing is performed only by the male, distinguished by his grey head and black gape, presumably in order to attract the female, recognisable by her olive crown, yellowish throat, yellow gape and smaller size. Towards the end of the Dry season, when breeding comes to a close, the gape of the male also turns yellow.[414]

Of all honeyeater species in the Top End, the Brown Honeyeater is the least fussy in its habitat choice.[424] Not only was it the most frequently recorded species during the Howard's Peninsula survey, but it was present in 16 of the 20 recognised habitats. Although most abundant in tall paperbark forest, it was also common in eucalypt forest and woodland, mangals and rainforest.[308] In a 14-month bird survey of 13 patches of mangals around Darwin, from Buffalo Creek to East Arm, the Brown Honeyeater was the most frequently encountered species, and second only to the Red-headed Myzomela in abundance. Well adapted to the urban environment, it was the second most frequently recorded species, and the fourth most abundant, during a bird survey of Palmerston in 2007.[307] It appears to be less successful in Darwin's suburbs, probably due to competition with the Rufous-banded Honeyeater, which is much less common in Palmerston.

In woodland at the Territory Wildlife Park, Berry Springs, the Brown Honeyeater was the second most abundant honeyeater species, but with a distinct preference for dense thickets of small non-eucalypt trees. A banding study here suggested that Brown Honeyeaters were largely sedentary within a relatively small home range.[391] Of 137 individuals banded between 1994 and 1997, 28 (20%) were recaptured up to three times, mostly less than 200 m from their original banding site, although three birds moved over a kilometre within two years.[420] One bird that had been banded in 1989 was recaptured six years later only 800 m away from its original banding site. Similarly a male banded in the mangal adjacent to Palmerston SP was recaptured at its banding site exactly seven years after being banded.[47]

A Dry season breeder, the Brown Honeyeater lays eggs from April to August, with a peak in May and June.[412] This breeding activity, and the singing that accompanies it, probably explains the increase in reporting rates during the mid-Dry season. Nests are invariably built in shrubs, typically only 0.5 to 2 m above the ground, and in suburbia pot plants are sometimes utilised as nest plants.[47] Incubating females seem fearless, sitting tight on their eggs even when humans brush past the nest, only 30 cm away. A study of Brown Honeyeaters nesting in low mangroves at Nightcliff and Rapid Creek revealed that a high proportion of nests were flooded by high spring tides.[425] As the full lunar cycle is slightly less than the time it takes a pair to build a nest and raise young to fledging, all nests built below 7.4 m above sea level are destined to be flooded, yet these birds seem to repeat this mistake year after year, suggesting a limited capacity to learn.

Little Friarbird
Philemon citreogularis

RANGE: tropical and e. Australia, NG. HABITATS: mainly eucalypt forest and woodland. STATUS: common resident, probably partially migratory. BREEDING: Aug–May, peaking in Oct.

The status of the Little Friarbird in Darwin may have changed in recent decades. It was considered a spasmodic Dry season visitor to Darwin gardens during the 1970s and 1980s,[60] yet since at least the 1990s it has been known to nest in urban parks and along streets during the Wet season as well as the late Dry. Monthly reporting rates of the species from the 1970s to the present differ little from those of the Helmeted Friarbird, indicating that it is also resident, albeit probably making small-scale seasonal movements. It has been recorded in 80% of cells and breeds in many of them. A small number were present around Darwin Golf Course in all 12 months surveyed weekly during 2006–07.[309] It was also among the 12 most frequent and abundant bird species in Palmerston, and equally common in suburbs of all ages, during the Dry season of 2007, though its status in the Wet season was unknown.[307]

Just 13 km south of our region, however, the situation is radically different. In woodland at the Territory Wildlife Park the Little

Niven McCrie

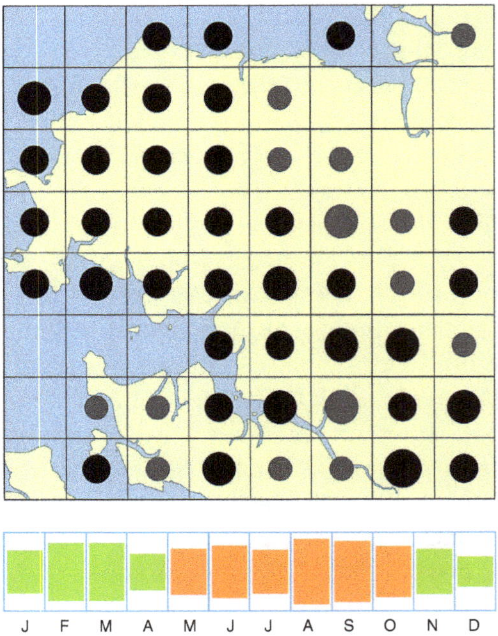

Friarbird was by far the most abundant of the 11 species of nectar-feeders,[391] with a mean density almost 10 times that of the Silver-crowned Friarbird. However, while it was present year-round its numbers fluctuated dramatically, with a sixfold increase between March and June. This increase was correlated with the amount of nectar available, principally in the flowers of the Darwin Woollybutt and Fern-leaved Grevillea during June, with the addition of the Yellow-barked Paperbark in July. This increase was very likely due to influxes of birds emigrating from areas outside the 100 ha of the park; of 63 birds banded here over three years only one was recaptured, and this was only one month after its initial capture.[420]

Large-scale movements are evident in eastern Australia, where one banded individual is known to have travelled 637 km.[414] Populations in south-eastern Australia appear to migrate north or north-west via inland routes[157] in March and April, after breeding during the late spring and summer. There is no evidence to suggest that these migrants move as far as the Top End, however, and it is interesting that at Yinberrie Hills, about 50 km north-west of Katherine, the species was present and breeding only during the Wet season,[311] not the Dry as would be expected if it emigrated from southern Australia. It is possible that such birds, breeding in semi-arid parts of the NT, are the source of immigrants to subcoastal areas like the Territory Wildlife Park.

In Darwin breeding has been reported in all months except June and July, but just over a half of all clutches were laid in September and October.[47] Like Helmeted Friarbirds, this species often nests in large 'nursery trees' that host nests of several species, usually Figbirds, and often Torresian Imperial Pigeons and Magpie-larks. Almost as soon as the chicks fledge, pairs may re-nest, one parent building the new nest, often within metres of the first, while the other feeds the fledglings.[119]

Helmeted Friarbird
Philemon buceroides

RANGE: largely coastal, tropical Australia and e. Indonesia. HABITATS: mangal, monsoon rainforests and suburbia. STATUS: common resident. BREEDING: Aug–May, but peaking Sep and Jan.

Niven McCrie

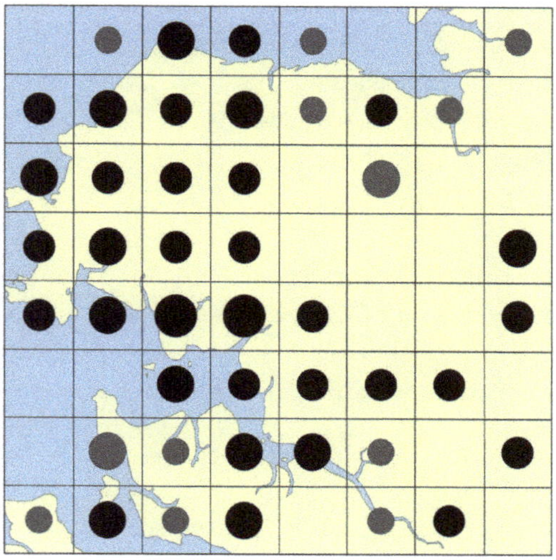

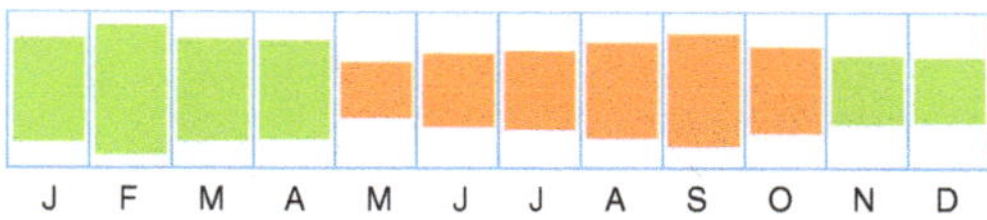

Within the Top End the Helmeted Friarbird has an unusually restricted distribution, and is represented by two distinct races. Confined to the coastline from Wadeye in the west to Cape Arnhem in the east, the race found in Darwin (*gordoni*) weighs 89 g on average,[47] being second only in size to the Blue-faced Honeyeater. In contrast, the other race (*ammitophila*), sometimes called the Sandstone Friarbird, is some 20 g heavier,[414] rivalling the Blue-faced Honeyeater, but it is confined to the sandstone plateaux and escarpments of western Arnhem Land, some 300 km east of Darwin. The Helmeted is the common large friarbird of inner Darwin city and the northern suburbs, often nesting in large trees on the perimeter of ovals or even in busy malls and car parks. Its adoption of the urban environment here doubtless relates to the close proximity of its preferred habitats, comprising mangal, monsoon rainforests and vine thickets, to many suburbs of Darwin. These habitat preferences also explain why the species rarely visits Palmerston,[307] which is far from the nearest rainforest or vine thicket. As well as consuming nectar and large insects, like mantids and cicadas, the Helmeted Friarbird, unlike the other friarbird species, eats figs and other rainforest fruits when they are available, and regularly raids pawpaws and bananas from backyards.[47] Flowers of the exotic Rain Tree and African Tulip are also plundered for nectar.

Banding studies at Nightcliff suggest that while some individuals of this species are sedentary, others make smallscale seasonal movements between habitats. During the Wet season many Helmeted Friarbirds move to subcoastal paperbark swamps, such as Marrara Swamp, to exploit the nectar of paperbarks alongside the other two species of friarbirds.[47] Soon afterwards, in the early Dry season, the nectar in the large, shaving-brush-shaped flowers of the Star Mangrove attracts hundreds of Helmeted Friarbirds to the seaward fringe of mangals, such as at Nightcliff.[421] One bird colour-banded at Nightcliff was sighted over two years later at the CSIRO Tropical Ecosystems Research Centre, 8.5 km away.

Nesting has been recorded locally in almost every month of the year, but there are two clear peaks that account for almost a half of all records, the first in September and the second in January.[47,412] The two peaks in reporting rates correspond remarkably well to these breeding peaks, though this may be fortuitous, since the other friarbirds show a similar pattern. African Mahoganies are favourite trees for nesting in surburbia, while in the mangals nests are often placed in Stilt Mangroves or on the Mangrove Mistletoes within them.[47]

In the past the Helmeted Friarbird was often confused with the Silver-crowned Friarbird, leading to some erroneous statements in the literature. However, it is readily distinguished from the latter species by the inconspicuousness of the casque on its bill and the smaller extent of bare skin on the face, which is leaden-grey in colour, as well as its larger size and more lyrical voice.

Silver-crowned Friarbird
Philemon argenticeps

RANGE: endemic to monsoon-tropical Australia. HABITATS: mainly eucalypt forest and woodland. STATUS: common resident. BREEDING: Sep–Mar (TE).

The Silver-crowned Friarbird replaces the Helmeted Friarbird as the common large honeyeater of Palmerston and rural parts of the Darwin region. The outer suburbs of Darwin, such as Malak, Karama and Woodleigh Gardens, are also frequently visited due to their close proximity to eucalypt forests and woodland, the principal habitat of this species.[365] In Palmerston a five-month survey showed that this species was most frequent and abundant in suburbs built in or since the 1990s, such as Bakewell, Gunn and Rosebery, but scarce in the oldest suburbs, like Driver and Moulden.[307] Similarly, in Darwin new suburbs appear to attract this species until they develop sufficient dense foliage to attract the Helmeted Friarbird.[60]

Unlike the Little Friarbird, which is also mainly found in woodland, the Silver-crowned Friarbird appears to be largely resident. The numbers of these birds showed little seasonal fluctuation in eucalypt woodland at the Territory Wildlife Park, 13 km south of the Darwin region.[391] During a three-year banding study there, three out of 37 banded individuals were recaptured twice, each time within 300 m of the site where they were

Niven McCrie

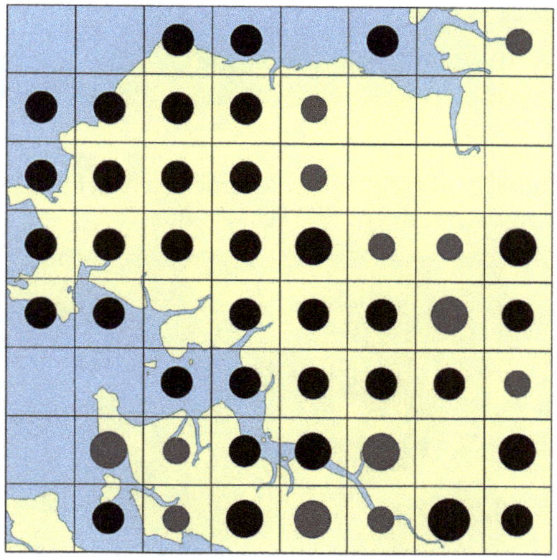

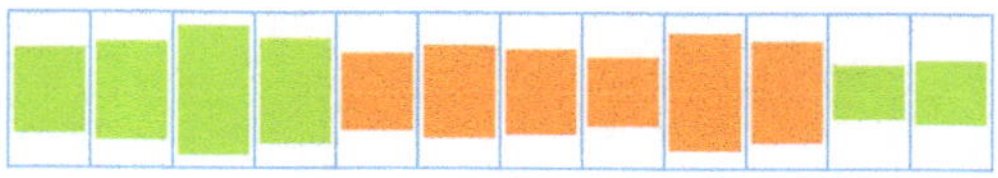

originally captured, and up to three years afterwards.[420] Observations of foraging at that site also suggested that this species spent about two-thirds of its time feeding on nectar and one-third on invertebrates.[426] Fruit is rarely eaten so it visits rainforests infrequently, unlike the Helmeted Friarbird. It also shuns mangals, except when the Star Mangrove is in bloom, at which time small numbers may be seen feeding on nectar alongside the other two friarbird species. All three species also converge on paperbark forests, such as at Marrara Swamp, during the Wet season when the Broad-leaved Paperbark flowers.[47]

Few nests have been observed in our region, though juveniles are regularly reported in Palmerston.[319] Elsewhere in the Top End, breeding records indicate that egg-laying starts in September and continues through to March, with peaks in those two months.[47] These peaks in breeding activity closely match those in reporting rates, though it is unclear if the two are causally related.

The Silver-crowned Friarbird is easily distinguished from other local species of friarbirds by the obvious casque or knob on the top of its bill and the large expanse of jet-black bare skin on the head, which extends well behind the eye and contrasts strongly with the narrow strip of silvery-white feathers on its crown. Its shrieking calls are also higher-pitched and less melodic than those of the Helmeted.

Blue-faced Honeyeater
Entomyzon cyanotis

RANGE: tropical and e. Australia, s. NG. HABITATS: mainly eucalypt forest and woodland. STATUS: common resident. BREEDING: Apr–Dec.

Niven McCrie

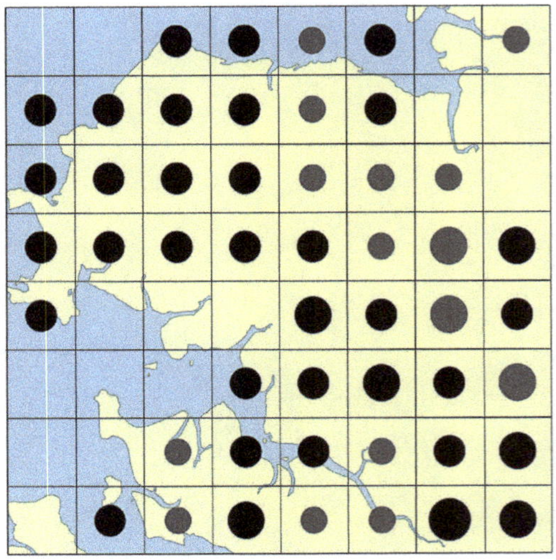

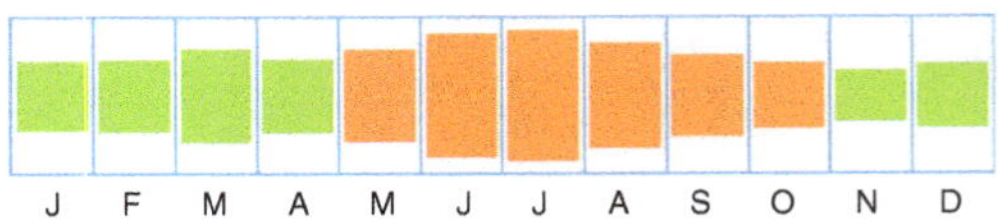

Weighing about 110 g on average,[391] the Blue-faced Honeyeater is the largest of all local honeyeaters. Given its wide distribution and conspicuous behaviour, the species is surprisingly little known in terms of its social organisation and breeding biology, but we do know that it has a dark side. Gregarious and pugnacious, it often usurps the nests of other species, especially those of the Grey-crowned Babbler, causing the death of the usurped brood. At Coomalie Farm, 45 km south of the Darwin region, a long-term study showed that 10% of Babbler nests were appropriated by the Blue-faced Honeyeater, and the latter was never found building its own nest.[427] Happily for the Babblers in Darwin, where they are much scarcer than at Coomalie Farm, there are no reports of their persecution by the pillaging honeyeater!

The local race of the Blue-faced Honeyeater differs from those of eastern Australia by the presence of a large white patch in the wing, conspicuous when in flight.[414] It has adapted well to the urban environment, but is perhaps commoner in Palmerston than Darwin. During a five-month bird survey in Palmerston it was the sixth most frequently recorded species, as well as being the fifth most abundant in the old suburbs of Moulden and Driver. In contrast, it was rarely recorded around Millner from 1988 to 1993.[296] During the Howard's Peninsula survey it was encountered in nine of the 20 habitats, but highest densities were in the tall paperbark forest, and eucalypt forest and woodland.[308]

Of 26 Blue-faced Honeyeaters banded over three years in woodland at the Territory Wildlife Park, 13 km south of our region, three birds were recaptured at last once, in all cases close to their original banding site, but in the case of one bird, after it had been recaptured 1.3 km away.[420] These findings indicate that the species is probably sedentary, but occupies a large home range. Its diet is broad, roughly two-thirds nectar, mainly from eucalypts, and one-third invertebrates. Its penchant for the nectar in banana flowers, noticeable in our region,[365] has earned it the nickname of 'banana-bird' in Queensland.[414] In some places it is also known as the 'pandanus-bird', alluding to its apparent affinity for these palm-like trees. Indeed, a study in Kakadu NP showed that these birds repeatedly fly from one pandanus to the next, searching the leaf-bases for invertebrates, and bypassing eucalypts when in mixed stands of these trees.[3880]

In the Top End, the long bimodal breeding season begins in April and ends in December, with peaks in May and September.[47,412] In suburbia, the species often builds its nest within the fronds of exotic palms, such as the Bismarck Palm. In Palmerston's CBD one pair used an old Magpie-Lark nest which was built on a vent inside a veranda on the second floor of a building.[319] Several nests in Darwin have been reported to have four or five attendants, providing evidence of cooperative breeding in this species.[414]

White-throated Honeyeater
Melithreptus albogularis

RANGE: tropical and subtropical Australia, NG. HABITATS: mainly eucalypt forest and woodland, and paperbark forest. STATUS: very common resident. BREEDING: Apr–Sep.

One of the commonest birds of the vast savannas across the Top End,[104] the White-throated Honeyeater is nevertheless patchily distributed within Darwin's suburbs, probably due to its strong association with eucalypts. Prior to tropical cyclone Tracy in 1974 the species was reportedly common in and around Darwin.[60] Shortly thereafter the local population is said to have declined due to the damage caused to eucalypts by the cyclone, and the burning of regrowth afterwards.[365] The clearing of this habitat during the late 1970s for new suburbs, such as Anula, Wanguri and Wulagi, exacerbated this local decline. However, since then the species has become a regular visitor to suburban parks, especially where close to remnant patches of eucalypt forest, such as those around the airport.

In Palmerston, a town surrounded by eucalypt woodland, this species was among the 10 most abundant bird species recorded in the oldest suburbs, such as Woodroffe and Driver, during the 2007 Dry season, though it was rarely recorded in suburbs built since 2000.[307] At the Territory Wildlife Park, the White-throated Honeyeater was resident, but of 67 individuals banded over 39 months only three were recaptured, all within 200 m of the original banding site and in one case after two years.[420] This indicates that at least some birds are sedentary, though territory size probably varies with the number of nectar sources available, since the species takes nectar as well as invertebrates throughout the year. These birds are bold and pugnacious, repeatedly harassing intruders (for example, humans)

Nolan Caldwell

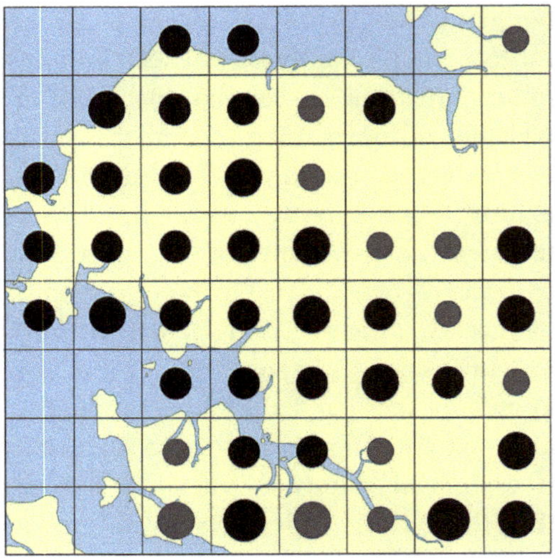

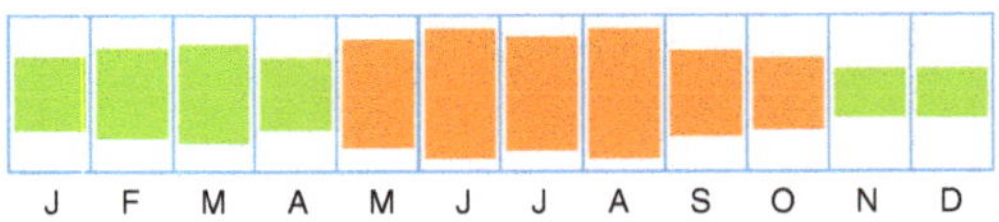

near their young, and sometimes attacking their own images in the side mirrors of cars parked in their territory.[47]

The White-throated Honeyeater is a Dry season breeder in the Top End, clutches being laid from April through to September, but peaking in July.[47,412] Increased calling during this period, especially by young birds begging for food, may explain the Dry season peak in reporting rates and their lull during the early Wet. Despite its wide distribution, the species is poorly known with respect to its social organisation and breeding biology. Observations of three adults provisioning the young in a nest in Darwin, and elsewhere, indicate that it breeds cooperatively, at least occasionally.

Yellow Chat
Epthianura crocea

RANGE: scattered localities in tropical and arid Australia. HABITATS: seasonally inundated floodplains and mangroves fringing tidal channels. STATUS: vagrant. BREEDING: Feb–May (TE).

The endangered Alligator Rivers subspecies of the Yellow Chat (*tunneyi*) is endemic to the Top End of the NT, and is largely restricted to a small area of coastal and subcoastal floodplains from the Mary River to East Alligator River. During a survey of the species in Kakadu NP in November 2004, 96 birds were recorded, from which the extrapolated population for the park was put at fewer than 300 individuals.[428] Numbers outside of the park have not been assessed, but are unlikely to be significant. Records of this species west of the Adelaide River have been sporadic.

Mike Reed

The largest recorded influx of Yellow Chats into the Darwin region occurred during 1980. Ten birds were found near Leanyer SP during May, and by October no fewer than 60 birds were present.[34] The species was also reported from Holmes Jungle in June 1985, but there is no record of the numbers involved.[34] Subsequently there were no records until the start of September 2005, when a single immature bird was found at Knuckey Lagoons. The bird was present until at least mid-October, by which time the lagoon had become almost dry. Less than a week before the bird being found at Knuckey Lagoons, a single immature bird was seen at Harrison Dam and, several days later, two 'uncoloured' birds and an adult male. The last sighting was of a single immature, at the end of October. Three birds were also reported from Harrison Dam in October the following year.[429]

Breeding apparently occurs during the late Wet season, from February or March to April or May.[428] Nesting was recorded on the South Alligator River floodplain, about 160 km east of the Darwin region, in early 1986, but there is otherwise very little information on breeding for the local subspecies.

Rufous-banded Honeyeater
Conopophila albogularis

RANGE: Coastal and subcoastal Top End, Cape York and NG. HABITATS: mainly paperbark forest, vine thicket, mangal and suburbia. STATUS: very common resident. BREEDING: year-round, except Jun–Jul.

It would be hard to find a better avian mascot for Darwin than the Rufous-banded Honeyeater, as not only does the Top End represent the bulk of its global range, but also the species has adopted this city in a way that is reminiscent of introduced species, such as sparrows and mynas, or the pugnacious Noisy Miner, in other Australian cities. It is one of the commonest resident bird species in many, if not most, suburbs of Darwin, where it occupies permanent territories as small as 1500 m^2, and nests in trees and shrubs in yards, sometimes within a metre of buildings or busy roads.[430] During the five years of the Millner survey (1988–93) it was the most frequently recorded species, detected in all 58 months sampled, and on average the fourth most abundant.[296] It can also be found in the older suburbs of Palmerston, such as Driver, where there are well-established gardens, but is scarce in the often bare environment of newer suburbs.[307] Similarly, around Darwin Golf Course, it was merely

Niven McCrie

the tenth most frequently recorded non-waterbird species over 52 weekly surveys during 2006–07.[309]

Outside the metropolitan areas of Darwin and Palmerston, however, the Rufous-banded Honeyeater is much less common and probably makes small-scale seasonal movements.[420] Like the Bar-breasted Honeyeater, it favours moist riparian habitats, especially seasonally inundated paperbark forest,[103,392] and its purse-shaped nests are often suspended above water.[47] As its short bill indicates, however, it is largely insectivorous, and although it takes nectar when available,[392] it is much less dependent on a continuous supply of nectar than most other local honeyeater species. Thus it also occupies coastal and subcoastal vine thickets and rainforests,[355] as well as mangals, habitats in which nectar is generally scarce.

While abundant in mangals around Darwin's foreshore,[47,313] such as at Nightcliff, it is seldom seen in the extensive mangal blocks surrounding Darwin Harbour.[312] Indeed, in the Grey Mangrove-dominated mangal surrounding Leanyer SP, it was easily the commonest species, accounting for 43% of all birds captured and banded, over a six-year study,[43] whereas in mangal adjacent to Palmerston SP only two birds were caught over 10 years.[312] As the former is now surrounded by suburbs but was originally vine thickets, while the latter abuts dry savannas, it seems likely that their presence in coastal mangals relates to the close proximity of complex alternative habitats. Another attribute that may have contributed to the success of the species in Darwin is its size, which is slightly larger than several potentially competing honeyeater species, enabling it to expel many intruders from its territories.

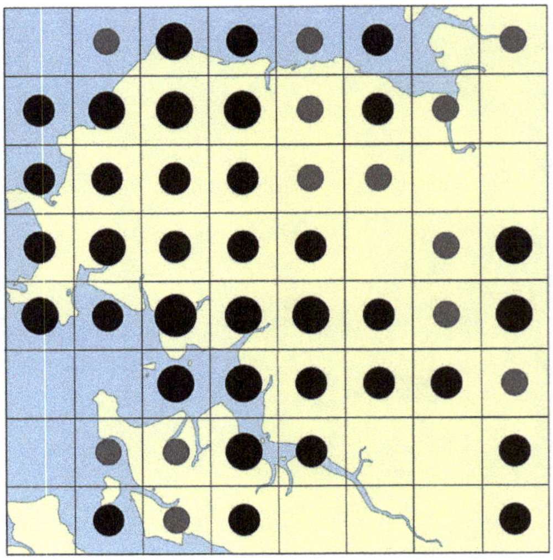

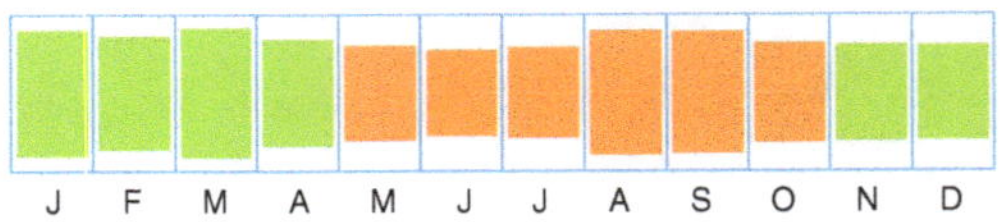

Urban Rufous-banded Honeyeaters enjoy the longest breeding season of any bird in the region, possibly facilitated by year-round watering of gardens, lawns and ovals, which in turn may increase the availability of insect prey through the year. Clutches are laid in every month, though very rarely in June or July. This short break in reproduction is used for that other essential component of the avian annual cycle, moulting. By August, when Brown Honeyeaters ease off with their strident singing as their relatively short breeding season comes to an end, Rufous-banded Honeyeaters power up with their more tuneful songs.[430]

Like Brown Honeyeaters, this species happily nests in pot plants, even when they are near major thoroughfares, but their favourite nesting tree in Darwin is the local Black Wattle, commonly planted around parks and ovals. The lack of many natural predators in suburbia probably contributes to its exceptional reproductive success. A study at Charles Darwin University Casuarina campus showed that pairs often nest up to four times per year so, despite usually laying only two eggs per nest, six young are raised, on average, annually.[430] One pair raised a remarkable 32 young over five seasons! As some individuals live up to nine years or more,[430] these birds must be among the most productive songbirds in the Australian tropics. The resultant surplus of youngsters probably wanders widely in search of vacant territories, explaining the high capture, but low recapture, rate in the Leanyer study.

Rufous-throated Honeyeater
Conopophila rufogularis

RANGE: endemic to monsoon-tropical Australia. HABITATS: open woodland. STATUS: rare visitor, largely in Dry season; moderately common to south. BREEDING: Dec–Apr (TE).

Mat and Cathy Gilfedder

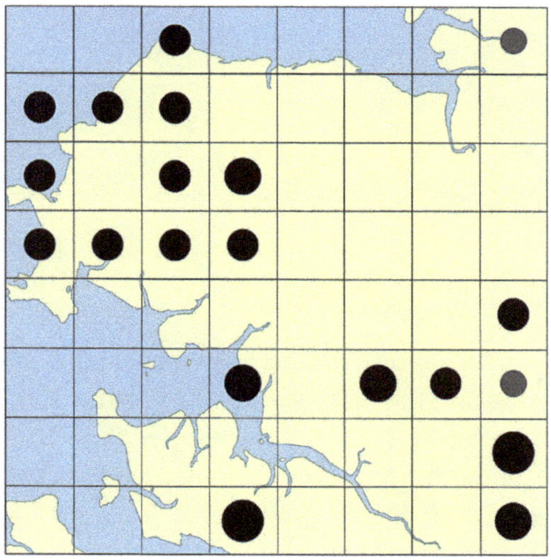

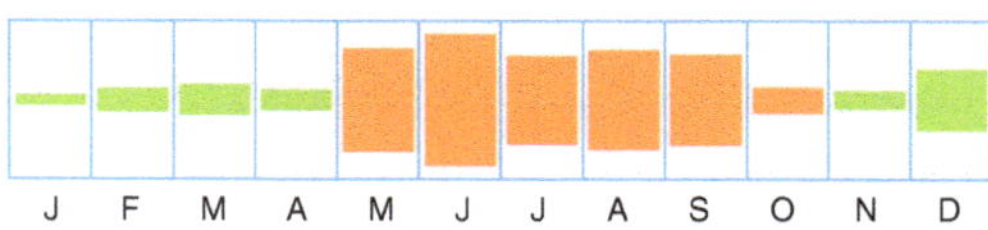

The feeble contact calls of the Rufous-throated Honeyeater belie its close genetic relationship with the garrulous Rufous-banded Honeyeater, but its nest, juvenile plumage and alarm call betray their affinity. Indeed, juveniles of these two species are almost identical in appearance, except for a greyish smudge on the breast of the Rufous-banded. While the Rufous-banded mainly occupies moist habitats of coastal and subcoastal environments, the Rufous-throated is a bird of inland woodland, though like the former species it is fond of vegetation around watercourses.[414]

Similar to the Banded Honeyeater in being largely confined to the monsoon tropics and showing a high degree of mobility, the Rufous-throated visits Darwin mainly during the Dry season,[1] arriving in most years during May and departing by October, presumably from breeding areas inland. Although the two species are often found together inland, their visits to the Darwin region do not appear to be synchronised. While 1998 and 2009 were 'good' years for the Banded Honeyeater in Darwin, the Rufous-throated was apparently absent in those years, and most numerous or widespread during 1999 and 2010.[21,47]

Rufous-throated Honeyeaters were recorded on only five out of over a thousand bird censuses involving 100 sites during the Howard's Peninsula survey of April–May 1986.[308] Further inland, however, the species appears to be resident in low densities. At Coomalie Farm, 45 km south of the Darwin region, it has been observed most months, mainly in the low paperbark woodland surrounding seasonally inundated floodplains, but sometimes in tall riverine trees.[47] It also appears to be present year-round in similar habitats in western Arnhem Land.[47]

Breeding records from across the Top End, though relatively few in number, indicate that the species breeds earlier than the Banded Honeyeater, from December to April.[47] This difference probably reflects disparities in their diets, as the short-billed Rufous-throated is less nectar-dependent than the longer-billed Banded,[417] so its breeding season may relate to the abundance of insects more than to the availability of nectar.

Bar-breasted Honeyeater
Ramsayornis fasciatus

RANGE: endemic to tropical Australia. HABITATS: all wooded habitats, especially paperbark forests. STATUS: moderately common resident. BREEDING: Oct–May.

Like the closely related Brown-backed Honeyeater of the Wet Tropics, the Bar-breasted Honeyeater has an obvious penchant for Paperbark trees. Apart from feeding on nectar from the flowers of paperbarks more than any other local honeyeater,[47] it constructs its large, pendulous nest almost entirely from the bark of paperbarks, and often attaches it to the branches of paperbarks.[414] Instead of being the usual open cup shape, however, the nests of both of these tropical honeyeaters are dome-shaped with a hooded side entrance, a design that seems to have more in common with gerygones than other honeyeaters.

Given its predilection for paperbarks, it is not surprising that, among Top End honeyeaters, the Bar-breasted shows the greatest affinity for riparian forests.[103,392,417] It is common in Broad-leaved Paperbark forests, such as at Marrara Swamp, and groves of Weeping and Cajaput Paperbarks fringing mangals, at least during the Wet season and early Dry[308] when these species bloom.[392] Outside this period the species appears to move to eucalypt forests and woodland, and even rainforests.[103] Although recorded in 50% of cells, 43% of all records came from the Casuarina CR–Royal Darwin Hospital area, and it was reported from urban

Niven McCrie

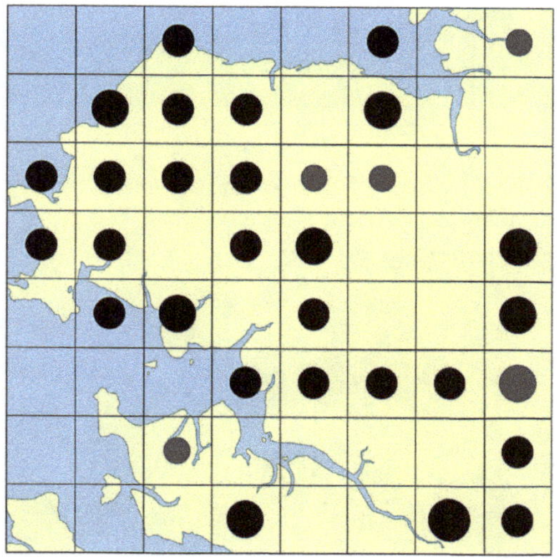

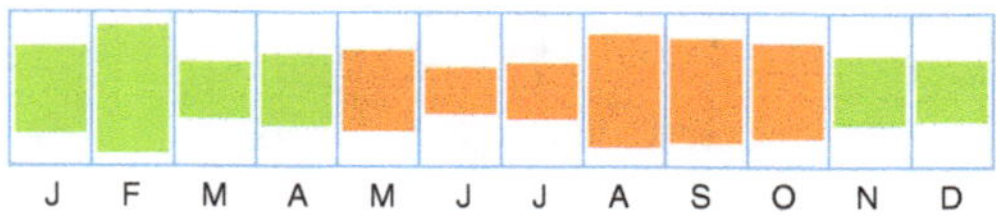

parks and gardens significantly more than rural and natural areas. At the Casuarina campus of Charles Darwin University, one pair was found nesting along a small drainage channel in 1987, but by 2009 at least six pairs were nesting in the tall paperbarks surrounding buildings, well away from standing water.[47] In Palmerston, by contrast, it is scarce, probably due to the lack of paperbark swamps in close proximity to the town.[307]

Nests have been found in all but two months of the year in our region, but 80% of clutches were started during the Wet season, with a peak in January and February, the two wettest months of the year.[47] This breeding season roughly coincides with the flowering period of most paperbark species,[392] suggesting that an abundant supply of nectar from this source is an important factor in the timing of reproduction. However, as their pendulous nests are often suspended low over the water on billabongs and creeks, nesting at a time when water levels are highest might be adaptive in escaping the attentions of ground-dwelling nest robbers that don't like swimming. The down side is that such nests risk being flooded due to sudden rises in water levels.

The resemblance of the Bar-breasted's nests to flood debris has led to the proposition that, despite their large size, their architecture may be cryptic in function, discouraging potential nest robbers from inspecting their contents. Indeed, experiments using artificial nests showed that domed nests were robbed less often than open cup nests in paperbark swamps, but not in other habitats.[431] Moreover, new nests are sometimes built within a metre of old nests, suggesting that some sites enjoy more protection from nest robbers than others.[432] The Bar-breasted Honeyeater is also a frequent host of the brood-parasitic Brush Cuckoo.[326]

Yellow-throated Miner

Manorina flavigula

RANGE: whole of Australian mainland, mostly inland. HABITATS: woodland and eucalypt forest, rural areas. STATUS: scarce resident. BREEDING: May–Sep (TE).

Niven McCrie

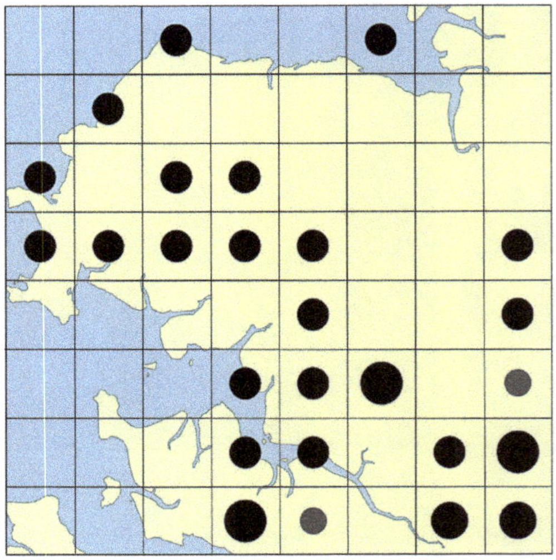

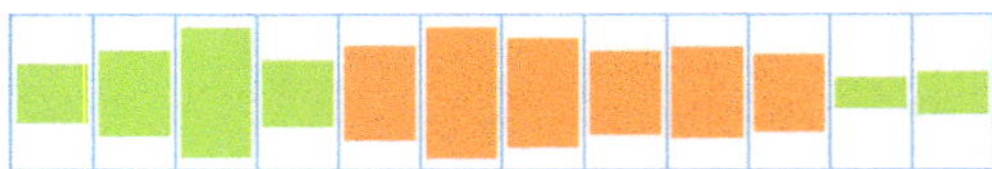

In stark contrast to the ubiquitous Noisy Miner in eastern Australian capital cities, the closely related Yellow-throated Miner is virtually absent from the metropolitan area of Darwin yet, curiously, occurs in the city of Palmerston. In a five-month bird survey of 10 Palmerston suburbs in 2007, this species was found to be abundant in the young suburbs of Rosebery and Bakewell, built after 2000, but less common in older suburbs and in the woodland on the edges of the city.[307] Absent from much of Kakadu NP[103] and generally scarce across the Top End,[58] it is patchily common in rural areas around Humpty Doo,[60] though mainly in woodland fragments rather than orchards and cleared areas.[433] It was found in 41% of cells, with significantly more records in rural cells than in urban and natural cells.

At the Territory Wildlife Park, 23 km south of Palmerston, Miners were regularly encountered on only four out of 13 sites surveyed in 1994.[391] Here the species appeared to be sedentary in moderately large territories in eucalypt woodland, feeding on nectar from blossoming eucalypts when available, as well as on bark-dwelling invertebrates. At Coomalie Farm, 45 km south of the Darwin region, it was resident in small numbers only in open forest dominated by Darwin Woollybutt, and absent from floodplains, rocky woodland and riparian forest.[47]

Like many other local honeyeaters the Yellow-throated Miner appears to breed mainly during the Dry season in the Top End.[412] Reporting rates in the Darwin region were lowest during the early Wet, possibly due to inconspicuous behaviour after breeding. In 1985 and 1986, one tame pair was noted nesting at the Noonamah pub,[60] where the species had been noted nesting in 1944, over 40 years earlier.[434]

White-gaped Honeyeater
Stomiopera unicolor

RANGE: endemic to tropical Australia. HABITATS: all wooded habitats. STATUS: very common resident. BREEDING: year-round, but mostly Wet season with peak in Jan.

Niven McCrie

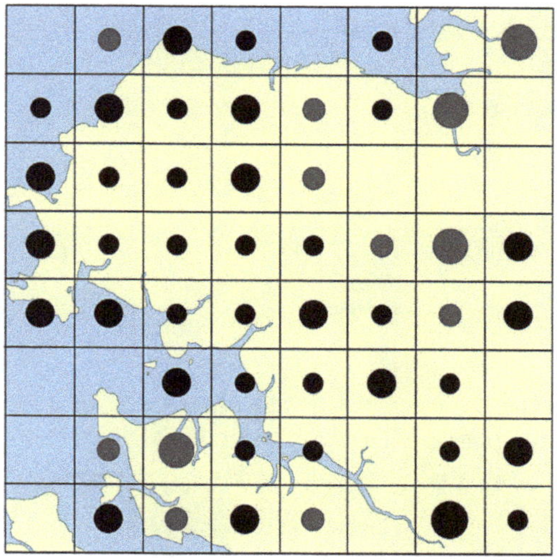

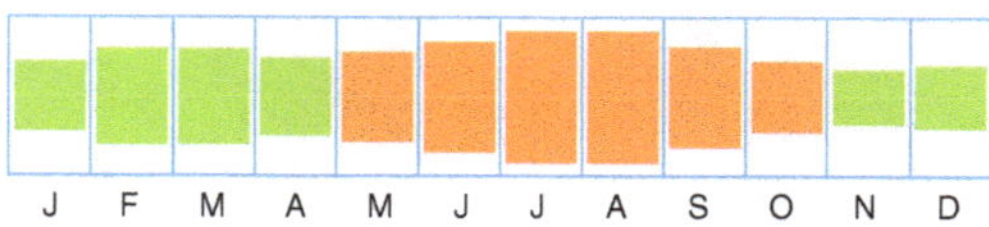

The eighth most frequently recorded bird in Darwin, the White-gaped Honeyeater is the only common medium-sized honeyeater in the Top End, which lacks the many honeyeater species weighing 22–50 g that are found in southern Australia. The characteristic rollicking song of this species is often given by two birds simultaneously. The male is thought to initiate the song, giving up to seven rapidly repeated phrases, each phrase consisting of five to seven chuckling notes, of which the last two or three sharply rise in pitch.[414] After a short pause, the female joins in either with the same song or with simple monotonal notes.

Common throughout the metropolitan area of Darwin, including the city, this species apparently prefers suburbs where gardens are well established.[60] In a five-month bird survey of 10 suburbs of different ages in Palmerston in 2007, this species was found to be abundant in the older suburbs of Driver and Moulden, but was rarely recorded in newer suburbs or in the woodland on the edges of the city.[307] In nature, this species favours the dense cover of rainforests of all types,[355] vine thickets and riparian woodland, as well as black wattle groves.[308] It also frequently visits coastal mangals abutting rainforests or older suburbs, such as at Nightcliff, where it is the most abundant honeyeater at times,[47,313] but is largely absent from mangals surrounding Darwin Harbour, which abut eucalypt woodland.[312]

Like the Helmeted Friarbird, the White-gaped Honeyeater is partial to figs and other fleshy fruits when they are available,[60] a fact that explains its affinity for rainforests. But its staple diet is insects and nectar, which it gathers from a wide variety of sources.[392] Its ability to adapt to suburban gardens is exemplified by the way it exploits the exotic Yellow Oleander, the nectar of which is enclosed in a chamber at the base of the bell-shaped flowers. The Honeyeater snips the flower off the branch, carries it to a horizontal perch, and while holding it down with one foot, makes an incision into the chamber to extract the nectar with its tongue, a process that leaves characteristic scars on the disposed flowers.[47]

Surprisingly few nests have been found in Darwin, possibly as they are often built high in large trees, such as African Mahoganies and mangos. One pair nested in a pot plant on the veranda of a third storey Nightcliff apartment.[47] However, observations of fledglings and food-begging juveniles indicate that while these birds can breed throughout the year,[412] the majority of clutches are started in the Wet season, almost a third of them in January alone.

Singing Honeyeater
Gavicalis virescens

RANGE: whole of Australian mainland, excluding e. and n. coastal areas. HABITATS: woodland and shrublands. STATUS: vagrant. BREEDING: Feb–Aug (NT).

Laurie Ross

Interestingly, all three honeyeater species recorded as vagrants to the Darwin region were discovered on the coast, within metres of the shoreline. The single record of Singing Honeyeater involves a bird seen on several occasions in mangroves on the shoreline at Nightcliff during September 2011.[21,40,320] It was often seen perching in taller mangroves at the edge of the shore, or feeding in mangrove saplings exposed at the low tides. The species is regularly found closer to Darwin than the other vagrant honeyeaters, but is generally rare in the Top End, especially north of Pine Creek.[58] During much of 1970 one to four birds were present in the Marrakai area,[1] about 40 km south-east of our region, and since at least the year 2000 one or two birds have been reported spasmodically here, mainly on the edges of floodplains.[47] To the west of Darwin several observers have found the species on the Dundee Beach Road.[301] It was not recorded during the Kakadu survey,[103] and was infrequently encountered at Yinberrie Hills, 50 km north-west of Katherine, during monthly surveys in 1986–87.[311]

Yellow-tinted Honeyeater
Ptilotula flavescens

RANGE: monsoon-tropical Australia and se. NG. HABITATS: eucalypt woodland and riparian forest. STATUS: vagrant. BREEDING: mostly Mar–Jul (TE).

Common on the Tiwi Islands to the north of Darwin,[357] around Maningrida in Arnhem Land to the east,[47] and Yinberrie Hills to the south,[311] the Yellow-tinted Honeyeater is curiously absent from the north-western NT mainland, including the Darwin–northern Kakadu region. The sole record in the Darwin region is from East Point, where an individual was observed in stunted mangroves in October 1970.[1] The bird was reported as showing characteristics of the Tiwi Islands subspecies (*melvillensis*), which has a yellower plumage than the mainland race, with distinct streaks on the breast. As island origin for the bird seems likely given that the nearest point of the Tiwi Islands to where it was found is only about 55 km, whereas on the mainland the species rarely occurs north of Pine Creek, some 200 km south of Darwin. At Yinberrie Hills,[311] densities in the Wet season were almost three times higher than in the Dry, suggesting some seasonal movement. Here and elsewhere in the Top End the species appears to breed principally during the late Wet and early Dry seasons.[47]

Grey-headed Honeyeater
Ptilotula keartlandi

RANGE: inland Australia, reaching coast in WA. HABITATS: eucalypt woodland, mulga. STATUS: vagrant. BREEDING: Jan–Aug (NT).

A common species of low open woodland in the arid and semi-arid parts of Australia,[414] the Grey-headed Honeyeater is not a bird one would normally expect to find anywhere near Darwin. However, in March 2009[435] one adult of this species was sighted at East Point, and very likely the same individual was seen there again in October 2009,[436] February 2010[437] and August 2010.[21] Darwin is well outside the normal range of the species, which is mainly south of Elliott,[414] about 600 km south-east of the Darwin region. The appearance of this species is all the more curious considering that the closely related Grey-fronted Honeyeater occurs regularly around Timber Creek and Mataranka,[300] only about 350 km from our region, but has not yet been recorded in the Darwin–Humpty Doo region or Kakadu NP.

Striated Pardalote
Pardalotus striatus

RANGE: whole of Australia. HABITATS: eucalypt forest and woodland, riparian forest. STATUS: very common resident and Dry season visitor. BREEDING: May–Sep.

Ashley Carlson

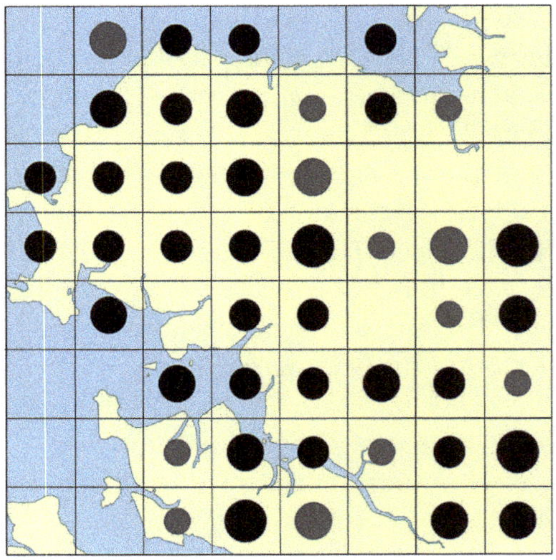

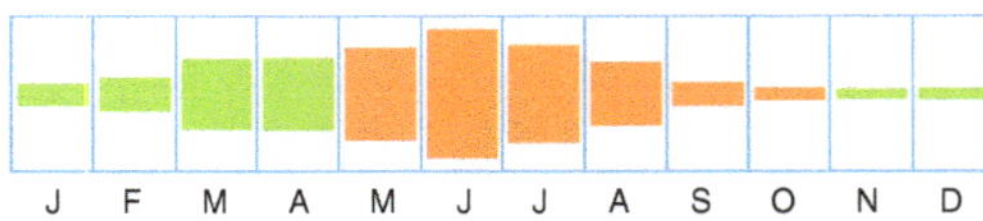

Having spent the first three weeks of their lives underground, Striated Pardalotes spend most of their adulthood in the canopies of eucalypt trees, until it is their time to breed. Then they will dig a hole in the earth, up to 45 cm deep and, after lining the nest chamber with grass, the female will lay her eggs.[438] In Darwin and Palmerston, this species is often observed around construction sites and road cuttings, where exposed or heaped soil provides suitable nest sites.[60,365]

Southern races of the Striated Pardalote feed mostly on lerp, the sugary exudate of certain bugs that are abundant on temperate eucalypts. But as lerp is very scarce in Top End savannas,[394] local Pardalote populations are less confined to eucalypt forests than their southern counterparts. Indeed, during the Howard's Peninsula survey they were found in 12 of the 20 habitats sampled, but were most abundant in riparian forest dominated by Swamp Box and Ghost Gum, tall paperbark swamp, and cyclone-damaged Woollybutt-Stringybark woodland.[308]

While some southern populations of this species are well known for their long-distance north–south migration, those in the tropics are thought to be resident or to make local movements only.[438] Indeed, local reporting rates show a steady increase during the late Wet (austral autumn), with a peak in June, followed by a faster decline until October. Its apparent scarcity from September through February may be due, in part at least, to inconspicuous behaviour at that time. The species gives its very audible disyllabic 'song' most frequently during the breeding season, which in the Top End is the early Dry season, while outside this period it mainly utters a soft trill.

Nevertheless there is some evidence of seasonal movements in the Top End at the local or even regional scale. In the bloodwood and Salmon Gum-dominated woodland of Yinberrie Hills, about 50 km north-west of Katherine, they were mainly Wet season visitors, largely absent from February to October,[311] when they are conspicuous in Darwin. It is thus plausible that towards the end of the Dry season, after breeding, a large proportion of the Darwin population moves from the low-lying coastal plains to higher country inland. At Coomalie Farm, 45 km south of the Darwin region, Striated Pardalotes are abundant throughout the year in open forest dominated by Woollybutt.[47] However, they appear to occupy the paperbark-dominated floodplains mainly while breeding during the Dry season, whereas the surrounding bloodwood-dominated stony hills are occupied only during the Wet. Presumably the black soil of the former habitat provides a suitable substrate for making nest burrows, while the lack of soil in the latter prevents it.

Breeding during the early Dry season is adaptive since nest burrows made during the Wet risk being flooded by rains, while later in the Dry season digging may become impossibly difficult in the hard-baked soil.[412] Within the suburbs they often commandeer heaps of builders' sand or topsoil for their nest burrows.[60] Outside the metropolitan area, nest tunnels are often dug in wheel-ruts along bush tracks, or drainage ditches, as well as creek banks.[47] On 11 July 2004 no fewer than 16 nest burrows, as little as 20 m apart, were found on the side of a dirt track near Humpty Doo.[3]

Weebill
Smicrornis brevirostris

RANGE: whole of Australian mainland. HABITATS: eucalypt forest and woodland. STATUS: very common resident. BREEDING: Mar–Sep.

Marc Gardner

Weighing a mere 5 g, the northern race of the Weebill is the smallest of all Australian birds, being smaller, as well as yellower, than its southern counterparts.[438] Though common within Darwin's metropolitan area where there are stands of eucalypts, such as those surrounding the airport, and on the urban fringes, it rarely enters parks or suburban backyards in either Darwin[60] or Palmerston.[307] During the Howard's Peninsula survey, it was among the six most frequently recorded species, with highest densities in open forests and woodland dominated by Woollybutt and Darwin

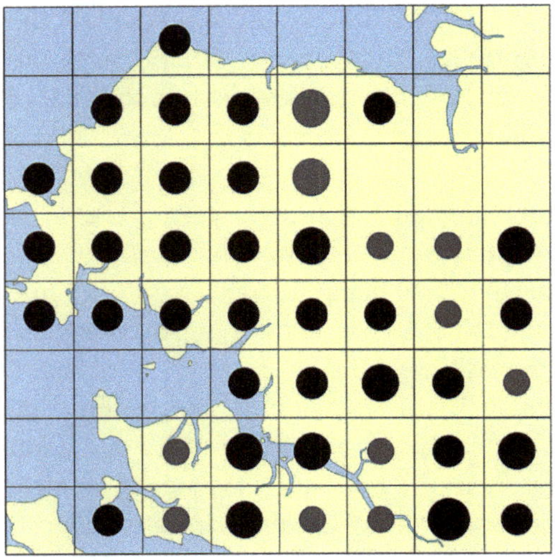

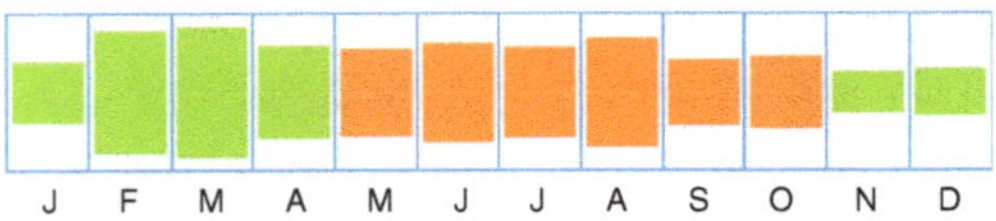

Stringybark, particularly those damaged by Cyclone Tracy or having a shrubby understorey or low canopy.[308] Not far outside the region, however, the species appears equally common in woodland dominated by bloodwoods and other eucalypt species.[47] Moreover, in Kakadu NP the species prefers to forage on the Kakadu Plum, when it is in leaf during the Wet season.[103] It was recorded in 73% of cells, the highest number of records coming from the Rapid Creek area.

Reporting rates peak during the latter half of the Wet season, followed by a steady decline over the Dry. Variation in singing rates is more likely the cause of these seasonal fluctuations than seasonal movements, as the species appears to be sedentary. Surprisingly, for a species that occurs so widely across the continent, little is known about the Weebill's social organisation and nesting behaviour. Often encountered in small groups of three to five birds, Weebills are suspected of breeding cooperatively at least occasionally, young from previous broods helping to provision their younger siblings. Top End birds breed largely during the Dry season,[412] with a peak in egg-laying in July.[47] Local nests are built in both trees and shrubs, at heights ranging from 1 to 12 m from the ground. This species is apparently sensitive to fire, being much more frequent in unburnt and biennially burnt areas than in areas burnt annually.[382]

Mangrove Gerygone
Gerygone levigaster

RANGE: coastal tropical and e. subtropical Australia, s. NG. HABITATS: mangal. STATUS: very common resident. BREEDING: Mar–Oct, but biannual with few nests May–Jul.

Jim Sneddon

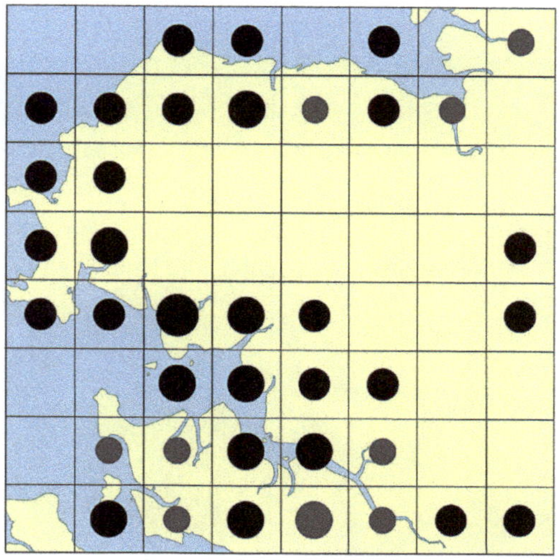

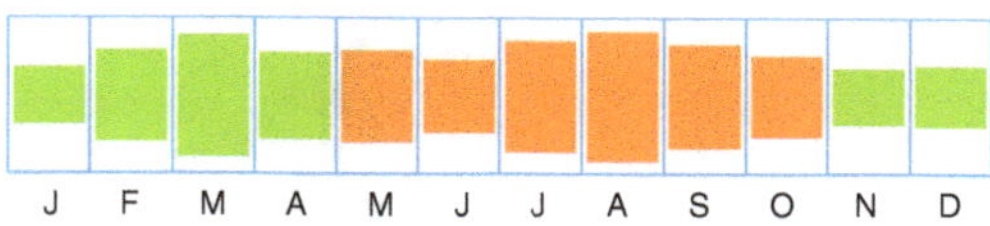

A more informative name for this bird might be the Grey-Mangrove Gerygone as its life revolves around the Grey Mangrove.[312,313] In the Darwin region, detailed studies have shown that it forages on Grey Mangroves for more than 80% of the time,[312] and nearly 90% of its nests are built in this plant.[439] Its penchant for this plant species explains why it has the widest distribution of all mangal-specialised birds in Australia, since the Grey Mangrove is the most widespread mangrove on the globe, let alone the Australian continent.

In contrast to the Large-billed Gerygone, which favours the tall, dense mangroves along tidal creeks, the Mangrove Gerygone prefers the open habitat formed by stunted Grey Mangroves, which are most abundant around bare saltflats near the landward edge of mangals, and along the muddy or rocky foreshore.[439] The species was recorded from 59% of cells in the region, of which, not surprisingly, most (81%) were coastal.

A study of these birds in the Ludmilla Creek catchment showed that the average distance between nests of adjacent pairs was about 70 m, though within a single 'saltpan', nests were often as close as 30 m.[334] Like the Large-billed Gerygone, the Mangrove Gerygone breeds biannually in the Darwin region, the first egg-laying peak occurring from late March to early May.[439,440]

The Mangrove Gerygone suffers almost as much as its larger cousin from brood parasitism by the Little Bronze Cuckoo. At the Ludmilla Creek site, over three years, an average of 34% of nests failed due to parasitism by this Cuckoo.[334] Like the Large-billed Gerygone, some pairs of this species have learned to pull the cuckoo hatchling out of their nests before it ejects their own eggs or nestlings.[441] However, like that species, individual parents sometimes make mistakes and pull out their own young, despite the hatchling Mangrove Gerygone looking quite dissimilar to the imposter. The local race of this species differs from the eastern Australian race in being somewhat smaller, having less black on the tail and no hint of grey on the clean white underparts.[438]

Large-billed Gerygone
Gerygone magnirostris

RANGE: tropical Australia and NG. HABITATS: mangal and monsoon rainforests. STATUS: very common resident. BREEDING: all year, but biannual with peaks in Mar–Apr and Sep–Nov.

Mat and Cathy Gilfedder

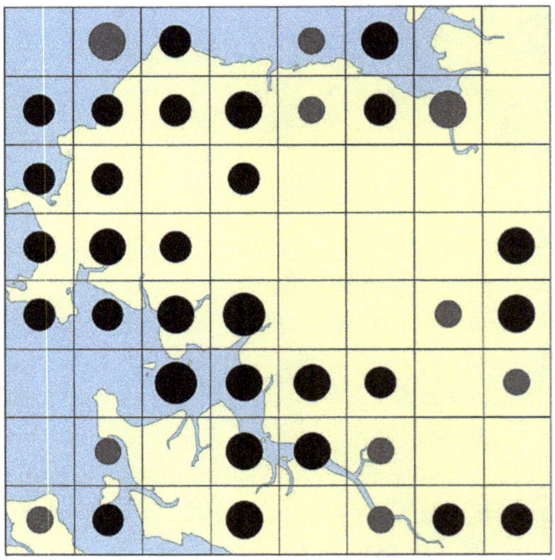

Of all Australian songbirds the Large-billed Gerygone builds the largest nest, relative to its body length. From top to bottom, an average nest measures about 50 cm, almost five times longer than its owner.[334] Suspended over a tidal creek or on the edge of a freshwater stream, this untidy, bulky structure looks more like flood debris than a nest, earning the species the name of 'flood bird'.[438] Much less well known is its habit of building new nests close to one or two old ones, sometimes within a metre. The eggs and young in such nests escape the attention of Black Butcherbirds and other predators more than isolated nests, but whether the old nests act as decoys remains to be demonstrated.[432]

Around Darwin Harbour, the Large-billed Gerygone is one of the four most abundant bird species in mangals,[312] but it also occurs in adjacent vine thickets and wet, spring-fed rainforests, such as Holmes Jungle, and occasionally even in tall Black Wattle or riparian forests.[308] In mangals, it shows a strong preference for Stilt and Rib-fruited

Mangroves often with an over-storey of tall Grey Mangroves that provide plenty of shade along tidal creeks. An intensive study of this species on the upper reaches of Ludmilla Creek showed that the average distance between nests of adjacent pairs was 65 m, suggesting territories of less than 0.5 ha.[334]

This species is one of several in the Top End that have a biannual breeding season.[412] Although nesting has been observed in every month of the year, there are clear peaks in egg-laying during the late Wet (March and April) and late Dry seasons (September to November), with very little breeding activity in the intervening periods.[334] This possibly explains the dip in the May reporting rate, but not the one in September. This unusual annual cycle appears to relate to their diet of small insects, which are most abundant during these transition seasons. Many eggs and young succumb to predators and drowning by high tides,[432] but the worst enemy of the species is the Little Bronze Cuckoo, which lays its eggs only in gerygone nests. When it hatches, the cuckoo chick ejects the gerygone eggs or hatchlings from the nest, so that it is the sole recipient of food brought by its foster parents. The cuckoo youngster requires feeding for at least four weeks after it fledges, thus preventing the gerygones from re-nesting as quickly as they could following predation or flooding of the previous nest.

But the Geryones of Darwin are fighting back, using behaviour that has not been observed in cuckoo hosts anywhere else in the world. At the Ludmilla Creek site, a staggering 41% of nests were parasitised by Little Bronze Cuckoos, on average, over three years - one of the highest rates of brood parasitism in the world.[334] Perhaps because of repeated parasitism, some pairs have learnt to recognise the 'ugly duckling' and pull it out of the nest.[339] But the cuckoo hatchling closely mimics the Large-billed hatchling in appearance, so these 'smart' parents might be using non-visual cues to distinguish the imposter. Sadly for their own young they don't always get it right, but natural selection will favour those that do.

Green-backed Gerygone
Gerygone chloronota

RANGE: monsoon tropical Australia, and NG. HABITATS: monsoon rainforests, vine thickets, riparian forest, mangal edge. STATUS: very common resident. BREEDING: year-round, but mostly Jan–Jul.

Though widespread in the rainforests of New Guinea, within Australia, the distinctive Green-backed Gerygone is confined to the Top End and Kimberley region of Western Australia. This distinctive species differs from the other gerygones of the Darwin region in its tuneless reeling song and in several aspects of its breeding biology, including its breeding season. One of the commonest species in Top End rainforests of all types,[355] the Green-backed Gerygone also occurs in forests dominated by Black Wattles and paperbarks. During the Howard's Peninsula survey, Green-backed

Adrian Boyle

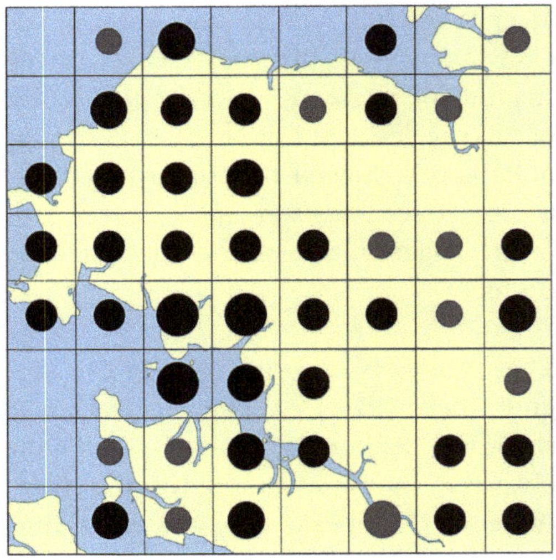

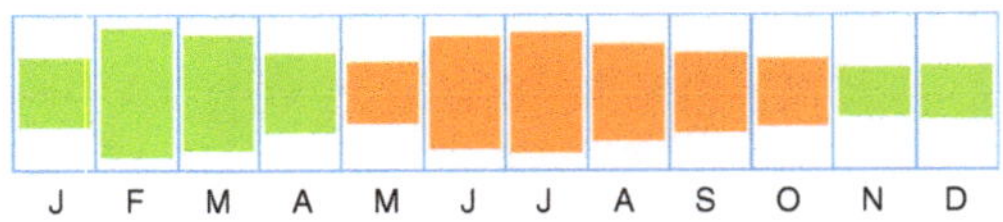

Gerygones were found to occupy eight out of 20 habitats, but densities were highest by far in dry rainforest and vine thickets.[308] They also inhabit the landward edge of mangals, where Black Mangroves form a dense canopy with little light penetrating to the ground. Here they rarely interact with the more mangal-specialised Large-billed and Mangrove Gerygones, which prefer tidal channels and open saltflats, respectively.[312] Reported from 72% of cells, this species was most often recorded around Casuarina CR and Buffalo Creek.

The reporting rates of this species do not appear to reflect its breeding activity. Though breeding has been recorded in every month of the year, egg-laying peaks in January then falls off gradually over successive months until September and October when breeding activity is at its lowest.[47] As with the two mangal-specialised gerygones, the Green-backed is parasitised by the Little Bronze Cuckoo[47,326] but to a much lesser extent, possibly because it is the smallest of local gerygone species and thus the least capable of raising the large cuckoo chick to independence. Nests of this species are characteristically decorated with caterpillar frass, and lined with plant down rather than feathers, as is the case with the two mangal-specialised gerygones.[47]

White-throated Gerygone
Gerygone olivacea

RANGE: tropical and e. Australia, and NG; HABITAT: mostly eucalypt forest and woodland. STATUS: scarce resident. BREEDING: Nov–Jan (TE).

David Webb

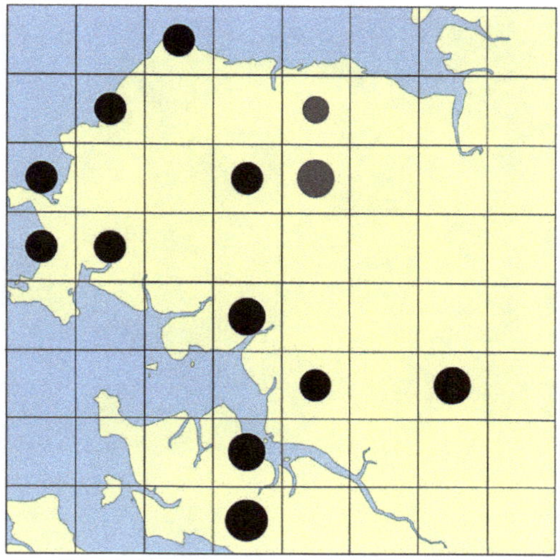

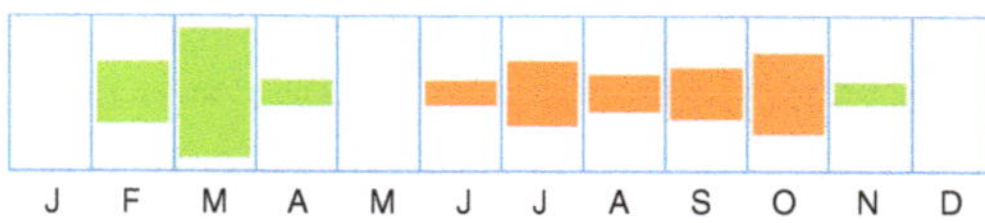

In tropical parts of Australia, the White-throated Gerygone is apparently resident, unlike its populations in south-eastern Australia, which migrate towards the tropics each year after breeding.[157] Compared to its south-eastern counterpart, the Top End and Kimberley race is quite small, and its tail has more extensive blackish-brown and less white.[438] In common with southern birds, however, local birds favour eucalypt forest and woodland,[1,103] which they usually share with the much more abundant Weebill, but no other gerygone species.

There are too few records of this species in our region to infer anything about its seasonal status, but there are one or more records for all months except May, December and January.[47] It was not recorded during the Howard's Peninsula survey of April–May 1986,[308] but was observed at Mickett Creek on three out of 60 visits between 2004 and 2009.[442] Within 50 km of the Darwin region, however, it is common along the Marrakai Road and at Coomalie Farm.[47]

Until recently there had been no breeding records for the species from the NT. However, at Coomalie Farm, observations of nest building and feeding of recently fledged young, including a juvenile Little Bronze Cuckoo, suggest that eggs are laid from November to January.[47] Further inland, however, they may breed later. At Yinberrie Hills, about 50 km north-west of Katherine, breeding was reported during April and May,[311] well outside its breeding season in southern Australia.[438]

Grey-crowned Babbler
Pomatostomus temporalis

RANGE: tropical and e. Australia. HABITATS: eucalypt forest and woodland. STATUS: moderately common resident, mainly outside urban areas. BREEDING: Jun–Nov (TE).

Niven McCrie

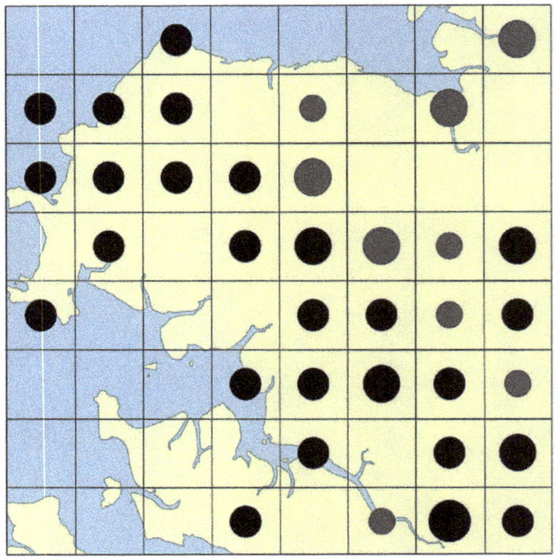

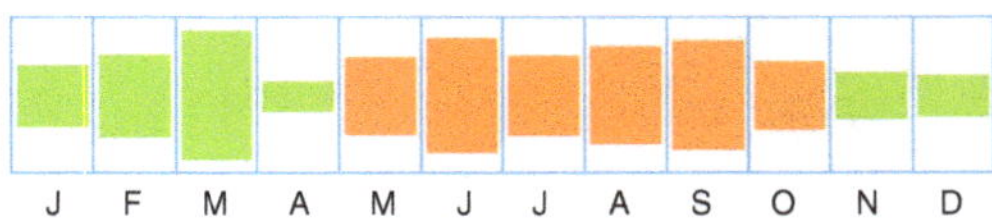

The bulky stick nests constructed by Grey-crowned Babblers are unmistakable, and sometimes the first clue to their presence in an area. Yet these structures are used for roosting as well as breeding, and although some 'roost-nests' are built for roosting alone, most are old 'brood nests' that have been refurbished for that purpose.[438] Babblers live communally, all members of the group sleeping together; as many as 14 birds have been reported sharing a single roost-nest.

During the late 1970s and early 1980s local populations of Babblers may have suffered some decline due to the destruction of forest caused by cyclone Tracy and subsequent urban development, but they persisted around Marrara Swamp, Holmes Jungle and Howard Springs.[60] Since then, groups have become established around ovals or parks in several suburbs, such as Wanguri, Moil and Rapid Creek. During the Palmerston survey in 2007, the species was recorded on only 6% of censuses, mostly in the suburbs of Gunn and Bakewell, and rarely in the oldest or youngest suburbs. Outside the metropolitan areas, it is a common resident of eucalypt woodland and forest, and rural areas, of the Darwin–Humpty Doo region.[1,60] There are significantly more subcoastal records than coastal records, yet surprisingly it was recorded only three times in over 1000 censuses during the Howard's Peninsula survey in 1986.[308]

Like many other species of Australian songbirds, babblers breed cooperatively, with all members of a group participating in the building of nests and the provisioning of young. At Coomalie Farm, 45 km south of the Darwin region, members of 20 groups were colour-banded to investigate their contributions towards raising young over six years.[443] Groups comprised a breeding pair and up to seven non-breeding helpers, but the average group consisted of just four birds. As groups increased in size the feeding contributions of helpers increased and those of the parents decreased, but this had little effect on breeding success. However, unaided pairs failed to raise young successfully, eventually splitting up and emigrating to nearby territories.[443] Older helpers contributed more than younger helpers, and females, often unrelated to the breeders, contributed more food than males.[444] Adults of both sexes moved frequently between groups, but male helpers became breeders more by staying at home, while females did so by emigration.

Top End Grey-crowned Babblers breed largely during the Dry season.[443] Egg-laying at Coomalie Farm started in July and ended in late October, extending into November only in two seasons that were preceded by low annual rainfall. Breeding success was low, mostly due to nest predation, and almost 10% of all active nests were usurped by the aggressive Blue-faced Honeyeater for their own breeding.[427] Although this species has declined dramatically in south-eastern Australia due to woodland clearing, Top End populations seem to be faring well despite widespread landscape changes due to increased burning; in fact, evidence from several studies suggests that not only are they fire-tolerant, but they may even favour burnt landscapes.[382,445] In a study of Top End woodland fragments, including sites around Humpty Doo and Berry Springs, the species was found to avoid cleared land and mango orchards, preferring woodland patches that had been burnt at 5–10-year intervals.[433]

Black Butcherbird
Cracticus quoyi

RANGE: coastal and subcoastal tropical Australia, NG and Aru Is. HABITATS: mainly mangal. STATUS: common resident. BREEDING: Sep–Jan.

There can be few bird songs that resonate like those of the Black Butcherbird in the mangals of the Top End during dawn, when they can reputedly be heard up to a kilometre away.[446] What sounds like a single voice, however, is often two birds singing alternately. The two birds typically perch 1–2 m apart in the same tree, one always higher than the other. The higher-perched bird invariably starts the sequence, with a low-pitched, rather muted and guttural phrase, to which the second bird replies with a higher-pitched, more melodious phrase of two or three syllables. The two main sequences may be rendered as 'croo-croo', answered with an upward-inflected 'oo-wa', while the other is a single 'ork', followed by 'oodle-oo', the second syllable lower and shorter than the first and third syllables. This antiphonal singing is mainly heard in the late Dry season, when birds are preparing to breed.

Black Butcherbirds are common in mangals of Darwin, both along the coast and around the harbour, favouring Stilt Mangroves and

Richard Noske

other tall mangroves along tidal creeks.[312,313] They are frequently observed perched on power lines beside Dick Ward Drive, where it crosses Ludmilla Creek. However, they also visit monsoon rainforests and vine thickets adjacent to mangals or on freshwater creeks, such as at Berry Springs NR.[47] Parks and gardens of suburbs close to mangals are also regularly visited, especially where there are large trees that can be used as calling posts, and sometimes, even for nesting.[47] Groups of four or five[96] are sometimes seen, presumably comprising breeding pairs with their brood of young birds, which evidently stay with their parents for several months. Immature birds have pale blue-grey bills,[47] in contrast to adults whose bills are predominantly black, with only the basal quarter to third pale blue-grey. Immatures of the local race *spaldingi* are always black, unlike those of the race *rufescens* of north-east Queensland in which there are both black and all-rufous forms. Local birds, both young and older, also sometimes show a thin pale grey 'spectacle', which continues as a short line behind the eye.[47]

By far the largest species of butcherbird, the Black Butcherbird appears to be at least as carnivorous as the much better-known

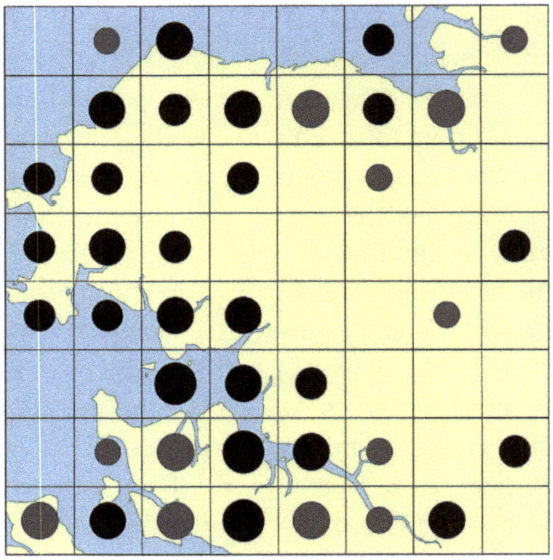

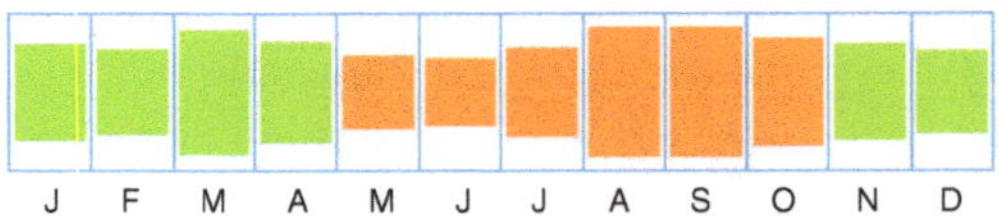

Silver-backed and Pied Butcherbirds. In Stuart Park Black Butcherbirds were seen killing a Bar-shouldered Dove sitting on a nest, eating the eggs, then dismembering the adult on the ground;[93] and a nestling of the same species was found impaled on a branch next to a building at Charles Darwin University Casuarina campus.[47] In Darwin mangals, Butcherbirds have been observed robbing the large pendant nests of the Large-billed Gerygone several times, in one case ripping open the nest, then consuming the two Gerygone eggs on the ground.[334] At Buffalo Creek one ate the two eggs in a nest of Shining Flycatchers, despite vigorous defence by their owners. At Nightcliff a nest of Broad-billed Flycatchers was being observed when a Black Butcherbird alighted on it, plucked out a small chick, then flew a few metres to a perch, where it bashed the chick three or four times before swallowing it. It then flew right back to the nest, plucked the other chick out and flew off with it across the reef, with one of the parents in hot pursuit.[447]

At East Point a Peaceful Dove was flushed off its nest by an adult Butcherbird before its egg was stolen and fed to one of the Butcherbird's dependent young.[119] Predation of nestling Rufous-banded Honeyeaters[411] and Double-barred Finches[21] has also been reported. Yet bird eggs and chicks are probably less frequent menu items than insects and marine creatures, such as crabs and mudskippers.[187] Indeed, one observer witnessed a 'tug-of-war' between a Black Butcherbird and a Frilled Dragon over a large stick insect![93]

In October 2009 at Charles Darwin University a Black Butcherbird was observed attempting to kill a female Pacific Koel, after the latter had been driven to the ground by three Little Friarbirds.[448] The Koel received a powerful blow to the head by the butcherbird, which then departed and, though the Koel was still alive when retrieved by the observers, it died soon afterwards. There is one potential record of the Butcherbird being parasitised by the brood-parasitic Pacific Koel.[330]

Ten local breeding records indicate that this species lays its eggs in the five months from September to January with equal frequency.[47] The high reporting rates in August–September may be due to increased singing at the start of the breeding season, while the secondary peak in March may be related to the presence of dependent young.

Silver-backed Butcherbird

Cracticus argenteus

RANGE: endemic to Top End and Kimberley. HABITATS: eucalypt forest and woodland. STATUS: sparse resident. BREEDING: Aug–Nov (TE).

Until recently considered an isolated monsoon-tropical race of the widespread Grey Butcherbird, the Silver-backed Butcherbird differs from the former in having a paler back, a white shoulder, and no white in front of the eyes.[411] The Grey Butcherbird occurs in the southern half of the NT, but is separated from the nearest Silver-backs by some 600 km.[96] Generally less common in the Top End than the Pied Butcherbird, this species is more habitat-specialised, preferring tall eucalypt forests[1] but also occurring in riparian woodland.[308,449] During the Howard's Peninsula survey it was found to be largely restricted to Stringybark- and Woollybutt-dominated forest that had not been damaged by tropical cyclone Tracy, and was more abundant at sites with a taller canopy than in forest of lower stature.[308]

Surprisingly, the species was rarely recorded in Kakadu NP during the 1980–83 Kakadu fauna survey,[103] but was among the 12 most frequently recorded birds in Woollybutt-dominated eucalypt forest at Coomalie Farm, only 45 km south of the Darwin region.[47] In our region it was significantly more frequent in rural and subcoastal cells than coastal or urban cells, and although it was rarely reported from metropolitan Darwin, 20% of records came from Palmerston, probably because this

Nolan Caldwell

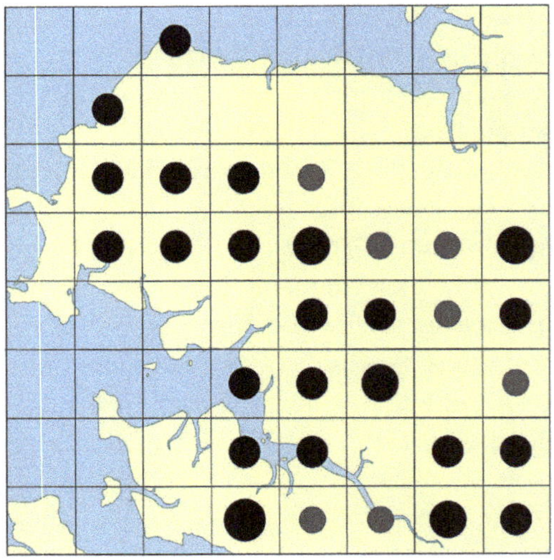

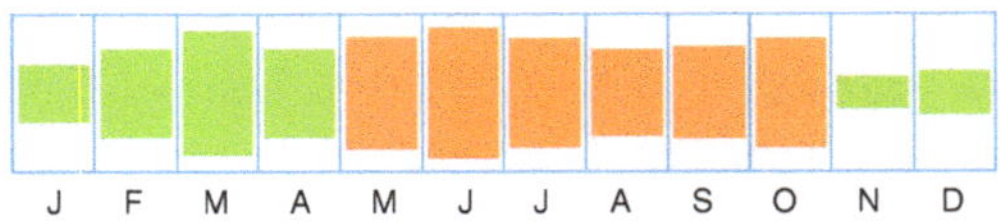

city is surrounded by eucalypt forest and woodland. During the 2007 Palmerston bird survey, it was recorded in five of 10 suburbs, most frequently in Rosebery.[307]

It is difficult to explain the fluctuations in reporting rates of the species, particularly the dip during the late Wet season. Nine breeding records from the Top End indicate that eggs are laid from late August to November. Like the Pied Butcherbird, this species is known to take advantage of recently burnt areas, hunting insects and lizards killed or narcotised by fire, especially during the first few days after the fire.[95] It also appears to favour frequently burnt habitat. At Solar Village, just outside the Darwin region, it was found to be two to three times more abundant at annually burnt sites than at long-unburnt sites.[399]

Pied Butcherbird
Cracticus nigrogularis

RANGE: whole of Australian mainland except much of SA and Vic. HABITATS: mainly eucalypt woodland and farmland. STATUS: common resident. BREEDING: Aug–Nov (TE).

Tolerating open habitats more than the Silver-backed Butcherbird, the Pied Butcherbird is generally commoner in the Top End, especially in Kakadu NP, where the former is apparently scarce.[103] More widespread in the Darwin region than the Silver-backed, being recorded in 63% (vs. 47%) of cells, it was similarly recorded more often in rural or subcoastal cells than urban or coastal ones in our region.[47] Nevertheless, the Pied

Adrian Boyle

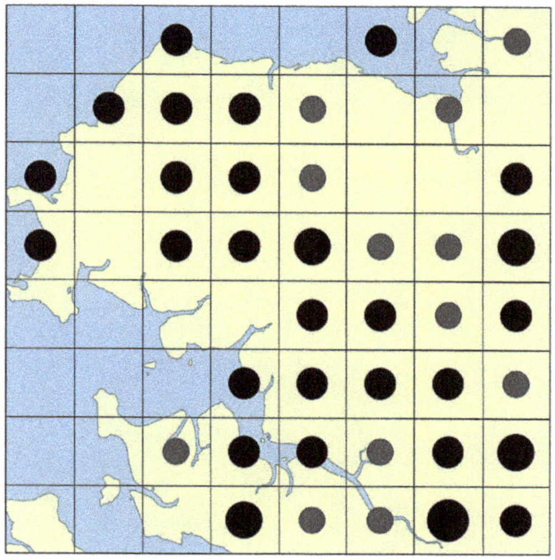

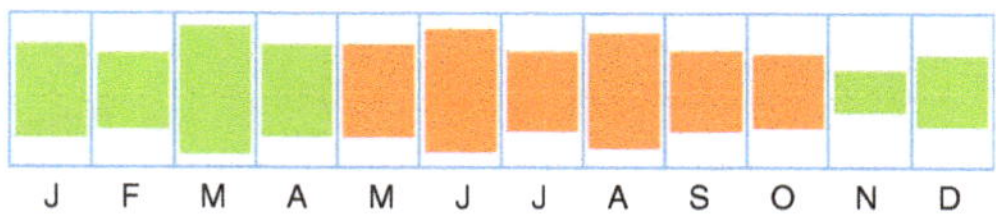

was reported more frequently from the metropolitan areas of both Darwin and Palmerston than the Silver-backed. During the Palmerston survey it was twice as abundant as the Silver-backed Butcherbird, but most records came from the newer suburbs of Durack and Rosebery, and few from the oldest, densely vegetated suburbs.[307]

During the Howard's Peninsula survey, it was found to be less restricted than the Silver-backed Butcherbird in habitat choice. Though most abundant in undamaged Woollybutt and Stringybark forest, it also occurred in cyclone-damaged eucalypt forest and low woodland, and occasionally in riparian forest, grassland and even rainforest.[308] Similarly, at Coomalie Farm, it was the only butcherbird on the floodplain and in bloodwood-dominated woodland, though it was less abundant in Woollybutt-dominated forest than the Silver-backed Butcherbird.[47] At Munmarlary, in Kakadu NP, and Yinberrie Hills, 50 km north-west of Katherine, this species showed an obvious preference for burnt areas, being most abundant just after a fire, and scarce to absent from unburnt areas.[382]

Only one nest has been reported from our region, but breeding records from elsewhere in the Top End suggest eggs are laid from August through to November.[31,43,47,103,311,397] Although the low reporting rates in the early Wet season are difficult to explain it is noteworthy that further inland, at Yinberrie Hills, the species showed significant seasonal variation in abundance,[311] possibly due to local movements within large home ranges.

White-breasted Woodswallow
Artamus leucorynchus

RANGE: tropical and e. Australia, NG, Indonesia and sw. Pacific. HABITATS: all except monsoon rainforests. STATUS: common Dry season migrant and scarce resident. BREEDING: Sep–Dec.

Kirri Hardy

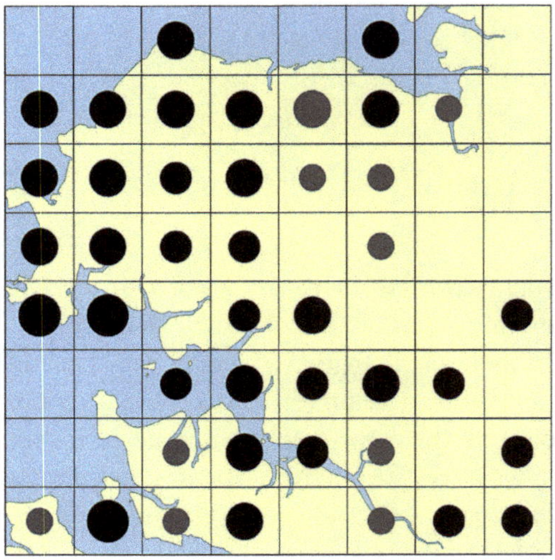

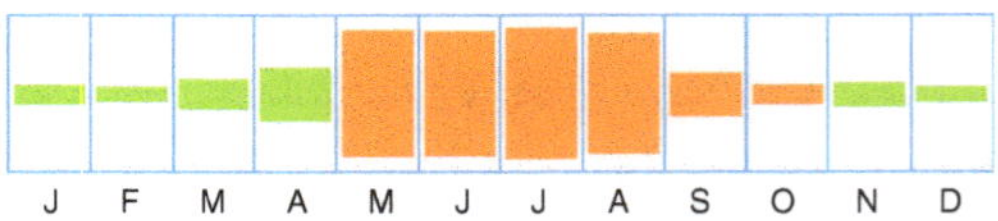

The annual arrival of large numbers of White-breasted Woodswallows in Darwin at the start of the Dry season is much less celebrated than the appearance of dragonflies, yet its timing is probably more predictable. Both the dragonflies and the birds are aerial hunters that screen the air continuously for small insect prey, the former hovering and darting relatively close to the ground and the latter gliding high in the sky, often betrayed only by their nasal calls. Together with Tree Martins and Rainbow Bee-eaters, their invasion at this time suggests that the Dry season skies abound with small aerial insects.

The arrival and departure dates of the migratory population are remarkably consistent from year to year. Around Millner, birds arrived in the last week of April in two years and the first week of May in three others during 1989-93.[296] Their departure took place in the last week of August in three years, and the third week of August in another year, but not until mid-October in 1990. Although the source of Darwin's 'wintering' migrants is uncertain, the species is known to be migratory in eastern Australia, leaving New South Wales and Victoria from March to May and returning during August and September. These months are roughly consistent with the times of their arrival and departure in Darwin, inferring that their migration is rapid. Moreover, an analysis of Atlas records indicated that this is one of several species that moves north-west from south-east Australia in the direction of the Gulf of Carpentaria and the Top End.[157] That the 'wintering' population does *not* originate from semi-arid inland parts of the NT is indicated by the lack of breeding records from that region except around permanent billabongs and rivers,[450] such as at Victoria River Downs[451] where they are resident. It is also noteworthy that the species was found in both Wet and Dry seasons on Cobourg Peninsula[143] and was recorded in February 1981 on South Goulburn Island,[452] suggesting that northern island populations are resident, as they are on Groote Eylandt.[82]

The species characteristically perches on power lines and radio masts, where it often congregates in large numbers. On the high power lines along Vanderlin Drive, for example, one count in early July 1995 yielded at least 300 birds over 2.5 km, mostly in scattered groups of three to 20 birds.[47] Almost 1500 Woodswallows were counted among eight suburban locations that were simultaneously sampled on the evening of 9 August 2008. The maximum number was 590 along 400 m of McMinns Street in the city.[3] Considering the small number of sites visited, it is possible that the total 'wintering' population in Darwin is several thousand.

But not all White-breasted Woodswallows vacate the region during the Wet season. The species was described as 'very plentiful in the Darwin area' between November 1944 and January 1945, when a nest was found at Larrakeyah Barracks.[434] Dependent juveniles were reported in November 1987 at Stoke's Hill Wharf,[34] and a few pairs were regularly sighted in Darwin city over the Wet seasons during the 1990s.[47] The first nest in the northern suburbs was recorded at Casuarina CR in 1999,[319] and the second breeding record at East Point two years later. In 2006, one pair nested in the mangal at Nightcliff, and since 2008 two pairs have nested there.[47] Finally, in 2010 a nest was discovered on the edge of the car park at Hibiscus shopping centre, with five adults in attendance.[47] These observations suggest the presence of a small, and apparently increasing, resident population. Based on 15 nests, clutches are laid from late September to December, though more than half of these in October.[47]

Roost sites are typically horizontal branches in trees or the horizontal cross-bars of power poles. The birds huddle closely to one another, usually facing the same direction, late birds squeezing into the middle, often landing on the backs of those already settled. In the car park at Hibiscus shopping centre, 245 individuals were counted as they slept in an African Mahogany on 6 August 2007, and a nearly identical number (240) were counted one year later, on 9 August 2008.[47] Such observations suggest that migratory individuals return to the same 'wintering' areas from year to year.

Masked Woodswallow
Artamus personatus

RANGE: whole of Australian mainland. HABITATS: woodland and grassland. STATUS: irregular Dry season visitor. BREEDING: none in Top End.

While often considered nomadic due to its irregular occurrence in many localities, Masked Woodswallows largely breed in the southern half of the continent during spring, after which many migrate towards the tropics where they overwinter during the Dry season.[411] Breeding populations in south-east Australia are known to move in

Nolan Caldwell

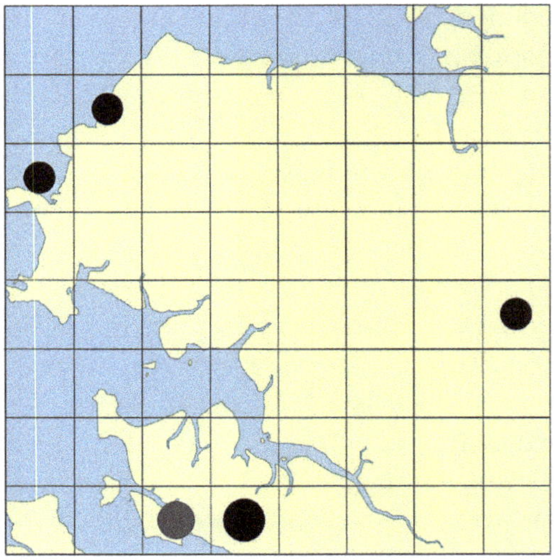

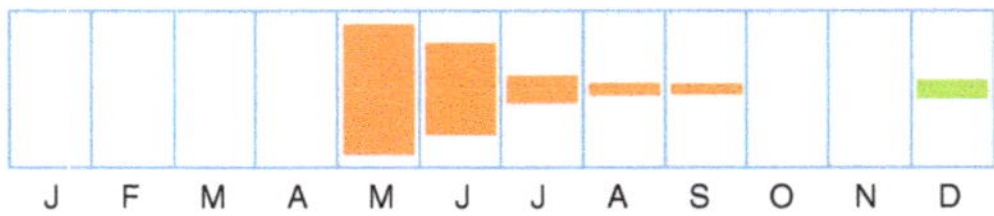

a north-westerly direction during late summer or autumn, but their destination is unknown.[157] Though it probably reaches inland parts of the Top End every year,[451] this species visits the Darwin region only once in every two to five years, invariably during the early Dry season, and usually in May or June. During the five years from 1967 through 1971, it was recorded only in 1970, when it was described as common during the last week of May and first three weeks of June.[1] This irruption followed a Wet season in which there had been less rain than usual in inland regions, suggesting that the birds may have moved further north than normal due to drought conditions further south.

A full quarter of a century passed before the Masked Woodswallow was noticed in the region again, yet it was recorded in nine of the following 19 years. One to three juvenile birds were sighted in May in three years (1995, 2000 and 2006) at the mouth of Rapid Creek or in the mangals close to the hospital.[21,47] Though the appearance of these birds on the coastline suggests they may have been on their way further north, there have been no confirmed records from the Tiwi Islands.[357] The species reappeared in 2008 and 2010 when many birds were seen over several successive days in the northern suburbs of Darwin, apparently part of a widespread influx across the region. Unfortunately these birds fly at such high altitudes at times that they are likely to be missed in some years.

Although the occurrence of these birds in the Top End is consistent with the north-westerly trajectory and timing of migration of south-eastern birds, it is also possible that our birds originate from breeding populations in southern NT, South Australia or even Western Australia. During a vehicular traverse of the NT from Alice Springs to Darwin in late July 1964, Masked Woodswallows were found in 'countless thousands' from Larrimah, about 430 km from Darwin, north through Mataranka to about halfway between Katherine and Pine Creek, and several hundred around Elliott (another 200 km further south).[96] Three years later, when the traverse was repeated in the same month, no birds were seen north of Larrimah, and several hundred only over the next 300 km south.[96] This change in the abundance of birds between the two years may relate to changes in environmental conditions. In 1964, when large numbers of birds had been observed closer to Darwin, the southern half of the NT was in the grip of an eight-year drought, perhaps forcing birds further north than normal, whereas in 1967, when they were absent north of Larrimah, the drought had just broken and the Top End had 'one of the most extensive wet seasons for many years'.[96] Thus there may be a relationship between ENSO and the extent of movements of this species.

White-browed Woodswallow

Artamus superciliosus

RANGE: whole of Australia except centre and most of WA. HABITATS: woodland and grassland. STATUS: vagrant. BREEDING: none in Top End.

Bastiaan Hensen

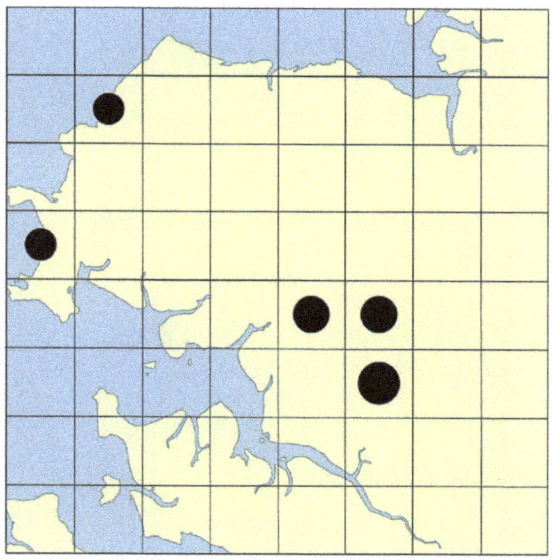

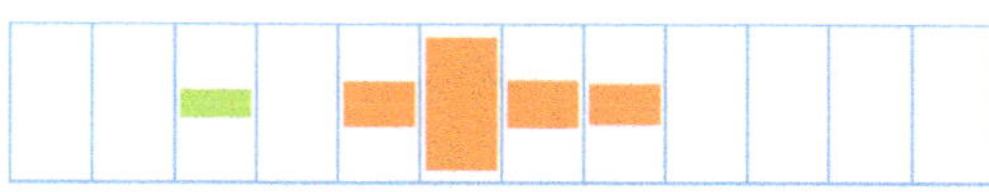

Typically occurring in mixed flocks with the Masked Woodswallow, the White-browed Woodswallow is the less common of the two in the Top End,[96,103] unlike eastern Australia where the situation is reversed. Both species were reported as being 'common' in the Darwin–Humpty Doo region during late May and June 1970.[1] Curiously, over the next 25 years only the White-browed was noted (in 1979, 1989, 1990), but in the 19 years from 1995 to 2013 it was recorded in only four years (2001, 2007, 2008, 2010). Thus over 46 years the average interval between records was almost six years. All records were from the Dry season, mostly June and July, except for one in March (1979). The June 1990 record coincides with a sighting of hundreds of birds in a mixed flock flying over Coomalie Farm, 45 km south of the Darwin region, where they were also present in July 2005.[47] On the latter occasion, however, over 200 White-browed and about 70 Masked, as well as one apparent hybrid bird, were seen feeding on bare ground, as well as on the blossoms of Darwin Woollybutts. Finally, at least one bird was present among numerous Masked at Casuarina CR in early June 2010[320] coincident with a sighting of a small number that included juveniles near Fogg Dam.[21]

Like the two previous species of woodswallows, the White-browed migrates in a northerly direction from its main south-east Australian breeding grounds during autumn but, unlike the others, takes a more inland route on its southward return journey, resulting in a circular, anticlockwise pattern of movement.[157] Thus birds visiting the Top End may originate from populations breeding in central Queensland or South Australia, rather than the south-east. Both species have been noted feeding on nectar from the flowers of eucalypts, though the importance of this energy source in their diet during winter is unknown.[411] Woodswallows have a brush-tipped tongue, which greatly facilitates nectar extraction, so it might be expected that they exploit the glut of nectar available in the Top End during the Dry season.[398]

Black-faced Woodswallow
Artamus cinereus

RANGE: whole of Australia except se. coast, s. NG and Timor. HABITATS: woodland and grassland. STATUS: vagrant, but moderately common resident just outside region. BREEDING: Oct–Nov (TE).

Ashley Carlson

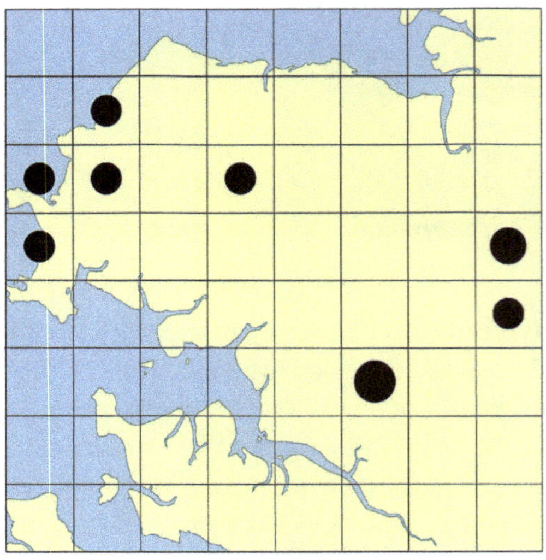

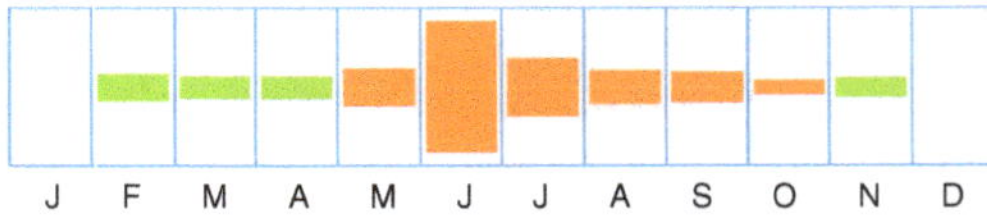

Despite being the common resident woodswallow of the Top End savannas, the Black-faced Woodswallow is rarely seen in the Darwin region. The nearest locality to Darwin where this species is regularly seen is the Naval Communications Station at Middle Point near Fogg Dam,[60] where small numbers can usually be seen perched on the power lines bordering the treeless paddocks. Though it was once considered to be migratory in the Top End,[1,110] with birds moving towards the coast during the Dry season and retreating inland during the Wet, data from the first Atlas suggested the reverse, with slightly higher reporting rates in coastal areas during the Wet.[4] However, they are present year-round at many sites including Coomalie Farm,[47] Yinberrie Hills,[311] and Victoria River Downs,[451] and more likely wander locally after breeding. In woodland dominated by Salmon Gum and Darwin Box at Yinberrie Hills, they were among the three most abundant species at times,[311] but were significantly more abundant in burnt areas, especially those burnt early, than in unburnt areas.[382] At Coomalie Farm and on the Marrakai track they build their cup-shaped nests among the branches of small paperbarks on dry floodplains during October and November.[47]

Little Woodswallow
Artamus minor

RANGE: whole of Australia except southern third. HABITATS: eucalypt forest and woodland. STATUS: rare Dry season migrant. BREEDING: Oct–Nov, and possibly Jun (TE).

Aiden Webb

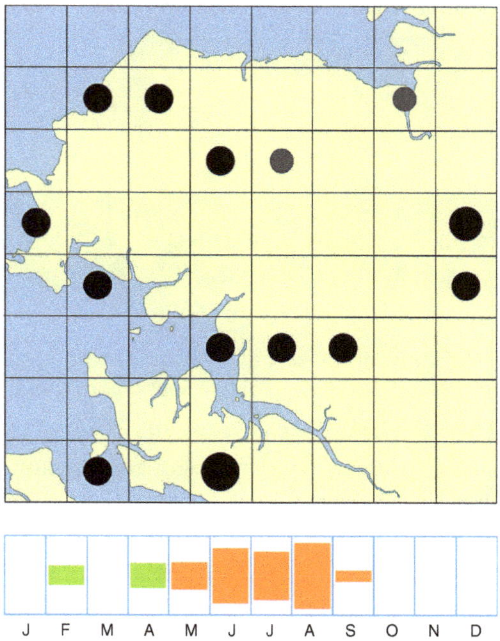

Though rare around Darwin itself, the Little Woodswallow is a moderately common early to mid-Dry season migrant to the Top End, usually arriving between mid-April and mid-May, and departing in the first two weeks of August.[1] Analyses of Atlas data suggested a north–south migration pattern for this species, in which populations from South Australia, northern and central New South Wales, southern Queensland and possibly southern NT move to northern Australia from March onwards, and return to breed during September.[157] However, populations to the west of Katherine[369,451] and on Groote Eylandt[82] are resident, and nesting has been reported in October among the sandstone cliffs near Gunbulanya in western Arnhem Land.[110] At Yinberrie Hills, where breeding was also recorded, highest densities were recorded in February–April and September–October, suggesting passage movements of migrants, possibly to and from areas to the north.[311] Numbers here evidently fluctuated between years, as they were significantly more abundant in October 1986 than in October of the following year. To complicate matters further, some birds may breed during the Dry season, as apparently occurred at Coomalie Farm in at least two years.[47]

This species generally occurs in more wooded habitats than the Black-faced Woodswallow, such as open forests dominated by Darwin Stringybark and Woollybutt, but it often visits the edges of floodplains and tidal saltflats,[47] and in many regions is associated with rocky outcrops and gorges.[96,451] In the latter habitat it builds its cup-shaped nests in clefts on cliff faces,[110,411] but more typically uses tree hollows. Like the Black-faced Woodswallows, the Little Woodswallows are sometimes attracted to intense fires in the mid-Dry season, when they screen the air at canopy height,[95] presumably scooping up fleeing insects that are caught in updrafts. At Munmarlary, near Jabiru, the species was present only during the Dry, and in almost all cases on unburnt or biennially burnt sites, rather than on annually burnt sites, whereas at Yinberrie Hills, it was more abundant on recently burnt sites than in unburnt sites.[382]

Black-faced Cuckooshrike
Coracina novaehollandiae

RANGE: whole of Australia, NG and e. Indonesia. STATUS: common Dry season visitor, and passage migrant. HABITATS: eucalypt woodland, grassland, suburbia. BREEDING: mostly Sep–Jan (TE).

The Black-faced Cuckooshrikes seen in Darwin during the Dry season are migrants that may have flown 2000 or possibly 3000 km from their breeding grounds in the south-east of the continent. Numbers arrive in Darwin in April, and peak in May,[453] one or two months after most migrants depart from New South Wales and Victoria, heading in a north-west direction.[157] Although reporting rates of this species remained high until August, observations at Leanyer from 1974 to 1984 revealed a marked drop-off in numbers after the May peak,[453] coinciding with its arrival in eastern Indonesia,[454] where it is also a non-breeding Dry season visitor.[86] Parties of these birds have been observed flying out to sea from Darwin in May,[453] though it is possible that they were on their way to the Tiwi Islands. As there is no corresponding peak in numbers at the end of the Dry, it is possible that southbound

Bastiaan Hensen

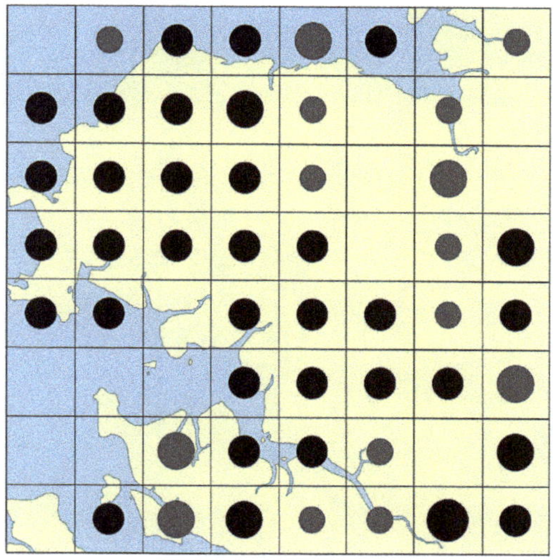

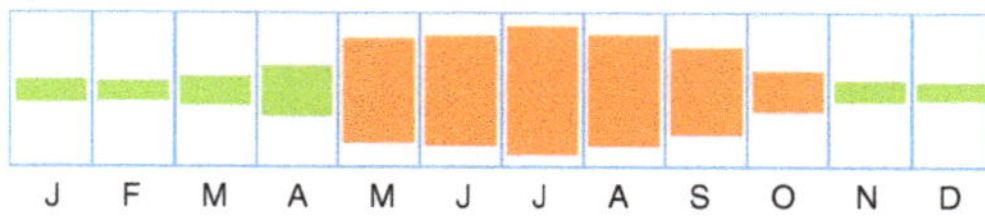

migrants from Indonesia return to northern Australia by a different route, or that they over-shoot Darwin.[453] Between 1967 and 1972 a few birds were recorded in the region throughout the Wet of 1969–70,[1] coinciding with poor rains inland. These and other Wet season records probably concern birds dispersing from subcoastal areas, where small numbers breed.

During their stay, these birds can be seen in the middle of suburbs, often perching on wires and clothes lines.[365] Outside suburbia, they frequent all wooded habitats except vine forests and mangals, preferring eucalypt woodland, especially where cyclone-damaged, and open grassland.[308] Until they depart, mostly in October or November, groups of migratory birds include immatures,[47] recognised by their black eye-stripe and pale throat, as they do not moult into adult plumage until about one year old.[411]

Until 2006 there were no breeding records for Darwin. In mid-November of that year a nest with eggs was found on the edge of Wanguri oval,[47] while another at McMinns Lagoon had young by mid-October.[136] In September 2010 a pair nesting in a Palmerston backyard raised at least one young, which stayed with its parents while they attended a second brood through November.[35] Presumably the same pair nested again in April 2011, but when its two chicks were already well feathered they were killed and eaten by a Brown Goshawk as the parents looked on. These records, and others from Coomalie Farm and Kakadu NP, suggest that most nesting in the Top End takes place from September to November, as it does in New South Wales, though clearly some also occurs in the late Wet season.[47,98,103]

White-bellied Cuckooshrike

Coracina papuensis

RANGE: tropical and e. Australian mainland, far e. Indonesia, and NG. STATUS: very common resident; HABITAT: all wooded habitats, suburbia. BREEDING: mainly Sep–Feb.

Colin Trainor

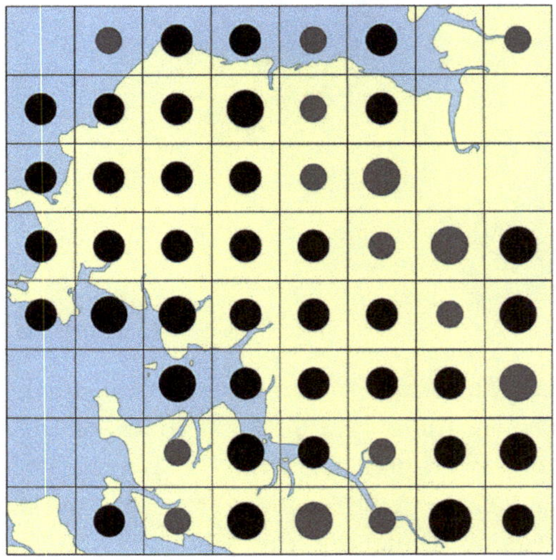

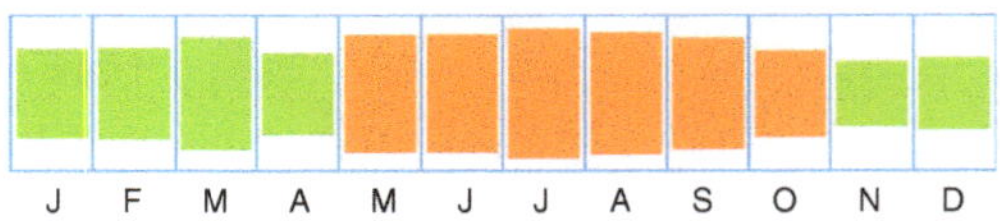

Unlike the larger Black-faced Cuckooshrike, the White-bellied Cuckooshrike is a resident of the Darwin region,[453] where it is among the 20 most frequently recorded species. It occurs in most suburbs of Darwin and Palmerston, but during a bird survey in the latter city was found to be commoner in suburbs built before 2000 than in those developed afterwards.[307] During the Howard's Peninsula survey, it was recorded in more sites than any other bird species, and was among the most generalised of all species in its habitat choice, occurring in 16 of the 22 vegetation types.[308] Although most abundant in cyclone-damaged eucalypt woodland, it was also common in undamaged eucalypt forest and paperbark woodland, and moderately common in rainforest and mangal.

Like most other members of its family it is mainly insectivorous, but it also takes fruits, such as figs,[47] the orange-coloured arils of Black Wattles,[455] and very occasionally nectar.[398] In open forests and woodland of Kakadu it was found to forage lower down during the Wet season than in the Dry, spending half of its time in shrubs and on the ground.[103] The species is not shy about where it builds its nest, which may be found in roadside trees and even beside city buildings.[365]

Although breeding has been reported in all but two months of the year in the NT,[411] data from Darwin suggest two peaks in egg-laying, one in September–October and the other in December–January,[47,453] possibly representing first and second broods. The nearly flat, slightly cup-shaped nest is built in the fork of a horizontal branch 2–12 m up in a tree, typically a eucalypt; but in urban areas exotic trees such as African Mahoganies are also used.

Common Cicadabird

Coracina tenuirostris

RANGE: coastal and subcoastal tropical and e. Australia, NG and e. Indonesia. STATUS: scarce resident. HABITAT: forests, including mangal. BREEDING: Nov–Dec (TE).

Ashley Carlson

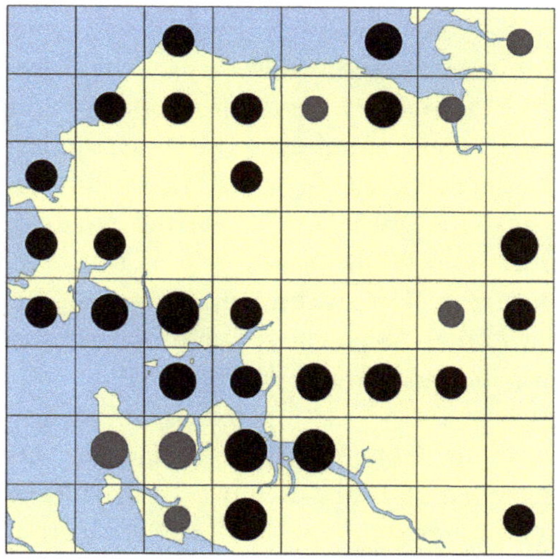

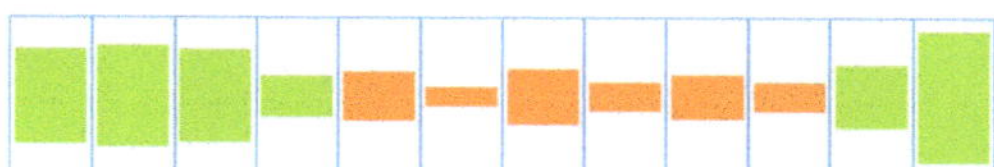

Identical in appearance to its eastern counterpart, the tropical race of the Common Cicadabird (*melvillensis*) is smaller, and the isolated population of the Top End and Kimberley gives an entirely different song.[411] While the song of eastern birds is reminiscent of cicadas, consisting of rapidly repeated notes on the same pitch, that of Top End birds consists of a series of loud, drawn-out upward-inflected whistles, slowly repeated up to 12 times, but usually four or five times. In contrast to the eastern race, which migrates to New Guinea in the winter, Top End birds occur year-round, but appear to undertake seasonal movements.

The Common Cicadabird is much less common in the Top End than in eastern Australia. During almost five years from 1967 through to 1971 it was recorded in the Darwin–Humpty Doo region on only six occasions, five of which were in the Wet season.[1] In a decade of monthly surveys of the area around Leanyer SP from 1974 to 1984 it was encountered only nine times,[453] and in 1986 it was not recorded at all during over 1000 censuses involving 20 habitats scattered throughout Howard's Peninsula.[308] However, sampling during the latter survey was confined to April and May, months in which reporting rates are low in the Darwin region. Indeed, reporting rates in Darwin during the Wet season are almost twice that of the Dry, suggesting either seasonal movements at a regional scale, or marked changes in behaviour, such as calling rates.

Top End birds inhabit monsoon rainforest,[103] paperbark and other riparian forests,[47] as well as mangal.[453] In the Kimberley, the species is said to be confined to rainforests in the Dry season, dispersing into drier open forests during the Wet.[321] However, during a Dry season bird survey of no fewer than 88 patches of monsoon rainforest in and around Kakadu NP, it was recorded in only two patches.[355] In mangal close to Palmerston SP the species was recorded 25 times over five years, 88% of records pertaining to the Wet season, suggesting the mangal and surrounding eucalypt woodland was vacated during the Dry.[312] In south-east Australia the species is known to be more vocal while nesting, in spring and early summer, than afterwards,[411] so it is likely that the low Dry season reporting rates in Darwin are at least partly attributable to inconspicuous behaviour at that time.

Only two nests have been found in the NT. The first was discovered at the Territory Wildlife Park in mid-December 1997[456] and the second at Girraween Lagoon in late December 2007.[457] In the 1970s local males shot for museum skins had enlarged testes in both November and January, so it is possible that breeding continues at least until February. The Common Cicadabird is one of the few species in Australia that lays a single egg in its nest, yet its incubation period (about 22 days) is identical to that of the similar-sized White-bellied Cuckoo-shrike, whose clutch is normally two.[411] However, both parents brood the eggs in the latter species, whereas in the Cicadabird only the female performs this role,[411] the male being free to defend the territory or attract other mates.

White-winged Triller
Lalage tricolor

RANGE: whole of Australia and NG. STATUS: moderately common Dry season migrant. HABITATS: mostly eucalypt forest, woodland and floodplains. BREEDING: Oct–Feb (TE).

Apart from the fairywrens, few Australian songbirds exhibit seasonal variation in plumage, but one such species is the White-winged Triller, which is also unusual among trillers in being migratory over much of its wide range. Soon after breeding, around December and January,[411] populations in south-east Australia move in a north-westerly direction towards the interior of Queensland and the NT to spend their 'winter' in warmer climes.[157] Whether these birds migrate as far as Darwin is not

Ashley Carlson

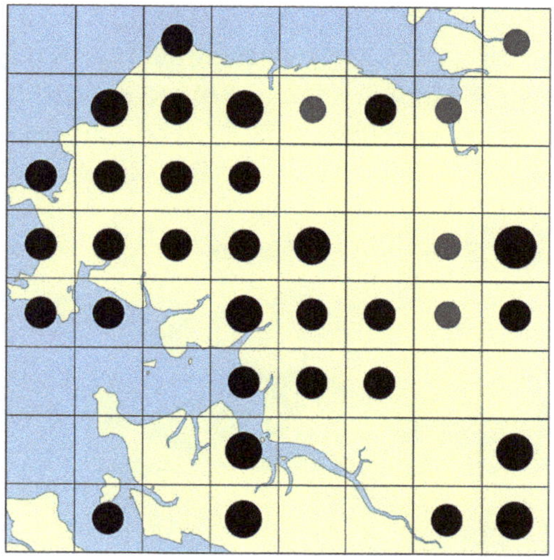

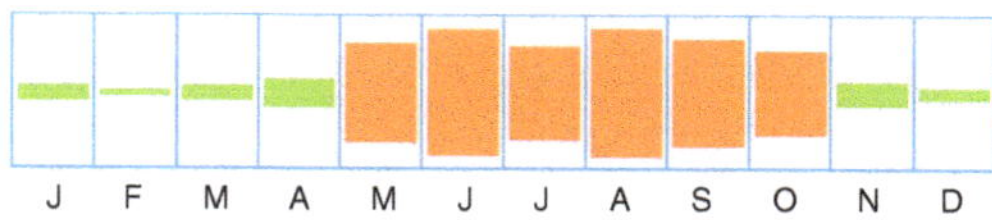

known, but the species is common and widespread across the Top End during the southern winter (mid-Dry season), and largely absent from northern parts of it during the Wet.[82,96,103,110] As reporting rates and historical data[1,453] show, the bulk of birds visiting Darwin arrive in May, long after the departure of south-eastern populations from their breeding grounds. Around Leanyer SP numbers dropped off soon after their arrival in Darwin, and gradually taper off through the Dry season until November when only a few remain for the Wet.[453] Where the majority of birds go after May is unknown, as the species occurs neither in the Moluccas nor Papua to the north. Assuming that they come from southern Australia, and that it takes them three to five months to reach Darwin, it is conceivable that some depart for the south soon after their arrival in May to allow sufficient time to return to their breeding grounds in August and September.

Another potential source for our visitors is the southernmost parts of the Top End. At Victoria River Downs and Keep River NP, the species breeds from December to March,[98,369] and while some birds remain in the region during the Dry season, the majority form post-breeding flocks and disperse. At Yinberrie Hills, its seasonal occurrence was opposite to that of Darwin, being moderately common from November to January,[311] when it presumably bred, and scarce during the Dry season. The main habitats of the species are eucalypt woodland and dry floodplains, where flocks of up to 25 may occur, but individuals are sometimes seen in mangal[313,453] and monsoon rainforest.[103] Insects are the main food of this species, mostly snatched from foliage, but sometimes from the ground while hopping. It also often feeds on the floral nectar of eucalypts, especially the Darwin Woollybutt and the Fern-leaved Grevillea, which blossom in the Dry season.[398]

On arrival in Darwin some adult males are still in full or at least partial breeding plumage,[453] with black on the head and back, but by June all have moulted into their non-breeding plumage which resembles the brown-coloured females, except for black on the wings and tail.[411] By October, however, many males have acquired breeding plumage, and can be heard singing, even in urban areas, as if attempting to attract mates. A nest found in the mangal at Nightcliff in November 1977[60,365] represents the only breeding record for Darwin, but in open forest near Fogg Dam three nests were found in October and another three in February.[453] At Coomalie Farm, a colony of about six pairs nested on the edge of a floodplain in December 2002,[47] but had disappeared by February. The post-breeding destination of these populations remains a mystery.

Varied Triller
Lalage leucomela

RANGE: tropical and subtropical e. Australia, and NG. STATUS: common resident. HABITATS: monsoon rainforest, paperbark forest and mangal. BREEDING: year-round, peaking Oct–Jan (TE).

Niven McCrie

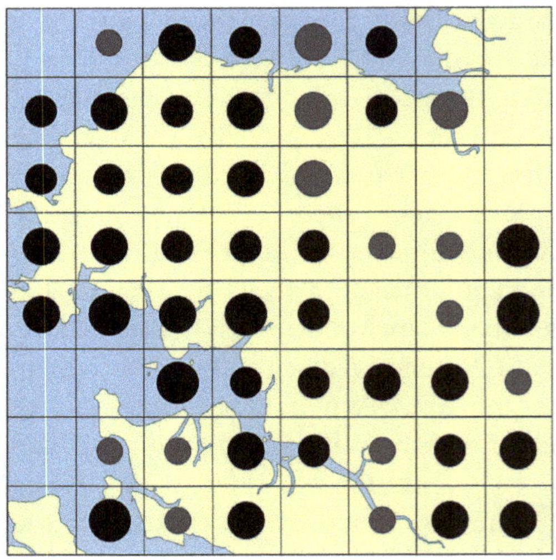

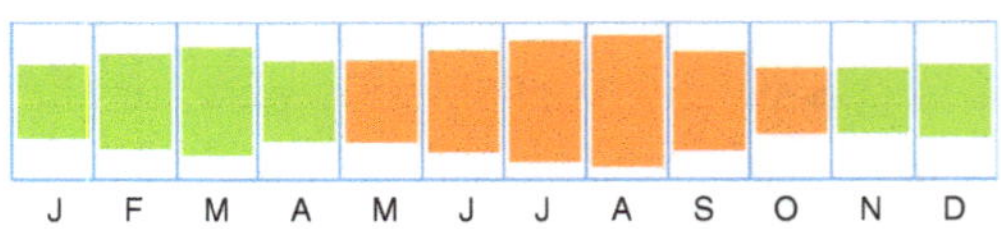

In stark contrast to the smaller, more streamlined White-winged Triller, the Varied Triller is sedentary, and favours rainforests and riparian forests rather than open habitats. It is a common bird of Darwin's suburban gardens, especially where there are mature trees with adequate cover,[60] but is less abundant in Palmerston, probably because of the lack of rainforest nearby. Its adaptability doubtless relates to its broad diet, which consists of both insects and fruit, and it is a regular visitor to fruiting fig trees in urban parks and schools.

During the Howard's Peninsula survey, it was among the four most frequently recorded species, occupying 12 of the 20 habitats sampled, with highest densities in monsoon rainforest, paperbark forest and Black Wattle forest, but also occurring in mangal and, occasionally, eucalypt forest and woodland.[308] Although frequently seen in mangals adjacent to rainforests or suburbs, such as at Nightcliff, where it has access to fruiting trees, it is scarce in mangals surrounded by savannas, such as most of Darwin Harbour.[312,313] Food items noted in Darwin include the 'looper' caterpillars that attack Poinciana trees[60] and the fleshy arils of Black Wattles seedpods.[455]

Astonishingly little is known of the breeding biology of this widespread species. It is one of a handful of songbirds in Australia that lays a single-egg clutch, a feature it shares with the Common Cicadabird. However, its incubation and nestling periods are totally unknown. Breeding records from Darwin reveal that it can lay eggs in any month of the year, though over a half of all records were from October through January.[47] Interestingly these are the months with the lowest reporting rate, suggesting that the species is less vocal or conspicuous when nesting.

Varied Sittella
Daphoenositta chrysoptera

RANGE: whole of Australian mainland and NG. HABITATS: eucalypt forests and woodland. STATUS: rare resident. BREEDING: Feb–Nov (TE).

Laurie Ross

Typically found swarming over a eucalypt tree, then flying to the next in a tight group of three to eight individuals, the Varied Sittella is the only bark-foraging specialist bird in the Top End apart from the Black-tailed Treecreeper.

Although moderately common in eucalypt forest and woodland in Kakadu NP[103] and parts of Arnhem Land,[47,143] the Varied Sittella is very sparsely distributed in the Darwin region. During almost six weeks of intensive surveys of Howard's Peninsula in 1986 it was recorded only once.[308] Then in August and September 1990 two nests were found on a rocky ridge on the Channel Island Road.[43] It was another 11 years before the species was recorded again at the latter site. Nine years later again, a group of six were encountered at Charles Darwin NP in May 2010.[21] However, the species has been sighted regularly on the Marrakai Road, about 40 km south-east of our region, since the late 1960s.[1,47]

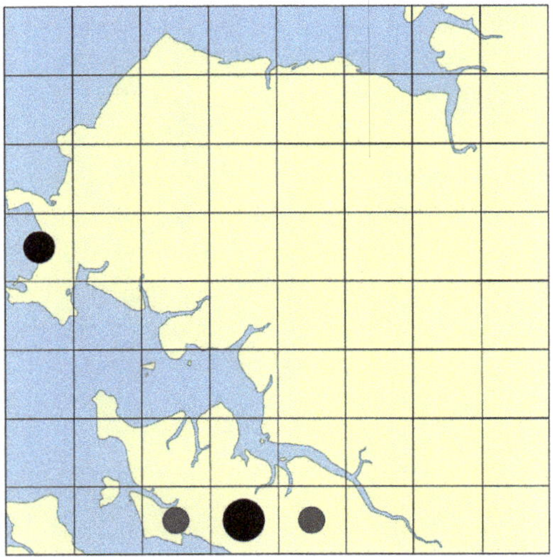

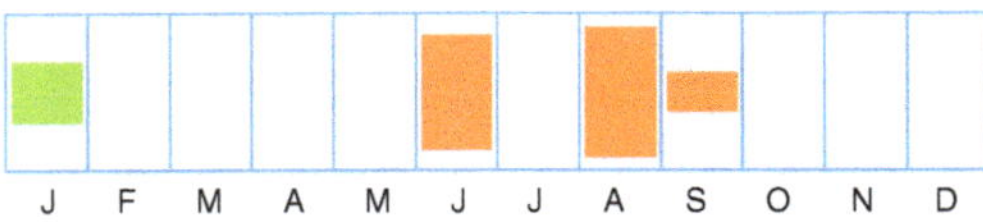

Breeding records in the Top End, mostly derived from observations of adults feeding fledglings or juveniles, suggest that egg-laying can occur in any month of the year, but over a half of all records are from the mid-Dry season months of June through August.[47] This contradicts much of the literature which states that breeding in the NT occurs from September to January.[438]

Grey Whistler
Pachycephala simplex

RANGE: endemic to Top End (race *simplex*); n. Qld and NG (other races). HABITATS: monsoon rainforests, vine thickets and mangal. STATUS: common resident. BREEDING: Mar–Aug.

Niven McCrie

Contradicting its name, the distinctive Top End race (*simplex*) of the Grey Whistler lacks any grey in its plumage, and may yet be found to represent a separate species, for which the common name 'Brown Whistler' is apt. The other Australian subspecies (*peninsulae*) is found only in north Queensland, and like all but one of the six subspecies in New Guinea, has yellow on the underparts and grey on the head.[438] Moreover, the two Australian races are separated from each other by the 500 km-wide Gulf of Carpentaria. One of the commonest birds in local monsoon rainforests and mangals, it was the most abundant species in semi-deciduous vine thickets during the Howard's Peninsula survey.[308]

In Top End mangals the Grey Whistler is far more widespread and abundant than the other two mangal-specialised whistlers,

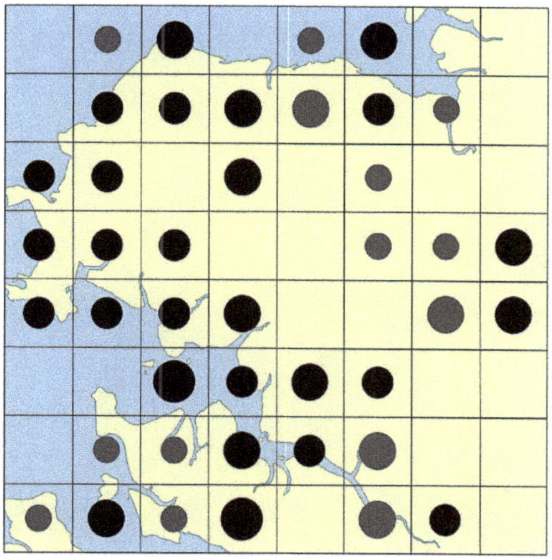

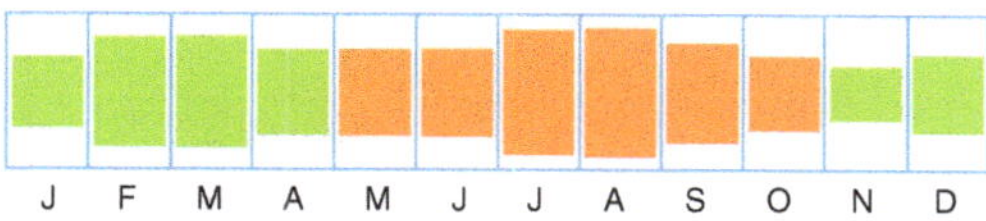

with an average density of about one bird per 2 ha.[312,313] In a bird survey of mangals in the Darwin region, this species was recorded at 10 of the 13 sites sampled, with highest densities at Mindil Beach, East Point and Casuarina CR, each of which is close to coastal rainforests or vine thickets.[313] During 10 years of banding on a 4 ha mangal site near Palmerston SP, 27 Grey Whistlers were banded, of which 10 (41%) were subsequently re-trapped on the site, one bird almost exactly six years after banding. Yet of the 26 individuals banded over nine years at Ludmilla Creek mangal, only 4 (15%) were re-trapped, up to three years after banding. Within mangals, the species prefers the Stilt and Rib-fruited Mangroves typically found along tidal creeks to the less-frequently inundated Spurred Mangrove thickets.[313]

As with the other mangal-inhabiting whistlers, very little is known about the breeding biology of the Grey Whistler. Indeed, only three clutches have ever been described.[438] However, recent records of nests and dependent juveniles in our region show that, contrary to the literature, this species breeds mostly in the Dry season with a peak in June, although nest building starts as early as March.[47] Supporting this conclusion is the fact that one early collector was unable to find their nests during four months of fieldwork between October 1915 and January 1916.[70]

Mangrove Golden Whistler
Pachycephala melanura

RANGE: coastal tropical Australia, NG and adjacent islands. HABITATS: mangal. STATUS: sparse, localised resident. BREEDING: Dec–Mar.

While the Mangrove Golden Whistler is largely confined to coastal mangals across the Top End, including the Darwin region, some populations occur in rainforests and bamboo thickets up to 55 km inland,[47] and on offshore islands it occupies a very broad range of habitats.[458] Even in mangals, however, it is patchily distributed and much less common than the Grey Whistler. During a bird survey of mangals in the Darwin region, it was found at seven sites, with an average density of about one bird per 6 ha.[313] The mangal at the mouth of Ludmilla Creek, including East Point Boardwalk, is home to several pairs. Here, and at Camerons Beach, Shoal Bay, the birds are associated with large Star Mangroves that occur in a 50–100 m wide band along the

Adrian Boyle

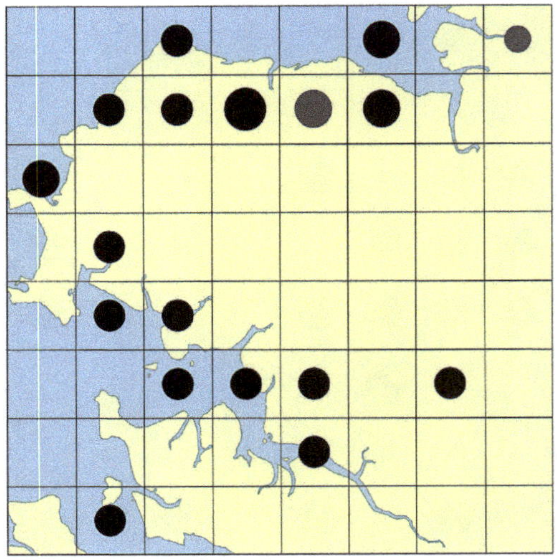

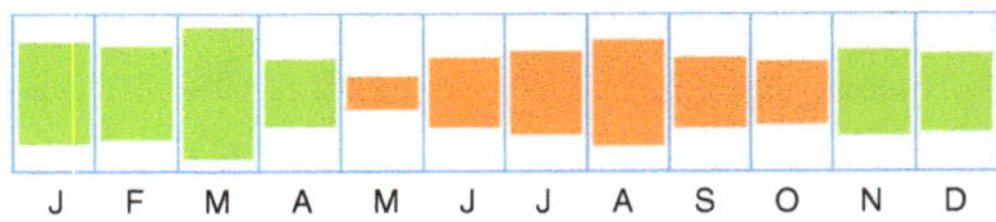

seaward fringe. This association may explain why these Whistlers are largely absent from the vast mangals surrounding Darwin Harbour, where this mangrove species, if present, forms an even narrower belt.

The species is also regularly sighted in mangal surrounding saltflats beside Palmerston SP, less than 400 m from the Star Mangrove-lined Elizabeth River, yet only 700 m inland, in a 4 ha site dominated by Stilt Mangroves and Spurred Mangroves, it was seen only four times during 10 years.[312] On the other hand, just south of Leanyer SP two or three pairs inhabit the Grey Mangroves growing along artificial canals and around the edge of large saltflats, 4 km from the mouth of Buffalo Creek.[313]

Very little is known of the breeding biology of this species.[438] Indeed, a nest found on the edge of Kulaluk Bay in early January 2009[47] was the first reported in the NT for a century. This nest contained one chick which, like the Australian Golden Whistler of eastern Australia, was chocolate-brown in colour.[47] Two months later, in late March, another nest, this time with eggs, was found near the East Point boardwalk. These observations suggest that breeding in the Top End occurs during the Wet season, as it does in the Kimberley region of Western Australia.[187]

Rufous Whistler

Pachycephala rufiventris

RANGE: whole of Australia and New Caledonia. HABITATS: all wooded habitats, but primarily eucalypt woodland. STATUS: sparse Dry season visitor. BREEDING: Oct–Feb (TE).

Jon Norling

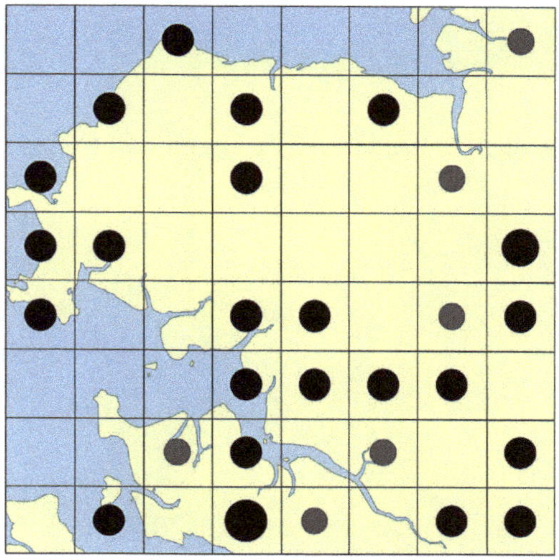

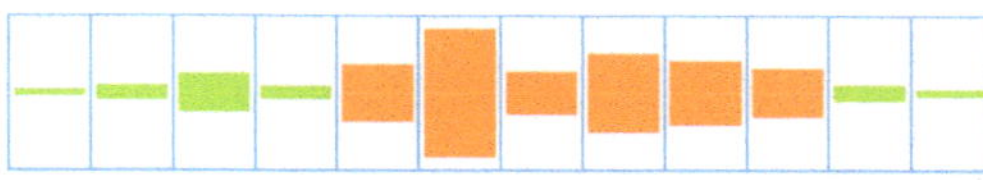

A well-known north–south migrant in eastern Australia, the Rufous Whistler is a common resident throughout much of the Top End savannas but a sparse visitor to the Darwin region, mostly during the Dry season.[60] Although migrants from south-eastern Australia move in a north-westerly direction towards inland Queensland and the NT after leaving their breeding grounds during March and April,[157,438] it is unlikely that any of these eastern birds reach the Top End, which hosts a different race, *falcata*. Males of the local race differ from those of southern Australia in being smaller, lacking a black mask, and having a paler belly and breast band, while in females the streaking of the underparts is finer and barely extends below the upper breast.[438]

Reporting rates suggest that most birds arrive in Darwin during May and June, but disperse soon afterwards. During the Howard's Peninsula survey the species was recorded only six times, twice each in mangals, vine forests and eucalypt woodland. On the landward edge of mangal near Palmerston SP, the species was recorded 10 times in monthly censuses over 36 months, eight of those during the Dry season.[312] Further inland, at Coomalie Farm it is a breeding resident in eucalypt open forest and woodland, as well as paperbark-fringed floodplains.[47] Interestingly, at Yinberrie Hills, much further inland, it was most abundant during the Wet season,[311] so it is possible that some visitors to Darwin are from southern parts of the Top End, although there are no published breeding records from that region.

While the breeding biology of this species is well known from studies in southern Australia, its annual cycle in the Top End is poorly known.[438] Breeding records from the last three decades suggest that it starts clutches between September and February, with a peak in October.[47] This is very similar to the breeding season of the species in south-eastern Australia, apart from a later start.

White-breasted Whistler
Pachycephala lanioides

RANGE: endemic to coastal monsoon-tropical Australia. HABITATS: mangal. STATUS: very rare resident. BREEDING: Mar–Jun.

Glenn Newton

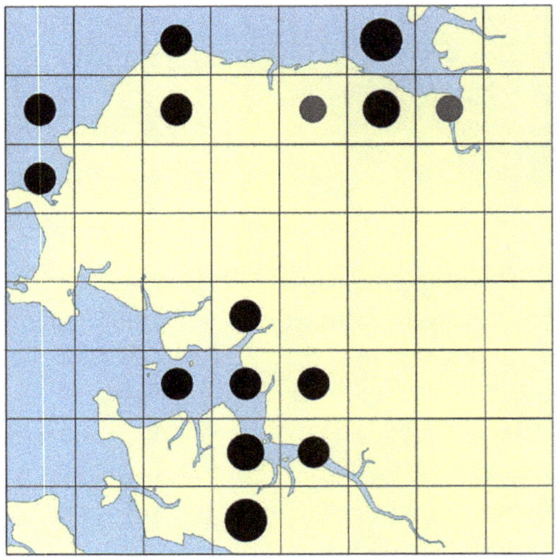

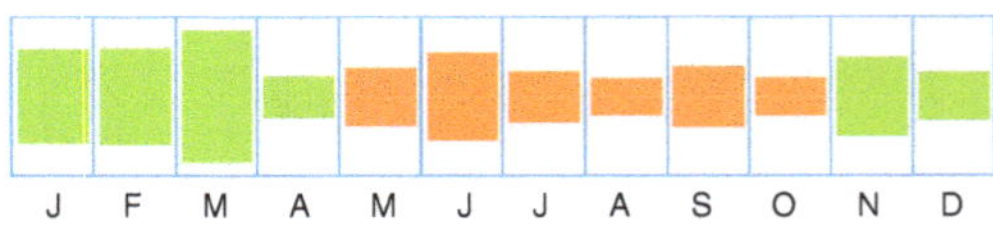

One of the Top End's rarest resident birds, the mangal-specialised White-breasted Whistler was reported from only two localities in the NT during the second nationwide bird atlas from 1998 to 2002: Darwin and the eastern tip of Melville Island.[5] Yet 20 years earlier, during the first Atlas project (1977–81), it was reported from at least six localities stretching between the Daly River, south of Darwin, and the Goomadeer River, in western Arnhem Land.[4] Indeed, during the late 1960s and early 1970s it was considered a common species in the dense low mangroves of the Darwin area, including Camerons Beach,[1,96] Mica Beach[96] on the western side of Darwin Harbour, and Middle Arm Jetty, which is 12 km south of Elizabeth River bridge.[1,34] Although it has not been recorded at the last site since around 1990, the species was still present at Camerons Beach in 2001.[47] Since 1981, the species has been reported occasionally from Buffalo Creek, Leanyer SP and Palmerston SP, although only one individual was recorded over 10 years of banding in mangals less than 300 m away from the latter site.[312]

In 2000, however, three White-breasted Whistlers were found in the narrow belt of mangals fringing Kulaluk Bay, about 1 km north of the mouth of Ludmilla Creek.[47] Here the birds foraged for small crabs on the sandy mud under the dense canopy of low Stilt Mangroves and mallee-like Grey Mangroves (2–6 m high), as well as under even lower (1–2 m) Grey Mangroves scattered on the mudflats towards the sea. The habitat here is remarkably similar to that at Camerons Beach, where the mangal is 100–300 m wide, and contains a mixture of low Stilt and Grey Mangroves. At both sites, the landward edge of the mangal is narrowly bordered by a sandy beach, fringing a vine thicket, which these birds occasionally visit.[1,96] Although a single male was recorded in monsoon forest on a tributary of the Howard River on 1 May 1967,[1] the species was not encountered 19 years later during the Howard's Peninsula survey in 1986.[308] Incredibly, the nearest recently reported population to that in Darwin lies about 160 km to the east, on the South Alligator River.[366]

Until recently the only NT breeding records concerned two nests with young found by William McLennan at the mouth of the Roper River in March 1916.[70] More than 90 years later, a pair at Kulaluk Bay was observed building a nest in December 2009 and May 2010, with an egg finally laid on 7 June 2010.[47] At the same site in 2013, two eggs were laid around 30 March and one hatched by 15 April, but whether the chick fledged or not is unknown.[119]

Little Shrikethrush
Colluricincla megarhyncha

RANGE: tropical and subtropical, largely coastal, Australia, NG, and adjacent islands. HABITATS: monsoon rainforests, riparian forest and mangal. STATUS: sparse resident. BREEDING: Jun–Feb (TE).

Niven McCrie

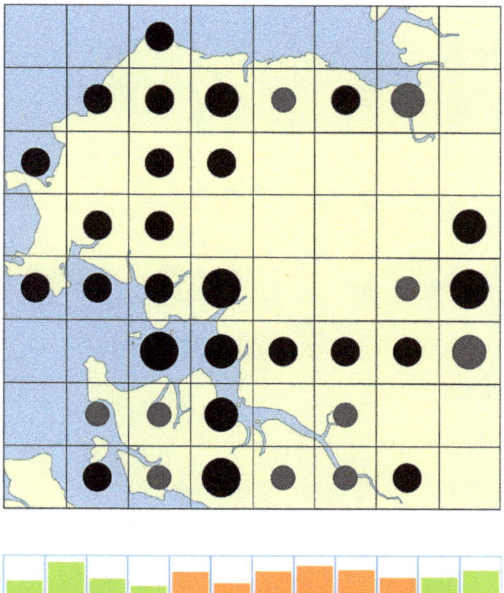

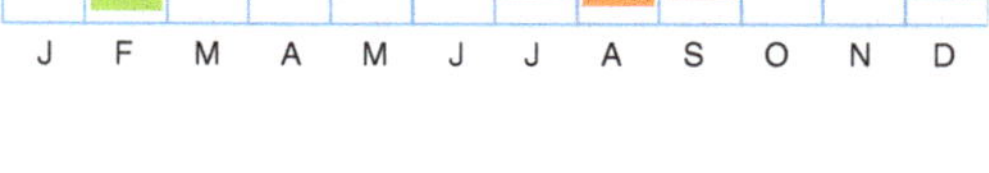

Given its wide range, surprisingly little is known of the ecology and breeding biology of the Little Shrikethrush. In Kakadu NP it is among the commonest species in monsoon rainforests,[103] found in all six types of this habitat, but most abundant around lowland springs.[355] In the Darwin region, however, it is rather scarce, and during the Howard's Peninsula survey it was recorded only four times in over 1000 censuses involving 100 sites. While recorded in 56% of cells in the region, over a quarter of all records came from the spring-fed monsoon rainforest at Howard Springs NR. Other sites with regular sightings include Buffalo Creek, East Point and Holmes Jungle.[96,300]

While it is well known to occur in mangals, it is much less abundant in this habitat than in monsoon rainforests. Of 13 patches of mangal around Darwin's harbour and coastline that were surveyed each month during 2008–09, only five contained this species, and they included the four largest patches.[313] The average density was approximately one bird per 11 ha, although at Buffalo Creek, the density was one bird per just over 2 ha. During monthly bird censuses in mangal near Palmerston SP over 4.5 years, it was recorded 11 times, mostly in the Wet season.[312] Only five birds were captured and banded in 77 banding sessions over 10 years at this site. None were captured in mangals during seven years of banding at Ludmilla Creek and three years at Rapid Creek.[47]

In 1902, three nests of the species were found in the Top End, all in January and February,[400] but it would be another century before another nest was reported. A bird was found sitting on its nest at Howard Springs NR on 18 October 2004.[21] Seven years later another nest, this time with an egg, was discovered at Lake Bennett, about 40 km south of our region, on 2 January 2011.[62] At Mary River Park[47] and in Kakadu NP[103] pairs have been observed feeding fledglings or juveniles in late July and early September, respectively, suggesting egg-laying in June and July. These observations indicate that the breeding season of local birds is longer than that of eastern races of the species.[438]

Grey Shrikethrush
Colluricincla harmonica

RANGE: whole of Australia and NG. HABITATS: mainly eucalypt forest and woodland. STATUS: scarce resident. BREEDING: Sep–Feb (TE).

Mike Reed

While it is a moderately common resident of eucalypt forests and woodland in areas to the south and east of Darwin,[47,103] within the Darwin region the Grey Shrikethrush appears to be sparsely distributed and mainly on the outskirts.[47,96] It was recorded from 27% of cells, but 40% of records came from Howard Springs NR and Channel Island Road. During the Howard's Peninsula survey it was recorded six times and, surprisingly, only from vine forest and degraded Black Wattle forest.[308] Unlike the

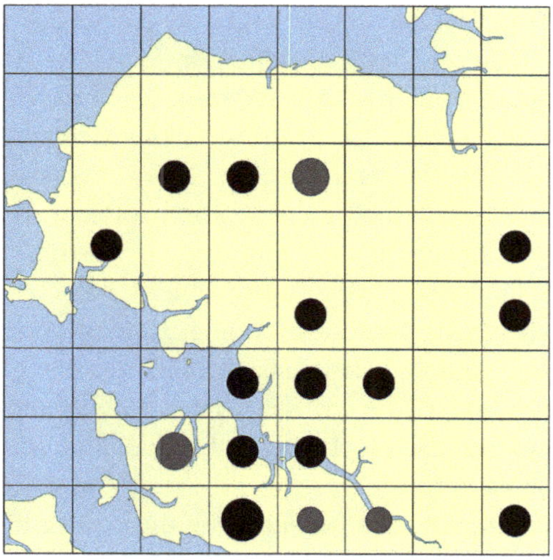

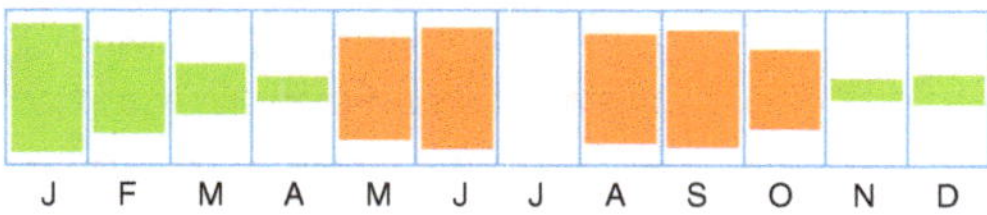

Little Shrikethrush, however, the species has no particular affinity for riparian vegetation. In a broad-scale analysis of Top End spatial data it showed a significant preference for woodland away from riparian areas, especially in lower rainfall regions.[450]

Grey Shrikethrushes apparently breed from September to February in the Top End,[41,47,103] though at Kapalga, in Kakadu NP, one nest was found in May.[43] Nests are usually built in tree hollows or on the top of stumps, but at Coomalie Farm a nest with two recently hatched young was located near the top of an ironwood sapling on 17 October 2006.[47]

Australasian Figbird
Sphecotheres vieilloti

RANGE: tropical and subtropical e. Australia, and s. NG. HABITATS: monsoon rainforests and suburbia. STATUS: very common resident. BREEDING: mainly Sep–Nov.

Niven McCrie

One of the most gregarious of all Australian songbirds, the Australasian Figbird not only forms flocks of up to 300 birds during the early and mid-Dry season in the Darwin region, but also nests in colonies of up to 20 pairs during the late Dry season. In Darwin itself, feeding flocks often congregate around fruiting Weeping Figs and Banyans,[459] when dozens of birds may be seen resting on nearby power lines between feeding bouts. So well adapted is the species to the urban environment that Darwin may support a higher density of birds than any natural coastal region of equivalent size in the Top End. Around Millner Figbirds were the second most abundant of all birds (after the Red-collared Lorikeet) during five years of monthly censuses, though their numbers varied dramatically over the year due to their high mobility.[296] In Palmerston, too, the Figbird was the third most abundant of all species counted during a Dry season survey, but older suburbs supported more birds than younger ones.[307]

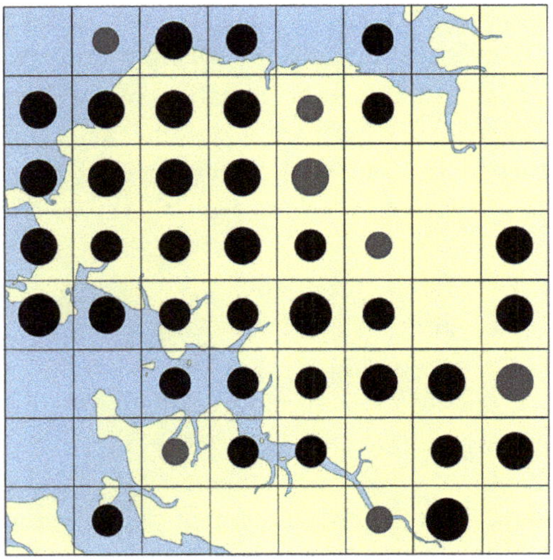

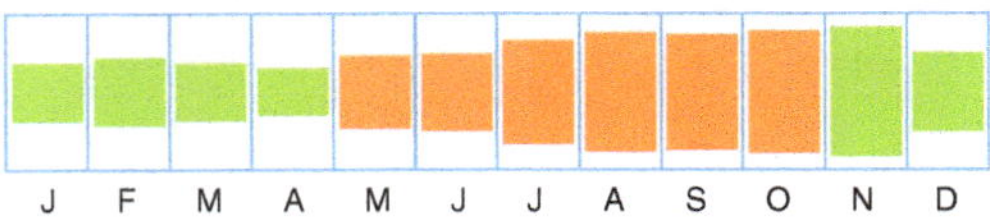

Outside suburbia the Figbird inhabits all types of rainforests[355] and riparian forests, but it appears to be less dependent on these habitats than other fruit-eating specialists.[316] Although the species eats the fruit of a wide range of rainforest plants in the Top End,[314] it has a particular penchant for figs and appears to seek them out across the landscape, even when they are 5 km from the nearest rainforest patch.[314,316] On Gunn Point peninsula, just outside the region, 14 birds fitted with radio-transmitters flew an average of 3.3 km between rainforest patches roughly every 5 days, the longest flight being 10.5 km.[315]

Breeding colonies can be found in most if not all of the older suburbs of Darwin, curiously often in locations where humans congregate. Examples are Raintree Park in the city, and Nightcliff and Parap malls where weekend markets see large crowds of people. Busy car parks, such as those at the Royal Darwin Hospital and at Hibiscus shopping centre, are often selected, as are school grounds, such as at Wanguri, and the edges of ovals, such as in Moil and Tiwi.[47] In such places large trees, both exotic and native, may support up to 10 nests, some as close as 3 m to each other. Surprisingly, the brightly coloured males incubate eggs and care for chicks almost as much as do their more sombrely plumaged, and hence more camouflaged, partners. When being fed, Figbird chicks not only give loud, chattering food-begging calls but also simultaneously shake their heads vigorously while gaping their bright red mouths. Two is the normal brood, and one chick often leaves the nest a day or two before the other, standing on the nest's rim or on adjacent branches, from which they sometimes fall to the ground, especially during stormy weather. Soon after the young have fledged, the colony members disperse to take advantage of the glut of rainforest fruits during the Wet season.[317] This may account for the lower reporting rates from December through May.

Within a colony, pairs build nests concurrently and even across many Darwin suburbs breeding is highly synchronised, with over 90% of 250 observed nests having eggs in September or October.[460] In such crowded conditions it is possible that some nests contain chicks that are fathered by males other than the brooding one. Indeed, females are guarded closely by their partners, though some have two male escorts. One female was seen mating with a male, only metres from its partner on the nest![47] Moreover, before and during breeding colony members sometimes gather in a large tree, where males sing continuously or chase one another, while watched by females,[47] reminiscent of the lekking behaviour of polygamous birds such as the birds-of-paradise. Like Figbirds, all members of the latter family eat fruits and are strikingly sexually dimorphic but, unlike Figbirds, their males take no part in nesting.[411]

Olive-backed Oriole
Oriolus sagittatus

RANGE: tropical and e. Australia, and s. NG. HABITATS: mainly eucalypt forests and woodland. STATUS: moderately common Dry season migrant. BREEDING: unknown (NT).

Bastiaan Hensen

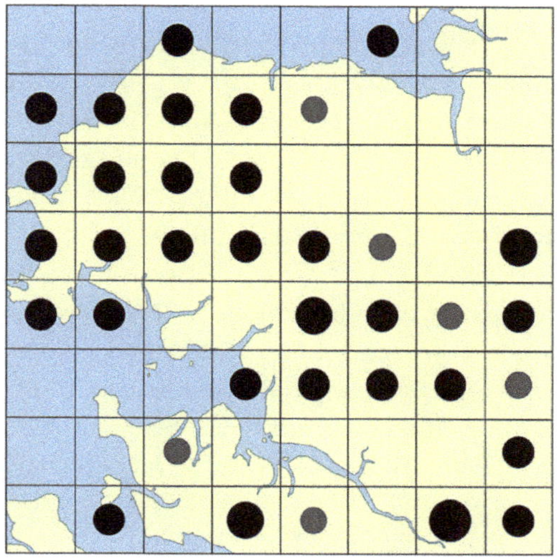

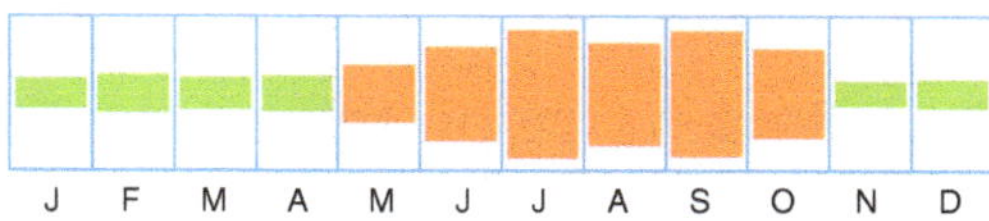

In contrast to the sedentary, fruit-eating Green Oriole, which is restricted to the tropics, the Olive-backed Oriole is a highly mobile, largely insectivorous bird whose range extends to southern Victoria. Populations in south-eastern Australia move north as far as Cape York and possibly New Guinea during early autumn, and return in late winter to breed in spring.[157,411] However, it is unlikely that these birds reach the Top End, where the local race *affinis* appears to exhibit more modest movements. In our region, as well as in Kakadu NP,[103] the species is scarce during the Wet season, but often seen from April or May through to October,[1] suggesting an influx from other regions. Conversely, at Coomalie Farm, only 45 km south of the Darwin region, it was recorded from October to February, and was apparently absent during the Dry. A Wet season influx was also reported for the Victoria River district.[451] These observations, though scant, suggest that subcoastal populations move to the coast during the Dry season and retreat inland during the Wet season,[41] when they probably breed.

Primarily a bird of eucalypt forests and woodland, and less frequently rainforests and riparian forests,[355] the Olive-backed Oriole probably eats fruit less frequently, and insects more frequently, than the Green Oriole. It also appears to take advantage of the Dry season surfeit of nectar in the woodland more than the latter species.[398] Both species are opportunistic carnivores, and Olive-backed Orioles have locally been observed taking eggs of a Peaceful Dove[47] as well as attempting to raid a Bar-breasted Honeyeater nest.[461]

Another clue to the origin of Darwin birds comes from their vocalisations. The species is a superb mimic of the calls of other bird species, individuals being heard to mimic up to 13 species in northern New South Wales.[411] In Nakara one adult bird regularly perched in a tall Black Wattle beside a main road, where it sang for long periods, especially during the middle of the day, for several months of the Dry season. Its vocal repertoire included calls of the Cockatiel, Masked Woodswallow, Magpie-lark, Grey-crowned Babbler, Yellow-throated Miner, White-bellied Cuckooshrike, Pied Butcherbird, Rainbow Bee-eater and Silver-crowned Friarbird, interspersed with the mellow 'ori-oriole' call of its own species.[47] As the Cockatiel and Masked Woodswallow rarely occur north of Katherine, especially in the Wet season, mimicry of their calls indicates that the Oriole was born some distance inland, where it had the opportunity to learn those calls.

There are few breeding records for the Top End, and none in recent times.[411] A large proportion of Top End birds possess either juvenile or immature plumage,[110] with greyish upperparts, rufous edges to the wing feathers, blackish bills and dark brown eyes, but this is not necessarily an indication of recent breeding, as individuals take about two years to attain adult plumage.[411]

Green Oriole
Oriolus flavocinctus

RANGE: tropical Australia and NG. HABITATS: mainly monsoon rainforest and riparian forest. STATUS: very common resident. BREEDING: Sep–Mar.

Niven McCrie

Given the lack of yellow in its plumage, and the fact that it belongs to a family in which 15 members are predominantly or partly bright yellow, the Green Oriole's traditional name of Yellow Oriole seems singularly inappropriate. Its song of three or four deep, resonating notes is one of the more characteristic sounds of the Australian tropics. Being a fruit-eater in the main, its primary native habitats are monsoon rainforests and vine thickets, which contain an abundance of fleshy-fruited plant species. Among the 10 most frequently encountered species during the Howard's Peninsula survey, it was recorded in eight habitats, though most birds were recorded

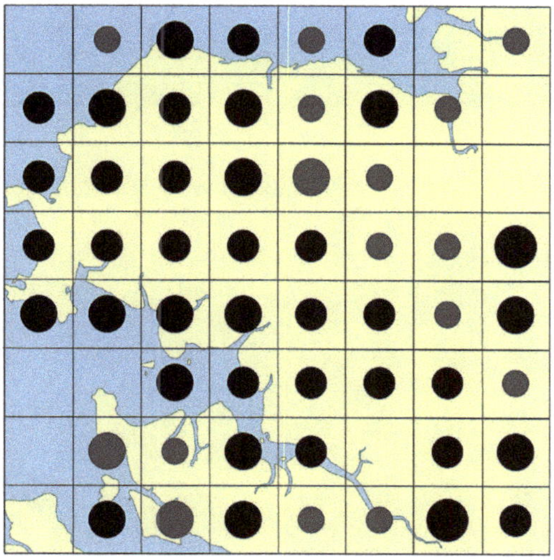

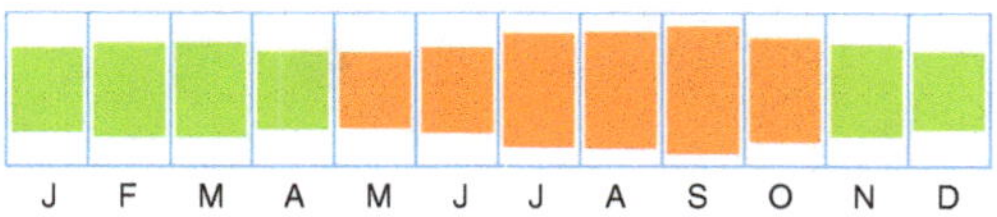

in vine thickets and monsoon rainforests, where it was the second most abundant species.[308] Indeed, during a study of fruit-eating birds in the Gunn Point region, just north-east of the Darwin region, the Green Oriole was the most abundant species in all but one of 10 patches of rainforest.[316] It was also the only one among the six species studied that regularly inhabited riparian rainforests.

The species is also well adapted to the urban environment, partly attracted by fruit trees such as pawpaw and bananas, and is especially common in older suburbs with well-established gardens.[60] In Palmerston, it was among the 15 most frequently recorded species in the old suburbs of Driver and Moulden during a bird survey in 2007, but was scarce or absent from suburbs built after 1990. In a survey of mangals around Darwin, the species was recorded in all but one of 13 patches and, surprisingly, was most abundant at those patches far from vine thickets.[313] As mangals are largely devoid of fleshy-fruited plants, the birds appear to visit this habitat for insect prey. Green Orioles also occasionally feed on nectar,[398] and there are several reports of them taking nestlings of small birds such as the Rufous-banded Honeyeater and Double-barred Finch.[366] Indeed, during an experiment using artificial nests and eggs under camera surveillance in the mangals at Nightcliff, a Green Oriole was filmed robbing the plasticine eggs![462] As part of a local study of movements of fruit-eating birds, six Green Orioles were radio-tracked but, unlike the Figbird which undertook long flights between rainforest patches, the maximum distance flown by an oriole was 850 m.[315] One adult male captured in Nightcliff in June 2008 was found dead 18 months later in the same suburb, apparently a victim of a pet cat.[47] These observations suggest that the bird is relatively sedentary.

The local breeding season of the species lasts from September to March, with peaks in October and February.[47] The bark of paperbark trees forms the bulk of its large cup-shaped nest, which is suspended from a fork in a horizontal branch of a tall tree. Exotic tree species, such as the African Mahogany, Rain Tree, Coffee Bush and mango, are often used. One nest was built on a low branch about 1 m above a parked car and, despite the noise and activity associated with the construction of a shed nearby, succeeded in raising one young to fledging.

Spangled Drongo
Dicrurus bracteatus

RANGE: tropical and e. Australia, NG, Moluccas and Philippines. HABITATS: all wooded habitats. STATUS: common resident, with seasonal movements. BREEDING: Sep–Dec, but mainly Oct.

Niven McCrie

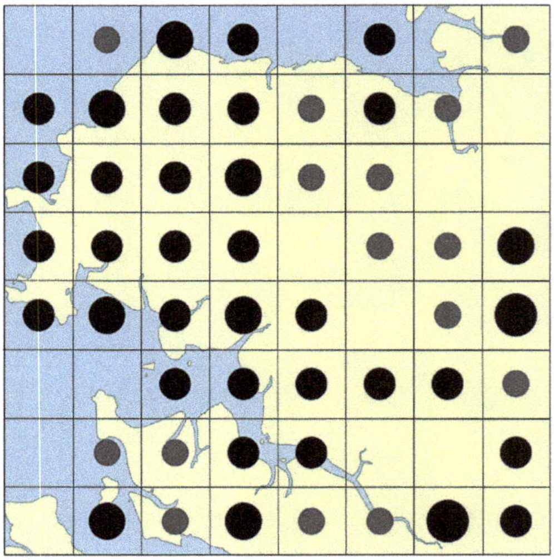

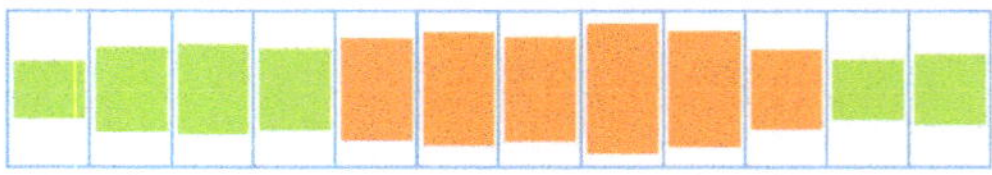

Described variously as raucous, rasping, metallic and brassy, the discordant calls of the Spangled Drongo reflect its audacious character. Like other members of the largely tropical family to which it belongs, it sometimes chases and bullies other birds in an attempt to steal their food.[411] The Top End and Kimberley race (*baileyi*) is isolated from those of eastern Australia, and while the latter migrate north as far as New Guinea in the cooler months, local birds are present year-round.[1,103] However, the sudden appearance of these birds, sometimes in small flocks, at sites where they are not normally present suggests seasonal movements. Reporting rates are lowest in the early Wet season immediately after the species breeds, but gradually increase over the remainder of the Wet season and much of the Dry, before dropping again. At Darwin Golf Course, an individual was present for six weeks of June and July 2007, but the species was not recorded in these months during the preceding year and in only two of the 44 weeks of the intervening months.[309]

The Drongo was recorded in 78% of cells, equally frequently in coastal and subcoastal cells, and significantly more in urban than forest or rural cells. During the Howard's Peninsula survey it was found in eight of the 20 habitats sampled, but was most abundant in monsoon rainforests, and moderately common in Black Wattle forests and eucalypt forest with a dense understorey.[308] Indeed, it was the sixth most frequently recorded bird species in an extensive survey of rainforests in Kakadu NP, present in five of the six rainforest types sampled.[355] In the Darwin region it also frequently visits mangals, being found in 12 out of 13 mangal patches surveyed every month for 14 months.[313]

The breeding season of the Spangled Drongo may be the shortest of any bird species in the Top End.[412] Over 90% of clutches are laid in October, and the remainder in the months immediately preceding and following this month.[47] The shallow basket-like nest is made mostly from vine tendrils, and in some cases is so flimsy that its contents can be seen from below. Both sexes build the nest, incubate the eggs and feed the young. Although insects are the main food item, usually captured while on the wing, the Spangled Drongo often visits the flowers of the Darwin Woollybutt and Fern-leaved Grevillea for nectar during the early Dry season. Indeed, of 208 observations in the Top End of opportunistic nectar-feeding by species other than honeyeaters and lorikeets, about 10% involved this species.[398]

Willie Wagtail
Rhipidura leucophrys

RANGE: whole of Australia, NG and e. Indonesia. HABITATS: eucalypt forest and woodland. STATUS: uncommon Dry season visitor. BREEDING: Sep–Jan (TE).

Kirri Hardy

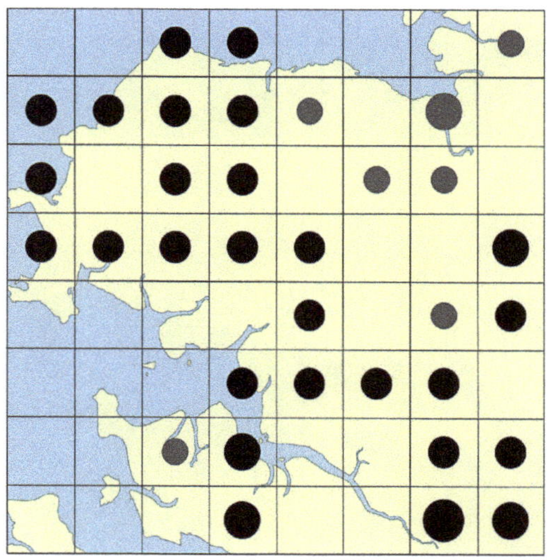

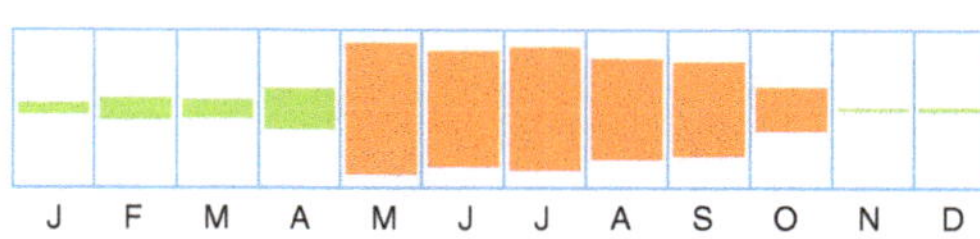

Although well adapted to urban settings in southern Australia, and fairly common in rural areas of the Darwin region, the Willie Wagtail is, curiously, largely absent from the suburbs of Darwin. Recorded in 54% of all cells, it was observed significantly more in subcoastal and rural-dominated cells than in coastal and urban or forest-dominated cells, respectively. During the Howard's Peninsula survey, it was recorded in six of the 20 habitats, but was scarce in all except cyclone-damaged eucalypt woodland.[308] It appears to be commoner in Kakadu NP, where it was recorded on almost all sites and in all habitats sampled during the Kakadu survey, but was most abundant at woodland sites.[103]

Despite being described in the late 1960s and early 1970s as a common resident of the Darwin–Humpty Doo region,[1] reporting rates show that the species is largely a Dry season visitor to Darwin, with only 10% of records falling in the Wet season. This is consistent with observations from Kakadu NP,[1] as well as other Top End coastal localities,[143] where the species was found to be more abundant and widespread in the Dry season than in the Wet. Around Darwin Golf Course, one to five individuals were seen each week from June to October 2006, but none were observed over the following nine months.[309] However, at Coomalie Farm, 45 km south of the Darwin region, Willie Wagtails appear to be sedentary,[47] so birds visiting the Darwin region may well emanate from much further south. Indeed, at Yinberrie Hills, 50 km north-west of Katherine, they were almost twice as abundant in December as in June, with highest densities in April–May and August–October,[311] suggesting a passage movement at that latitude.

Reporting rates indicate that most birds arrive in Darwin during May, but after July there is a gradual decline in records, possibly as birds move inland again. No breeding has been reported in our region, but elsewhere in the Top End the species lays its eggs from September to January.[41,47,197]

Northern Fantail
Rhipidura rufiventris

RANGE: tropical Australia, NG and e. Indonesia. HABITATS: all wooded habitats. STATUS: very common resident. BREEDING: May–Jan, peaking Oct–Nov.

Niven McCrie

Of the two resident species of fantails occurring in the region, only the Northern Fantail is likely to be encountered in eucalypt forest and woodland, though it is most abundant in monsoon rainforests and vine thickets.[308] It also occurs in paperbark forest and mangals, where it is much more abundant than the Mangrove Fantail. Indeed, during a study of mangal birds in 2008–09 this species was recorded in all 13 patches surveyed, from Buffalo Creek in the north to East Arm in the south.[313] The highest density of birds occurred at the smallest mangal patch, near Mindil Beach, while the second

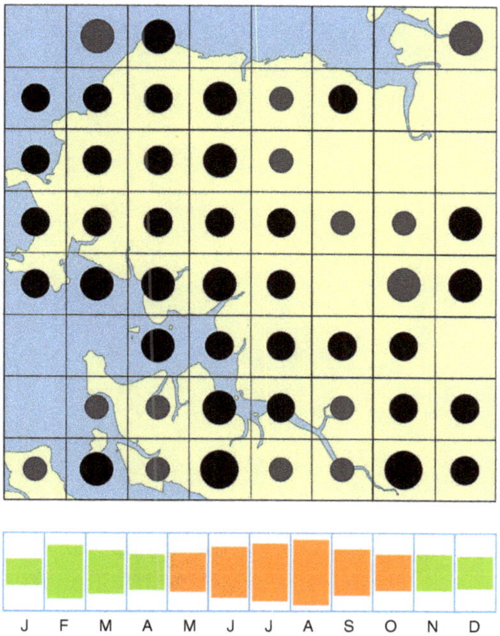

highest density was at East Arm, which was the largest patch. It also occasionally ventures into gardens of older suburbs of Darwin, especially in areas adjacent to mangals or vine thickets.

Larger and more sedate than the other fantails, it is much less acrobatic when foraging, and its posture is more upright as it perches, often for long periods, before darting into the air after aerial insects.[411] Despite its broad tropical distribution, the Northern Fantail remains poorly known, especially in terms of its territory size and breeding habits. In the Darwin region, breeding records span every month from May to January, with a dramatic peak in egg-laying during the Dry–Wet transition months of October–November.[412] The August peak in reporting rates is probably due to increased calls and territorial defence activity just before breeding, but the secondary peak in February is difficult to interpret.

As with other fantails, the nest of the Northern is shaped like a wine glass without the base. The eggs are rarely left uncovered as both adults incubate, swapping every 7 to 15 minutes.[47] In the mangals at Nightcliff, observations revealed that at least four nests failed due to flooding by high spring tides, while another two nests were parasitised by the Brush Cuckoo.[47]

Grey Fantail
Rhipidura albiscapa

RANGE: whole of Australia. HABITATS: all wooded habitats. STATUS: rare Dry season migrant. BREEDING: not in NT.

In stark contrast to its status in many other parts of Australia, the Grey Fantail is a rare and poorly known Dry season migrant to the Top End.[411] Most birds migrating to the Top End are assumed to belong to the subspecies *alisteri*, which breeds in south-east Australia. However, historical records suggest that the subspecies *preissi* and *albicauda* from south-western Western Australia and southern NT, respectively, might also reach the Top End.[41,411] Although there are only six records for the Darwin region, the species was recorded on seven of the 100 sites sampled during the Howard's Peninsula survey, conducted in April and early May 1986.[308] Of the seven

Kirri Hardy

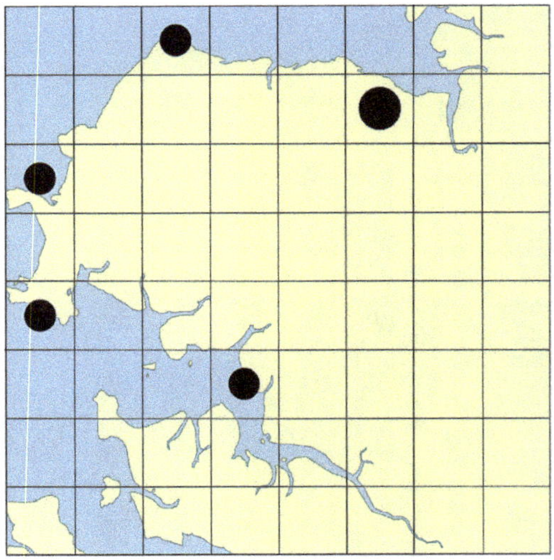

sightings, three were made in monsoon rainforest, two in paperbark forest, and one each in vine thicket and Black Wattle forest. Similarly, during the three-year Kakadu survey the species was recorded six times, five of which were in monsoon rainforest and one in woodland, but all sites were adjacent to the South Alligator River,[103] supporting evidence that it prefers riparian habitats.[450] Two of these records fell in the Wet season, when the species would be expected to be absent from the Top End.

Mangrove Fantail
Rhipidura phasiana

RANGE: coastal monsoon-tropical Australia, and se. NG. HABITATS: mangal. STATUS: uncommon resident. BREEDING: Jan–Mar.

Belinda Rafton

First described in 1884, the Mangrove Fantail was treated as a subspecies of the much more widely distributed Grey Fantail for the next century.[411] However, the Mangrove Fantail differs from the Grey Fantail not only in its smaller size, dark grey 'necklace', pale buff underparts and pinkish-yellow bill base, but also in its distinctive call, which is a twittering descending whistle that starts at an almost imperceptibly high pitch. Restricted to mangals, it is generally uncommon and, despite its name, is much less abundant in this habitat than the Northern Fantail. Indeed, among the 12 local mangal-dependent species, the Mangrove Fantail has the second lowest average density after the White-breasted Whistler.[312,313] This is partly due to its

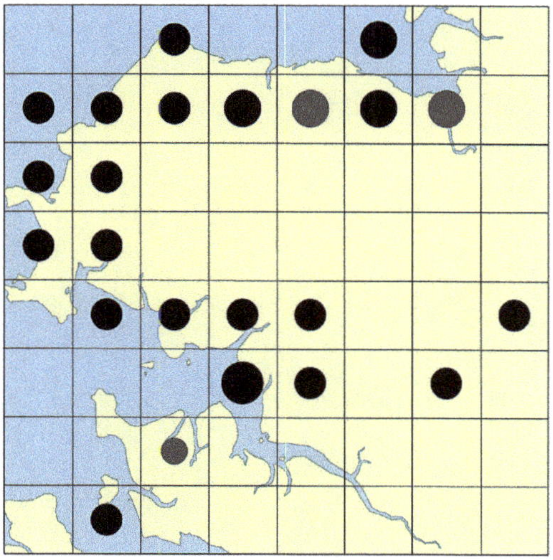

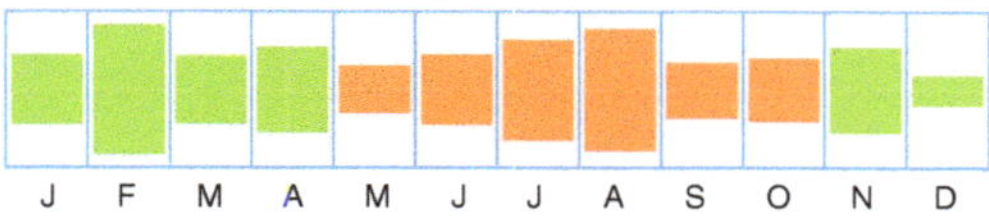

preference for the wetter, more frequently inundated zones of the mangal, where it often forages among the aerial roots of Stilt Mangroves beneath the dense canopy of mangroves surrounding tidal creeks.[312]

During a survey of mangal birds in the Darwin region, the species was found in five of 13 patches surveyed.[313] The site with the highest density was Ludmilla Creek, where several pairs appear to occupy linear territories.[47] They were not, however, in the mangal at Nightcliff, possibly due to the absence of tidal creeks. During the Howard's Peninsula survey they were recorded several times in mangal dominated by Stilt Mangroves, but not in landward mangal dominated by Spurred Mangrove.[308] Not surprisingly they were recorded significantly more from coastal cells in the Darwin region than from subcoastal ones.

Apart from an alleged record of breeding in January in 1903,[41] no nests had been found in the Top End until one containing young was found in mid-February 1999 and, 10 years later, another in which the eggs had been laid during the last two weeks of March 2009.[47] Both nests were in the Ludmilla Creek estuary. A juvenile bird reported from Sandy Creek, Casuarina CR, in early December 2010 was also possibly this species.[320]

Arafura Fantail
Rhipidura dryas

RANGE: monsoon-tropical Australia, and e. Indonesia. HABITATS: monsoon rainforests, vine thickets and mangal. STATUS: uncommon, largely Dry season visitor. BREEDING: Jul–Nov (TE).

Originally described in 1843,[411] the Arafura Fantail was treated as a subspecies of the much more widespread Rufous Fantail until 2006. However, it differs morphologically from the latter species in being smaller, and having more white but less rufous on the tail. In addition, while the Rufous Fantail of eastern Australia is a long-distance migrant, breeding in the south-east and wintering as far north as New Guinea,[411] the Arafura Fantail is thought to be a resident in at least some parts of its range, which extends from the tip of Cape York west to the Kimberley, and includes eastern Indonesia.[86] In the Top End it is generally uncommon and confined to monsoon rainforests, mangals and riparian forests.[41,103,355]

Laurie Ross

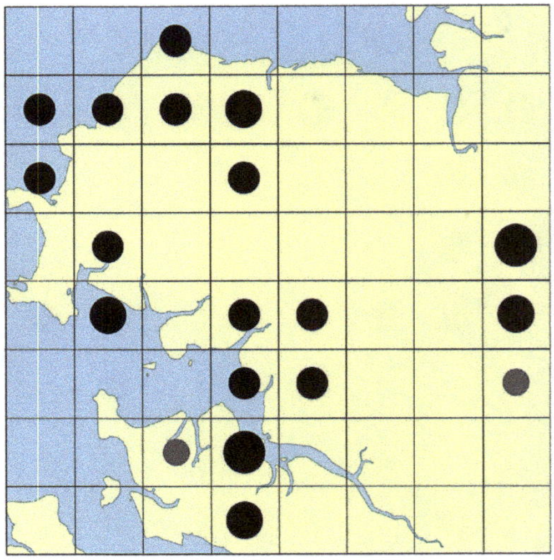

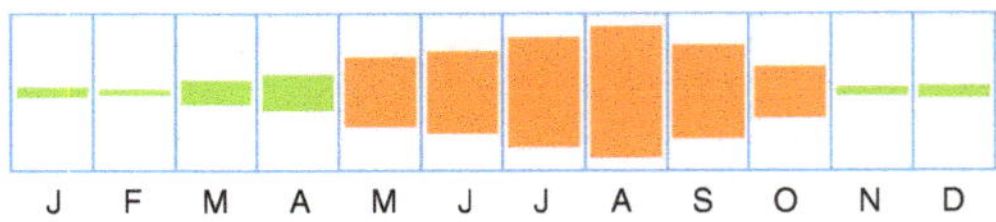

During the Howard's Peninsula survey in April and May 1986 it was recorded 12 times in over 1000 censuses, mostly in semi-deciduous vine thicket, with only a few records in semi-evergreen rainforest and mangal.[308] Within mangals this species is much less abundant than the Northern Fantail. During a bird survey of mangals in the Darwin region, it was found in only four of 13 sites surveyed, being most abundant in Charles Darwin NP, where both the Mangrove Fantail and Northern Fantail were also found, but in lower densities.[313] At each of the other three sites where Arafura Fantails were present, Northern Fantails were more abundant and Mangrove Fantails were absent. Reporting rates show that the species largely vacates our region for much of the Wet season, from late November to March. It is also absent from western Kakadu NP during the Wet season.[103] These observations contradict a claim that Top End birds move inland during the Dry season,[41] an unlikely scenario for any insectivorous bird.

Very little is known of the breeding season and biology of this species. No nests have been found in Western Australia,[187,411] and only four nests have been reported from the Top End: one in July, two in September and one in mid-November.[361,411] A juvenile was observed in November on Groote Eylandt, suggesting nesting before or during October.[82] Thus the purported breeding season overlaps the period when birds are present in the Darwin region, yet no nests have ever been found here. As its eastern counterpart, the Rufous Fantail, often nests near water, it is feasible that local birds retire to large floodplains east of the Darwin region during the Wet season.

Magpie-lark
Grallina cyanoleuca

RANGE: whole of Australian mainland. HABITATS: grassland and woodland. STATUS: very common Dry season migrant and resident. BREEDING: Aug–Apr.

Niven McCrie

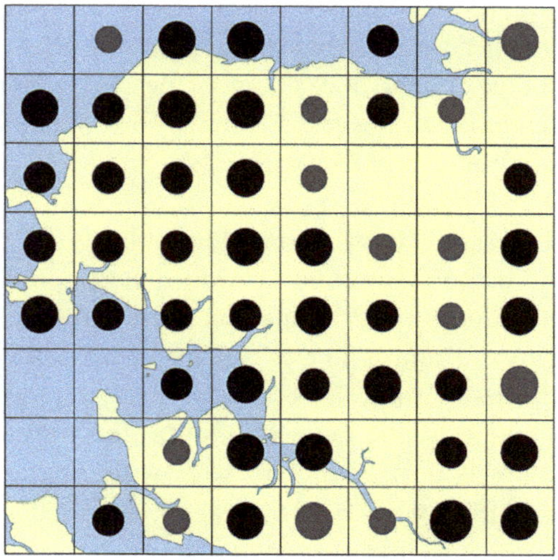

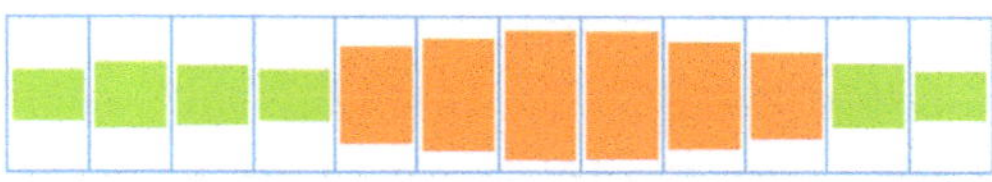

There are few better examples of a partially migratory bird species in Darwin than the Magpie-lark, whose relatively small sedentary population is augmented by a much larger one in the Dry season. However, the ratio of these two populations may be changing. During the late 1960s and early 1970s it was rarely recorded during the Wet season,[1] and by 1978 was still regarded as a non-breeding visitor to the 'coastal Darwin area'.[365] Even in 1987, it was considered to be largely absent from Darwin during the Wet season, though possibly resident only a few kilometres inland.[60] Indeed, during almost daily surveys around Millner from 1988 to 1993, the species was rarely recorded between November and April, but in May several birds arrived and numbers increased until August.[296] Mean monthly numbers tripled over those six years, indicating that there may have also been an increase in the population visiting Darwin.

Breeding by Magpie-larks was first noticed in Darwin in 1988, but over the next two decades nests were found in many suburbs, usually in trees bordering ovals and school grounds, but also in trees in the city and above marketplaces, the birds apparently unconcerned by weekend crowds.[47] Reporting rates also confirm a well-established resident population and suggest that the migrant population arrives quite suddenly in May, but departs much more gradually. The two populations may be difficult to distinguish. Small groups of six to 10 birds foraging on lawns and ovals may represent aggregations of two or three resident families, as young of this species stay with their parents for two to three months.[411] Flocks of migrants are typically composed of larger numbers, which can be easily counted when they roost on power lines along major arterial roads, such as McMillans Road. On 25 May 2008, for instance, four groups of 20–60 were counted on power lines beside the Stuart Highway near Palmerston at 06:45 hours just before they moved off to forage.[47] Large flocks numbering 150 to 250 have been observed in all Dry season months except October.[21]

The annual Dry season influx of birds is by no means unique to Darwin, being noted in Kakadu NP[103] and western Arnhem Land[49,163] and as far away as Groote Eylandt[82] in the NT, as well as the Kimberley region to the west and Cape York to the east.[411] Yet there is no evidence of large-scale migration from south-east Australia,[157] where the species is largely sedentary, except for non-breeding flocks which are thought to wander locally. Thus despite the wide-spread nature of the influx to the Top End, its source remains a mystery.

In the Darwin region, the species has been recorded in 81% of cells. During surveys of Palmerston from May to September 2007 it was the fifth most abundant species, but while common in suburbs built in the 1980s and 2000s, it was strangely absent from those of intermediate age.[307] During the Howard's Peninsula survey, it was recorded on 10% of survey sites and in seven habitats, but was most abundant in cyclone-damaged eucalypt woodland and riparian forest.[308]

Although local birds have been recorded breeding from August to April, the peak months for egg-laying in Darwin, as in populations in southern Australia, are September and October. Of 71 clutches, nearly half were probably laid in these two months alone.[47] Nests are often built in the introduced African Mahogany and Rain Tree, often in association with other medium-sized species of songbird. In the car park of Hibiscus shopping centre, one such pair shared an isolated Yellow Flame Tree with six pairs of Figbirds and a pair of Helmeted Friarbirds, and in a Nightcliff park one nest was situated only 3 m from that of a Little Friarbird. Such mixed-species breeding colonies probably benefit from improved defence of nests from nest robbers, and indeed Magpie-larks often mob predatory birds, even outside the breeding season.[21,411]

Leaden Flycatcher
Myiagra rubecula

RANGE: tropical and e. Australian mainland, and s. NG. HABITATS: all wooded habitats. STATUS: common Dry season visitor and scarce resident. BREEDING: Sep–Jan (TE).

While Leaden Flycatchers are long-distance migrants in much of eastern Australia, vacating the south during autumn to spend winter in the warmer climes of north Queensland and New Guinea,[411] those in the tropical parts of Australia appear to be resident or to undertake short-distance movements only.

Reporting rates in the Darwin region show a marked seasonality in records, being lowest at the start of the Wet season, and

Ashley Carlson

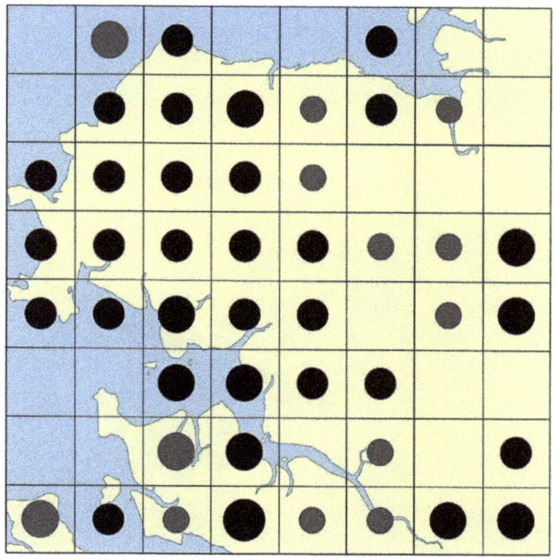

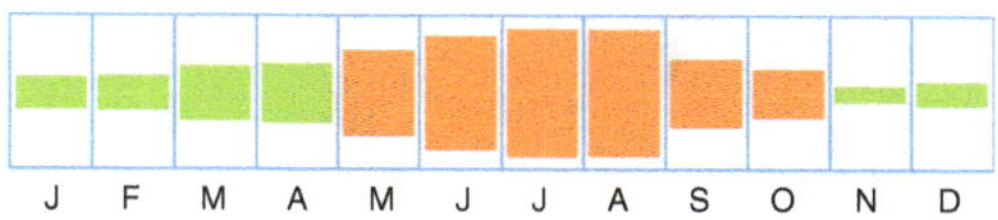

gradually increasing to a peak in the mid-Dry season. In Kakadu NP, too, the species was more numerous and widespread during the Dry season than in the Wet.[103] Subcoastal populations may be more sedentary. At Coomalie Farm, for instance, they were recorded in eucalypt forest in every month, but their occurrence in other habitats was sporadic, presumably due to local movements.[47] During the Howard's Peninsula survey this species was encountered in six habitats, but was most abundant in eucalypt forest with a dense shrubby understorey and low mangal.[308] In a survey of birds of mangals in the Darwin region, Leaden Flycatchers were recorded in nine out of 13 patches, the highest density being found at Casuarina CR.

Breeding records for the Top End suggest that egg-laying starts in September and finishes in January, with a peak in October-November.[47] In late October 2008 a pair was watched as it vigorously defended its nest against a potentially nest-robbing Green Oriole in a small patch of eucalypt forest at the Casuarina campus of Charles Darwin University.

Broad-billed Flycatcher
Myiagra ruficollis

RANGE: coastal monsoon-tropical Australia, NG and e. Indonesia. HABITATS: mangal. STATUS: common resident. BREEDING: Mar–Nov.

Niven McCrie

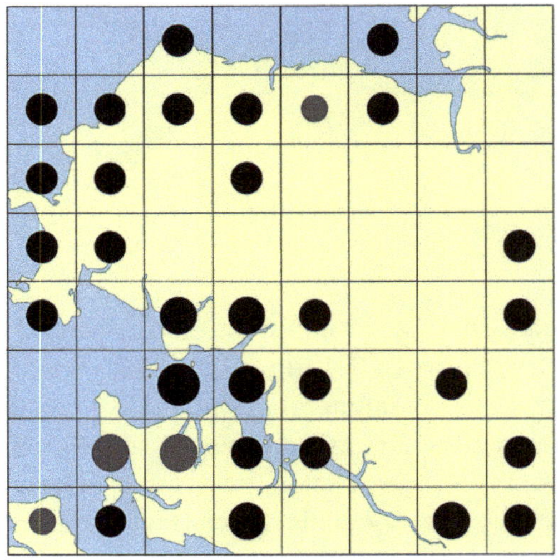

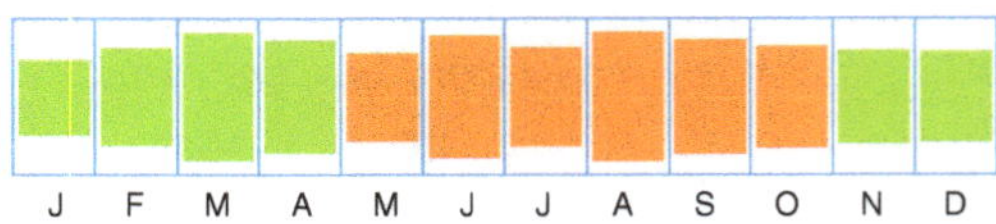

Given its virtual confinement within Australia to coastal mangals of the monsoon-tropics, it is not surprising that the Broad-billed Flycatcher is poorly known compared to the widespread Leaden Flycatcher, with which it often shares its habitat.[47] While adult male Broad-bills are readily distinguished by their dark, glossy blue-grey upperparts, the duller females closely resemble the Leaden Flycatcher, but differ in having white or greyish-white rather than mid-grey lores (the feathers between the bill and eye) and conspicuous white sides to their tails.[411] Contrary to some sources, however, the amount and intensity of the rufous-orange on the throat and breast appear to vary as much between individuals as between the two species.[47] Both species make buzzing notes and two-note whistles, with the second note higher or lower than the first, but the Broad-bill's calls include a repeated three-note whistle, with the third note lower than the second, and its notes are often down-slurred.[47]

In a survey of birds of mangals in the Darwin region, Broad-billed Flycatchers were recorded in all but the smallest one of 13 patches, the highest density being found at Nightcliff.[313] During an intensive study of the mangal near Palmerston SP this species was encountered in half of the 54 months sampled.[312] In both studies it showed a distinct affinity for the frequently inundated Stilt Mangrove zone. Outside mangals the species also occurs in monsoon rainforests[355] and paperbark forest.[1,96]

Very little is published on the breeding biology of the Broad-billed Flycatcher[411] but observations in Darwin show that, like the Leaden Flycatcher, both sexes build the relatively large cup-shaped nest and incubate the eggs.[47] Here, the species breeds biannually. The first period of egg-laying occurs from March to May, while the second, more intense, egg-laying period occurs mostly in September and October. A study at Nightcliff from 2006 to 2010 revealed that three to four pairs occupied an 800 m stretch of mangal, each pair occupying a territory of about 1 ha.[47] Some of their nests failed due to flooding by high spring tides, and one brood of two chicks was taken by a Black Butcherbird.[447] Two colour-banded birds resided there for at least three years, while at Ludmilla Creek another bird was recaptured at the same site where it had been captured and banded eight years previously.[47]

Shining Flycatcher
Myiagra alecto

RANGE: tropical Australia, NG and e. Indonesia. HABITATS: mangal, monsoon rainforest and riparian forest. STATUS: common resident. BREEDING: Sep–Apr.

Niven McCrie

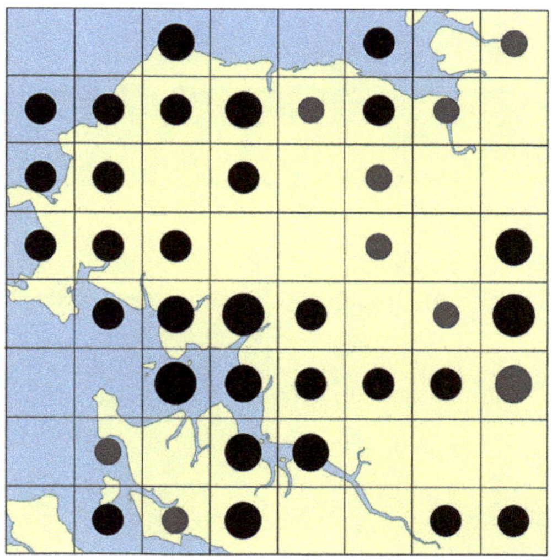

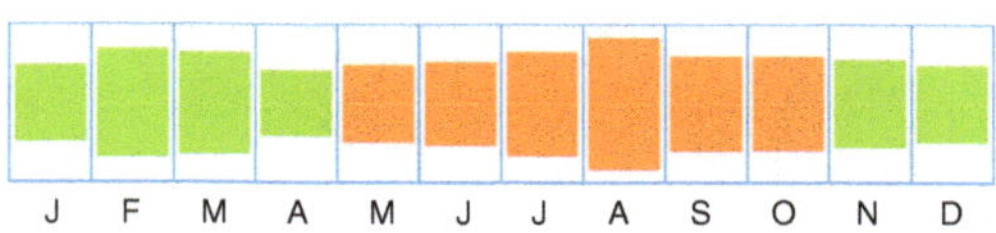

Compared to the Leaden and Broad-billed Flycatchers, the Shining Flycatcher is large (18 g), and unusual in that the female is arguably the more beautiful of the sexes. It is also unusual in being common in both mangal and monsoon rainforest,[1,103] but also occurs in vine thickets and riparian habitats. During the Howard's Peninsula survey, it was most abundant in rainforest,[308] and in Kakadu NP it was found in four of six rainforest types.[355] During a bird survey of mangals in the Darwin region, Shining Flycatchers were recorded in all 13 patches sampled, with the highest density on Ludmilla Creek.[313] It was also among the 10 commonest bird species in a three-year study of mangal-dwelling birds near Palmerston SP.[312] In both studies, it was found to show a strong preference for the frequently inundated Stilt Mangrove zone, where it spends most of its foraging time within 2 m of, or even on, the muddy ground.[312,463]

This species often indulges in impressive displays, most involving loud calls, crest raising, tail arching and gaping of the bill to expose the bright salmon-pink palate.[411] In Kakadu NP one aggressive interaction between two males consisted of high-speed circuit chases alternating with stationary displays, in which each bird raised its head and body while at right angles to its opponent, and emitted a whirring 'creeek' call.[464] In courtship displays males stretch the neck and tilt the head back to a sharp angle, spread their tail and wings, and alternately raise and lower the body.[411]

The breeding biology of the Shining Flycatcher is poorly known, but breeding records around Darwin show that it nests from September to April, with a major peak in October (41% of records) and a gradual decline in activity over the Wet season.[47] The deep cup-shaped nest is invariably built over water, such as along a creek or on the edge of a billabong. Nests built too low in mangroves are sometimes flooded by high spring tides, which drown the young or carry the eggs away. Young birds resemble the female, and when begging for food give a high-pitched rapid chatter quite unlike the adults. In the mangal at Nightcliff one female was recaptured at the same site where she was captured and banded 7.5 years previously, while a colour-banded male at Palmerston SP was at least six years old when last seen.[47]

Paperbark Flycatcher
Myiagra nana

RANGE: endemic to monsoon-tropical Australia and sc. NG. HABITATS: mainly riparian forest. STATUS: moderately common resident. BREEDING: Sep–Dec (TE).

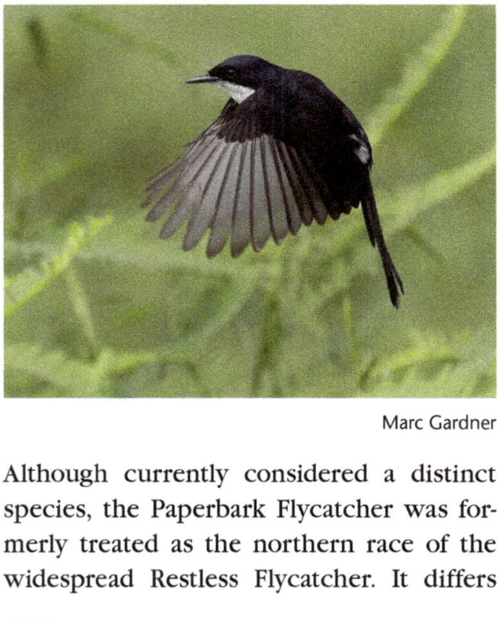

Marc Gardner

Although currently considered a distinct species, the Paperbark Flycatcher was formerly treated as the northern race of the widespread Restless Flycatcher. It differs from the latter species in having a much smaller body and slightly broader bill, and in lacking the characteristic 'scissors-grinding' call, but its behaviour is otherwise much the same.[411] While it is usually described as a resident in the Top End,[369,411] reporting rates in the Darwin region show marked seasonality, most records occurring in the Dry season, with a noticeable peak in July. In both Kakadu NP and the Victoria River Downs area, this species appeared to concentrate in riparian habitats, such as paperbark swamps, during the Dry season, then move to eucalypt forest and woodland for the Wet season.[98,103] The Dry season peak in reporting rates may thus be due to birds moving into well-watered areas, and

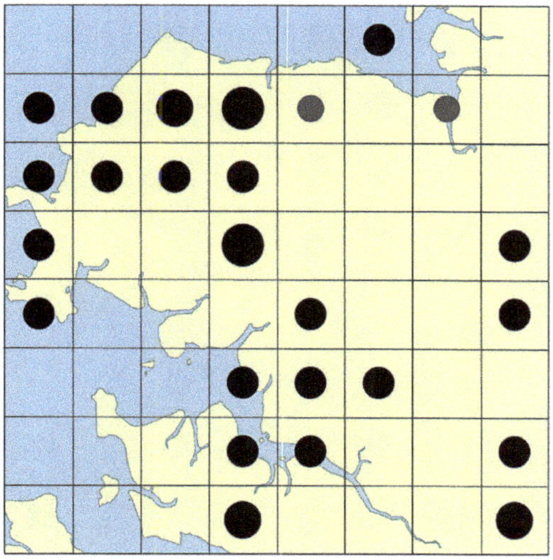

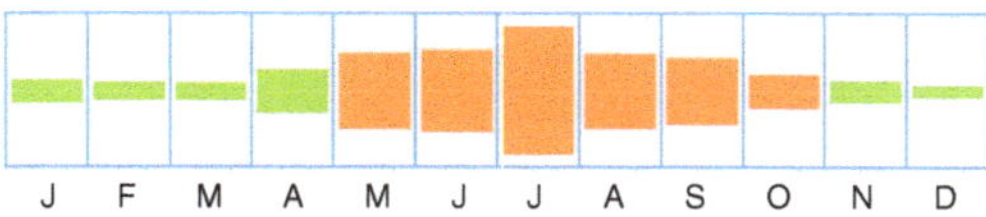

indeed almost two-thirds of all records came from Knuckey Lagoons and Leanyer SP. Surprisingly, it was recorded only once during the Howard's Peninsula survey in April–May 1986.[308]

Apart from paperbark forests, the species often frequents the stands of Freshwater Mangroves or pandanus that fringe many floodplains and billabongs. In Kakadu NP it was one of six bird species that showed a strong affinity for pure stands of pandanus.[388] It is also sometimes seen on the edges of ovals in suburbs on the outskirts of Darwin,[47] such as Leanyer, possibly because of its proximity to the floodplain adjacent to Holmes Jungle. One individual was present around Darwin Golf Course from late April to the end of July 2007.[309] Single birds were also resident in the mangal at Nightcliff for at least five months from April to September 2005, and from September to November 2009.[47]

Only one nest has been reported from the Darwin region, but elsewhere in the Top End egg-laying appears to occur mainly from September to December,[47] though there are reports of nestlings and fledglings as late as February and March.[411]

Torresian Crow
Corvus orru

RANGE: tropical and subtropical Australia, NG and far e. Indonesia. HABITATS: mainly woodland and rubbish tips. STATUS: scarce resident. BREEDING: Jan–May (TE).

Well known as a scavenger in cities and towns across the Top End,[1,103] the Torresian Crow is surprisingly scarce in the Darwin region.[365] Its attraction to rubbish tips has been noted in several Top End regions, such as South Goulburn Island,[452] Gove[365] and Groote Eylandt.[82] Indeed, 29% of records in the Darwin region emanate from the vicinity of Shoal Bay Waste Management Facility, where up to 50 birds have been reported at times. However, the avian scavenger community at this site is dominated, especially during the Dry season, by Black Kites, Whistling Kites and Pied Herons, which together comprise up to 1000 birds, sometimes more.[47] As all three species are very efficient at seizing edible morsels from the refuse, even as it is being unloaded from the dump trucks, it is possible that they out-compete the Crows.

Although the Torresian Crow has been reported from 73% of the cells in our

Niven McCrie

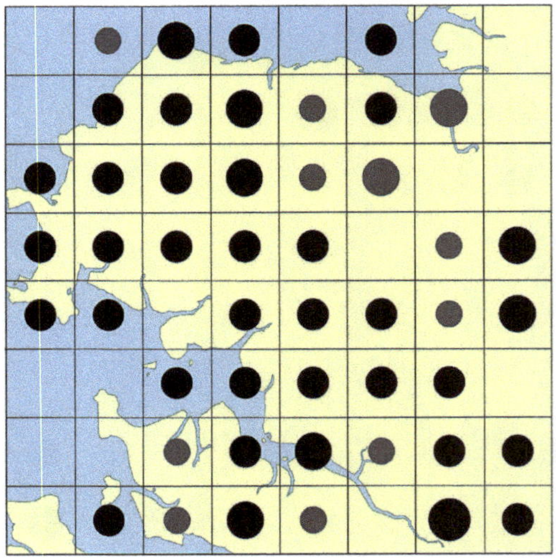

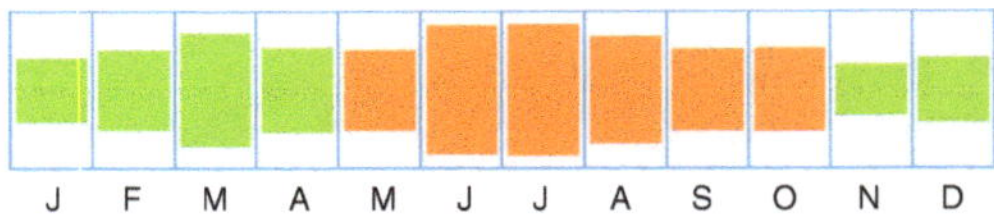

region, sightings away from the waste disposal site are infrequent. Around Darwin Golf Course one or two birds were recorded in almost a third of the weekly censuses over 12 months in 2006–07, with no obvious seasonality, while in Palmerston it was recorded only three times during weekly surveys throughout the Dry season of 2007.[307] Only three birds were sighted during the six weeks of the Howard's Peninsula survey.[308] In the late 1960s and early 1970s it was suggested that the species is commoner in the Dry season than in the Wet.[1,365] This is partly supported by the reporting rates, which rise to a peak in mid-June, then drop to minimal values in November and December. As Crows are often observed foraging in burnt areas immediately behind a fire front[95] they may be attracted to fires during the Dry season.

While breeding purportedly occurs over nine months of the year in the NT,[411] observations of six nests in the Darwin-Fogg Dam region indicate that clutches are mainly laid from January to May.[47] One nest was built on a tall power-line tower on the Channel Island road, remote from urban areas, while another was built in a non-native palm in the grounds of Wagaman School.[21] The species is also the only known local host of the Channel-billed Cuckoo, and there are several reports of Crows chasing and mobbing these Cuckoos.

The learning ability and memory of crows is well known. One of the few reported cases of this species caching food took place at Edith Falls, about 200 km south-east of the Darwin region, where a bird was observed collecting pieces of bread, burying them under a small pile of twigs and soil, and retrieving them the following day.[465] Torresian Crows have also discovered a way to eat the potentially deadly, introduced Cane Toad with no ill effects, by rolling the Toads onto their backs and consuming the intestines and thighs, thus avoiding the poison glands on the head.[466] Such behaviour was observed in three separate regions of the Top End during 2002.[466] Considering that this predation technique was learned within two years of the Toad's invasion, it is likely that the behaviour was culturally transmitted among Crows.

Buff-sided Robin

Poecilodryas cerviniventris

RANGE: endemic to monsoon-tropical Australia. HABITATS: riparian forest. STATUS: vagrant, but localised resident to s. BREEDING: Sep–Nov, possibly to Mar (TE).

Micha V. Jackson

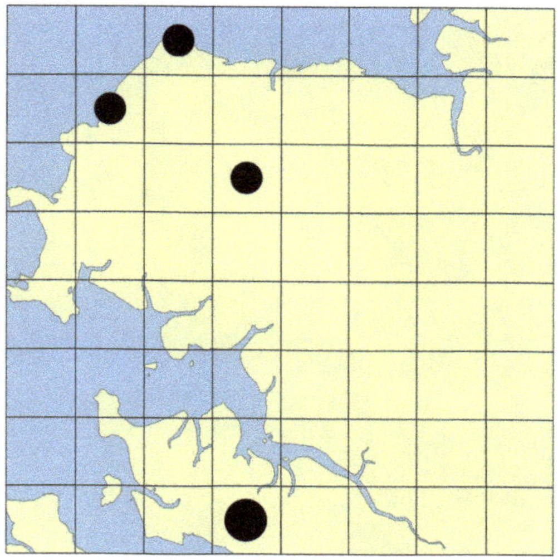

Until recently considered a race of the White-browed Robin of northern Queensland, the Buff-sided Robin has been recorded only rarely in the Darwin region, but has resident populations less than 40 km away.[300] Although the typical habitat of this species is riparian vegetation, invariably including paperbarks and pandanus, and often bamboo, it was first reported in our region in 1964 when a bird was sighted in the monsoon rainforest at Holmes Jungle.[96] The species was not recorded in the region again until December 2006, when an individual was found on the landward edge of the Rapid Creek mangal adjacent to the Casuarina campus of Charles Darwin University. In August 2008, over 1.5 years later, another one (possibly the same individual) was discovered in vine thickets at Lee Point, where it stayed for at least eight weeks before disappearing. Potential sources areas for these vagrants include Manton Dam, 29 km south of the Darwin region, where at least one pair resides along the creek leading from the dam wall. The species also occurs at Marrakai Crossing, on the Adelaide River, and suitable riparian habitat appears to extend for at least 20 km downstream towards the Adelaide River bridge. Six breeding records, including four from over a century ago, indicate clutches are laid mainly from September through to November, though the species allegedly breeds in the Top End until March.[411]

Mangrove Robin
Peneoenanthe pulverulenta

RANGE: tropical Australia, NG, and adjacent islands. HABITATS: mangal. STATUS: common resident. BREEDING: Sep–May.

Glenn Newton

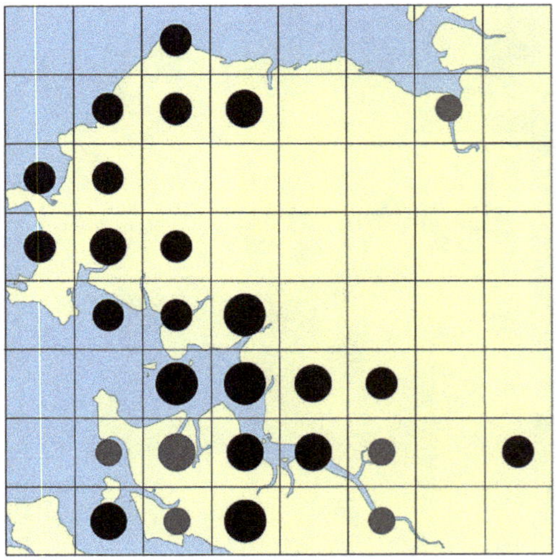

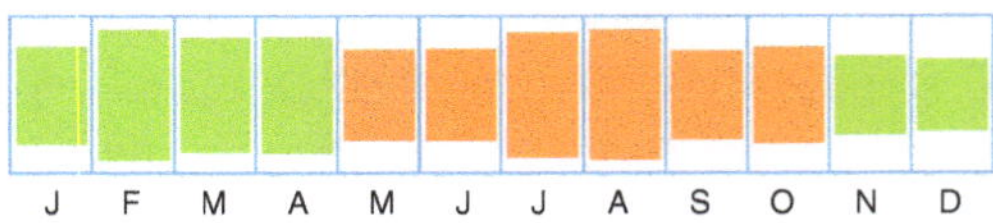

The first sign of the secretive Mangrove Robin is usually its mournful two- or three-note whistle. It is strictly confined to mangals, where it typically hops slowly on the muddy ground as it hunts for small crabs,[187] or pounces on them from an aerial root. On very high tides, however, when such substrates are covered by water, it gleans ants and other insects from the branches of mangroves.[312,463] In the Darwin region it is common in large tracts of mangal with well-developed zonation, such as those around the Elizabeth River, where it prefers dense thickets of Spurred and Rib-fruited Mangroves often close to the landward edge.[312,313] Along Darwin's shoreline, however, it is patchily distributed, and was recorded in only six of the 13 mangal patches sampled during an intensive survey in 2008-09.[313] On Ludmilla Creek, for example, it occurs only on the lower reaches, where dense stands of low Spurred Mangrove grow.[47]

Mangrove Robins were reported from 42% of cells in the region and, not unexpectedly, significantly more of the cells were coastal than subcoastal. East of Buffalo Creek they may be more scarce as none were recorded in any of the 10 mangal sites sampled during the Howard's Peninsula survey.[308] On a 4 ha mangal site near Palmerston SP, 25 Mangrove Robins were banded over 10 years,[312] while a similar banding effort over 8-9 years on Ludmilla Creek and Rapid Creek produced only two Robins.[47] Five colour-banded juveniles at the Palmerston site stayed in their presumed natal territory for up to eight months, indicating that the species may breed cooperatively, young birds helping their parents to raise later broods.[47] Although some robin species are known to be long-lived,[411] Mangrove Robins at this site were not particularly long-lived, the oldest colour-banded individual being only 2.3 years old when last recaptured at its banding location.

Almost nothing is known of the breeding biology of this tropical species, few nests having been reported in the last 100 years. However, local records of fledglings over the last three decades suggest that the breeding season is long, spanning from September to May, with possible egg-laying peaks in October and February. Data on moult indicate a biannual pattern, with some birds moulting their wing feathers in both March–April and August–October.[47]

Lemon-bellied Flyrobin
Microeca flavigaster

RANGE: tropical Australia and NG. HABITATS: all wooded habitats, favouring mangals and paper-bark forests. STATUS: very common resident. BREEDING: biannual, peaking Mar–Apr and Aug–Oct.

Niven McCrie

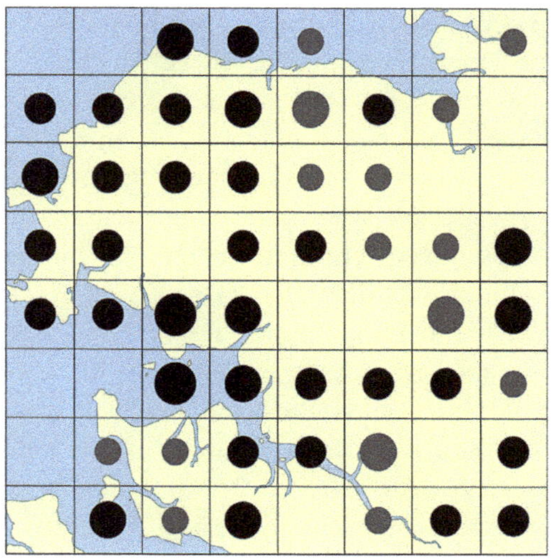

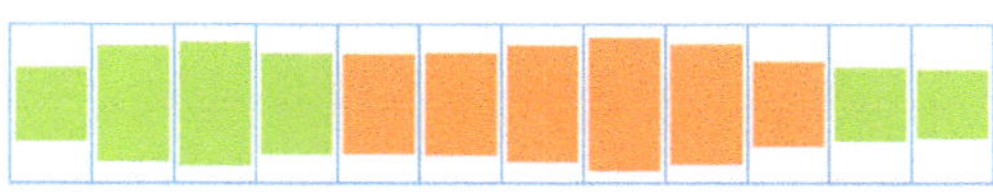

With a song flight display that has been compared to that of the celebrated Eurasian Skylark, a pre-dawn song that finishes before most birds even start their dawn calls, and the ability to mimic the calls of more than a dozen species, the Lemon-bellied Flyrobin is surely one of Australia's finest songbirds. Yet until recent studies in the Darwin suburb of Nightcliff its vocal prowess went practically unnoticed, and its unusual breeding behaviour was unknown.[467] Song flights are performed by males before and during nesting, and begin with the bird launching into the air and rising slowly to 30–40 m. It then flies in tight circles for 1–3 minutes, all the while singing, before descending to a favourite perch at the top of a prominent tree. The pre-dawn song starts up to one hour before sunrise and continues for about 15 minutes, consisting of up to 250 song phrases, repeated at intervals of 4–5 seconds. Moreover, individuals can mimic the calls of at least 14 species, including Beach Stone-curlews, Figbirds and Mistletoebirds. Yet this rich and varied vocal repertoire belongs to the males alone, females having a simple and invariant song. Indeed, as males and females of the species look identical, their voices are the only means by which the sexes can be distinguished.[467]

In the Darwin region Lemon-bellied Flyrobins were recorded in 75% of cells, significantly more of which were coastal than subcoastal. This species was recorded in seven of the 20 habitats sampled during the Howard's Peninsula survey, including monsoon rainforest, vine thicket and mangal, but was commonest in paperbark forest, where it was one of the most abundant species, with an average density of almost three birds per hectare.[308] During an intensive 14-month survey of mangals, it was found in all 13 patches sampled, but was most abundant at Mindil and Nightcliff, the two smallest patches.[313] It also occurs around many suburban ovals and parks of Darwin, but is not normally a backyard bird. In the Star Mangrove-dominated mangal along the southern foreshore of Nightcliff, seven territories of Lemon-bellied Flyrobins averaged 4 ha in area and were mostly occupied by pairs, but some by trios.

Among Australian birds, the Lemon-bellied Flyrobin has the double distinction of being the smallest species with a single egg clutch, and of having the smallest nest, which is a shallow open cup about 3.5 cm wide. The study at Nightcliff revealed that the large egg is incubated by the female alone, yet she spends less time on the nest than possibly any other Australian songbird – around 30% of the day, and typically only 3 minutes at a sitting! She also flees at the first sight of danger, relying on the minuteness and camouflage of the nest and its contents to avert discovery. The egg takes a full 18 days to hatch and, though both sexes feed it, the chick spends a whopping 18–20 days on the nest, by which time the latter is barely visible.[467] The youngster is fed for at least another seven weeks, and stays with its parents for three to six months after fledging. It sometimes assists its parents in their next breeding attempt by feeding its sibling. This behaviour, known as cooperative breeding, was unknown in this species until the study at Nightcliff. Colour-banding also showed that these birds can live for at least 13 years.[47] Like the mangal-dwelling gerygones and Canary White-eye, the Lemon-bellied Flyrobin is a biannual breeder in Darwin, eggs being laid in both March–April and August–October.[412,440]

Horsfield's Bush Lark
Mirafra javanica

RANGE: tropical and e. Australia, NG and Indonesia. HABITATS: grassland. STATUS: moderately common resident. BREEDING: Nov–Mar.

Bastiaan Hensen

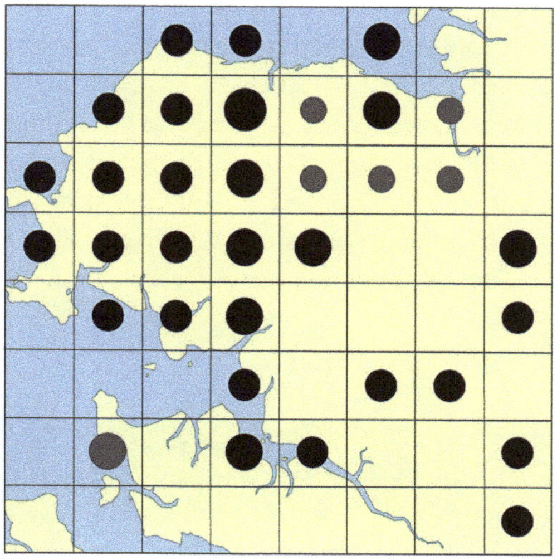

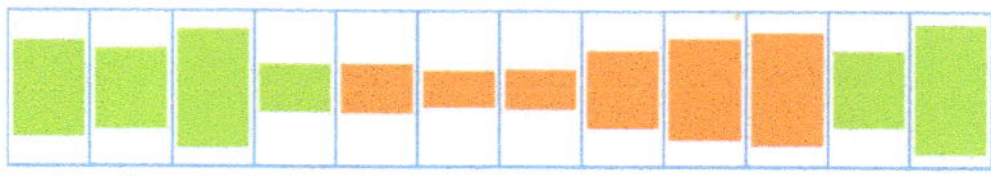

The only true lark in Australia, the Horsfield's Bush Lark is not, however, the only Australian bird that indulges in impressive song flights. Male cisticolas, songlarks, Tawny Grassbirds, White-winged Trillers and Lemon-bellied Flyrobins also sing while performing display flights to advertise their territory or attract females, though perhaps none match the Bush Lark in tonality and duration and, with the exception of the Flyrobin, in their ability to mimic the calls of other species. Before the start of the breeding season, which coincides with the Wet, males begin their displays, ascending into the sky on quivering wings until almost out of sight, then hovering while singing for 5 minutes or more, before dropping to the ground.[411] They also sing from perches, such as bushes and fence wires, and may even be heard during moonlit nights.[60] During the non-breeding season, however, they are quiet and inconspicuous, typically encountered only when flushed from cover. This partly explains why reporting rates are much lower in the early Dry season than in the Wet season.

Like the cisticolas and Tawny Grassbird, the Bush Lark is a bird of open grassland, but unlike those species generally avoids moist or flooded areas during the Wet season. It also appears to avoid suburban lawns and ovals where the grass is short and dense, but is common on the sparsely grassed areas of Darwin International Airport, where up to a dozen scattered individuals have been counted during surveys.[47] Other cleared areas where the species has been regularly noted are the RAAF aerial farm near Lee Point, Fannie Bay racecourse and Berrimah Farm.[47,60] It has been recorded in 53% of cells, significantly more of which were subcoastal than coastal. Over half of all records were from the Knuckey Lagoons and Leanyer Swamp area.

The two known nesting records of the species from the Darwin region were of eggs in late November 1970 and early March 1996,[1,415] which accords with its purported breeding season of November to early April for the whole of the Top End.[411] Thus the drop in reporting rates coincides with the end of the breeding season, suggesting a cessation of singing and possibly post-breeding dispersal. At Darwin International Airport few birds were seen during May 2011 compared to September and November 2010.[47] On the Barkly Tableland the species was also found to be more abundant in the latter half of the Dry than in the early half. That similar trends are apparent over widely separated regions suggests that the species does not undertake any large-scale movements. During the Dry season at Victoria River Downs, some birds were observed to form flocks that moved locally, while others remained on their breeding grounds.[98]

Barn Swallow
Hirundo rustica

RANGE: breeds in n. hemisphere and migrates to s. hemisphere; e. Asian race *gutturalis* migrates to Philippines, Indonesia, NG, tropical Australia. HABITATS: aerial. STATUS: rare Wet season visitor. BREEDING: not in Australia.

Dr Chan Ah Lak

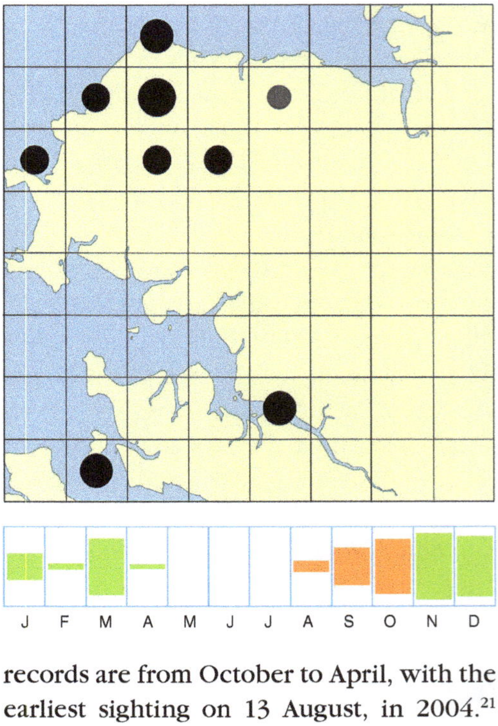

The cosmopolitan Barn Swallow was first recorded in Darwin in October 1971[1] and is now known to be a nearly annual visitor to the region, with records in 26 of the years from 1977 to 2014. There are June records from Leanyer SP in 2003[3] and Lake Bennett, 40 km south of Darwin,[21] in 1990, but most records are from October to April, with the earliest sighting on 13 August, in 2004.[21]

Though most frequently seen individually or in small numbers, there are occasionally larger influxes, such as at least 15 birds at Lee Point in December 2003 and 20 or more at Leanyer SP in December 2004.[21]

Records are widespread, as may be expected for an aerial species, including Casuarina CR, Lee Point, Knuckey Lagoons, Dinah Beach, Nightcliff and Palmerston SP, but the majority are from Leanyer SP, where birds feed over the ponds and tracks, frequently in loose association with Tree Martins.

Welcome Swallow
Hirundo neoxena

RANGE: e. and s. Australia, largely excluding very arid areas. HABITATS: aerial. STATUS: vagrant. BREEDING: not in NT.

Julie Sarna

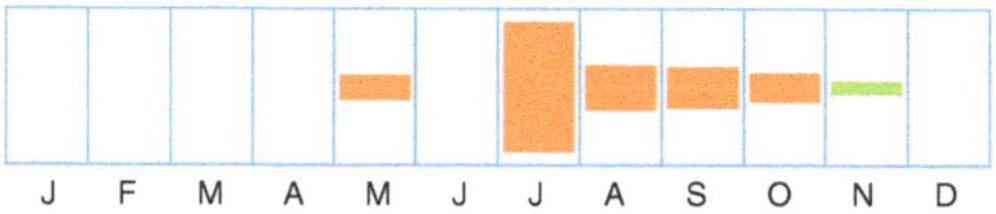

The Welcome Swallow is an occasional and erratic Dry season visitor to the Darwin region, and all records have been of three or fewer birds. The species has been confirmed in only eight of the years from 1978,[34] when it was first reported, until 2013. There have since been observations from the Darwin Botanic Gardens and suburban Parap, but most records have been from Leanyer SP, where the most recent sighting was in June 2013.[118]

Red-rumped Swallow
Cecropis daurica

RANGE: breeds in Eurasia and migrates s. to Africa and Asia, and se. to Lesser Sundas. HABITATS: aerial. STATUS: vagrant. BREEDING: not in Australia.

Ashley Carlson

Records of the Red-rumped Swallow from the Darwin region are all from Leanyer SP, all of single birds, and have all occurred in the late Wet season, from February to April. The timing of sightings suggests birds on return migration to south-east Asia, but in the absence of corresponding observations either at the time of southward migration or during the overwintering period, no conclusions can be drawn as to how long, or where, the birds occur in the Top End.

The first Darwin region record involved a bird seen intermittently from 20 February until 1 March 2003, sometimes feeding over the ponds, and at other times perched on power lines with up to 100 Tree Martins.[21] On 11 March 2005,[21,40] a Red-rumped Swallow was observed hawking insects over the sewage ponds, among Tree Martins and at least one Barn Swallow. The bird's off-white, slightly buff underparts, marked rather indistinctly with dark longitudinal streaks, confirmed identification as Red-rumped rather than as the somewhat similar Striated Swallow, which has not yet been recorded in Australia. Subsequent records were in mid-March 2006, 23 February and 1 March 2008, and 18 April 2010.[468]

Fairy Martin
Petrochelidon ariel

RANGE: whole of Australian mainland. HABITATS: aerial, mostly over wetlands. STATUS: uncommon Dry season visitor. BREEDING: Sep–Dec (TE).

Although the Fairy Martin breeds in southern parts of the Top End, it is neither as regular nor as numerous as the Tree Martin in the Darwin region. Indeed, there are no records for some years, including 1982, 1988-89, 1992-95 and 2003-04. During 1967-71, the first date it was observed in the Darwin-Humpty Doo regions varied from 29 March to 28 May, but the final day of records was during the last week of June in all years except 1969 when it was recorded as late as 11 October.[1] Reporting rates since then confirm that it is rarely recorded before March, but the major drop

Ashley Carlson

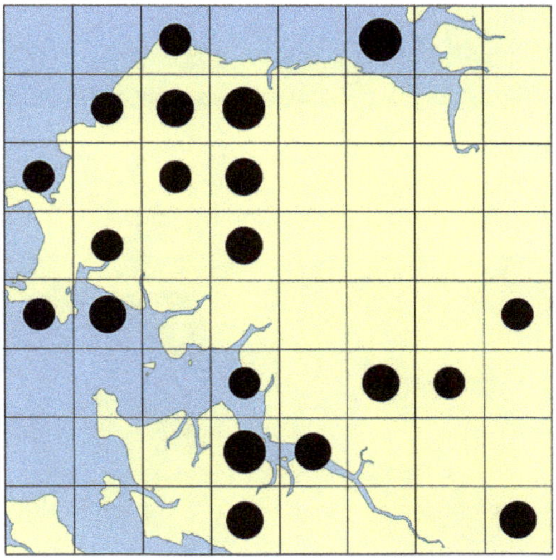

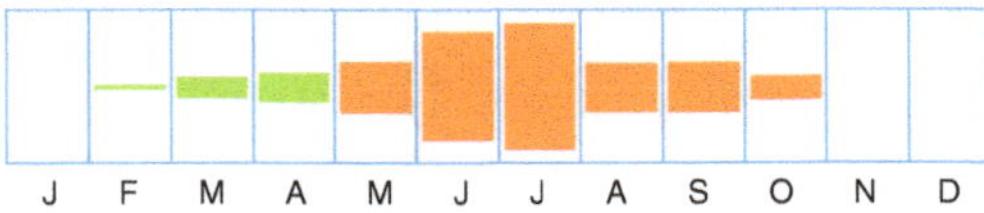

in records occurs in August. However, monthly reporting rates mask considerable annual variation, and in some years the species was first recorded no earlier than June, such as in 2002, or even September, as in 1998. The largest number counted was 50 birds at Knuckey Lagoons on 17 August 2002.[21]

During the Howard's Peninsula survey, in April–May 1986, it was recorded in the air above 22% of sites, half as many as the Tree Martin, and the latter was as much as six times more abundant in most shared sites.[308] The disparity in seasonality evident in the charts of the two species suggests that their origins are separate, though it has been suggested that Fairy Martins appear simply as a result of becoming 'caught up' in flocks of Tree Martins[1] as mixed flocks of these two species are well known.[411]

In eastern Australia both species are strongly migratory, moving north from breeding grounds in autumn, though they do not appear to reach the northern coastline, and there is no evidence of movement towards the Top End.[157] Fairy Martins are known to breed in Katherine Gorge (Nitmiluk NP),[1,60] Jasper Gorge[98] and Keep River NP, where they are present throughout the year, and considered more common than the Tree Martin.[369] As this species also breeds in many localities in the Kimberley region,[411] it seems more likely that visitors to Darwin originate from the west than the east.

Tree Martin
Petrochelidon nigricans

RANGE: whole of Australia, NG and e. Indonesia. HABITATS: aerial over all habitats. STATUS: common largely Dry season migrant. BREEDING: none in Top End.

Ashley Carlson

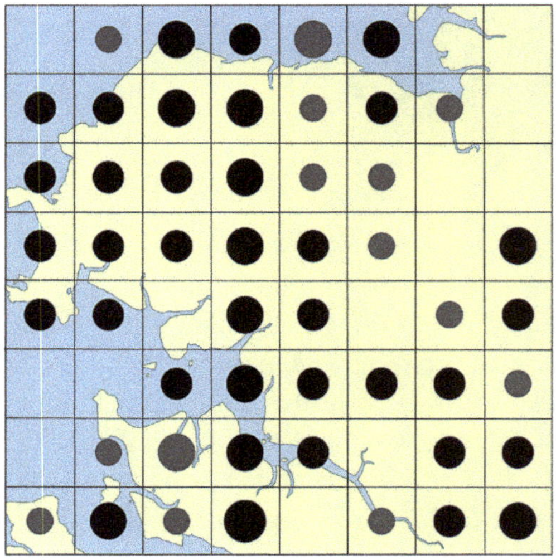

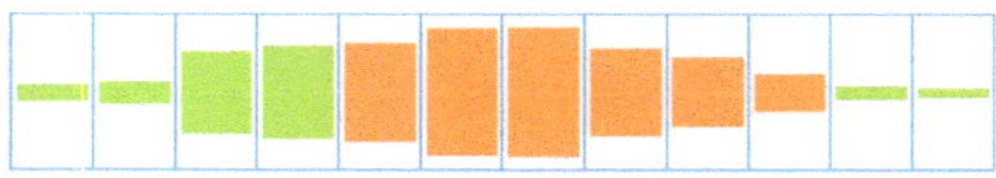

One of the few Australian birds that migrate across the Timor Sea, the Tree Martin is a regular 'winter' visitor to Timor[86] and the Moluccas, as well as the Top End. Yet like most migratory birds visiting the Darwin region during the Dry season, their place of origin remains a mystery. Unlike the Fairy Martin, this species does not breed in the NT, except around Alice Springs and to the south.[4,5] Analysis of Atlas data suggested that Tree Martins breeding in south-east Australia migrate directly north, but there was no evidence that these populations move north-west.[157] This suggests that birds visiting the Top End originate from other regions of southern Australia, such as South Australia or even south-west Western Australia. Tree Martins breeding in the south-west, for example, start leaving as early as January, arriving in coastal parts of the Kimberley region in late March,[411] with some birds staying until October, much the same as those visiting the Top End. Indeed, of the many species that migrate to the Darwin region for the Dry season, this is the earliest to arrive, with reporting rates increasing sharply after February.

While there are records for the species through the early and mid-Wet season (November–February) in most years, the size of flocks during this period rarely reaches 100. From March to September, however, flocks sometimes number a thousand or more. On 1 August 1994, between 07:00 and 07:50 hours, an estimated 10 000 birds, based on a passage rate of about 200 birds per minute, were observed flying in an easterly direction over a site near Palmerston.[47] In April–May 1986, this species was the second most widespread species recorded during the Howard's Peninsula survey, observed screening the air over 46 out of 101 sites representing all 20 surveyed habitats.[308] Monthly counts at Leanyer SP over four years (1998–2001) showed that flocks of 100 or more birds usually arrived by mid-March and all birds had gone by mid-October, except in 1999 when a small number was seen in late November.[21] From 2002 to 2004, the species was recorded at several sites as late as December.[21]

Oriental Reed Warbler
Acrocephalus orientalis

RANGE: breeds e. Asia, migrating as far s. as Indonesia, NG. HABITATS: wetlands. STATUS: Wet season vagrant. BREEDING: not in Australia.

The Oriental Reed Warbler was first recorded in Australia in 1912,[41] when a bird was collected on Melville Island, north of Darwin. There have since been records from the Barkly Tableland[26,55] and northern Western Australia,[256] as well as from the Top End, and all are from the Wet season, when the species might be expected to 'overwinter' in Australia. The first record for the Darwin region was in March 1979, when a bird was observed and heard calling at Leanyer SP.[469] The species was recorded there again, identified on the basis of its call, in March 1982. The last Darwin record, also identified by call, appears to have been in January 1983, when a bird was present near the Shoal Bay Waste Management Facility.[469]

Outside the Darwin region, two specimens, one in each of January and February, were collected from Harrison Dam, near Fogg Dam, in 1982.[469] One was later observed at Fogg Dam in March, and two in April, 1994.[21,26] The following year there were reports there of two birds in mid-March, and one or two in mid-November. A bird was seen and photographed at Fogg Dam on 16 December 2012[263] and, although daily visits for the two weeks following failed to relocate it, an Oriental Reed Warbler, possibly the same individual, was again seen at the site on 11 February 2013 and was present there until at least mid-March 2013. Another was found at the same locality during December 2013.[263]

The close similarity between this species and Australian Reed Warbler makes confirmation of reports somewhat difficult, and it may be that some claims of Oriental Reed Warbler are misidentifications. Nevertheless, the Oriental Reed Warbler may be more a rare visitor rather than a vagrant to the Top End.

Australian Reed Warbler
Acrocephalus australis

RANGE: e. and w. Australia excluding much of tropics and arid regions. HABITATS: reedbeds in wetlands or along watercourses. STATUS: scarce resident and Dry season visitor. BREEDING: Jan (TE).

Kirri Hardy

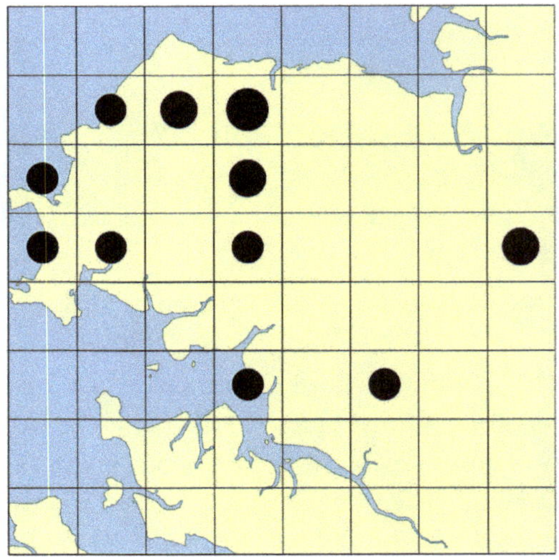

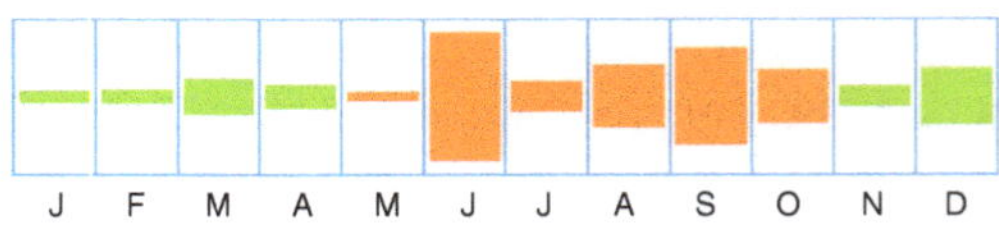

Arguably the least known of the resident understorey-dwelling warblers in the Top End because of its skulking habits, this species has a restricted and patchy distribution from Darwin to western Arnhem Land.[411] During 1967–71 it was recorded each year from June to January at Fogg and Harrison Dams, but appeared to vacate the region from February to May.[1] Indeed, 85% of all records for the Darwin region since then pertain to these seven months, but the majority fall in June and September. There were few reports of the species from March to May, but the species has been observed at Fogg Dam in mid-April in both 1999 and 2006.[21] Almost two-thirds of all local records came from Knuckey Lagoons and Leanyer SP.

This species is largely migratory in south-east Australia, departing from its breeding areas during February–April and returning in August or September.[411] Its pattern of movement is similar to that of the Black-faced Cuckooshrike, White-winged Triller and White-breasted Woodswallow,[157] which are well-known Dry season visitors to the Darwin region. However, migratory populations of these three species depart well before the Reed Warbler 'disappears'. Moreover, this species is invariably detected by its loud melodious calls, and singing is much less frequent outside the breeding season, which in southern Australia lasts from September to January. As the latter period overlaps the months in which they are recorded in the Top End, it is possible that Darwin birds also breed locally. The only confirmed breeding record for the Top End pertains to a clutch found on the Daly River on 1 February 1902, presumably laid in January,[41,411] but in the Kimberley region of Western Australia, where the species is considered a resident,[321] it is said to breed in 'autumn'.[359]

As Top End birds may be heard calling from June to December it is likely that the local population is largely resident, and that its apparent absence during the late Wet season is due to lower calling rates and, possibly, local movements out of flooded habitats after breeding. It is also plausible that this population is augmented by migrants from southern latitudes, as suggested by the 'spikes' in reporting rates, but additional data are required to test such hypotheses.

Rufous Songlark
Megalurus mathewsi

RANGE: whole of Australian mainland. HABITATS: grassy woodland and floodplain edges. STATUS: rare Dry season visitor. BREEDING: unknown in tropics.

Mat and Cathy Gilfedder

Regarded as one of the most exuberant songsters among the bush-birds of southern Australia, the Rufous Songlark has an altogether different demeanour in the Top End. Throughout the spring, when the species breeds in southern Australia, males are almost continuously vocal, pouring out their tuneful song whether perched or while ascending into the air during their song flights.[411] Yet during the autumn and winter, when these migratory birds visit the tropics, they are silent and rarely encountered unless flushed from the dense grass in which they typically forage. Surprisingly, while this species is a regular Dry season visitor to many parts of the Top End, including the Marrakai track, just 40 km from the Darwin region, there are only two confirmed records of it in the Darwin region. The first was of a male singing on 20 July 1974 in the area now occupied by Leanyer Recreation Park, and the second was of a bird at Leanyer SP on 11 May 1985.[470]

As the Rufous Songlark is not known to breed north of about Alice Springs,[411] it is generally considered a non-breeding Dry season visitor to the Top End, though its status varies from place to place. In the Fogg Dam region it was reported as being locally common from April through August during 1969 and 1970, when southern parts of the NT received roughly half of their normal rainfall,[124] but absent in 1967, 1968 and 1971, when the interior had good rains.[1] On the other hand, at Coomalie Farm, one or two individuals were recorded each year from 1986 to 2004, with extreme dates being 20 April and 23 November.[47] On the latter date, a bird was heard singing, indicating that breeding was imminent. Similarly, near Maningrida in western Arnhem Land, small numbers were recorded in woodland between May and September during three successive years.[47] At Victoria River Downs it was present throughout the year,[98] yet further west at Keep River NP, on the NT–WA border, it was recorded during the Dry season only.[369]

It seems improbable that visitors to the Top End emanate from south-east Australia, from which birds migrate directly north in December or January, and return from August onwards, using an inland route.[157,411] More likely, small populations breed through subcoastal NT, possibly augmented by migrants from further inland.

Tawny Grassbird
Megalurus timoriensis

RANGE: largely coastal, tropical and subtropical e. Australia, NG and e. Indonesia. HABITATS: grassland and sedgelands, often in or near wetlands. STATUS: common resident. BREEDING: Feb (TE).

Marc Gardner

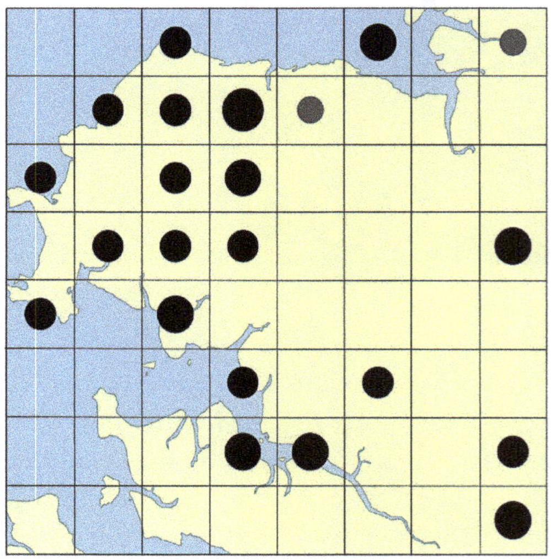

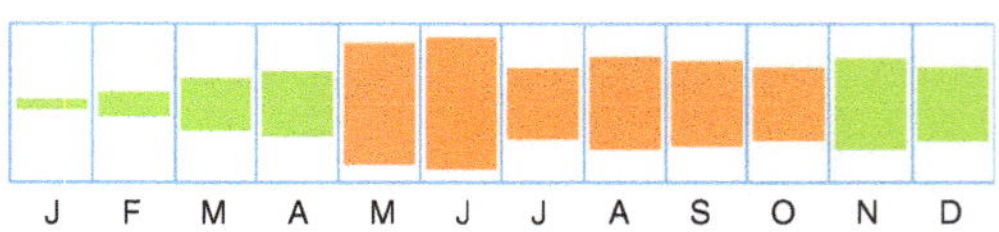

Like most of the grass-dwelling warblers, the Tawny Grassbird has two personas. Quietly skulking about in dense grass for much of the year, these birds become bold and vociferous during the breeding season, the males declaring their territories with sweet descending reels while quivering their wings during short display flights or from a conspicuous elevated perch. Given their wide distribution from Australia to the Philippines, astonishingly little is known of their biology, though like the cisticolas they are probably polygamous. The only documented record of breeding in the Top End pertains to a nest found on the McArthur River in February 1914.[360]

In the Darwin region the species has been recorded in 34% of cells, though almost a third of all records were from the Knuckey Lagoons area. During the Howard's Peninsula survey it was found to be most abundant in mixed, short grassland and sedgeland, but was also recorded in tall bulrush swamps and dense sedgelands that were flooded to a depth of 70 cm.[308] Occasionally the species wanders into suburban gardens, presumably from nearby grassy areas.[60] During the late 1960s the species was noted in almost every month from April through November in the Fogg Dam area.[1] Reporting rates in the Darwin region drop dramatically after December and rise slowly to a peak in May–June. The pattern of seasonality contrasts with that of birds in eastern Australia, where singing and displaying rates are said to be high during the breeding season, from September to February, and low during the non-breeding season.[411]

In early March 2007 at least four males were observed displaying and singing frequently at Berrimah Farm,[47] suggesting that breeding was imminent if not already under way. This timing is commensurate with the breeding seasons of other local grass-dwelling warblers, which coincide with the mid to late Wet season. The May–June peak in reporting rates may therefore reflect post-breeding dispersal, as flooded grasslands start to dry up, while the lack of records from January to March may be partly due to the difficulty, and inherent dangers, of surveying wetlands when water levels are highest.

Brown Songlark
Megalurus cruralis

RANGE: whole of mainland Australia, but rare or absent in much of tropics. HABITATS: grassland, including floodplains. STATUS: scarce, irregular Dry season migrant. BREEDING: not in tropics.

Like the Rufous Songlark, the Brown Songlark is a ground-dwelling species that breeds in southern Australia during the spring, then migrates to northern Australia in late summer, before returning south via the inland in late winter.[157] Unlike the former species, however, the occurrence of Brown Songlarks in the Top End is best described as irruptive, as their numbers and timing vary from year to year, probably depending on the amount and distribution of rainfall inland. The first known irruption of this species in the Darwin region took place in 1970, when one or two birds were seen at Leanyer Swamp in mid-October, though earlier in the year up to 12 birds had been counted in the Humpty Doo and Marrakai areas between 12 April and 16 May.[1] However, the species was not recorded in these areas during the previous three years or in the following year. Significantly, the irruption took place

David Webb

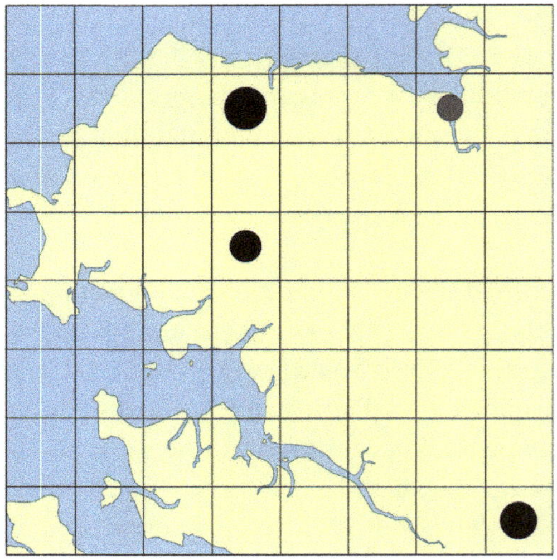

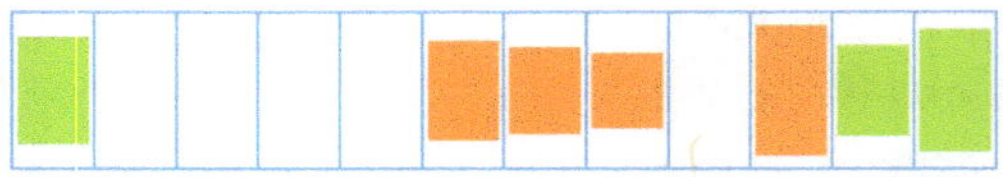

during a period of drought in southern parts of the Top End and Barkly Tableland. Total rainfall for the preceding Wet season was about half that of each of the previous three Wet seasons for Katherine and Tennant Creek, and about a quarter in the case of Alice Springs.[124] During the following year, however, Wet season rainfall returned to pre-1970 levels in Katherine and Tennant Creek.

Since then, there have been single sightings of Brown Songlarks at Leanyer Swamp in December 1983, August 1985, July 1986 and June 2001.[21] Although few regions of the continent experienced serious rainfall deficiencies in either 1983 or 2001, drought was widespread in inland eastern Australia in early 1985, and in the Gulf and Channel Countries in early 1986.[124] The Barkly Tablelands suffered low rainfall in both years. The next widespread irruption of these birds was not until 2008, when single birds were seen from May to November at Knuckey Lagoons, McMinns Lagoon[21] and Darwin International Airport.[47] Just outside the region, at least 35 birds were counted on the Adelaide River floodplain along the Arnhem Highway,[47] while around Maningrida in western Arnhem Land, about 350 km east of the Darwin region, flocks of up to 25 individuals were seen on several floodplains in June and July of the same year.[47] Interestingly, the 2007–08 Wet season was again exceptionally poor in the Barkly Tableland.[124]

Zitting Cisticola
Cisticola juncidis

RANGE: coastal tropical Australia, NG and Indonesia, as well as Eurasia and Africa. HABITATS: short grass floodplains. STATUS: uncommon resident. BREEDING: Dec–Feb.

Tony Ashton

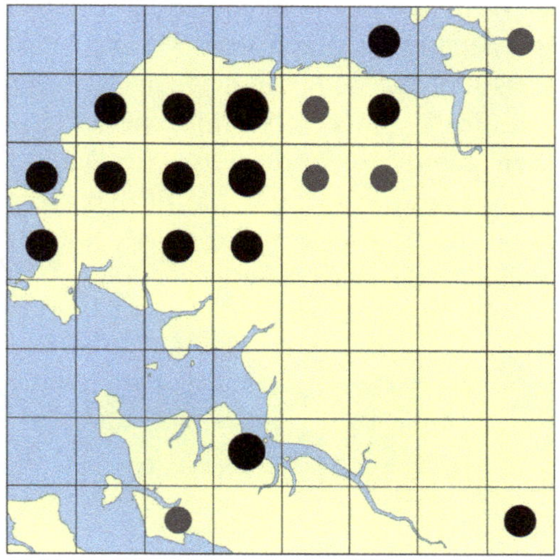

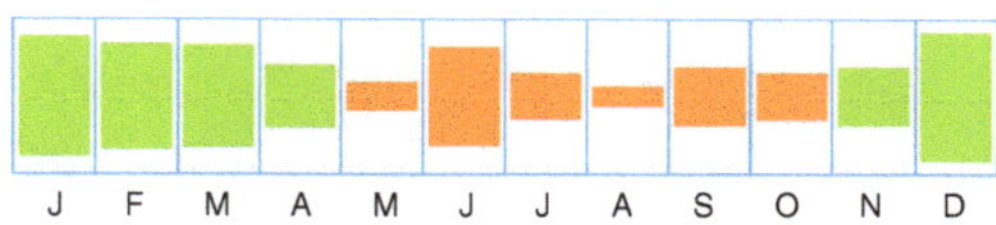

With 18 subspecies scattered between Europe and Australia, including Africa and India, the Zitting Cisticola is possibly the only Australian native songbird that is better known outside than inside Australia. Yet it was one of the first bird species to be studied in any detail in the Darwin region. Indeed, the local race was named *leanyeri* after Leanyer Swamp, where it was first collected in 1952,[471] though its range was later found to extend from the East Kimberley region of Western Australia to the south-west head of the Gulf of Carpentaria. The simple call from which the species takes its first name is given only by males during their song flights, which in Darwin commence in late November or early December and finish by April.[471] The bird launches into the air from ground level with a rapidly repeated 'lik-lik-lik-lik-lik', climbing to a height of 15–30 m before it levels off and starts circling with a bouncing motion, repeating a 'lik-lik' call at the top of each undulation, until it dives towards the ground.[411] The species is polygynous, and studies outside Australia have revealed that a single male may have up to 11 female partners, but more usually four, nesting on his territory.

The habitats of the two cisticola species during the Wet season are mutually exclusive yet closely abut. At this time Zitting Cisticolas are restricted to plains covered with short grass about 30 cm high,[471] either on coastal saltflats adjacent to mangal or black-soil floodplains further inland, whereas Golden-headeds occupy the long grass on higher ground. However, during the Dry season the two species overlap, especially when their preferred habitats are burnt.[1] During the Howard's Peninsula survey in April–May 1986 the Zitting was found to be most abundant in dense sedgeland (50 cm high) that was flooded to a depth of 70 cm, but it was also recorded in tall (150 cm high), flooded bulrush swamps and patches of mixed short (10 cm high) grassland and sedgeland, albeit at lower densities.[308] Since 1977, the species has been recorded in 31% of cells in the Darwin region, though three-quarters of all records came from Knuckey Lagoons, Holmes Jungle and Leanyer Swamp.

Based on observations of 10 nests at Leanyer swamp during the Wet seasons of 1950–52, and more recent records from the Top End,[411,471] clutches are laid from late December to mid-February. Male songs and displays associated with breeding activity probably account for the January peak in reporting rates, but the 'spike' in June is difficult to interpret. Some Dry season records may refer to the commoner Golden-headed as the two species are practically indistinguishable in non-breeding plumage, except for the whiter tip to the tail in the Zitting.

Golden-headed Cisticola
Cisticola exilis

RANGE: tropical and e. Australia. HABITATS: floodplains and other wetlands, as well as woodland with tall grass. STATUS: very common resident. BREEDING: Dec–Apr.

Mat and Cathy Gilfedder

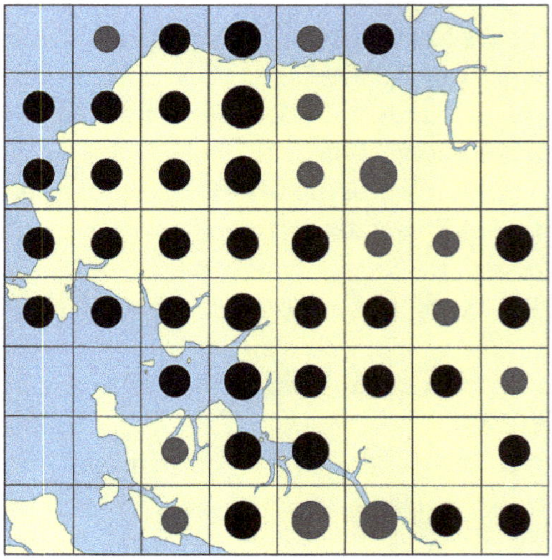

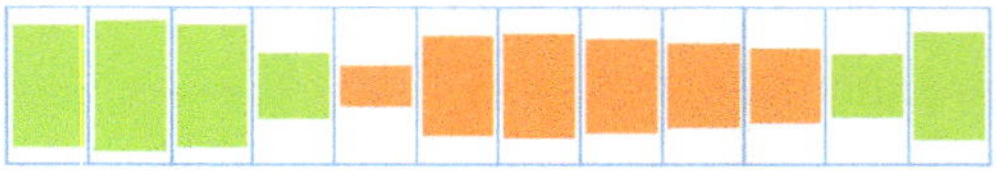

While the tiny Golden-headed Cisticola is one of the commonest birds of grassy areas in the Darwin region, there is nothing common about its bizarre mating and nesting habits. Just before the breeding season males acquire their bright 'golden' head coloration and moult their tails, then start singing, first from a prominent perch, then while performing spectacular aerial song displays, in which they ascend as high as 35 m before diving to within 2 m of the ground, re-ascending either immediately or after a 20 m zig-zag flight around the territory.[411] Each territory owner then builds the 'shells' of several nests in the tall grass, each constructed by sewing leaves together. This is done by making small holes with the bill along the leaf margins and pulling spider-web through them, a stitching technique that is unique to the family Cisticolidae, which includes the Asian Tailorbirds. Having attracted up to 14 females to his territory to finish construction of the nests that he started, the male plays no further role in breeding other than mating.[472]

The Golden-headed Cisticola has been recorded in 80% of cells in the Darwin region. It was the second most frequently recorded species during the 1986 Howard's Peninsula survey, where it was found in seven of the 20 sampled habitats. As with the Zitting Cisticola, it was most abundant in flooded bulrush swamps and dense sedgelands, although its density in these habitats was five to seven times higher than that of the Zitting. It was also three times more abundant than the Zitting in mixed grassland-sedgeland. Habitats occupied by Golden-headed Cisticolas alone included open pandanus–grevillea woodland and tall grasslands of Spear Grass.[308]

Nests are well concealed and difficult to locate among the grass, which partly explains why so few nests have been reported from the Darwin region. Observations of nest building or recently fledged young, however, suggest clutches are laid from December through April.[47] High reporting rates from December through to March probably reflect increased levels of singing and displaying by breeding males at that time. The drop in reporting rates over April and May coincides with the abandonment of territories soon after nesting has finished, and the subsequent increase in rates may be related to the formation of loose flocks of up to 10 birds on floodplains during the Dry season.[47]

The reason why males replace their tail feathers at the start of the breeding season with new ones that are 30–40% shorter was a mystery until a series of elegant experiments was conducted on a colour-banded population near Townsville in north Queensland.[472] By artificially lengthening and shortening the tails of males, these experiments showed that males with short tails attracted more nesting females and produced more offspring than those with longer tails, despite the latter being aerodynamically superior.[472]

Canary White-eye
Zosterops luteus

RANGE: endemic to coastal tropical Australia. HABITATS: mainly mangal. STATUS: very common resident. BREEDING: biannual, mostly Mar–Apr and Sep–Oct.

Like its well-known and much more widespread cousin, the Silvereye, the Canary White-eye is omnivorous, yet while the former may be found in any vegetated habitat, the latter is largely restricted to mangals. One of 11 bird species that are dependent on mangals in the Australian monsoon tropics,[312,313] it is also among the three most abundant species in this habitat within the Darwin region.[312] Contrary to

Micha V. Jackson

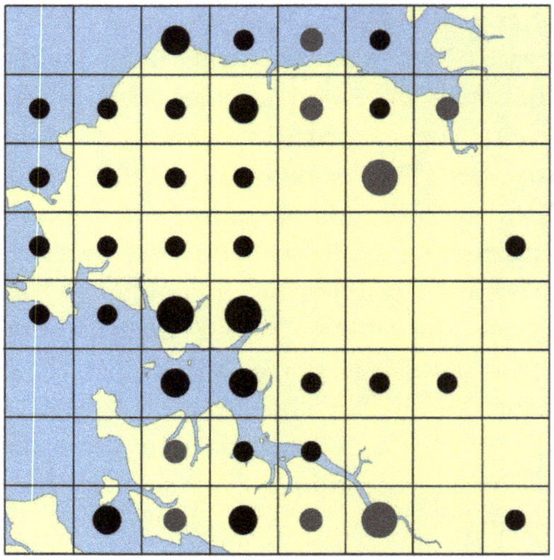

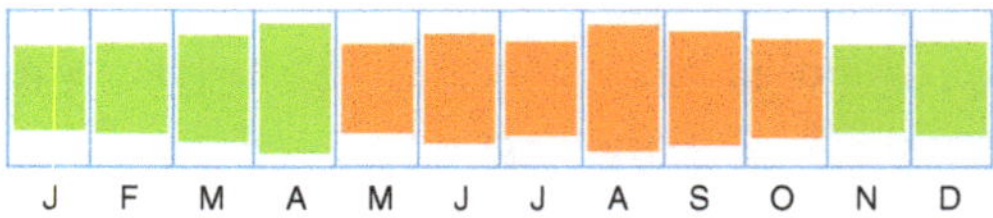

claims that local populations are nomadic,[1] colour-banding at three sites revealed that these birds are sedentary, though when they are not breeding, pairs amalgamate to form small flocks of up to 15 birds that move around a home range of 5 ha or more. Seven birds were at least four years old when last recaptured, the oldest being at least six years old.[47]

Like several other mangal-dependent species, this species often visits coastal monsoon rainforests and vine thickets,[308] as well as paperbark forests that adjoin mangals,[1,96] but it has also been seen as far inland as Koolpin Gorge[43] on the South Alligator River, some 140 km from the coast. It often visits gardens in suburbs beside the coast,[60] and in the future may well become a city bird, as very similar species have done in Indonesia. Although primarily an insectivore with a penchant for beetles and bugs,[421] it also consumes nectar and fruit.[312] As in other white-eyes, the tips of its forked tongue are fringed with fine hairs, which are used to mop up nectar from flowers of most mangrove species and paperbarks.[411] It also has the ability to prise open small flowers, like that of the Black Mangrove, by inserting its somewhat wedge-shaped bill into the floral tube, then gaping, behaviour known as *zirkelning*.[411] Fruits, such as figs and Native Passionfruit, are obtained outside mangroves. All mangal zones are used by Canary White-eyes, but when nesting they prefer the Spurred Mangrove zone.[312] Most nests are built in the short Grey Mangroves that grow on and around bare saltflats within that zone.[421,473] Prior to and during nesting, individuals sing canary-like medleys from selected perches for 10 minutes or more.

This species breeds biannually.[440] Although nests have been found in all months, studies at several sites in Darwin, involving more than 150 nests, show that nearly 80% of clutches are laid during two peaks, the first in March–April and the second in September–October, coinciding with the late Wet and late Dry seasons, respectively.[421,473] These periods coincide with peaks in the abundance of small insects in the mangroves. This species also has one of the shortest nest cycles of any Australian songbird, the eggs taking about 10 days to hatch and the nestlings another 10 days to fledge.[473] Unlike most small Australian songbirds, females of which incubate the eggs and brood the young alone, Canary White-eye parents share these duties equally.[47] Despite being fast breeders, about two-thirds of nests monitored in the above study succumbed to nest predators, whose identity was unknown.

Metallic Starling
Aplonis metallica

RANGE: e. Indonesia, NG, Solomon Is, wet-tropical Australia. HABITATS: rainforest, mangal, suburbia. STATUS: vagrant. BREEDING: not in NT.

Jim Sneddon

Two Metallic Starlings, an adult and an immature, were observed in a suburban garden in Millner in August 2002.[474] The nearest population of the species occurs in south Maluku, Indonesia, some 500 km north of Darwin, while the nearest population within Australia is about 1000 km east, on Cape York. While on proximity grounds the birds may have come from Indonesia, the subspecific status of the birds in Darwin could not be ascertained, and their origin remains speculative.

Common Starling
Sturnus vulgaris

RANGE: native Europe, e. to Mongolia, nw. China; introduced to s. and e. Australia. HABITATS: urban, rural and cultivated areas. STATUS: vagrant. BREEDING: not in NT.

A single adult Common Starling in breeding plumage was observed on several occasions at Leanyer SP from 22 June to 29 June 1996.[475] While there is some possibility that the bird originated north of Australia, where the closest records are from Thailand, it is more feasible that the bird arrived as a transported migrant, by ship or road, from southern or eastern Australia.

Mistletoebird
Dicaeum hirundinaceum

RANGE: whole of mainland Australia, and far e. Indonesia. HABITATS: all wooded habitats. STATUS: very common resident. BREEDING: mainly Nov–Apr.

Jon Norling

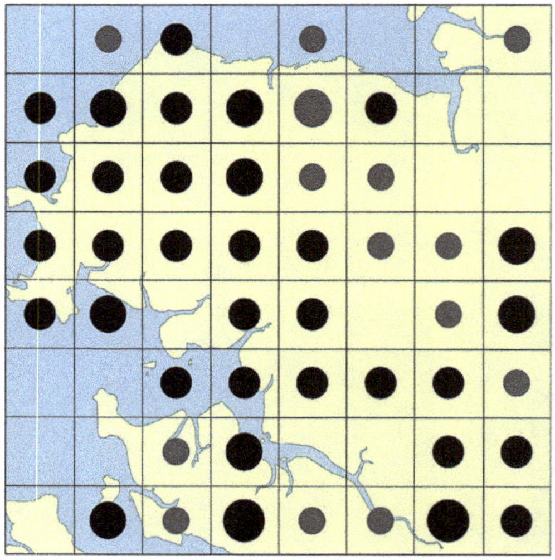

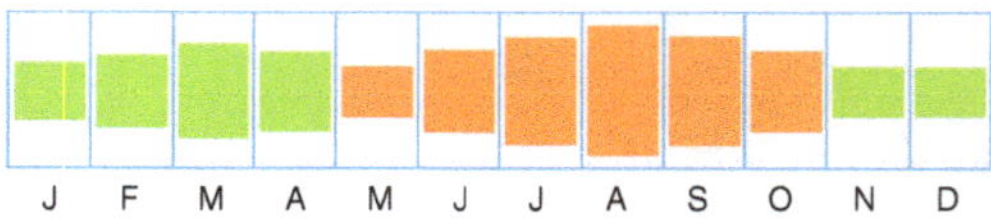

A prime example of mutualistic relationships between birds and plants, the tiny Mistletoebird is the principal dispersal agent of many of Australia's 90 or so species of mistletoes, the fruits of which form its staple diet. While mistletoes date back to the time of the dinosaurs, the Mistletoebird probably colonised Australia relatively recently from the Indonesian archipelago, where there are over 20 species of flowerpeckers (Dicaeidae), the family to which it belongs. Found in all habitats, from rainforest to desert, over much of the Australian mainland,[411] Mistletoebirds occur wherever mistletoes are able to grow. In the Top End, these birds have also been seen eating the fruits of several rainforest plant species,[47,314] as well as insects and occasionally even nectar.[398]

In Darwin the species is abundant, possibly partly due to the widespread tree planting of Milkwoods and Cheesewoods, which are sometimes heavily infested with Matchstick Mistletoes, a popular source of fruit for the bird.[47,418] Mistletoebirds were reported from 73% of cells in the Darwin region, significantly more in urban than rural or forested cells. During the Howard's Peninsula survey the species was recorded in six of 20 sampled habitats, mainly in eucalypt forest and woodland, but also in paperbark forest.[308] In a bird survey of mangals in the Darwin region it was recorded in 10 of 13 sites sampled, but was most abundant at Mindil, the smallest site, where the density was over two birds per hectare.[313] Although it has been recorded in some Darwin suburbs in every month,[47,60] reporting rates for the region are strongly bimodal, with a major peak in August and a minor one in March. In Kakadu NP, too, it was more abundant in the late Dry season than in the Wet and early Dry seasons, especially in monsoon rainforests.[103,355] Such fluctuations in reporting rates and abundance probably relate to local movements associated with the reproductive timetable of various mistletoes. In the mangal at Nightcliff, for example, it is most frequent in February and March, when the Mangrove Mistletoe is fruiting.[421]

The pendant, purse-shaped nest of the Mistletoebird surely rates as one of the neatest and softest built by any Australian bird. Composed almost entirely of plant down and silk from spider web and suspended from a twig,[411] it hangs limply when empty. The nest is so flaccid that during a storm one was blown across an adjacent twig, causing the entrance to be blocked and the two trapped chicks inside to die of starvation.[47] This is a Wet season breeder, with a third of local clutches being laid in February, although one clutch was found in September 1999.[47] Fledglings and dependent juveniles are instantly recognisable from other young birds by their bright reddish-orange bills.

House Sparrow
Passer domesticus

RANGE: native Europe to n. Japan; introduced and naturalised e. half of Australia. HABITATS: settled areas incl. towns, farms. STATUS: transported vagrant. BREEDING: Sep (NT).

Niven McCrie

The House Sparrow was first reported in the NT in October 1977, when several birds were found at Lake Nash, about 400 km south-east of Tennant Creek.[411] Somewhat prematurely, it was reported in 1980 that all sparrows in the NT had been eradicated,[476] as the species has subsequently become established in Tennant Creek and areas of the Barkly Tableland. Occurrences in the Darwin region probably relate to birds transported by road, either from there or from elsewhere in southern or eastern Australia. The first report in the Darwin region concerns a bird found at Fannie Bay in January 1980, and which was later destroyed by officers of the Parks and Wildlife Commission. The only records since 1980 were from Palmerston and Stuart Park, both in 1992.[477]

Eurasian Tree Sparrow

Passer montanus

RANGE: native Eurasia, s. to Malaysia and Indonesia; introduced se. Australia. HABITATS: urban areas. STATUS: transported vagrant. BREEDING: not in NT.

Dr Chan Ah Lak

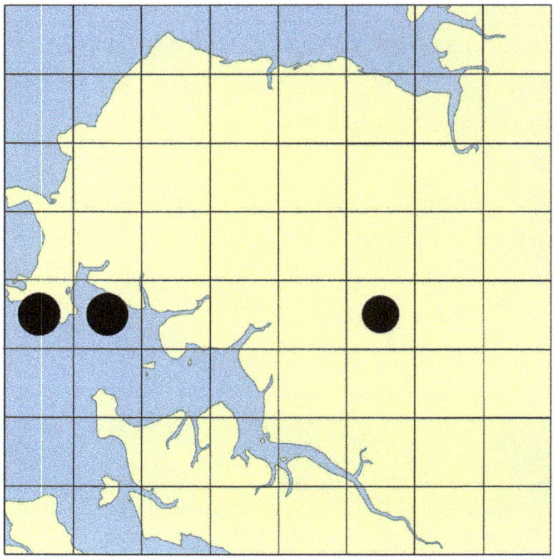

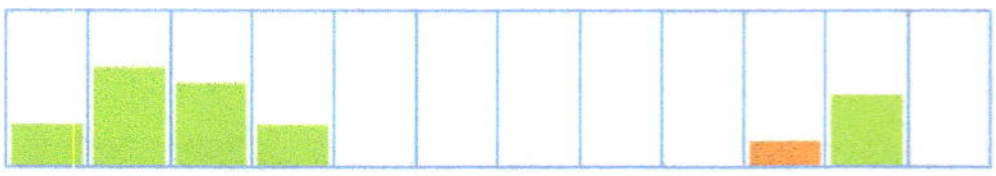

The Darwin region's first Tree Sparrow was found in 1966 at Stokes Hill Wharf,[477] where it was subsequently also seen in 1981,[34] 2001[21] and 2013.[478] These and other records in the coastal suburbs of Cullen Bay in 1998, Nightcliff in 2002[21] and Fannie Bay in 2006[47] clearly suggest arrival in Darwin by ship, probably from south-east Asia. In June 1983, a Tree Sparrow was found at Darwin International Airport and, though it may have been coincidental, it is possible that the bird was transported in the hold of a plane.

The range of the Tree Sparrow has expanded eastwards through Indonesia over the last century.[86] Already well established in Timor and Ambon, its potential to become established in the Darwin region is high. The birds found at Cullen Bay included three fledglings, clearly from local nesting, and there was an additional nest, containing four eggs, in the vicinity.[477]

Crimson Finch
Neochmia phaeton

RANGE: endemic to tropical Australia. HABITATS: woodland, grassland, and wetland edges, especially where pandanus present. STATUS: common resident. BREEDING: Dec–Jul.

Niven McCrie

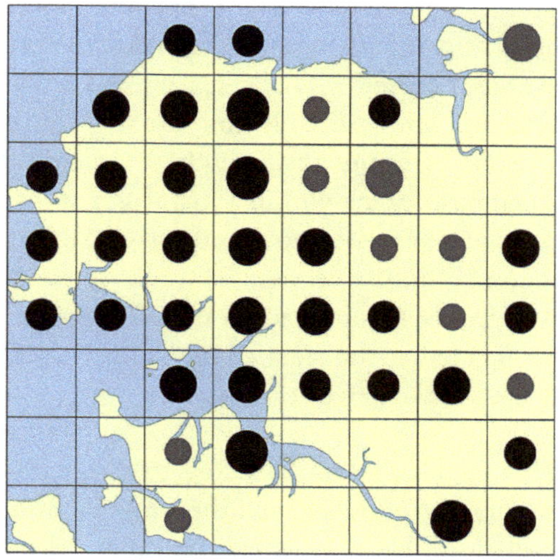

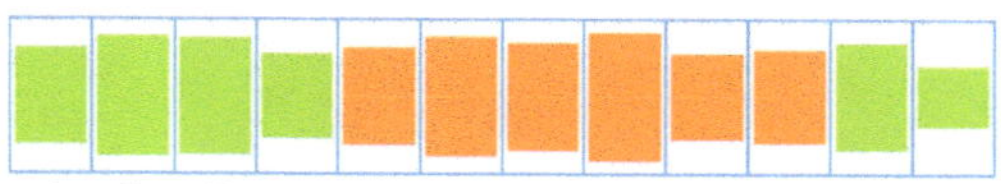

Were it not for its obviously colourful plumage, the Crimson Finch could be called the Pandanus Finch for no other bird species shows such a strong affinity for these palm-like trees.[1,388] In a detailed study of this species in the Kimberley region of Western Australia, 98% of the 870 nests found were situated in the River Pandanus, a third of them over water.[479] In Kakadu NP Crimson Finches were found to regularly use the spiny-edged leaves of Common Pandanus not only as sites for nests, but also for roosting, foraging for insects and protection from raptors.[388] While habitats other than those dominated by pandanus are used, they are invariably associated with wetlands or rivers.[450] Of the five grassfinch species recorded during the Howard's Peninsula survey, the Crimson was the most widespread, being found in no fewer than 19% of sites, representing nine habitats.[308] Though most abundant in bulrush swamps, it was moderately abundant in tall paperbark swamps and riparian forest containing Swamp Box and Ghost Gum.

In the early 1970s the Crimson Finch was regarded as generally scarce in Darwin, except near the entrance to Leanyer SP and in the area of pandanus to the north of Vanderlin Drive.[60] Since then, however, it has been recorded in two-thirds (66%) of cells in the Darwin region, significantly more of which were in urban-dominated environments than forest or rural ones. As reporting rates show, the species is a resident, though it may make small-scale seasonal movements between habitats. In the Victoria River district, they were restricted to riverside habitats during the Dry season, but following the first rains dispersed to breed around the margins of billabongs and creeks.[451] In the Kimberley, they were non-territorial year-round, congregating in flocks of 13 individuals, on average, during the Dry season and splitting up into pairs during the breeding season.[479] Despite nesting in high densities, some nests being within 1 m of each other, pairs were monogamous and usually stayed together until one died.[479] Adult survival was relatively high and, of the 64 individuals banded at the start of the study, 12.5% were still alive five years later.

This species starts breeding in the early Wet season, earlier than other tropical Australian grassfinches, a fact that may relate to the availability of insects and spiders which form a greater part of the diet than in the other species.[480] Breeding records in our region suggest that clutches are laid from December to July. As early nests coincide with the first rains, they are often washed away with the first floods.[481] Nest building is performed almost exclusively by males, and while each male in the Kimberley study was found to construct four to five nests per year, on average, half of these were abandoned before the female laid her eggs.[479] Clutches were composed of five eggs, on average, and were incubated by both sexes, but the failure rate was high, with 60% of clutches being depredated. Artificial structures are used for nest sites more by Crimson Finches than by any other Australian grassfinch.[481] Of 74 nests across northern Australia, 30% were sited in artificial structures, mainly roof rafters, but one was built on a regularly used tourist-operator's boat!

Masked Finch
Poephila personata

RANGE: endemic to tropical Australia. HABITATS: eucalypt woodland and grassland. STATUS: scarce visitor. BREEDING: Mar–Jul (TE).

Peter Kyne

As elsewhere in coastal and subcoastal areas of the Top End,[103,105,143] the Masked Finch is much less common in the Darwin region than the Long-tailed Finch, with which it often associates. Nevertheless, it was described as common in the Humpty Doo and Marrakai areas, east of the Darwin region, during 1967–71.[1] Since then the species has been recorded in over a third of the cells in the Darwin region. Apart from the 'spike' of records in February, reporting rates suggest that this species is most frequent during the early Dry season. In

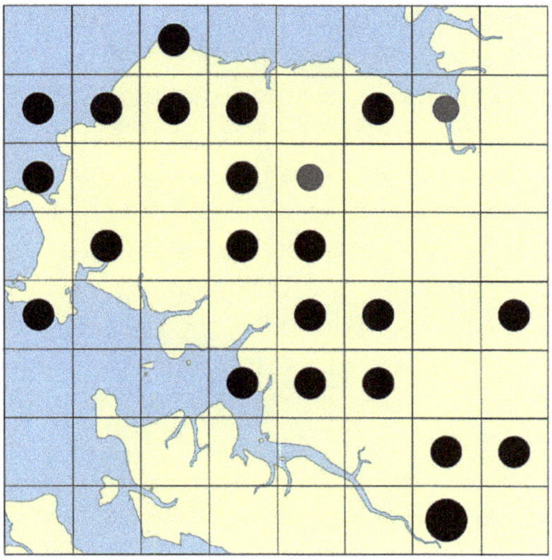

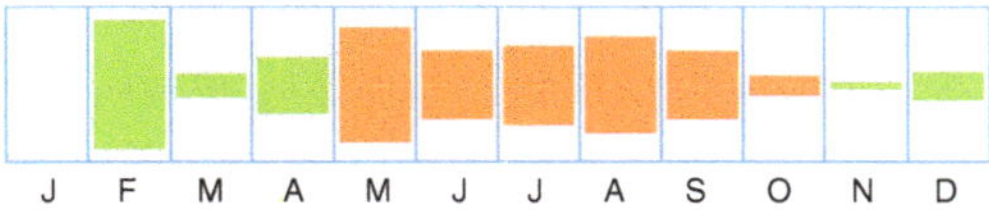

weekly bird surveys of Palmerston from May to September 2007, the species was found only in suburbs built in the 1990s, with up to five birds seen in Durack.[307]

In contrast to the Long-tailed Finch, this species builds its nests in grass or on the ground more than in trees.[411] However, the only potential breeding record from the Darwin region concerns a pair that was seen near Thorak Cemetery in mid-March 2007, carrying grass stems to a half-built nest about 9 m up in the upper foliage of a Woollybutt.[47] Most clutches in the Top End are laid from March to July.[47,143,177,451]

Long-tailed Finch
Poephila acuticauda

RANGE: endemic to monsoon-tropical Australia. HABITATS: eucalypt woodland and forest, and grassland. STATUS: moderately common resident. BREEDING: Feb–Nov.

Niven McCrie

Abundant in savannas across much of the Top End, the Long-tailed Finch was considered scarce or absent in coastal parts of the Darwin region until at least the mid-1980s.[1,60] Indeed, it was recorded at only two out of 101 sites spread across 20 habitats during the Howard's Peninsula survey in 1986.[308] Since then, however, the species has been recorded in 63% of cells in our region, coastal cells being as well represented as subcoastal cells. Up to 15 birds, including dependent juveniles, were seen around Darwin Golf Course each month from July 2006 to June 2007.[309] Small numbers occur along the entire length of Rapid Creek, including its mouth, where they have been seen foraging in vegetation fringing the beach.[21] During the Palmerston bird survey in 2007, the majority of records of this species came from the then undeveloped suburbs of Mitchell Creek and Bellamack.[307]

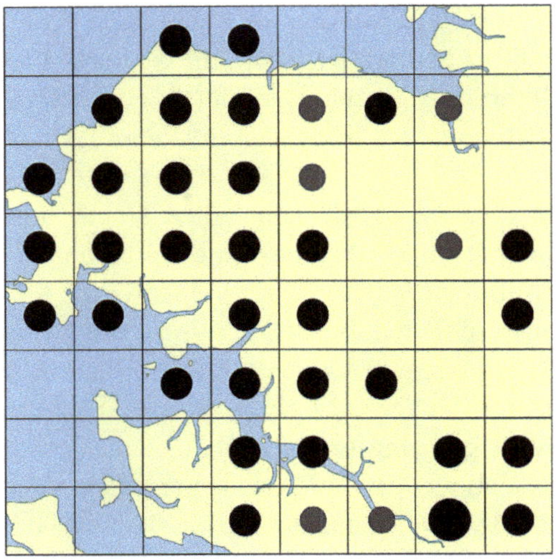

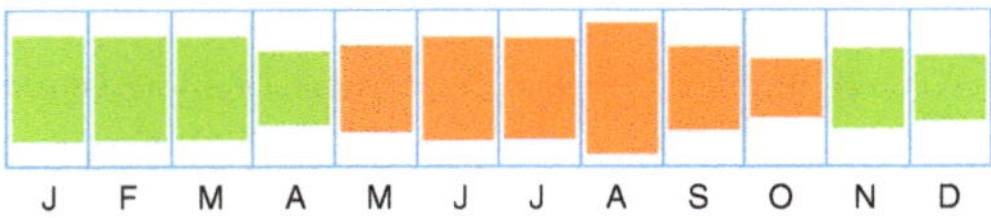

Across the Top End, Long-tailed Finches are often seen in mixed flocks with the Masked Finch, as the two have similar habitat preferences and diets.[310,482] Both breed primarily during the early Dry season, when seeds of Annual Speargrass become abundant, but whereas Masked Finches usually construct their nests in the grass, Long-tailed Finches build their grass nests in the canopy of eucalypts or in tree hollows.[483,484] Annual grasses remain important in the diets of both species during the late Dry and early Wet seasons but, unlike the Masked, the Long-tailed consumes many insects, especially termites at this time.[482] During the first few weeks of the Wet season, when grass seed from the previous season has germinated and the coming crop is not yet ripe, these birds are forced to feed on both old and sprouting seed, for which they spend at least six hours per day foraging.[310] Later in the Wet season, however, they exploit the seeds of perennial grasses, such as the Golden Beard Grass, which become abundant.

In the Darwin region nest building has been recorded from January to October, and in woodland at Holmes Jungle no fewer than 10 pairs were observed building nests in the latter half of August 1996, though it is not known whether eggs were laid in these nests.[415] At Yinberrie Hills, near Katherine, clutches were laid from February to November.[483] A three-year study in the Kimberley region of Western Australia showed that pair-bonds between adult birds were very strong, and the same birds were seen nesting together at the same location during successive breeding seasons.[484]

Zebra Finch
Taeniopygia guttata

RANGE: whole of Australian mainland except coastal areas. HABITATS: woodland and grassland. STATUS: occasional aviary escapee and vagrant. BREEDING: Jan–Sep (s. NT).

Mat and Cathy Gilfedder

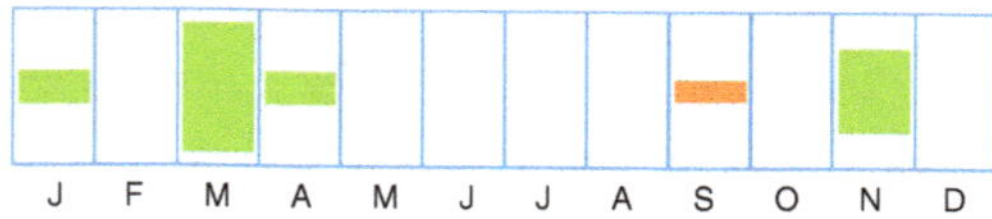

The desert-adapted Zebra Finch is generally rare north of the latitude of Larrimah,[41,96,300] about 500 km south-east of Darwin. It was first reported in Darwin in September 1952[41] when much of the northern half of the NT, including most of the Barkly Tableland and part of the Tanami Desert, was gripped with drought.[124] It is therefore possible that this report refers to wild birds driven towards the coast in search of food and possibly water. Since then the species has been reported at least nine times, but many of these records almost certainly refer to aviary escapees. For instance, the next two records concerned two birds sighted at East Point in July 1967[96] and one bird feeding with Double-barred Finches near Fort Hill wharf in February 1971,[1] years in which there were no droughts in northern Australia.[124] In the 1990s, the species was reported in five years, only one of which (1992) coincided with drought conditions in the northern half of the NT. The species is resident in the Victoria River Downs[451] and Keep River NP.[369]

Double-barred Finch
Taeniopygia bichenovii

RANGE: tropical and e. Australia. HABITATS: eucalypt forests and woodland, mangal, wetlands. STATUS: very common resident. BREEDING: mainly Jan through Jul.

Of the six grassfinch species found in the Top End, the Double-barred Finch has the widest geographical range, reaching as far south as Victoria. While the eastern race *bichenovii* of this species is distinguished by its white rump patch, which contrasts with its black tail, the local race *annulosa* has a black rump, which in turn distinguishes it from the white-rumped Long-tailed and Masked Finches. During the late 1960s it was described as common in the Humpty Doo–Marrakai region, but 'scarce elsewhere', including Darwin.[1] By the mid-1970s, however, it was regarded as 'fairly common' in coastal suburbs from Parap to Nightcliff,[365] suggesting an

Niven McCrie

increase in abundance in the Darwin region, as has occurred in northern Australia generally.[485] Since then it has been recorded in 81% of cells in the region, making it the most widespread of our grassfinches.

Unlike the tropical savanna-specialised grassfinches that undertake seasonal movements according to the availability of grass seeds, the Double-barred Finch appears to be largely sedentary. For example, up to 25 were counted every week over 12 months during 2006–07 around the eastern half of the Darwin Golf Course.[309] It is the most urbanised finch species in Darwin, regularly nesting in pot plants on verandas. In Palmerston, too, it was among the 20 commonest species recorded during a survey in 2007, when it was noted in suburbs of all ages. During the Howard's Peninsula survey it was found to be most abundant in tall paperbark swamp, but also occupied eucalypt forest with a shrubby understorey, low eucalypt woodland, and a wet mosaic of grassland and sedgeland.[308] In Kakadu NP it was recorded in all major habitats, most

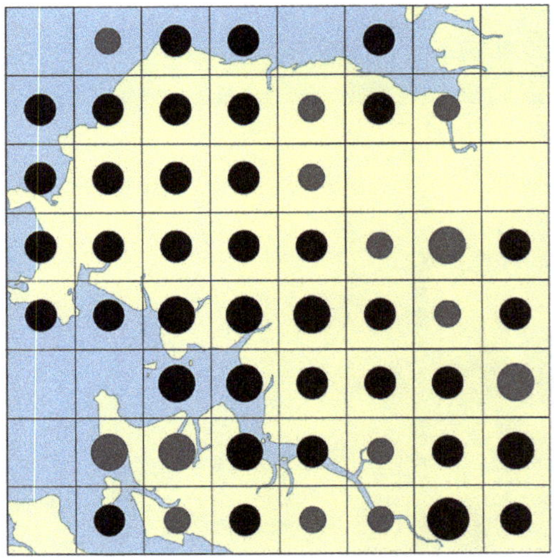

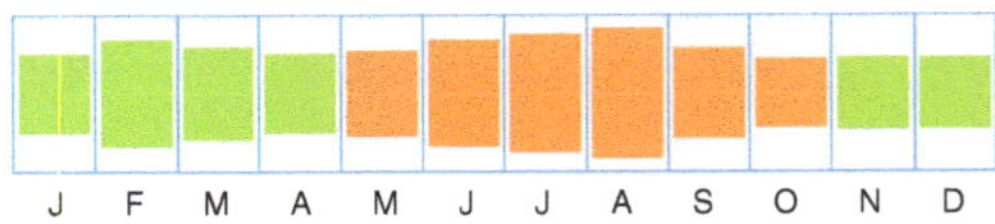

frequently in woodland, but it was also common at some monsoon rainforest and eucalypt forest sites.[103] Analyses of bird survey data from across the Top End showed that this species is strongly associated with riparian habitats.[450] Like the Crimson Finch, the Double-barred has a strong affinity for pandanus 'palms', though in Kakadu NP, it was not as abundant as the former species in either pure or mixed stands of the plant.[388] In a bird survey of mangals, Double-barred Finches were recorded in all 13 patches sampled, with the highest density at Mindil, the smallest patch.[313]

Breeding takes place from January to July in the Darwin region, with a peak in egg-laying in April.[47] Like other grassfinches, this species is multi-brooded, and nests from previous breeding attempts are often reused. One pair at Charles Darwin University Casuarina campus even commandeered an old nest of a Crimson Finch from which fledglings had emerged only four weeks earlier.[47]

Yellow-rumped Mannikin
Lonchura flaviprymna

RANGE: endemic to w. Top End and e. Kimberley Division of WA within monsoon tropics. HABITATS: grassland, woodland, and tall grass along rivers. STATUS: rare late Dry–early Wet season visitor, invariably associated with following species. BREEDING: unknown (NT).

Adrian Boyle

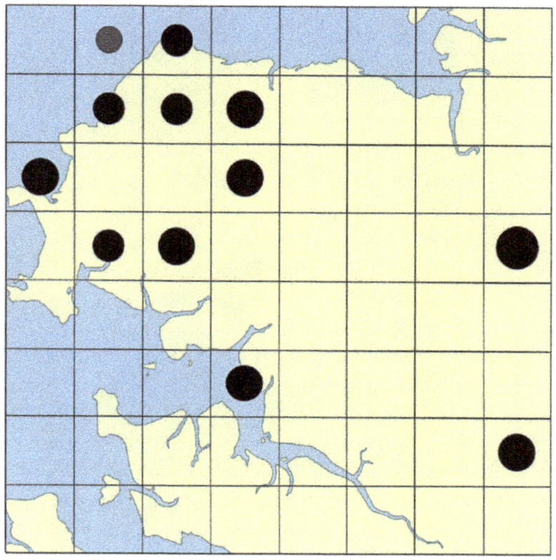

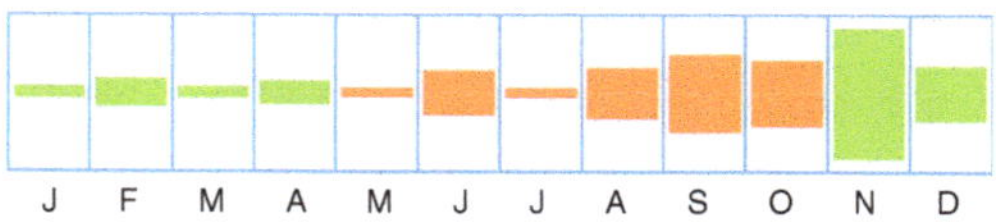

Although the Yellow-rumped Mannikin occurs in single-species flocks in the south of its exceptionally restricted range, in the Darwin region it is invariably associated with the much commoner Chestnut-breasted Mannikin. Small numbers of the former have been reported in the Darwin region since the late 1960s.[1,365] Even before 1990, mixed flocks were estimated to contain about one Yellow-rumped to every hundred Chestnut-breasted.[60] Yellow-rumped Mannikins have been recorded in 19% of cells in the Darwin region, nearly a third of all records coming from the East Point-Nightcliff area. Reporting rates show a general increase over the Dry season and start of the Wet, with a sudden drop in December. Since 1992 observations for up to seven consecutive years in five well separated suburbs (Larrakeyah, Parap, Leanyer, Durack and Nightcliff) have shown that Yellow-rumps typically arrive with flocks of Chestnut-breasted Mannikins during the first two weeks of October, but sometimes as early as late August, and depart in December or January.[47,319] The appearance of up to four Yellow-rumps among flocks of Chestnut-breasted in multiple locations every year suggests that the two species disperse together from shared breeding grounds. In the East Kimberley, where both species are common, the two roost and breed together, and often hybridise,[411] indicating that the Yellow-rumped may merely be a colour variant of the Chestnut-breasted, somewhat like the various forms of the Gouldian Finch. Indeed, immature Chestnut-breasted are often mistaken for Yellow-rumps, because of their pale-headed appearance and lack of distinct breast bands.

Chestnut-breasted Mannikin
Lonchura castaneothorax

RANGE: tropical and e. Australia s. to s. coast NSW, and NG. HABITATS: grassland, sedgeland and disturbed areas. STATUS: common resident and possibly Dry season visitor. BREEDING: Dec–Apr.

Jon Norling

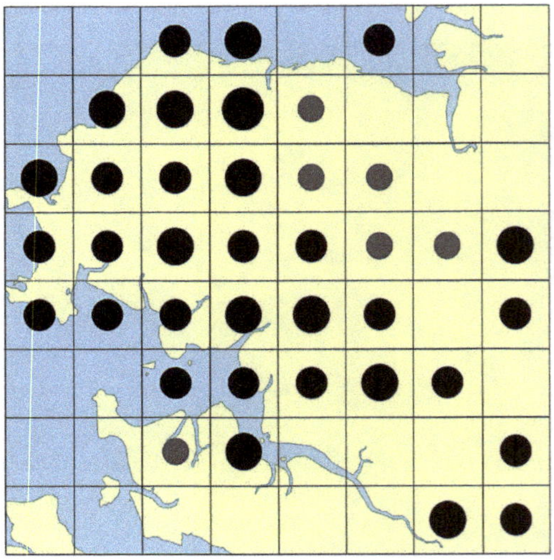

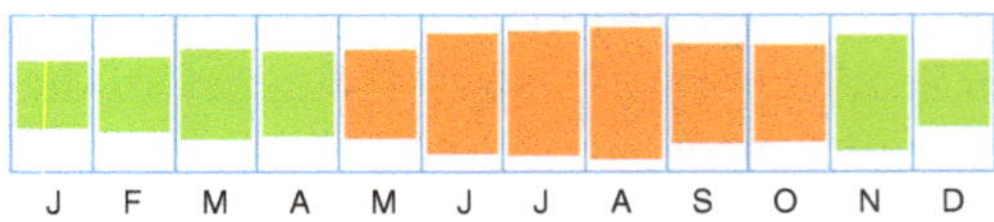

Unlike many other grassfinches, which build small roost-nests for sleeping outside the breeding season, the grassland-dwelling Chestnut-breasted Mannikin roosts in long grass or reeds, often in very large flocks.[411] One perennial roost, situated in reed beds near Holmes Jungle, contained several hundred birds.[60] From 1967 to 1971, the abundance of this species in the Darwin–Humpty Doo regions was said to fluctuate markedly from year to year, with few records from the coast.[1] By 1987, however, it was considered a regular visitor to the coastal fringes of Darwin, such as at East Point and the Wharf precinct, as well as the edges of the northern suburbs.[60] Since 1977 the species has been recorded in 59% of cells in our region, equally often in coastal and subcoastal areas. During the Howard's Peninsula survey, it was encountered in five habitats, but was most abundant in wet sedgelands with Wild Rice.[308]

Although reporting rates over the whole region show a slight increase during the Dry season, year-round observations in several suburbs suggest that this species makes seasonal movements. Around Millner it was largely absent from January through August each year from 1988 to 1993. Up to 183 birds, but usually 20 to 55 birds, appeared in late August or September and remained until December in all years.[296] Similarly, each year from 2004 to 2008 a flock of between 90 and 180 birds appeared during July or August in beach vegetation adjacent to mangal at Nightcliff and remained until December, regularly visiting a backyard bird feeder.[47] A very similar late Dry season influx was reported in the Palmerston suburb of Durack.[319] In contrast, around Darwin Golf Course the species was absent from July to December 2006, but up to 20 were present from January to June 2007, with dependent young present in early April.[309]

Few active nests have been found in the region. In 1948 a nest was found in Nightcliff during late March,[110] and almost 50 years later three nests with eggs were located in rank grass behind an abandoned car park in the same suburb during December and January.[47] Three pairs were observed building nests within 10 m of one another at Berrimah Farm in early March 2007.[47] The above observations suggest that local Chestnut-breasted Mannikins breed during the Wet season, sometimes in small colonies, after which they aggregate into flocks that move towards the coast in the late Dry season. As the Wet approaches they disperse once again to breed, though how far is unknown. Interestingly, at Victoria River Downs, over 200 km from the nearest coast, the species was present during the Wet season but largely absent during the Dry season.[451]

Gouldian Finch
Erythrura gouldiae

RANGE: endemic to monsoon-tropical Australia. HABITATS: woodland. STATUS: vagrant. BREEDING: Jan–Aug (TE).[484]

Nolan Caldwell

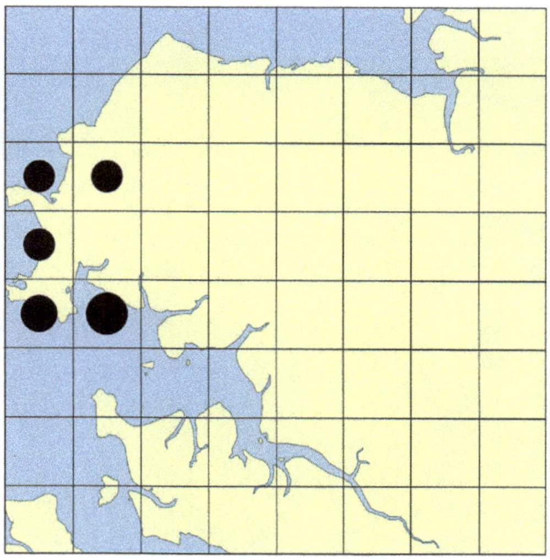

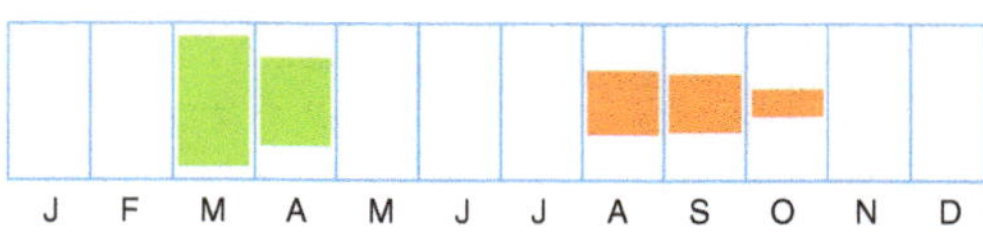

After suffering many decades of range contraction and population decline throughout its tropical range, the Gouldian Finch appears to be recovering in the Top End, and there has been an increasing number of sightings at subcoastal locations. Its status in the Darwin region, however, is difficult to assess, as separating individuals that originated in aviaries from wild birds is problematical, and there are some specimens in museum collections labelled 'Port Darwin District' that actually come from localities as distant as Pine Creek, some 200 km from Darwin.

While more than 20 birds were recorded at Fogg Dam in September 1967, there appear to be no specific reports of birds occurring in the Darwin region at that time.[1] A bird seen carrying nesting material near the old Darwin hospital in 1969[34] cannot be ruled out as an escapee, as Gouldian Finches will breed in captivity and it is not unlikely that caged birds will breed, or attempt to breed, if set free, as evidenced by feral populations in Australia of, for example, Scaly-breasted Munias, which derived from aviary stock. The majority of the post-1969 Darwin records are from the mid to late 1970s, when numbers of the species were still critically low and Darwin was recovering from tropical cyclone Tracy. It is quite likely therefore that many, if not all, of these records, which include adults and juveniles seen in suburban Fannie Bay in 1976,[2] represent escaped aviary stock.

Since 2005, there have been regular reports of Gouldian Finches during the Dry season at waterholes near the Mary River, where up to 400 or more birds have been observed coming to drink, as well as along the Marrakai Road, about 40 km east of the Darwin region, where the species was first recorded in 1970.[1] It seems likely that any wild Gouldian Finches reaching the Darwin region would derive from such areas to the east. A sighting in July 2008 at Humpty Doo, on the edge of our region, for example, may well represent wild birds. Although bird research has been carried out at Coomalie Farm, 45 km south of the Darwin region, since 1989, the species was first recorded there in 2012.[47]

Australian Pipit
Anthus australis

RANGE: whole of Australia, and c. NG. HABITATS: grassland and saltflats. STATUS: uncommon resident and Wet season visitor. BREEDING: Mar–Aug (TE).

David Webb

The distinctive tropical race of the Australian Pipit (*rogersi*) is both smaller and much more boldly marked on the head and underparts than birds of southern Australia.[411] In addition southern birds, apart from populations in the high country and Tasmania, are largely sedentary whereas Top End birds appear to make seasonal movements, possibly between habitats. Although generally scarce across the Top End,[41,103] the Pipit was described during the late 1960s and early 1970s as moderately common in all open habitats of the Darwin-Humpty Doo regions, but confined to bare or short-grass saline flats in the Wet season.[1,41]

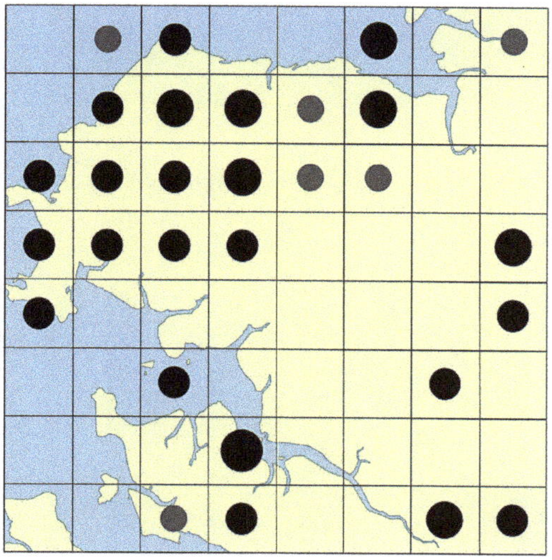

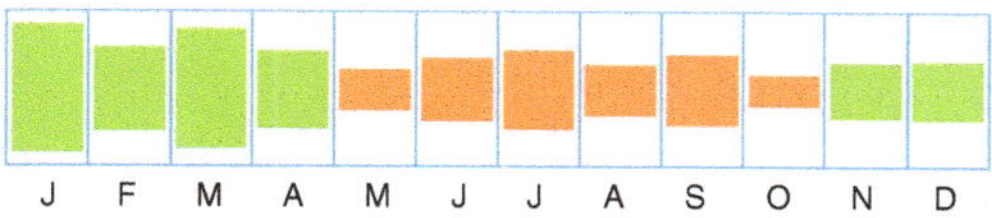

Reporting rates in the Darwin region since then suggest that the species is commoner in the mid to late Wet season, perhaps indicating a movement towards the coast during that period. Consistent with this, counts over 2.5 years (2003–05) at Darwin International Airport showed that it was most abundant in February and March, when average numbers ranged from five to 16 birds, while combined Dry season months averaged 1.5 birds.[385] Similarly, up to 22 birds were counted at the airport in late November 2010, while in September 2010 and May 2011 the maximum number was six and two, respectively.[47] Around Millner it was recorded only in mid-January 1989, when three birds were seen, despite almost daily bird censuses from 1988 to 1993.[296]

While the species was reported from 45% of cells, over a half of all records came from the Casuarina CR–Rapid Creek mouth area. Indeed, in the mid-1970s it was reported to be an occasional visitor to suburban grassy areas such as Casuarina Drive in Rapid Creek.[365] Surprisingly, it was not recorded at all during the Howard's Peninsula survey of April–May 1986 despite intensive sampling of seven grassland habitats.[308]

There are few published records of Pipits breeding in the Top End, but unpublished records suggest that unlike birds of southern Australia, which breed mainly in spring, local birds breed largely during the late Wet and Dry seasons.[411] The earliest record was of a nest with eggs in Leanyer on 9 April 1970.[1] On a melon farm at Lambell's Lagoon, about 18 km east of the Darwin region, two nests, one with eggs and the other with young chicks were found on 30 June 2012.[263] Breeding in the Dry season may be an adaptation by local birds to avoid Wet season flooding of their cup-shaped nests, which are typically built slightly below ground level.[411] Moreover, during the Wet season grasses and other ground cover plants tend to be high and dense, limiting feeding opportunities for this ground foraging species.

Eastern Yellow Wagtail
Motacilla tschutschensis

RANGE: breeds in e. Russia, se. Siberia, n. Japan and Alaska; winters in China, Taiwan to se. Asia, NG, Indonesia, Australia. HABITATS: wetlands and grassland. STATUS: moderately common late Dry, Wet season migrant. BREEDING: not in Australia.

Niven McCrie

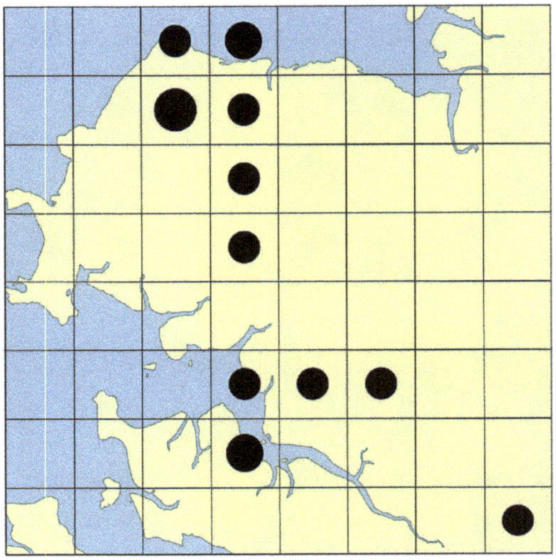

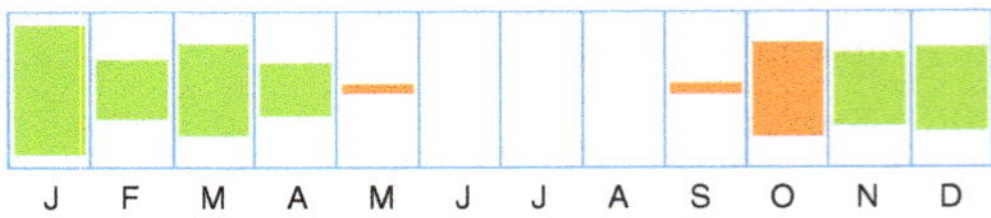

The Eastern Yellow Wagtail was first recorded in the NT in February 1968, when a single bird was seen near Fogg Dam. Up to five birds were present in December that year at Harrison Dam, where a specimen was collected in December 1969.[486] Now known to be an annual visitor to the Darwin region, this species typically arrives in October and November, though sometimes as early as September. For some time after their arrival, Eastern Yellow Wagtails appear wary and flush readily, whereas later in the season they can be approached more closely.

There appear to be few records in the Top End of multiple birds outside of the Darwin region; however, there were 31 observed at Lake Finniss in November 1993 and 27 there in October 1994.[26] Within the Darwin region there are reports of Eastern Yellow Wagtails from various locations but, other than a record of 25 birds at Shoal Bay Waste Management Facility in mid-March 1998,[21] these are generally only in small numbers, with the exception of Leanyer SP. Here, the maximum count has been 140 birds, in early March 1998.[21] Other high counts have all been in March and April, and probably represent pre-migration concentrations. Fluctuations in numbers present during these months, when most birds have acquired extensive yellow plumage, suggest that return migration takes place in waves rather than as a single event. The latest sighting was of a bird in full breeding plumage at Marlow Lagoon, Palmerston on 17 May.

Late afternoon observations of wagtails, singly and in groups, flying towards Leanyer from Shoal Bay and Knuckey Lagoons have been shown to coincide with birds arriving at the sewage ponds,[21] suggesting that birds foraging over a wide area during the day use the sewage works as a gathering area where they forage before roosting, probably in the mangroves.[487] The small numbers to be seen at the sewage ponds during the day, relative to the peak daily counts, support this. On those occasions when birds have been present in the middle of hot days, they are generally seen resting in shade, either on the ground or perched in trees.

While Yellow Wagtails are reported to feed mainly on flying insects in their wintering grounds,[488] observations in the Darwin region, mostly from Leanyer SP, indicate that the wagtails mainly forage by walking or sometimes darting after prey, picking items from the short grass, gravel, concrete boulders or mud, as well as from matted floating vegetation on the surface of sewage ponds. In longer grass, birds glean prey at low levels, or stretch to their limit with both feet on the ground, sometimes jumping to take prey from higher levels. The birds also hover at about a metre or more above ground to pick prey from grass heads, or jump-fly to take insects from the air, at heights of 2–3 m. When feeding in this manner wagtails have been seen to chase off and pursue Tree Martins and White-winged Terns, which were also feeding down to that height. Wagtails have also been observed to chase off Common Sandpipers, Sharp-tailed Sandpipers and Little Ringed Plovers. While there is usually little aggression between Wagtails, even when foraging closely, some birds have been observed grappling each other, followed by short flight-chases, mostly in March.

The taxonomy of the Yellow Wagtail group is still debated, and in some classifications the subspecies *taivana* is considered a separate species, the Green-headed Yellow Wagtail. Separating the two forms in the field can be difficult, even with close scrutiny. While birds showing prominent yellow brows and green crown, mantle and back may confidently be considered *taivana*, some individuals of this form show a distinct whitish supercilium. Additionally, first-winter *taivana* may be dull grey and white,[489] and as a result some may be overlooked. Conversely, *tschutschensis* may show some, even extensive, yellow on the brow, and the impression of its extent to an observer can be greatly exaggerated because of the human eye's heightened sensitivity to yellow. Nevertheless, there have been several sightings of *taivana*, confirmed by the combination of all diagnostic features, at Leanyer SP, and one was seen and photographed at Palmerston SP in October 2011.[490]

Grey Wagtail
Motacilla cinerea

RANGE: breeds Eurasia; winters Africa, e. Asia to Indonesia, NG. HABITATS: wetlands, flowing streams. STATUS: vagrant. BREEDING: not in Australia.

Adrian Boyle

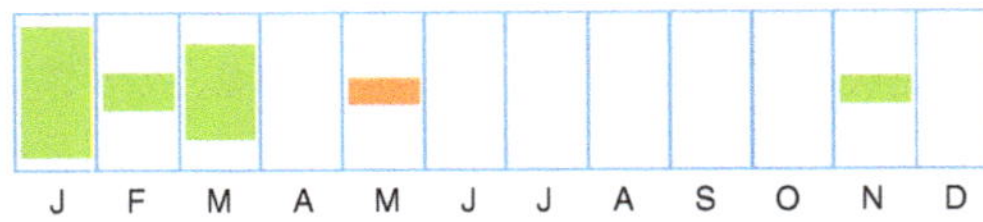

The first report of the Grey Wagtail for the Darwin region was of a single bird, probably an adult male, seen at Leanyer SP in March 1980.[491] The bird was observed for a short time but disappeared during a 30 minute absence by the observer and was not seen again, and it may be that the bird was found immediately before northward migration.[491] From January to March 1998, up to three birds were seen at Leanyer SP,[21] mainly as they foraged along a rocky stream, but they were also observed feeding along the gravel tracks in the sewage ponds, at times together with Eastern Yellow Wagtails. A sighting at Palmerston SP during November and December 2008 followed an October report of the species at Twin Falls, in Kakadu NP.[354] A Grey Wagtail was also seen and photographed foraging in a drain at Charles Darwin University Casuarina Campus during December 2014.[366] The bird was seen irregularly in the same area until early January 2015.

There have been reports of up to 10 Grey Wagtails in the sandstone uplands of Kakadu NP,[411] and it may be that the species occurs regularly there, or in similar habitat in Arnhem Land. However, this is difficult to confirm, as these areas are largely inaccessible to observers during the Wet season when Wagtails are likely to be present.

White Wagtail
Motacilla alba

RANGE: breeds Eurasia to Japan, s. and se. Asia; subsp. *leucopsis* winters s. Asia; subsp. *lugens* winters Myanmar, se. China, Taiwan, Japan. HABITATS: grassland, wetlands, suburbia. STATUS: vagrant. BREEDING: not in Australia.

The White Wagtail was first reported in the Darwin region during June 1993, when an adult male of the subspecies *leucopsis* was found at Leanyer SP. The species was subsequently seen during each of the following three years, all at the same site and all referable to the same subspecies. While the first three records were of adult

Adrian Boyle

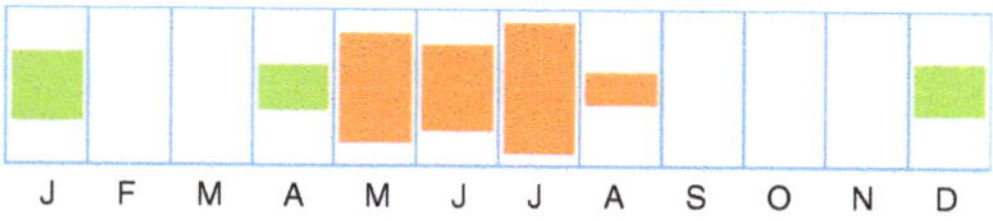

birds, perhaps the same individual, the bird present during 1996 was identified on plumage characters to be a first-year male, thus showing that at least two individuals were involved.[21,26] In each of the occurrences, the birds were present for up to four months, arriving – with one exception – in April and leaving in August or September, their stays covering their traditional northern hemisphere breeding seasons.

Another *leucopsis* White Wagtail was present at Leanyer SP for about six weeks from December 2003 to January 2004. In January 2009,[207] a White Wagtail of the *lugens* subspecies, commonly referred to as Black-backed Wagtail, was found at Leanyer SP, though in this case the bird was seen for only about a week. Perhaps coincidentally, a White Wagtail was seen at Pine Creek in October of that year, although the subspecies to which it belonged was not established. In March 2014, a White Wagtail of subspecies *lugens* was also found at Leanyer SP.[354]

European Goldfinch
Carduelis carduelis

RANGE: native Europe to Middle East; introduced se. Aust, including se. SA and Tas. HABITATS: grassland, woodland, suburbia. STATUS: probably escaped or released aviary stock. BREEDING: unknown (NT).

There have been irregular reports of European Goldfinches in the Darwin region since 1974, in widely scattered localities including Casuarina, Tiwi, Holmes Jungle, Mindil Beach and Berrimah.[3,21,34,101] The records are all likely to represent birds that had escaped from captivity, but feral breeding was suspected in the Nightcliff area where the species was seen from the early 1980s until 1992.[21] There have been sporadic reports of the species around other parts of Darwin since then, with the most recent sightings including three or four birds at Rapid Creek in March 2006,[3] a single bird near Charles Darwin University in February 2012, and a single bird feeding among a mixed group of finches at Lee Point in August 2013.[119]

Unconfirmed species

Stubble Quail
Coturnix pectoralis

RANGE: temperate and subtropical Australia. HABITATS: grassland. STATUS: unconfirmed. BREEDING: not in TE.

The first record of the Stubble Quail in the Top End was in June 1967, when a pair, including a male in breeding condition, showing bright orange facial skin, was seen at Middle Point,[1] near Fogg Dam. Five years later, in September 1972, a specimen was collected from nearby Harrison Dam.[34] In the Darwin region, the species was reported from Leanyer Swamp in September 1980,[135] but no details are available to substantiate the identification. Thus while the two earliest records can be considered confirmed, the bird's status in the Top End, particularly in northern areas, remains obscure.

Common Ringed Plover
Charadrius hiaticula

RANGE: breeds ne. Canada, Greenland, Scandinavia, n. Russia; winters Europe, Africa, sw. Asia. HABITATS: coast and wetlands. STATUS: unconfirmed. BREEDING: not in Australia.

The single report of the Common Ringed Plover in the Darwin region concerns a bird found at Leanyer SP in February 1980.[492] While mention was made in the published account of the need to compare this species with the Semipalmated Plover of North America, which has since been confirmed in Australia,[241] the only feature used to do so was the breadth of the breast band. Although the breast band may average slightly broader on Common Ringed Plover,[493] this character is of limited value in differentiating between the two species. On the basis of Australian records, the latter species is certainly the more likely, but there is insufficient evidence in the description to confirm its identity.

Caspian Plover
Charadrius asiaticus

RANGE: breeds in w., n. and e. Caspian Sea to e. Kazakhstan, nw. China; winters in e. and s. Africa. HABITATS: grassland, floodplains. STATUS: unconfirmed. BREEDING: not in Australia.

Reports of a lone Caspian Plover in late September 1974 at Leanyer Swamp, then of two birds among Oriental Plovers at the same site a week later,[494] lack sufficient details to be considered substantiated. The origin of a specimen purported to have been collected from the Pine Creek area on 26 September 1896[495] has also been questioned. A single confirmed record, from Lake Finniss, about 60 km east of Leanyer SP, in October 1994,[496] nevertheless suggests that the species may well visit Darwin occasionally.

Eurasian Curlew
Numenius arquata

RANGE: breeds n. Europe, central Asia, s. Urals and Kazakhstan; winters Africa, s. and se. Asia. HABITATS: tidal mudflats. STATUS: unconfirmed. BREEDING: not in Australia.

Purported Eurasian Curlews were reported as being seen 'on several occasions' at Nightcliff in late March and early April 1948.[110] The birds were described as being as large and long-billed as Far Eastern Curlews, but having white rumps. While the description certainly suggests Eurasian Curlew, the records have not been appraised by the BirdLife Australia Rarities Committee. A confirmed record of Eurasian Curlew in north-west Western Australia in November 2007,[497] however, offers support to its occurence in the Darwin region.

Spotted Redshank
Tringa erythropus

RANGE: breeds from n. Europe to e. Russia; winters from w. Europe and Africa to se. Asia. HABITATS: wetlands, occasionally muddy shores. STATUS: unconfirmed. BREEDING: not in Australia.

An alleged sighting of a Spotted Redshank on 16 November 1983 in 'Darwin' is the only published claim of this species for the Darwin region.[498] The record was considered as 'very likely' by the Birds Australia Rarities Committee, which concluded however that there was insufficient documentation to accept it as the first for Australia.[256]

Nordmann's Greenshank
Tringa guttifer

RANGE: breeds in the Sakhalin region of e. Russia; winters in s. and se. Asia. HABITATS: tidal mudflats, occasionally wetlands. STATUS: unconfirmed. BREEDING: not in Australia.

Observations of a purported Nordmann's Greenshank at Leanyer SP on two days in March 1974[494] are inadequately documented. Confirmed records of the species in north-west Australia in December 2006 and December 2009[499] nevertheless add some support to the possibility of its occurrence in the Darwin region.

Wandering Tattler
Tringa totanus

RANGE: breeds in ne. Siberia, Alaska; winters in sw. USA, S. America, Pacific islands, ne. Australia, NZ. HABITATS: coast; rocky shores, reefs and intertidal mudflats. STATUS: unconfirmed. BREEDING: not in Australia.

Problems of distinguishing the Wandering Tattler from the Grey-tailed Tattler make its status in the NT difficult to assess. It is unlikely to occur as more than a vagrant due to its generally Pacific Ocean wintering distribution and its probable entry into Australia from directly across that ocean.[209] Reports of the species from Darwin, identified on the basis of calls, are from Lee Point, December 1978, and Camerons Beach, April 1979.[251]

Latham's Snipe
Gallinago hardwickii

RANGE: breeds in n. Japan and e. Asia; majority of population winters in se. Australia. HABITATS: wetlands. STATUS: unconfirmed. BREEDING: not in Australia.

A snipe collected at Nightcliff in 1948[110] was identified by the collector as Latham's Snipe, on the basis of its having 16 tail feathers whereas Swinhoe's has 18 or more.[206] Subsequent study of the specimen showed it to be Swinhoe's Snipe.[500]

The only confirmed record of Latham's Snipe in the Top End was of a 'freshly dead' bird found in a vegetated drain under power lines at Jabiru in August 1985.[501] Given its normal wintering range of south-eastern Australia, the species is unlikely to be anything other than a vagrant to the Top End. However, any snipe seen early in the season, before October, is particularly worth scrutinising with this species in mind, as Swinhoe's Snipe tends to arrive later.

Pin-tailed Snipe
Gallinago stenura

RANGE: breeds in nc. and e. Russia, winters in s. and se. Asia to w. Indonesia, and occasionally nw. Australia. HABITATS: freshwater wetlands. STATUS: unconfirmed. BREEDING: not in Australia.

Breeding well north of the range of Swinhoe's Snipe and Latham's Snipe,[228] and over a much wider area, the Pin-tailed Snipe is the common snipe wintering in peninsular Malaysia and western Indonesia.[184] The two confirmed Australian records are from near Broome, Western Australia.[502,503]

On the evening of 1 January 1985, mist-nets were erected to capture snipe at Palmerston SP, where up to 45 birds had been observed foraging on the muddy ground at the bottom of an empty pond.[43] Of the 26 snipe captured, 25 were identified as Swinhoe's, but one was believed to be Pin-tailed. However, subsequent analysis of the measurement data of the suspected Pin-tailed show that the critical measurements fit within the range of the sampled Swinhoe's, and the photographs of the bird do not rule out the latter.

Gray's Grasshopper Warbler
Locustella fasciolata

RANGE: breeds from s. Siberia to n. Japan; winters from se. Asia to NG. HABITATS: vine thickets, forest edges, tall grass. STATUS: unconfirmed. BREEDING: not in Australia.

There is a published report of Gray's Grasshopper Warbler at Holmes Jungle in December 1979 and at Harrison Dam in January 1982.[504] The description provided is scant, and was considered by the Birds Australia Rarities Committee as insufficient to confirm the bird's identity.[256]

References

1. Crawford DN (1972) Birds of Darwin area, with some records from other parts of Northern Territory. *Emu* **72**, 131–148. doi:10.1071/MU972131
2. Thompson HAF (1977) Notes on the birds in the Darwin and northern areas of the Northern Territory. *The Sunbird* **8**, 83–91.
3. Keates A, Keates S Personal data.
4. Blakers M, Davies SJJF, Reilly PN (1984) *The Atlas of Australian Birds.* Melbourne University Press, Carlton.
5. Barrett G, Silcocks A, Barry S, Cunningham R, Poulter R (2003) *The New Atlas of Australian Birds.* Royal Australasian Ornithologists Union, Hawthorn East.
6. Taylor R, Woinarski J (2006) Emu *Dromaius novaehollandiae.* Threatened Species of the Northern Territory, PWCNT <http://www.lrm.nt.gov.au/_data/assets/pdf_file/0005/10859/emu_vu.pdf>.
7. Johnsgard PA (1965) *Handbook of Waterfowl Behavior: Tribe Anseranatini (Magpie Goose).* <http://digitalcommons.unl.edu/bioscihandwaterfowl/9>.
8. Whitehead PJ, Bayliss P, Yeomans K (1987) *The Distribution of the Nests of the Magpie Goose (*Anseranas semipalmata *Latham) at Kapalga NT: A Four-year Study.* Conservation Commission NT, Technical Report No 35, Palmerston NT.
9. Wetlands International (2013) Waterbird population estimates. Available at: http://www.wpe.wetlands.org (posted 19 November 2013).
10. Tulloch DG, McKean JL (1983) Magpie Goose populations on the coastal plains of the Northern Territory (1958–1980). *Corella* **7**(2), 32–36.
11. Whitehead PJ, Saalfeld K (2000) Nesting phenology of magpie geese (*Anseranas semipalmata*) in monsoonal northern Australia: responses to antecedent rainfall. *Journal of Zoology* **251**, 495–508. doi:10.1111/j.1469-7998.2000.tb00805.x
12. Traill LW, Bradshaw JA, Brook BW (2010) Satellite telemetry and seasonal movements of Magpie Geese (*Anseranas semipalmata*) in tropical northern Australia. *Emu* **110**, 160–164.
13. Shurcliff KS, McKean JL (1990) The status of ducks (Anatidae) in the Darwin area, Northern Territory, 1974–1986. *Australian Bird Watcher* **13**, 243–248.
14. Whitehead P, Bayliss P, Fox R (1988) Recreational waterfowl hunting activity and harvests in Northern Territory. *Australian Wildlife Research* **15**, 625–631. doi:10.1071/WR9880625
15. Brook BW, Whitehead PJ (2005) Plausible bounds for maximum rate of increase in magpie geese (*Anseranas semipalmata*): implications for harvest. *Wildlife Research* **32**, 465–471. doi:10.1071/WR02105
16. Vardon MJ, Missi C, Cleary M, Webb GJW (1997) Aboriginal use and conservation of wildlife in northern Australia: a cultural necessity. In *Conservation Outside Nature Reserves* (Hale PT, Lamb D eds), pp 241–245. Centre for Conservation Biology, University of Queensland, Brisbane.

17. Delaney R, Fukuda Y, Saalfeld K (2009) *Management Program for the Magpie Goose (Anseranas semipalmata) in the Northern Territory of Australia 2009–2014*. Parks and Wildlife Service of the Northern Territory.
18. Brann M (2012) Mango growers crying 'fowl' over bird damage <http://www.abc.net.au/site-archive/rural/content/2012/s3622518.htm>.
19. Mollah WS (1982) *Humpty Doo: Rice in the Territory*. Monograph. Australian National University North Australia Research Unit, Darwin.
20. Frith HJ (1977) *Waterfowl in Australia*. AH & AW Reed, Sydney.
21. McCrie N Personal data.
22. Niland D (1996) First record of Spotted Whistling Duck in Australia. *The Sunbird* **26**, 39–41.
23. Kyne P (2013) First record of Spotted Whistling Duck *Dendrocygna guttata* for the NT. *NT Naturalist* **24**, 50–54.
24. eBird Basic Dataset. Version: EBD_relAug-2013. Cornell Lab of Ornithology, Ithaca, New York. August 2013.
25. Jarvis M (2014) <https://groups.yahoo.com/neo/groups/ntbirds/conversations/messages/7631>.
26. Jaensch R *pers. comm.* (Wetlands International unpublished data).
27. Chatto R (2006) The Distribution and Status of Waterbirds around the Coast and Coastal Wetlands of the NT. Technical Report 76. Parks & Wildlife Commission NT, Palmerston.
28. Morton SR, Brennan KG, Armstrong MD (1991) *Distribution and Abundance of Waterbirds in the Alligator Rivers Region, Northern Territory*. Supervising Scientist for the Alligator Rivers Region, Jabiru.
29. Frith HJ, Davies SJJF (1961) Breeding seasons of birds in subcoastal Northern Territory. *Emu* **61**, 97–111. doi:10.1071/MU961097
30. Marchant S, Higgins PJ (eds) (1991) *Handbook of Australian, New Zealand and Antarctic Birds. Vol. 1: Ratites to Ducks*. Oxford University Press, Melbourne.
31. Thompson HAF Personal data.
32. Johnsgard PA (2010) *Ducks, Geese, and Swans of the World, Revised Edition*. University of Nebraska, Lincoln.
33. Mitchell D *pers. comm.*
34. Thompson HAF, Goodfellow DK (no date) *Annotated List of the Birds of the Top End*. Unpublished manuscript, Darwin, NT.
35. Jarvis M *pers. comm.*
36. Jaensch R (2003) Breeding by Freckled Duck *Stictonetta naevosa*, and a review of recent records in the Barkly Wetlands, Northern Territory. *NT Naturalist* **17**, 38–42.
37. Garland C *pers. comm.*
38. Chatto R (2000) *Waterbird Breeding Colonies in the Top End of the Northern Territory*. Technical Report. Parks and Wildlife Commission of the Northern Territory, Palmerston.
39. Kyne P (2012) Southern ducks hit the Top End. *Nature Territory* **June 2012**, 4.
40. Hensen B *pers. comm.*
41. Storr GM (1977) *Birds of the Northern Territory*. Western Australian Museum, Perth (Special Publication Number 7).
42. Rawsthorne J (2013) <https://groups.yahoo.com/neo/groups/ntbirds/conversations/messages/6830>.
43. Hertog AL Personal data.
44. Baker B Personal data.
45. Crawford D (1979) Effects of drought on Green Pygmy-Geese and Comb-crested Jacanas in Northern Territory. *Emu* **79**, 37–39. doi:10.1071/MU9790037

46. Ward MJ, Smith JG (2003) Brown Goshawk *Accipiter fasciatus* attacking a Green Pygmy-goose *Nettapus pulchellus* over and in water. *Australian Field Ornithology* **20**, 105–106.
47. Noske R Personal data.
48. Roshier DA, Klomp NI, Asmus M (2006) Movements of a nomadic waterfowl, Grey Teal *Anas gracilis*, across inland Australia-results from satellite telemetry spanning fifteen months. *Ardea* **94**, 461–475.
49. Smith A *pers. comm.*
50. Daly G *pers. comm.*
51. Bellio M *pers. comm.*
52. Thompson HAF (1978) Further records of palaearctic species in Darwin. *The Sunbird* **9**, 54–59.
53. Jerram M *pers. comm.*
54. Reis T *pers. comm.*
55. Jaensch R (2003) Records of northern waterbirds in the Barkly wetlands, Northern Territory, 1993–2002. *Northern Territory Naturalist* **17**, 20–30.
56. Jaensch RP (1989) *Birds of Wetlands and Grasslands in the Kimberley Division, Western Australia: Some Records of Interest, 1981–88.* Royal Australasian Ornithologists Union, Melbourne.
57. Birdlife Australia (no date) Hardhead <http://www.birdsaustralia.com.au/bird-profile/hardhead>.
58. Storr GM (1967) *List of Northern Territory Birds.* Western Australian Museum, Perth.
59. Noske RA, Van Gessel F (1987) First record of the Blue-billed Duck for the Northern Territory. *Northern Territory Naturalist* **10**, 13.
60. Thompson H, Goodfellow D (1987) *Common Birds of the Darwin Area.* Sandpiper Productions, Winnellie.
61. Rothwell N (2012) Love and war in our own backyard. <http://www.theaustralian.com.au/news/features/love-and-war-in-our-own-backyard/story-e6frg6z6-1226544278293?nk=a9376756dbefc8561ec62c1d1b33bbb2>.
62. Wright J *pers. comm.*
63. Jones D, Goth A (2008) *Mound-builders* CSIRO Publishing, Melbourne.
64. Franklin DC, Baker B (2005) The Orange-footed Scrubfowl *Megapodius reinwardt* as an urban bird in Darwin, NT. *Australian Field Ornithology* **22**, 48–50.
65. Palmer C, Christian KA, Fisher A (2000) Mound characteristics and behaviour of the Orange-footed Scrubfowl in the seasonal tropics of Australia. *Emu* **100**, 54–63. doi:10.1071/MU9853
66. Gillis M, Noske R (2007) Orange-footed Scrubfowl in Darwin – horticultural pest or partner? *Northern Territory Naturalist* **19**, 76–80.
67. Douglas F *pers. comm.*
68. Pettit M *pers. comm.*
69. Ainsworth G *pers. comm.*
70. White HL (1917) North Australian birds observed by William McLennan. *Emu* **16**, 205–231.
71. Eclipse L *pers. comm.*
72. Reed M (2004) <https://groups.yahoo.com/neo/groups/ntbirds/conversations/messages/1444>.
73. Walbridge P (1995) First Australian record of Streaked Shearwater? *Australian Birding* **2**(2), 7.
74. Cheshire N (2010) Procellariiformes observed around Papua New Guinea including the Bismarck Archipelago from 1985 to 2007. *South Australian Ornithologist* **36**, 9–24.
75. Carter M (1983) Streaked Shearwaters in Northern Australia. *Australian Bird Watcher* **10**(4), 113–121.
76. Percival D (1984) Streaked Shearwater *Calonectris leucomelas* in the Northern Territory. *Northern Territory Naturalist* **7**, 10–11.

77. Jessop R, Collins P (2000) The effects of cyclonic weather conditions on the bird life around Broome, Western Australia. *Stilt* **36**, 11–15.
78. McBride A *pers. comm.*
79. McKean JL, Gray D (1973) Unusual seabird records from the Northern Territory. *Emu* **73**, 184. doi:10.1071/MU973184a
80. Clarke RH (2010) *The Status of Seabirds and Shorebirds at Ashmore Reef and Cartier and Browse Islands: Monitoring Program for the Montara Well Release – Pre-impact Assessment and First Post-impact Field Survey.* Prepared on behalf of PTTEP Australasia and the Department of the Environment, Water, Heritage and the Arts, Australia (now the Department of Sustainability, Environment, Water, Population and Communities).
81. Kyne PM, Jackson MV (2011) Short-tailed Shearwater *Ardenna tenuirostris* in the Northern Territory. *Northern Territory Naturalist* **23**, 54.
82. Noske RA, Brennan GP (2002) *The Birds of Groote Eylandt.* Northern Territory University Press, Darwin.
83. Poole C, Brickle N, Bakewell D (2011) Pelagic birdwatching: South-East Asia's final frontier? *Birding Asia* **16**, 26–31.
84. Mlíkovský J (2010) Systematic notes on Asian birds: 74. A preliminary review of the grebes, family Podicipedidae. *British Ornithologists Club Occasional Publications* **5**, 125–131.
85. Trainor CR (2011) The waterbirds and coastal seabirds of Timor-Leste: new site records clarifying residence status, distribution and taxonomy. *Forktail* **27**, 63–72.
86. Coates BJ, Bishop KD (1997) *A Guide to the Birds of Wallacea: Sulawesi, the Moluccas and Lesser Sunda Islands, Indonesia.* Dove Publications, Alderley.
87. Trainor CR (2005) Waterbirds and coastal seabirds of Timor-Leste (East Timor): status and distribution from surveys in August 2002–December 2004. *Forktail* **21**, 61–78.
88. Hobcroft D *pers. comm.*
89. Shelley L *pers. comm.*
90. Andrew D (ed) (1996) This birding life. *Australian Birding* **Summer 1995–96**, 24.
91. Johnstone RE, Storr GM (1998) *Handbook of Western Australian Birds.* Western Australian Museum, Perth.
92. Gardner M (2011) <https://groups.yahoo.com/neo/groups/ntbirds/conversations/messages/5982>.
93. Estbergs J Personal data.
94. Jaensch R, Bellchambers K (1997) *Waterbird Conservation Values of Ephemeral Wetlands of the Barkly Tableland, Northern Territory.* Unpublished report to Australian Heritage Commission and Parks and Wildlife Commission of the Northern Territory.
95. Braithwaite RW, Estbergs J (1987) Fire-birds of the Top End. *Australian Natural History* **22**, 298–302.
96. Rix CE (1970) Birds of the NT. *South Australian Ornithologist* **25**, 147–191.
97. Nichols P *pers. comm.*
98. Boekel C (1980) Birds of the Victoria River Downs Station and of Yarralin, NT. Part 1. *Australian Bird Watcher* **8**(6), 171–193.
99. Webb D (2012) <https://groups.yahoo.com/neo/groups/ntbirds/conversations/messages/6697>.
100. Paterson L (2013) <https://groups.yahoo.com/neo/groups/ntbirds/conversations/messages/6742>.
101. Adlam N Personal data.
102. Chatto R *pers. comm.*
103. Brooker MG, Parker BS (1985) The birds of Kakadu National Park. *The Kakadu Fauna Survey: An ecological survey of Kakadu National Park* (Braithwaite R ed.), pp. 327–388.

104. Rosove M *pers. comm.*
105. Lendon A (1966) Notes on Northern Territory birds. *Australian Bird Watcher* **2**, 191–206.
106. Braithwaite R, Estbergs J (1982) Discovery of a Rufous Night Heron rookery in the Northern Territory. *Northern Territory Naturalist* **5**, 20.
107. Hancock J, Kushlan JA (2010) *The Herons Handbook*. Christopher Helm, London.
108. Pyle J (2000) Fish-baiting behaviour observed in Striated Heron *Butorides striatus. South Australian Ornithologist* **33**, 85–86.
109. Reynolds S *pers. comm.*
110. Deignan HG (1964) Birds of the Arnhem Land Expedition. In *Records of the American Australian Scientific Expedition to Arnhem Land* (Speight RL ed.), pp 345–425. Melbourne University Press, Melbourne.
111. Maddock M (1990) Cattle Egret: south to Tasmania and New Zealand for the winter. *Notornis* **37**, 1–2.
112. Bridgman HA, Maddock M (1994) Meteorological conditions and cattle egret migration: an update. *Notornis* **41**, 189–204.
113. McKilligan N (2005) *Herons, Egrets and Bitterns: Their Biology and Conservation in Australia*. CSIRO Publishing, Melbourne.
114. Heron Conservation (2011) White-necked Heron *Ardea pacifica* (Latham) <http://www.heronconservation.org/styled-5/styled-64/>.
115. Recher HF, Holmes RT, Davis WE, Jr, Morton S (1983) Foraging behavior of Australian herons. *Colonial Waterbirds* **6**, 1–10. doi:10.2307/1520961
116. Schulz M (1989) *The Importance of Wetlands in Kakadu National Park to Selected Waterbirds*. A Report to the Australian National Parks and Wildlife Service, May 1989. Australian National Parks and Wildlife Service, Canberra.
117. Seton DHC (1973) Observations on breeding of the Great-billed Heron in northern Queensland. *Emu* **73**, 9–11. doi:10.1071/MU973009
118. May I *pers. comm.*
119. Rawsthorne J *pers. comm.*
120. Carter M, Menkhorst P (2006) Nominate subspecies of the Little Egret *Egretta garzetta garzetta* in Australian territory. *Australian Field Ornithology* **23**, 104.
121. Gilfedder M *pers. comm.*
122. Garnett S (2013) <https://groups.yahoo.com/neo/groups/ntbirds/conversations/messages/6981>.
123. Michael B (2007) <https://groups.yahoo.com/neo/groups/ntbirds/conversations/messages/3018>.
124. Commonwealth of Australia Bureau of Meteorology (2014) Climate statistics for Australian locations <http://www.bom.gov.au/climate/averages/tables/cw_014015.shtml>.
125. Birch D (2000) Feeding ecology of Greater Frigatebirds *Fregata minor* and Lesser Frigatebirds *F. ariel* on Aride Island. *Phelsuma* **8**, 29–33.
126. Whiting SD, Noske RA, Guinea M, Chatto R, Donato D (1997) Observations of birds on islands in Northern Fog Bay, Northern Territory. *Northern Territory Naturalist* **15**, 27–40.
127. Chatto R (2006) *The Distribution and Status of Waterbirds around the Coast and Coastal Wetlands of the Northern Territory*. Technical Report. Parks & Wildlife Commission NT, Palmerston.
128. Clarke RH, Carter M, Swann G, Thomson J (2011) The status of breeding seabirds and herons at Ashmore Reef, off the Kimberley coast, Australia. *Journal of the Royal Society of Western Australia* **94**, 365–376.

129. McKean JL, Bartlett MC, Perrins CM (1975) New records from the Northern Territory. *Australian Bird Watcher* **6**, 45–46.
130. McCrie T *pers. comm.*
131. Dostine PL, Morton SR (1988) Notes on the food and feeding habits of cormorants on a tropical floodplain. *Emu* **88**, 263–266. doi:10.1071/MU9880263
132. Bartlett M (1981) A Great Cormorant *Phalacrocorax carbo*, in the Darwin area, Northern Territory. *Australian Bird Watcher* **9**(1), 27.
133. Boekel C (1978) Notes on the Pied Cormorant and Black Cormorant in the Northern Territory. *Australian Bird Watcher* **7**, 209–211.
134. Christidis L, Boles W (2008) *Systematics and Taxonomy of Australian birds.* CSIRO Publishing, Melbourne.
135. Marchant S, Higgins PJ (eds) (1994) *Handbook of Australian, New Zealand and Antarctic Birds. Vol. 2: Raptors to Lapwings.* Oxford University Press, Melbourne.
136. Binns D *pers. comm.*
137. Madsen T, Ujvari B, Shine R, Buttemer W, Olsson M (2006) Size matters: extraordinary rodent abundance on an Australian tropical flood plain. *Austral Ecology* **31**, 361–365. doi:10.1111/j.1442-9993.2006.01564.x
138. Madsen T, Ujvari B, Shine R, Olsson M (2006) Rain, rats and pythons: climate-driven population dynamics of predators and prey in tropical Australia. *Austral Ecology* **31**, 30–37. doi:10.1111/j.1442-9993.2006.01540.x
139. Pettit M (2012) <https://groups.yahoo.com/neo/groups/ntbirds/conversations/messages/6241>.
140. Rawsthorne J (2012) <https://groups.yahoo.com/neo/groups/ntbirds/conversations/messages/6236>.
141. O'Brien G, O'Brien M (2008) <https://groups.yahoo.com/neo/groups/ntbirds/conversations/messages/3837>.
142. Bezuijen M *pers. comm.*
143. Frith HJ, Hitchcock WB (1974) Birds. In *Fauna Survey of the Port Essington district, Cobourg Peninsula, Northern Territory of Australia.* CSIRO, Melbourne.
144. Boekel C (1976) Ten months on Gove Peninsula. *Australian Bird Watcher* **6**, 231–246.
145. Garnett S (2005) <https://groups.yahoo.com/neo/groups/ntbirds/conversations/messages/1773>.
146. Donato D *pers. comm.*
147. Popple S (2003) <https://groups.yahoo.com/neo/groups/ntbirds/conversations/messages/900>.
148. Aumann T, Baker-Gabb D (1991) *The Ecology and Status of the Red Goshawk in Northern Australia.* Report 75. Royal Australasian Ornithologists Union, Moonee Ponds.
149. Favaloro NJ (1981) The Red Goshawk. *Australian Bird Watcher* **9**, 44–53.
150. Robertson S *pers. comm.*
151. Riddell W (2011) The juvenile plumage of the Grey Goshawk *Accipiter novaehollandiae* in tropical Australia. *Australian Field Ornithology* **28**, 180–185.
152. Gardner M (2007) <https://groups.yahoo.com/neo/groups/ntbirds/conversations/messages/3194>.
153. Riddell W (2013) Double-brooding and other observations of Grey Goshawks in Darwin, Northern Territory. *Australian Field Ornithology* **30**, 152–156.
154. Bell A (2009) <https://groups.yahoo.com/neo/groups/ntbirds/conversations/messages/4403>.
155. Potts R (2005) <https://groups.yahoo.com/neo/groups/ntbirds/conversations/messages/1968>.
156. Riddell W (2012) <https://groups.yahoo.com/neo/groups/ntbirds/conversations/messages/6310>.
157. Griffioen PA, Clarke MF (2002) Large-scale bird-movement patterns evident in eastern Australian atlas data. *Emu* **102**, 99–125. doi:10.1071/MU01024

158. Riddell W (2011) Double-brooding by the Brown Goshawk *Accipiter fasciatus* in the Northern Territory. *Australian Field Ornithology* **28**, 92–93.
159. McKay R (2002) <https://groups.yahoo.com/neo/groups/ntbirds/conversations/messages/2353>.
160. Garnett S (2012) <https://groups.yahoo.com/neo/groups/ntbirds/conversations/messages/6145>.
161. Potts R (2002) <https://groups.yahoo.com/neo/groups/ntbirds/conversations/messages/156>.
162. Ainsworth G (2011) <https://groups.yahoo.com/neo/groups/ntbirds/conversations/messages/5620>.
163. Riddell W (2014) <https://groups.yahoo.com/neo/groups/ntbirds/conversations/messages/7722>.
164. Sedgwick EH (1946) Northern Territory bird notes. *Emu* **46**, 294–308. doi:10.1071/MU946294
165. Wright J, Wright J (2010) <https://groups.yahoo.com/neo/groups/ntbirds/conversations/messages/5037>.
166. Brown G (2012) <https://groups.yahoo.com/neo/groups/ntbirds/conversations/messages/6165>.
167. Sedgwick EH (1947) Northern Territory bird notes. Part 2. *Emu* **47**, 349–378.
168. Bell A (2011) <https://groups.yahoo.com/neo/groups/ntbirds/conversations/messages/5983>.
169. Lindsay BY, Waliwararra K, Miljat F (2001) Malakmalak and matngala plants and animals. Aboriginal flora and fauna knowledge from the Daly River area, northern Australia. *Northern Territory Botanical Bulletin* **26**, 1–128.
170. Estbergs JA (1987) Unusual feeding observations of some Top End birds. *Northern Territory Naturalist* **10**, 16.
171. Rawsthorne J (2012) <https://groups.yahoo.com/neo/groups/ntbirds/conversations messages/6334>.
172. Riddell W (2013) Raptor observations in Darwin, Northern Territory. *Australian Field Ornithology* **30**, 160–163.
173. Wright J, Wright J (2003) <https://groups.yahoo.com/neo/groups/ntbirds/conversations/messages/666>.
174. Rawsthorne J (2011) <https://groups.yahoo.com/neo/groups/ntbirds/conversations/messages/5841>.
175. Gardner M (2013) <https://groups.yahoo.com/neo/groups/ntbirds/conversations/messages/7026>.
176. Corbett L, Hertog T (2011) Diet and breeding of White-bellied Sea-Eagles *Haliaeetus leucogaster* in subtropical river habitats in the Northern Territory, Australia. *Corella* **35**, 41–48.
177. Rhodes LJ (1944) Birds of the Adelaide River District, Northern Territory. *Emu* **44**, 87–93. doi:10.1071/MU944087
178. Ziembicki M (2010) *Australian Bustard.* CSIRO Publishing, Melbourne.
179. Funnell D *pers. comm.*
180. McKean JL (1983) Lewin's Rail *Rallus pectoralis*, a new addition to the avifauna of the Northern Territory. *Australian Bird Watcher* **10**(4), 137.
181. McKean JL, Reed M (1979) Further sightings of the bush-hen *Gallinula olivacea* from sub-coastal Northern Territory. *The Sunbird* **103–4**, 73–74.
182. Mason I, Wolfe T (1975) First record of the Marsh Crake for the Northern Territory. *Emu* **75**, 235.
183. McFadden C (2004) <https://groups.yahoo.com/neo/groups/ntbirds/conversations/messages/1661>.
184. Wells D, Round PD, Treesucon U (1999) *The Birds of the Thai-Malay Peninsula: Covering Burma and Thailand South of the Eleventh Parallel, Peninsular Malaysia and Singapore.* Academic Press, San Diego.
185. Bell A (2011) <https://groups.yahoo.com/neo/groups/ntbirds/conversations/messages/5588>.

186. Barnes TA, Franklin DC (1997) Notes on nests of the Chestnut Rail. *Northern Territory Naturalist* **15**, 6–11.
187. Johnstone RE (1990) *Mangroves and Mangrove Birds of Western Australia.* Western Australian Museum, Perth.
188. Sonnenburg R (2009) <https://groups.yahoo.com/neo/groups/ntbirds/conversations/messages/4507>.
189. Hadden D (2004) <https://groups.yahoo.com/neo/groups/ntbirds/conversations/messages/1100>.
190. Barnes C *pers. comm.*
191. Whitehead B *pers. comm.*
192. Jaensch RP (1994) *An Inventory of Wetlands of the Sub-humid Tropics of the Northern Territory.* Conservation Commission of the Northern Territory, Palmerston.
193. Hall BP (ed.) (1974) *Birds of the Harold Hall Australian Expeditions, 1962–1970: A Report on the Collections Made for the British Museum of Natural History.* British Museum (Natural History), London.
194. Hance I *pers. comm.*
195. Chatto R (2003) *The Distribution and Status of Shorebirds Around the Coast and Coastal Wetlands of the Northern Territory.* Technical Report. Parks and Wildlife Commission of the Northern Territory, Palmerston.
196. Keates A, Collins T, Franklin DC (2008) Interspecific aggression in the Beach Stone-curlew *Esacus magnirostris. Northern Territory Naturalist* **24**, 34–37.
197. Franklin D *pers. comm.*
198. AWSG count data, Darwin area Dec 2008–Mar 2014, supplied by Gavin O'Brien.
199. Eclipse L (2010) <https://groups.yahoo.com/neo/groups/ntbirds/conversations/messages/5138>.
200. Hansen B, Minton C, Jessop R, Collins P (2009) Biometrics, sexing criteria, age-structure and moult of Sooty Oystercatchers in south-eastern and north-western Australia. *Emu* **109**, 25–33. doi:10.1071/MU07045
201. del Hoyo J, Elliott A, Sargatal J (eds) (1996) *Handbook of the Birds of the World. Vol. 3. Hoatzin to Auks.* Lynx Edicions, Barcelona.
202. Dostine PL, Morton SR (1989) Food of the Black-winged Stilt *Himantopus himantopus* in the Alligator Rivers Region, Northern Territory. *Emu* **89**, 250–253. doi:10.1071/MU9890250
203. Garnett S *pers. comm.*
204. Keates A, Keates S (2004) <https://groups.yahoo.com/neo/groups/ntbirds/conversations/messages/1685>.
205. BirdLife International (2014) The BirdLife checklist of the birds of the world: Version 7. Downloaded from <http://www.birdlife.org/datazone/userfiles/file/Species/Taxonomy/BirdLife_Checklist_Version_70.zip>.
206. Geering A, Agnew L, Harding S (2007) *Shorebirds of Australia.* CSIRO Publishing, Melbourne.
207. Bell A. *pers. comm.*
208. Minton C, Wahl J, Jessop R, Hassell C, Collins P, Gibbs H (2006) Migration routes of waders which spend the non-breeding season in Australia. *Stilt* **50**, 135–157.
209. Lane BA (1987) *Shorebirds in Australia.* Nelson, Melbourne.
210. Nebel S, Rogers KG, Minton CDT, Rogers DI (2013) Is geographical variation in the size of Australian shorebirds consistent with hypotheses on differential migration? *Emu* **113**, 99–111. doi:10.1071/MU12076
211. Rogers DI (2012) Females fly further – extreme differential migration in the Grey Plover. *Stilt* **62**, 65.

212. Shurcliff KS, Thompson HAF (1993) Seasonal patterns of wader populations in Darwin, Northern Territory, 1974–1987. *Northern Territory Naturalist* **14**, 10–18.
213. Bellio M (2002) <https://groups.yahoo.com/neo/groups/ntbirds/conversations/messages/220>.
214. Bamford M, Watkins D, Bancroft W, Tischler G, Wahl J (2008) *Migratory Shorebirds of the East Asian-Australasian Flyway: Population Estimates and Internationally Important Sites.* Wetlands International - Oceania. Canberra.
215. Schodde R (1973) *Birds in CSIRO Division of Wildlife Research Alligator Rivers Region Fact Finding Study.* Wildlife Canberra, CSIRO Publishing, Melbourne.
216. Jaensch RP (1982) Little Ringed Plover at Bool Lagoon. *SA Ornithologist* **28**, 201–204.
217. McCrie N (1995) First record of the Kentish Plover *Charadrius alexandrinus* in Australia. *Australian Bird Watcher* **16**, 91–95.
218. Kennerley PR, Bakewell DN, Round PD (2008) Rediscovery of a long-lost *Charadrius* plover from South-East Asia. *Forktail* **24**, 63–79.
219. Minton C (2006) The history of wader studies in north-west Australia. *Stilt* **50**, 224–234.
220. Minton C (2008) Report on field-work in the Darwin area NT, 26 October to 2 November 2008.
221. Minton C, Jessop R, Collins P, Hassell C, Ewing A, Gibbs H (2005) Sightings of waders and terns leg-flagged in north-west Australia: report number 9 *Stilt* **47**, 47–57.
222. Michael B (2004) <https://groups.yahoo.com/neo/groups/ntbirds/conversations/messages/1684>.
223. Lane B, Jessop R (1985) Report on the 1985 north-west Australia Wader Studies Expedition. *Stilt* **6**, 2–16.
224. Rawsthorne J (2013) <https://groups.yahoo.com/neo/groups/ntbirds/conversations/messages/6830>.
225. Jaensch R (2003) Recent records and breeding of Painted Snipe *Rostratula benghalensis* in the Mitchell Grass Downs and Sturt Plateau, Northern Territory. *Northern Territory Naturalist* **17**, 31–37.
226. Dostine PL, Morton SR (2000) Seasonal abundance and diet of the Comb-crested Jacana. *Irediparra gallinacea* in the tropical NT. *Emu* **100**, 299–311. doi:10.1071/MU9833
227. Crawford DN (1972) First Australian record of Asiatic Dowitcher. *Emu* **72**, 112–113.
228. Higgins PJ, Davies SJ (1996) *Handbook of Australian, New Zealand and Antarctic Birds. Vol. 3: Snipe to Pigeons.* Oxford University Press, Melbourne.
229. Battley PF, Warnock N, Tibbitts TL, Gill RE, Piersma T, Hassell CJ *et al.* (2012) Contrasting extreme long-distance migration patterns in bar-tailed godwits *Limosa lapponica. Journal of Avian Biology* **43**, 21–32. doi:10.1111/j.1600-048X.2011.05473.x
230. Jaensch R (1994) Lake Finniss: an internationally significant site for Little Curlew. *Stilt* **25**, 21.
231. Garnett S, Minton C (1985) Notes on the movements and distribution of Little Curlew *Numenius minutus* in Northern Australia. *Australian Bird Watcher* **11**(3), 69–73.
232. McKean JL, Shurcliff KS, Thompson HAF (1986) Notes on the status of the Little Curlew *Numenius minutus* in the Darwin Area, Northern Territory, 1974–1985. *Australian Bird Watcher* **11**, 258–260.
233. Collins P, Jessop R (2001) Arrival and departure dates and habitat of little curlew *Numenius minutus* at Broome, north-western Australia. *Stilt* **39**, 10–12.
234. Bellio MG, Bayliss P, Morton S, Chatto R (2006) Status and conservation of the Little Curlew *Numenius minutus* on its over-wintering grounds in Australia. In *Waterbirds Around the World: A Global Overview of the Conservation, Management and Research of the World's Waterbird Flyways* (Boere G, Galbraith C, Stroud D eds), pp. 346–348. The Stationery Office, Edinburgh.

235. Minton C, Jessop R, Collins P, Standen R (2011) The migration of Eastern Curlew Numenius madagascariensis to and from Australia. *Stilt* **59**, 6–16.
236. Wilson JR (2000) The northward movement of immature Eastern Curlews in the Austral winter as demonstrated by the Population Monitoring Project. *Stilt* **36**, 16–19.
237. Crawford DN (1972) First record of Redshanks in Australia. *Emu* **72**, 112.
238. Crossland AC, Sinambela SA, Sitorus AS, Sitorus AW (2006) An overview of the status and abundance of migratory waders in Sumatra, Indonesia. *Stilt* **50**, 90–95.
239. McCrie N (2000) A sighting of the Green Sandpiper *Tringa ochropus* at Darwin, Northern Territory. *Australian Bird Watcher* **18**, 229–232.
240. McKean JL (1980) A sight record of the Green Sandpiper *Tringa ochropus* from the Northern Territory. *Australian Bird Watcher* **8**, 165–166.
241. Birds Australia Rarities Committee (2014) *BARC_Index_of_Cases_v2014Jan.xls* <http://www.tonypalliser.com/barc/barc-home.html>.
242. Cramp S, Simmons K (eds) (1983) *The Birds of the Western Palaearctic, Vol 3, Waders to Gulls.* Oxford University Press, Oxford.
243. Fry G (1990) Biometrics, moult and migration of Terek Sandpipers *Tringa terek* spending the non-breeding season in North West Australia. *Stilt* **17**, 24–28.
244. Martindale J (1982) National Winter Count 1981. *Stilt* **2**, 2–9.
245. Watkins D (1993) *A National Plan for Shorebird Conservation in Australia.* Australasian Wader Studies Group RAOU Report No 90. Royal Australian Ornithologists' Union, Moonee Ponds.
246. Keates A, Keates S (2006) Possible territorial behaviour of Common Sandpiper on non-breeding grounds. *Stilt* **49**, 44.
247. Lane B, Jessop R (1983) *National Wader Count, Summer 1983, Report to Participants.* Royal Australian Ornithologists' Union, Melbourne.
248. Barter M, Tonkinson D, Tang SX, Yuan X, Qian FW (1997) Staging of Great Knot *Calidris tenuirostris*, Red Knot *C. canutus* and Bar-tailed Godwit *Limosa lapponica* at Chongming Dao, Shanghai: jumpers to hoppers. *Stilt* **31**, 2–11.
249. Hassell C, Southey I, Boyle A, Yang HY (2011) Red knot *Calidris canutus*: subspecies and migration in the East Asian-Australasian flyway-where do all the red knot go. *BirdingASIA* **16**, 89–93.
250. Rogers DI, Hassell CJ, Boyle A, Gosbell K, Minton C, Rogers KG, Clarke RH (2011) Shorebirds of the Kimberley Coast – populations, key sites, trends and threats. *Journal of the Royal Society of Western Australia* **94**, 377–391.
251. McKean JL, Hertog AL (1981) Some further records of uncommon migrant waders near Darwin, NT. *Northern Territory Naturalist* **4**, 10–13.
252. McKean JL (1984) A Northern Territory sighting of the Baird's Sandpiper. *Australian Bird Watcher* **10**, 169.
253. Garnett S, Szabo J, Dutson G (2011) *The Action Plan for Australian Birds 2010.* CSIRO Publishing, Melbourne.
254. Crawford DN (1997) Wader censuses from Darwin, Northern Territory, Australia, 1970 to 1972. *Stilt* **30**, 32–34.
255. McKean JL, Hertog AL, Marr N (1982) An Australian record of the Stilt Sandpiper (*Micropalama himantopus*) *Northern Territory Naturalist* **5**, 22–23.
256. Patterson RJ (1996) *RAOU Records Appraisal Committee: Opinions and Case Summaries 1992–1995.* Royal Australian Ornithologists Union, Hawthorn East.
257. McCrie N (1987) Recent NT record of Red-necked Phalarope. *Northern Territory Naturalist* **10**, 17.
258. Kyne P (2014) <https://groups.yahoo.com/neo/groups/ntbirds/conversations/messages/7357>.
259. Minton C *pers. comm.*

260. Ward VL (2006) Austral winter records of the Grey Phalarope *Phalaropus fulicarius* in southern Africa. *Wader Study Group Bulletin* **109**, 123.
261. Taufiqurrahman I, Sujatmiko D, Utomo MA (2011) First record of Red Phalarope *Phalaropus fulicaria* for Indonesia. *Kukila* **15**, 106–108.
262. Bamford MJ (1988) *Kakadu National Park: a Preliminary Survey of Migratory Waders, October/November 1987.* Royal Australian Ornithologists' Union, Melbourne.
263. Webb D *pers. comm.*
264. Jaensch RP (1994) Internationally significant numbers of Oriental Plover and Oriental Pratincole at ephemeral inland wetlands in the Northern Territory, Australia. *Stilt* **24**, 12–13.
265. McKean JL, Gray D (1973) Common Noddy in the NT. *Emu* **73**, 184. doi:10.1071/MU973184a
266. Jackson M (2014) <https://groups.yahoo.com/neo/groups/ntbirds/conversations/messages/7298>.
267. Bellio MG, Bayliss P, Williams AJ, van Dam R, Fox GJ, Moulden JH (2007) *A preliminary ecological risk assessment of the impact of tropical fire ants* (Solenopsis geminata) *on colonies of seabirds at Ashmore Reef.* Department of the Environment and Water Resources Supervising Scientist, Canberra.
268. Stenhouse IJ, Robertson GJ (2005) Philopatry, site tenacity, mate fidelity, and adult survival in Sabine's gulls. *The Condor* **107**, 416–423. doi:10.1650/7627
269. Shannon G, McKean JL (1983) First record of Sabine's Gull *Xema sabini* from Australia. *Australian Bird Watcher* **10**, 82–83.
270. Mathews GM (1914) A list of the birds of Melville Island, NT, Australia. *The Ibis* **2**, 91–132.
271. McCrie N, McCrie T (1999) Sight records of a Black-headed Gull *Larus ridibundus* in the Northern Territory. *Australian Bird Watcher* **18**, 87–92.
272. Birds Australia Rarities Committee (undated) Submission No 703: Black-headed Gull *Chroicocephalus ridibundus* Darwin, NT, 30 December 2009 – 6 March 2010. <http://www.tonypalliser.com/barc/summaries/SUMM703.htm>.
273. McKean JL, Thompson HAF (1983) Northern Territory record of the Japanese Gull *Larus crassirostris. Australian Bird Watcher* **10**, 84–85.
274. Schodde R (1991) The Asian subspecies of Gull-billed Tern *Sterna nilotica affinis* in Australia. *Bulletin of The British Ornithologists' Club* **111**, 215–217.
275. Rogers DI, Collins P, Jessop RE, Minton CD, Hassell CJ (2005) Gull-billed Terns in north-western Australia: subspecies identification, moults and behavioural notes. *Emu* **105**, 145–158. doi:10.1071/MU04045
276. McKean JL (1981) The status of gulls and terns (Laridae) in the Darwin area 1974–80. *Australian Seabird Group Newsletter* **15**, 11–16.
277. Harrison L, McGuire L, Ward S, Fisher A, Pavey C, Fegan M, Lynch B (2009) *An Inventory of Sites of International and National Significance for Biodiversity Values in the Northern Territory.* Department of Natural Resources, Environment, The Arts and Sport, Darwin.
278. Jaensch RP, Vervest RM (1990) *Waterbirds at Remote Wetlands in Western Australia 1986–8. Part Two, Lake MacLeod, Shark Bay, Camballin Floodplain and Parry Floodplain.* Royal Australasian Ornithologists Union, Melbourne.
279. Noske RA, Watanabe Y Unpublished data.
280. Watanabe Y (1999) *The Breeding Biology of Crested Tern* Sterna bergii *on Seagull Island, Northern Territory, in Relation to Egg Harvest by Aboriginal People.* Unpublished Graduate Diploma (Science) School of Biological & Environmental Sciences, Northern Territory University, Darwin.
281. Conservation Commission of Western Australia (2010) *Status Performance Assessment: Biodiversity Conservation on Western Australian islands Phase II – Kimberley Islands.* Final report.

282. Nicholson L (1998) A new breeding record for the Lesser Crested Tern *Sterna bengalensis* in the Lowendal Islands, northwestern Australia. *Records of the Western Australian Museum* **19**, 241.
283. Collins P, Jessop R (1997) Little Terns *Sterna albifrons* in the West Kimberley Division of Western Australia. *Stilt* **26**, 26–31.
284. O'Brien G, O'Brien M (2012) <https://groups.yahoo.com/neo/groups/ntbirds/conversations/messages/6784>.
285. Stewart A (2011) <https://groups.yahoo.com/neo/groups/ntbirds/conversations/messages/5376>.
286. Gardner M (2014) <https://groups.yahoo.com/neo/groups/ntbirds/conversations/messages/7321>.
287. Anon (no date) Ringing recoveries. Seychelles Bird Records Committee. <http://www.seychellesbirdrecordscommittee.com/ringing-recoveries.html>.
288. Guinea M, Ryan P (1990) Observations of turtles and birds on Bare Sand Island. *Northern Territory Naturalist* **12**, 26–27.
289. Schodde R, Mason IJ (1974) Common Tern in the Northern Territory. *Emu* **74**, 103–104. doi:10.1071/MU974103b
290. Morton SR, Brennan KG, Armstrong MD (1993) Distribution and abundance of grebes, pelicans, darters, cormorants, rails and terns in the Alligator Rivers Region, Northern Territory. *Wildlife Research* **20**, 203–217. doi:10.1071/WR9930203
291. Kyne P *pers. comm.*
292. Huyser O (1997) Skuas – pirates of the oceans. *Africa – Birds and Birding* **2**, 57–63.
293. Thompson HAF (1978) First Northern Territory record of Arctic Skua. *The Sunbird* **9**, 9.
294. Gosford B *pers. comm.*
295. Read R (2002) <https://groups.yahoo.com/neo/groups/ntbirds/conversations/messages/185>.
296. Healey C Personal data.
297. PWCNT (1998) *Feral Pigeon Eradication Program: Darwin 1997.* Parks & Wildlife Commission NT, Darwin.
298. Heaton S *pers. comm.*
299. Frith HJ (1982) *Pigeons and Doves of Australia.* Rigby, Adelaide.
300. McCrie N, Watson J (2003) *Finding Birds in Darwin, Kakadu and the Top End, Northern Territory, Australia.* Niven McCrie, Casuarina.
301. Franklin DC *pers. comm.*
302. Parker SA (1970) Critical notes on the status of some Northern Territory birds. *South Australian Ornithologist* **25**, 115–125.
303. Garnett ST, Crowley GM (2000) *The Action Plan for Australian Birds.* Environment Australia, Canberra.
304. Fraser F (2000) Species profile: Patridge Pigeon *Geophaps smithii* (Columbidae) *Northern Territory Naturalist* **16**, 38–39.
305. Fraser F, Lawson V, Morrison S, Christopherson P, McGreggor S, Rawlinson M (2003) Fire management experiment for the declining Partridge Pigeon, Kakadu National Park. *Ecological Management & Restoration* **4**, 94–102. doi:10.1046/j.1442-8903.2003.00142.x
306. Donato D (2005) <https://groups.yahoo.com/neo/groups/ntbirds/conversations/messages/1849>.
307. Noske RA, Spiers A, Carmichael Z, Ryan H, Scott T (unpublished) Dry season bird survey of Palmerston suburbs.
308. Woinarski JCZ, Tidemann SC, Kerrin S (1988) Birds in a tropical mosaic: the distribution of bird species in relation to vegetation patterns. *Australian Wildlife Research* **15**, 171–196. doi:10.1071/WR9880171
309. Carlson AJ Personal data.

310. Garnett S, Crowley G (1994) Wet-season feeding by four species of granivorous birds in the Northern Territory. *Australian Bird Watcher* **15**, 306–309.

311. Woinarski JCZ, Tidemann SC (1991) The bird fauna of a deciduous woodland in the wet-dry tropics of northern Australia. *Wildlife Research* **18**, 479–500. doi:10.1071/WR9910479

312. Noske RA (1996) Abundance, zonation and feeding ecology of birds in mangroves of Darwin Harbour, Northern Territory. *Wildlife Research* **23**, 443–474. doi:10.1071/WR9960443

313. Mohd-Azlan J, Noske RA, Lawes MJ (2012) Avian species-assemblage structure and indicator bird species of mangroves in the Australian monsoon tropics. *Emu* **112**, 287–297. doi:10.1071/MU12018

314. Woinarski JCZ (2000) *Unpublished Notes on Frugivory by Birds in Monsoon Vine Forests of the NT.* Conservation Commission of NT, Darwin.

315. Price OF (2006) Movements of frugivorous birds among fragmented rainforests in the Northern Territory, Australia. *Wildlife Research* **33**, 521–528. doi:10.1071/WR06029

316. Price OF (2004) Indirect evidence that frugivorous birds track fluctuating fruit resources among rainforest patches in the Northern Territory, Australia. *Austral Ecology* **29**, 137–144. doi:10.1111/j.1442-9993.2004.01331.x

317. Bach CS (2002) Phenological patterns in monsoon rainforests in the Northern Territory, Australia. *Austral Ecology* **27**, 477–489. doi:10.1046/j.1442-9993.2002.01209.x

318. Wright A, Jaensch RP, Woinarski J, Soulos P (1995) First sighting of the elegant Imperial Pigeon *Ducula concinna* in Australia. *Australian Bird Watcher* **16**, 110–114.

319. Ryan H *pers. comm.*

320. O'Brien G *pers. comm.*

321. Storr, GM (1980) *Birds of the Kimberley Division, Western Australia.* Western Australian Museum, Perth (Special Publication number 11).

322. Healey C (1992) Abundance, diet and roosting defecations of the Torresian Imperial-Pigeon *Ducula spilorrhoa* in Darwin. *Corella* **16**, 107–110.

323. Maurer G (2008) Who cares? Males provide most parental care in a monogamous nesting cuckoo. *Ethology* **114**, 540–547. doi:10.1111/j.1439-0310.2008.01498.x

324. Maurer G, Double MC, Milenkaya O, Süsser M, Magrath RD (2011) Breaking the rules: sex roles and genetic mating system of the pheasant coucal. *Oecologia* **167**, 413–425. doi:10.1007/s00442-011-2002-4

325. Maurer G, Smith C, Süsser M, Magrath RD (2008) Solo and duet calling in the pheasant coucal: sex and individual call differences in a nesting cuckoo with reversed size dimorphism. *Australian Journal of Zoology* **56**, 143–149. doi:10.1071/ZO08049

326. Thompson HAF (1982) The status of cuckoos Cuculidae in the Darwin area Northern Territory, 1974–1980. *Northern Territory Naturalist* **5**, 13–19.

327. Higgins PJ (ed.) (1999) *Handbook of Australian, New Zealand and Antarctic Birds. Vol. 4: Parrots to Dollarbird.* Oxford University Press, Melbourne.

328. Healey C, Healey E (2007) Diet and roost-site fidelity in the Common Koel *Eudynamys scolopacea* in suburban Darwin. *Australian Field Ornithology* **24**, 184–186.

329. Price OF, Woinarski JCZ, Robinson D (1999) Very large area requirements for frugivorous birds in monsoon rainforests of the Northern Territory. *Biological Conservation* **91**, 169–180. doi:10.1016/S0006-3207(99)00081-6

330. Brooker MG, Brooker LC (1989) Cuckoo hosts in Australia. *Australian Zoological Reviews* **2**, 1–67.

331. Langmore NE, Kilner RM (2007) Breeding site and host selection by Horsfield's bronze-cuckoos *Chalcites basalis. Animal Behaviour* **74**, 995–1004. doi:10.1016/j.anbehav.2007.02.028

332. Langmore NE, Maurer G, Adcock GJ, Kilner RM (2008) Socially acquired host-specific mimicry and the evolution of host races in Horsfield's Bronze-Cuckoo *Chalcites basalis*. *Evolution* **62**, 1689–1699. doi:10.1111/j.1558-5646.2008.00405.x
333. McKean JL (1980) Oceanic ornithological survey between Derby, W.A. and Darwin, N.T. *Northern Territory Naturalist* **3**, 8–13.
334. Mulyani Y (2004) Reproductive ecology of tropical mangrove-dwelling warblers. PhD thesis, Charles Darwin University, Darwin.
335. Brooker LC, Brooker MG, Brooker AMH (1990) An alternative population/genetics model for the evolution of egg mimesis and egg crypsis in cuckoos. *Journal of Theoretical Biology* **146**, 123–143. doi:10.1016/S0022-5193(05)80048-7
336. Langmore NE, Stevens M, Maurer G, Kilner RM (2009) Are dark cuckoo eggs cryptic in host nests? *Animal Behaviour* **78**, 461–468. doi:10.1016/j.anbehav.2009.06.003
337. Sato NJ, Mikami OK, Ueda K (2010) The egg dilution effect hypothesis: a condition under which parasitic nestling ejection behaviour will evolve. *Ornithological Science* **9**, 115–121. doi:10.2326/osj.9.115
338. Langmore NE, Stevens M, Maurer G, Heinsohn R, Hall ML, Peters A, Kilner RM (2011) Visual mimicry of host nestlings by cuckoos. *Proceedings. Biological Sciences* **278**, 2455–2463. doi:10.1098/rspb.2010.2391
339. Sato NJ, Tokue K, Noske RA, Mikami OK, Ueda K (2010) Evicting cuckoo nestlings from the nest: a new anti-parasitism behaviour. *Biology Letters* **6**, 67–69. doi:10.1098/rsbl.2009.0540
340. Rosenfeld EJ, Maurer G, Longmore NE (2010) Breaking free: Gould's Bronze-cuckoo nestlings can forge a second exit to fledge from domed nests. *Northern Territory Naturalist* **22**, 81–87.
341. King B (2005) The taxonomic status of the three subspecies of *Cuculus saturatus*. *Bulletin of the British Ornithologists' Club* **125**, 48–55.
342. Payne RB (2005) *The Cuckoos*. Oxford University Press, New York.
343. Heywood MR, Pavey CR (2002) Relative importance of plague rodents and dasyurids as prey of Barn Owls in central Australia. *Wildlife Research* **29**, 203–207. doi:10.1071/WR01104
344. Estbergs J, Garstone R, Hertog A (1978) Observations of the Eastern Grass Owl near Darwin, Northern Territory. *Emu* **78**, 93. doi:10.1071/MU9780093
345. Smith G *pers. comm.*
346. Beste H (1982) Observations at a nest of Eastern Grass Owls. *Australian Bird Watcher* **9**, 154–157.
347. Estbergs JA, Braithwaite RW (1985) The diet of the rufous owl *Ninox rufa* near Cooinda in the Northern Territory. *Emu* **85**, 202–205. doi:10.1071/MU9850202
348. Pettit M (2011) <https://groups.yahoo.com/neo/groups/ntbirds/conversations/messages/5823>.
349. Hance I (2012) <https://groups.yahoo.com/neo/groups/ntbirds/conversations/messages/6541>.
350. Goulding W, Venables B (2012) Rufous owl (*Ninox rufa*) diet and breeding in an urban forest remnant in north-eastern Queensland. *The Sunbird* **42**, 1–10.
351. Forsstrum S Personal data.
352. Sawyer G *pers. comm.*
353. Anon (2013) Rare owl sighted at Darwin botanic gardens. <http://www.abc.net.au/news/2013-11-21/rare-rufous-owl-returns-to-darwin-botanical-gardens/5108656>.
354. Paterson L *pers. comm.*
355. Woinarski JCZ (1993) A cut-and-paste community: birds of monsoon rainforests in Kakadu National Park, Northern Territory. *Emu* **93**, 100–120. doi:10.1071/MU9930100
356. Kaplan G (2007) *Tawny Frogmouth*. CSIRO Publishing, Melbourne.

357. Woinarski JCZ, Brennan K, Hempel C, Armstrong M, Milne D, Chatto R 2003. *Biodiversity Conservation on the Tiwi Islands, Northern Territory: Part 2. Fauna*. Dept of Infrastructure, Planning and Environment, Darwin.
358. Le Souëf D (1899) Ornithological notes from the Northern Territory. A list of the birds, with the nests and eggs obtained by Mr. E. Olive on the Katherine River. *Victorian Naturalist* **16**, 61–72.
359. Slater P (1959) Breeding periods of birds in the Kimberley Division, Western Australia. *Western Australian Naturalist* **7**, 35–41.
360. Barnard HG (1914) Northern Territory birds. *Emu* **14**, 39–57. doi:10.1071/MU914039
361. Le Souëf D (1902) Descriptions of birds' eggs from the Port Darwin District, Northern Australia. Part 2. *Emu* **2**, 139–159. doi:10.1071/MU902139
362. Jarman HEA (1945) Birds of Elliott and Newcastle Waters, Northern Territory. *South Australian Ornithologist* **17**, 74–78.
363. del Hoyo J, Elliott A, Sargatal J (eds.) (1996) *Handbook of the Birds of the World. Vol. 5. Barn Owls to Hummingbirds*. Lynx Edicions, Barcelona.
364. Thompson HAF (1984) The status of kingfishers and their allies (Coraciiformes) in the Darwin area, N.T, 1974 to 1982. *Northern Territory Naturalist* **7**, 18–29.
365. Thompson HAF (1978) Common birds of the Darwin suburbs. *Northern Territory Naturalist* **1**, 7–12.
366. Gardner M *pers. comm.*
367. Legge S (2004) *Kookaburra: King of the Bush*. CSIRO Publishing, Melbourne.
368. Curl DA (2005) The call of the Kookaburra: the natural history of the Blue-winged Kookaburra and other kingfishers. Unpublished PhD thesis. Monash University, Melbourne.
369. McKean JL (1985) Birds of the Keep River National Park (Northern Territory), including the Night Parrot. *Australian Bird Watcher* **11**(4), 114–130.
370. Bellis GS, Profke AM (2003) Roosting behaviour of the Rainbow Bee-eater (*Merops ornatus*) in suburban Darwin. *Corella* **27**, 75–80.
371. Gosper D, Holmes G (2008) A survey of diurnal raptors in the wet-dry tropics, northern Australia. *Corella* **32**, 71–75.
372. Riddell W *pers. comm.*
373. Rawsthorne J (2012) <https://groups.yahoo.com/neo/groups/ntbirds/conversations/messages/6379>.
374. Carter M (2012) An irruption of Nankeen (= Australian) Kestrels into areas north of Australian <http://bioacoustics.cse.unsw.edu.au/birding-aus/2012-05/msg00233.html>.
375. Fraser F (2000) Unusual predator avoidance behaviour by a Peaceful Dove Geopelia striata. *Northern Territory Naturalist* **16**, 34–35.
376. Schoenjahn J (2013) A hot environment and one type of prey: investigating why the Grey Falcon (*Falco hypoleucos*) is Australia's rarest falcon. *Emu* **113**, 19–25. doi:10.1071/MU12049
377. Schoenjahn J (2010) Field identification of the Grey Falcon *Falco hypoleucos. Australian Field Ornithology* **27**, 49.
378. Kowalick D *pers. comm.*
379. Estbergs J (2008) <https://groups.yahoo.com/neo/groups/ntbirds/conversations/messages/3649>.
380. Estbergs JA, Garstone R (1979) An observation of piracy in the Black Falcon (*Falco subniger*) at Darwin, Northern Territory. *Northern Territory Naturalist* **2**, 6.
381. Kurucz N (2000) The nesting biology of the Red-tailed Black Cockatoo (*Calyptorhynchus banksii macrorhynchus*) and its management implications in the Top End of Australia. Unpublished MSc thesis, School of Biological Sciences, Northern Territory University, Darwin.

382. Woinarski JCZ (1990) Effects of fire on the bird communities of tropical woodlands and open forests in northern Australia. *Australian Journal of Ecology* **15**, 1–22. doi:10.1111/j.1442-9993.1990.tb01016.x
383. Vardon M, Noske R, Moyle B (1997) Harvesting Black Cockatoos in the Northern Territory: catastrophe or conservation? *Australian Biologist* **10**, 84–93.
384. Ryan H (2005) <https://groups.yahoo.com/neo/groups/ntbirds/conversations/messages/2139>.
385. Brady C, Smith J (2005) *Bird Hazard Management Review of Darwin International Airport.* Report to Darwin International Airport by Indicus Biological Consultants Pty. Ltd. Nightcliff, Darwin.
386. Beeton RJS (1985) The Little Corella: a seasonally adapted species. *Proceedings of the Ecological Society of Australia* **13**, 53–63.
387. Lim TK, Bowman L, Tidemann S (1993) *A Report on the Survey of Winged Vertebrate Pest Damage on Crops in the NT.* Unpublished report. Conservation Commission of the Northern Territory, Darwin.
388. Verbeek NAM, Braithwaite RW, Boasson R (1993) The importance of *Pandanus spiralis* to birds. *Emu* **93**, 53–58. doi:10.1071/MU9930053
389. Franklin DC (2007) Sulphur-crested Cockatoo *Cacatua galerita* feeds on nectar. *Northern Territory Naturalist* **19**, 46–47.
390. Hasebe M, Franklin DC (2004) Food sources of the Rainbow Lorikeet *Trichoglossus haematodus* during the early wet season on the urban fringe of Darwin, NT. *Corella* **28**, 68–74.
391. Franklin DC, Noske RA (1999) Birds and nectar in a monsoonal woodland: correlations at three spatio-temporal scales. *Emu* **99**, 15–28. doi:10.1071/MU99003
392. Franklin DC, Noske RA (2000) Nectar sources used by birds in monsoonal north-western Australia: a regional survey. *Australian Journal of Botany* **48**, 461–474. doi:10.1071/BT98089
393. Woinarski JCZ, Connors G, Franklin DC (2000) Thinking honeyeater: nectar maps for the Northern Territory. *Pacific Conservation Biology* **6**, 61–80.
394. Franklin DC (1999) Lerp: an enduring mystery of Australian ecology. *Nature Territory* **September 1999**, 3–5.
395. Franklin DC (2003) Vertebrate pollinators visit flowers of an Umbrella Tree *Schefflera actinophylla* exclusively in the afternoon. *Northern Territory Naturalist* **17**, 1–6.
396. Franklin DC (1996) Massive aggregation of the Varied Lorikeet. *Eclectus* **1**, 6–7.
397. Thomas DAG (1947) Birds seen in the Northern Territory, part 2. *South Australian Ornithologist* **18**, 62–65.
398. Franklin DC (1999) Opportunistic nectarivory: an annual dry season phenomenon among birds in monsoonal northern Australia. *Emu* **99**, 135–141. doi:10.1071/MU99016
399. Woinarski JCZ, Risler J, Kean L (2004) Response of vegetation and vertebrate fauna to 23 years of fire exclusion in a tropical Eucalyptus open forest, Northern Territory, Australia. *Austral Ecology* **29**, 156–176. doi:10.1111/j.1442-9993.2004.01333.x
400. Le Souëf D (1902) Descriptions of birds' eggs from the Port Darwin District, Northern Australia. Part 1. *Emu* **2**, 85–96. doi:10.1071/MU902139
401. Glover B (1952) Movements of birds in South Australia. *South Australian Ornithologist* **20**, 82–91.
402. Zimmermann UM (1996) Ecology of the monsoon-rainforest endemic Rainbow Pitta *Pitta iris.* PhD thesis, Northern Territory University, Darwin.
403. Zimmermann U, Noske RA (2003) The breeding ecology of the monsoon rainforest-endemic Rainbow Pitta *Pitta iris. Emu* **103**, 245–254. doi:10.1071/MU02005
404. Zimmermann U, Noske RA (2004) Why do Rainbow Pittas *Pitta iris* place dung at the entrance to their nests? *Australian Field Ornithology* **21**, 163–165.
405. Frith CT, Frith DW (2004) *The Bowerbirds.* Oxford University Press, Oxford.

406. Katsuno Y, Okida T, Yamaguchi N, Nishiumi I, Eguchi K (2010) Bower structure is a good predictor of mating success in the Great Bowerbird. *Journal of the Yamashina Institute for Ornithology* **42**, 19–33. doi:10.3312/jyio.42.19
407. Okida T, Katsuno Y, Eguchi K, Noske RA (2010) How interacting multiple male sexual signals influence female choice in the Great Bowerbird. *Journal of the Yamashina Institute for Ornithology* **42**, 35–46. doi:10.3312/jyio.42.35
408. Katsuno Y, Eguchi K, Noske RA (2013) Preference for, and spatial arrangement of, decorations of different colours by the Great Bowerbird *Ptilonorhynchus nuchalis nuchalis. Australian Field Ornithology* **30**, 3–13.
409. Haruyama N, Eguchi K, Noske RA (2013) Experimental evidence on the local variation in the Great Bowerbird's colour preferences for bower decorations. *Emu* **113**(4), 367–373. doi:10.1071/MU13006
410. Mikami OK, Katsuno Y, Yamashita DM, Noske RA, Eguchi K (2010) Bowers of the Great Bowerbird (*Chlamydera nuchalis*) remained unburned after fire: is this an adaptation to fire? *Journal of Ethology* **28**, 15–20. doi:10.1007/s10164-009-0149-9
411. Higgins PJ, Peter JM, Cowling SJ (eds) (2006) *Handbook of Australian, New Zealand and Antarctic Birds. Vol. 7: Boatbill to Starlings.* Oxford University Press, Melbourne.
412. Noske RA, Franklin D (1999) Breeding seasons of land birds in the Australian monsoon tropics: diverse responses to a highly seasonal environment. *Australian Biologist* **12**, 72–90.
413. Nakamura M, Takaki Y, Mori S, Ueda K, Nishiumi I, Takagi M, Noske RA, Eguchi K (2010) Impacts of fire on the group composition of the Red-backed Fairy-wren *Malurus melanocephalus cruentatus* in the non-breeding season. *Journal of the Yamashina Institute for Ornithology* **42**, 47–64. doi:10.3312/jyio.42.47
414. Higgins PJ, Peter JM, Steele WK (eds) (2001) *Handbook of Australian, New Zealand and Antarctic Birds. Vol. 5: Tyrant-flycatchers to Chats.* Oxford University Press, Melbourne.
415. Ueda K Personal data.
416. Murphy SA, Legge SM, Heathcote J, Mulder E (2010) The effects of early and late-season fires on mortality, dispersal, physiology and breeding of red-backed fairy-wrens (*Malurus melanocephalus*). *Wildlife Research* **37**, 145–155. doi:10.1071/WR09007
417. Brooker MG, Braithwaite RW, Estbergs JA (1990) Foraging ecology of some insectivorous and nectarivorous species of birds in forests and woodlands of the wet-dry tropics of Australia. *Emu* **90**, 215–230. doi:10.1071/MU9900215
418. Woinarski JCZ, Press AJ, Russell-Smith J (1989) The bird community of a sandstone plateau monsoon forest at Kakadu National Park, Northern Territory. *Emu* **89**, 223–231. doi:10.1071/MU9890223
419. Noske RA, Carlson AJ (2011) The breeding biology of the Dusky Honeyeater *Myzomela obscura* in the Northern Territory, and the importance of nectar in the diet of nestling honeyeaters. *Australian Field Ornithology* **28**, 97–113.
420. Franklin DC, Noske RA (1998) Local movements of honeyeaters in a sub-coastal vegetation mosaic in the Northern Territory. *Corella* **22**, 97–103.
421. Noske RA (2003) The role of birds in mangroves – pollination and insect predation. In *Proceedings: Darwin Harbour Region: Current Knowledge and Future Needs* (Working Group for the Darwin Harbour Advisory Committee ed.), pp. 74–86. Department of Infrastructure, Planning and Environment, Darwin.
422. Noske RA, Andrew D (1992) Birds in mangroves. *Wingspan* **8**, 16–18.
423. Carlson A *pers. comm.*
424. Morton SR, Brennan KG (1991) Birds. In *Monsoonal Australia: Landscape, Ecology and Man in Northern Lowlands* (Haynes CD, Ridpath MG, Williams MAJ eds), pp. 133–149. A.A. Balkema, Rotterdam.

425. Franklin DC, Noske RA (2000) The nesting biology of the Brown Honeyeater *Lichmera indistincta* in the Darwin Region of northern Australia with notes on tidal flooding of nests. *Corella* **20**, 38–44.
426. Franklin DC (1997) The foraging behaviour of avian nectarivores in a monsoonal Australian woodland over a six-month period. *Corella* **21**, 48–54.
427. Eguchi K, Yamaguchi N, Ueda K, Noske RA (2013) The effects of nest usurpation and other interference by the Blue-faced Honeyeater on the reproductive success of the Grey-crowned Babbler. *Emu* **113**, 77–83. doi:10.1071/MU12044
428. Armstrong M (2005) <https://groups.yahoo.com/neo/groups/ntbirds/conversations/messages/2055>.
429. Michael B (2006) <https://groups.yahoo.com/neo/groups/ntbirds/conversations/messages/2637>.
430. Noske RA (1998) Breeding biology, demography and success of the rufous-banded honeyeater, *Conopophila albogularis*, in Darwin, a monsoonal tropical city. *Wildlife Research* **25**, 339–356. doi:10.1071/WR97070
431. Noske A, Fischer S, Brook BW (2008) Artificial nest predation rates vary among habitats in the Australian monsoon tropics. *Ecological Research* **23**, 519–527. doi:10.1007/s11284-007-0403-y
432. Noske RA, Mulyani YA, Lloyd P (2013) Nesting beside old nests, but not over water, increases current nest survival in a tropical mangrove-dwelling warbler. *Journal fur Ornithologie* **154**, 517–523. doi:10.1007/s10336-012-0922-3
433. Rankmore BR, Price OF (2004) Effects of habitat fragmentation on the vertebrate fauna of tropical woodlands, Northern Territory. In *The Conservation of Australia's Forest Fauna* (second edition) (Lunney D ed.), pp. 452–473. Royal Zoological Society of New South Wales, Mosman, NSW, Australia.
434. Thomas DAG (1947) Birds seen in the Northern Territory, part 1. *South Australian Ornithologist* **18**, 55–60.
435. Steele P *pers. comm.*
436. Fisher D *pers. comm.*
437. Kyne PM, Jackson MV (2010) An extralimital record of Grey-headed Honeyeater *Lichenostomus keartlandi* from Darwin, Northern Territory. *Northern Territory Naturalist* **22**, 75.
438. Higgins PJ, Peter JM (Eds.) (2002) *Handbook of Australian, New Zealand and Antarctic Birds, Vol. 6: Pardalotes to Shrike-thrushes.* Oxford University Press, Melbourne.
439. Noske RA (2001) The breeding biology of the Mangrove Gerygone *Gerygone laevigaster* in the Darwin region, with notes on brood parasitism by the Little Bronze-cuckoo *Chrysococcyx minutillis. Emu* **101**, 129–135. doi:10.1071/MU00013
440. Noske RA (2003) Monsoon madness: the upside-down breeding seasons of Top End birds. *Wingspan* **13**(3), 16–20.
441. Tokue K, Ueda K (2010) Mangrove Gerygones *Gerygone laevigaster* eject Little Bronze-cuckoo *Chalcites minutillus* hatchlings from parasitized nests. *The Ibis* **152**, 835–839. doi:10.1111/j.1474-919X.2010.01056.x
442. Reynolds SJ (2010) Birdlife of Mickett Creek: factors influencing frequency of occurrence and detection. *Northern Territory Naturalist* **22**, 45–59.
443. Eguchi K, Yamaguchi N, Ueda K, Nagata H, Takagi K, Noske RA (2007) Social structure and helping behaviour of the Grey-crowned Babbler *Pomatostomus temporalis. Journal fur Ornithologie* **148**, 203–210. doi:10.1007/s10336-007-0196-3
444. Eguchi K Unpublished reports.

445. Woinarski JCZ, Armstrong M, Price O, McCartney J, Griffiths AD, Fisher A (2004) The terrestrial vertebrate fauna of Litchfield National Park, NT: monitoring over a 6-year period and response to fire history. *Wildlife Research* **31**, 587–596. doi:10.1071/WR03077

446. Healey C (1993) Antiphonal song in the Black Butcherbird *Cracticus quoyi* in Darwin. *Northern Territory Naturalist* **14**, 30–32.

447. Noske RA (2010) Black Butcherbird depredating nestlings of Broad-billed Flycatcher. *Nature Territory* **August 2010**, 4.

448. Jackson MV, Kyne PM (2010) Potential host species fatally attack female Eastern Koel *Eudynamys orientalis*, a brood parasite. *Australian Field Ornithology* **27**, 133–136.

449. Crawford DN (1979) Effects of grass and fires on birds in the Darwin area, Northern Territory. *Emu* **79**, 150–152. doi:10.1071/MU9790150

450. Woinarski JCZ, Brock C, Armstrong M, Hempel C, Cheal D, Brennan K (2000) Bird distribution in riparian vegetation in the extensive natural landscape of Australia's tropical savanna: a broad-scale survey and analysis of a distributional data base. *Journal of Biogeography* **27**, 843–868. doi:10.1046/j.1365-2699.2000.00439.x

451. Boekel C (1980) Birds of the Victoria River Downs Station and of Yarralin, Northern Territory. Part 2. *Australian Bird Watcher* **8**(7), 205–211.

452. Thompson HAF (1983) Birds recorded on a visit to South Goulburn Island, Northern Territory. *The Sunbird* **13**(2), 21–30.

453. McKean JL (1986) The status of cuckoo-shrikes (Campephagidae) in the Darwin area, Northern Territory, 1974–1984. *Northern Territory Naturalist* **9**, 8–14.

454. Verhoeye J, Holmes DA (1999) The birds of the islands of Flores – a review. *Kukila* **10**, 3–59.

455. Noske RA (1990) Seed dispersal of Black Wattles *Acacia auriculiformis* by birds. *Northern Territory Naturalist* **12**, 24–25.

456. Noske RA, Barnes T, Barnes V (1997) First breeding record of the Cicadabird for the Northern Territory. *Northern Territory Naturalist* **15**, 43–44.

457. Carlson AJ (2008) Nesting event of the Cicadabird *Coracina tenuirostris* in the Northern Territory. *Northern Territory Naturalist* **20**, 38–40.

458. Woinarski JCZ, Fisher A, Brennan K, Morris I, Chatto R (2001) Patterns of bird species richness and composition on islands off Arnhem Land, Northern Territory, Australia. *Austral Ecology* **26**, 1–13.

459. Kowalick D (1986) Agonistic behaviour of figbirds at a fruiting banyon fig. *Northern Territory Naturalist* **9**, 23.

460. Noske RA (1997) Short breeding season of Figbirds *Sphecotheres viridis* in Darwin, Northern Territory. *Corella* **21**, 44–47.

461. Keates A, Keates S. (2010) Olive-backed Oriole as likely would-be nest predator. *Nature Territory* **April 2010**, 4.

462. Sato NJ, Morimoto G, Noske RA, Ueda K (2010) Nest form, colour, and nesting habitat affect predation rates of Australasian warblers (*Gerygone* spp.) in tropical mangroves. *Journal of the Yamashina Institute for Ornithology* **42**, 65–78. doi:10.3312/jyio.42.65

463. Mohd-Azlan J, Noske RA, Lawes MJ (2014) Resource partitioning by mangrove bird communities in North Australia. *Biotropica* **46**, 331–340. doi:10.1111/btp.12108

464. Stokes T (1981) An agonistic display in male Shining Flycatchers. *Northern Territory Naturalist* **4**, 24.

465. Ley AJ (1995) Recovery of cache food by Torresian Crow. *Australian Bird Watcher* **16**, 77.

466. Donato DB, Potts RT (2004) Culturally transmitted predation and consumption techniques by Torresian Crows *Corvus orru* on Cane Toads *Bufo marinus*. *Australian Field Ornithology* **21**, 125–126.

467. Noske RA (2011) Lemon-essence. An aerial-displaying, single-egg laying, biannual, tropical cooperative-breeder: introducing the humble Lemon-bellied Flycatcher. *Wingspan* **21**(4), 36–39.
468. Jackson MV, Kyne PM (2013) Red-rumped Swallow *Cecropis daurica* in the Northern Territory. *Northern Territory Naturalist* **24**, 55.
469. McKean JL (1983) Some notes on the occurrence of the Great Reed Warbler *Acrocephalus arundinaceus* in the Northern Territory. *Northern Territory Naturalist* **6**, 3–8.
470. Shurcliff KS *pers. comm.*
471. Givens TV, Hitchcock WB (1953) *Cisticola juncidis* (Raf.) in the Northern Territory. *Emu* **53**, 193–200. doi:10.1071/MU953193
472. Balmford A, Lewis MJ, Brooke M de L, Thomas ALR, Johnson CN (2000) Experimental analyses of sexual and natural selection on short tails in a polygynous warbler. *Proceedings. Biological Sciences* **267**, 1121–1128. doi:10.1098/rspb.2000.1117
473. Noske RA (1999) Notes on the breeding biology of the tropical mangrove-dwelling Yellow White-eye *Zosterops luteus. Australian Bird Watcher* **18**, 3–7.
474. Healey C (2003) First record of the Metallic Starling *Aplonis metallica* in the NT. *Northern Territory Naturalist* **17**, 53–55.
475. McCrie N (2000) A record of the Common Starling *Sturnus vulgaris* in Darwin, Northern Territory. *Northern Territory Naturalist* **16**, 26–27.
476. Parsons D, Adams S, Berman D (1995) *Sparrows on the Barkly Tableland: Distribution and Abundance*. Conservation Commission of the Northern Territory, Darwin.
477. Chapman S (2000) Occurrence and eradication of a small population of the Eurasian Tree Sparrow *Passer montanus* in Darwin. *Northern Territory Naturalist* **16**, 32–34.
478. Parker C *pers. comm.*
479. Milenkaya O, Legge S, Walters JR (2011) Breeding biology and life-history traits of an Australasian tropical granivore, the Crimson Finch (*Neochmia phaeton*). *Emu* **111**, 312–320. doi:10.1071/MU10088
480. Todd MK, Felton A, Garnett ST (2003) Morphological and dietary differences between common and uncommon subspecies of Crimson Finch, *Neochmia phaeton*, and Star Finch, *Neochmia ruficauda*, in northern Australia. *Emu* **103**, 141–148. doi:10.1071/MU02015
481. Todd MK (2002) Nest-site and breeding-season data for the Crimson Finch *Neochmia phaeton* in Australia. *Australian Bird Watcher* **189**, 161–171.
482. Dostine PL, Franklin DC (2002) A comparison of the diet of three finch species in the Yinberrie Hills area, Northern Territory. *Emu* **102**, 159–164. doi:10.1071/MU01034
483. Tidemann SC, Boyden J, Elvish R, Elvish J, O'Gorman B (1992) Comparison of the breeding sites and habitat of two hole-nesting estrildid finches, one endangered, in northern Australia. *Journal of Tropical Ecology* **8**, 373–388. doi:10.1017/S0266467400006702
484. van Rooij EP, Griffith SC (2011) Breeding ecology of an Australian estrildid, the Long-tailed Finch (*Poephila acuticauda*) *Emu* **111**, 297–303. doi:10.1071/MU10092
485. Franklin DC (1999) Evidence of disarray amongst granivorous bird assemblages in the savannas of northern Australia, a region of sparse human settlement. *Biological Conservation* **90**, 53–68. doi:10.1016/S0006-3207(99)00010-5
486. Crawford DN, Parker SA (1971) First record of the Yellow Wagtail in the Northern Territory. *Emu* **71**, 142. doi:10.1071/MU971140g
487. McCrie N (1987) Roosting of Yellow Wagtails near Darwin. *Northern Territory Naturalist* **10**, 18.
488. Owen DF (1969) The migration of the yellow wagtail from the equator. *Ardea* **57**, 77–85.
489. Alström P, Mild K (2003) *Pipits and Wagtails of Europe, Asia and North America*. Christopher Helm, London.

490. Roderick M *pers. comm.*
491. McKean JL (1980) A sight record of the Grey Wagtail *Motacilla cinerea* in the Northern Territory. *Australian Bird Watcher* **8**, 237.
492. McKean JL (1980) A sight record of the Ringed Plover *Charadrius hiaticula* in the Northern Territory. *Australian Bird Watcher* **8**, 236–237.
493. Dunn JL (1993) The identification of Semipalmated and Common Ringed plovers in alternate plumage. *Birding* **25**, 238–243.
494. McKean JL, Thompson HA, Estbergs JA (1976) Records of uncommon migrant waders near Darwin. *Australian Bird Watcher* **6**, 143–148.
495. Condon HT (1961) The Caspian Plover – a new record for Australia. *Emu* **61**, 191–195. doi:10.1071/MU961191
496. McCrie N, Jaensch R (1999) Sighting of a Caspian Plover *Charadrius asiaticus* at Lake Finniss, Northern Territory. *Australian Bird Watcher* **18**, 81-86.
497. Birds Australia Rarities Committee (2011) Submission No 545: Eurasian Curlew Numenius arquata, 80 Mile Beach, WA, 26 November 2007. <http://www.tonypalliser.com/barc/summaries/SUMM545.htm>.
498. McKean JL, Dampney AR (1984) First sighting of the Spotted Redshank *Tringa erythropus* in Australia. *Northern Territory Naturalist* **7**, 8–9.
499. Birds Australia Rarities Committee (2007) Submission No 519: Nordmann's Greenshank *Tringa guttifer* 80 Mile Beach, Near Broome WA, 2nd December 2006. <http://www.tonypalliser.com/barc/summaries/SUMM519.htm>.
500. Frith HJ, Crome FHJ, Brown BK (1977) Aspects of the biology of the Japanese snipe *Gallinago hardwickii*. *Australian Journal of Ecology* **2**, 341–368. doi:10.1111/j.1442-9993.1977.tb01150.x
501. Bywater J, McKean JL (1987) A record of Latham's Snipe *Gallinago hardwickii* in the Northern Territory. *Australian Bird Watcher* **12**, 65.
502. Birds Australia Rarities Committee (2007) Submission No 473: Pin-tailed Snipe *Gallinago stenura*, Broome, WA, 31st January 2004. <http://www.tonypalliser.com/barc/summaries/SUMM473.htm>.
503. Birds Australia Rarities Committee (2007) Submission No 478: Pin-tailed Snipe *Gallinago stenura* Broome, WA, 19th February 1998. <http://www.tonypalliser.com/barc/summaries/SUMM478.htm>.
504. McKean JL (1984) The occurrence in Australia of Gray's Grasshopper Warbler *Locustella fasciolata*. *Australian Bird Watcher* **10**(5), 171–172.

Species list: common names

Species list: scientific names

www.ingramcontent.com/pod-product-compliance
Lightning Source LLC
LaVergne TN
LVHW070846210926
848482LV00016B/877
9781486300341